Chance
change
& challenge

THE EVOLVING EARTH

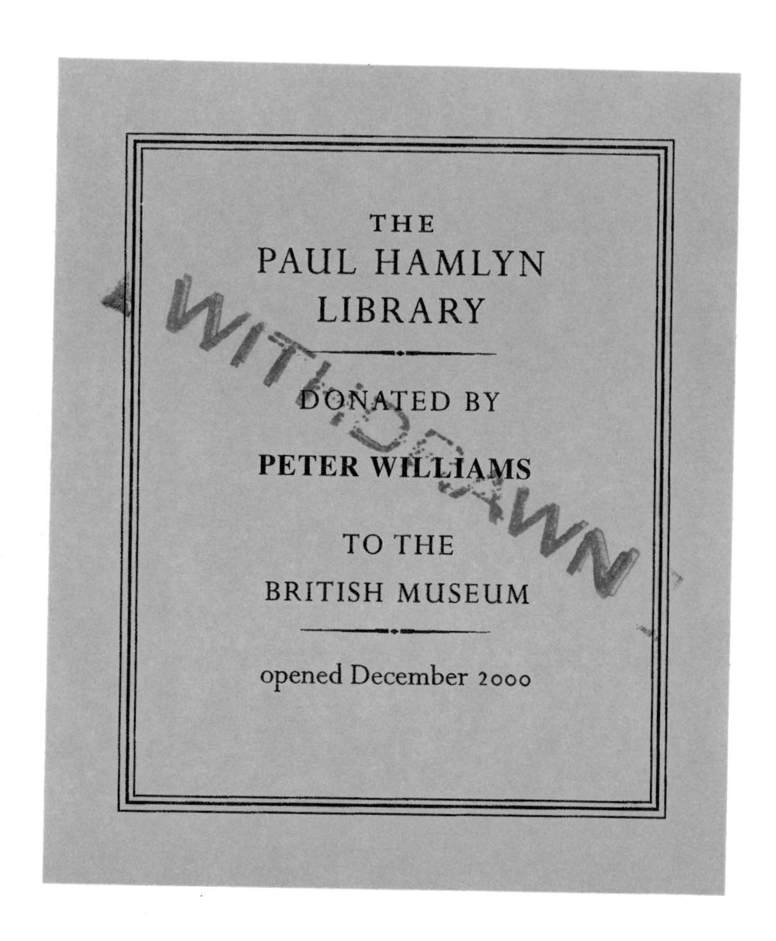

Meteorite with original planetary matter? (see chapter 1).

Orgueil carbonaceous chondrite. Beneath the dark fusion-crust (caused by frictional heating during atmospheric flight) lies a mixture of dark, hydrous silicates and white carbonates and sulphates, often in veins. The veins are indicative of the former presence of water. Such meteorites contain a variety of carbon compounds which may have been involved in the origin of life.

Chance change & challenge

General Editor: *P. H. Greenwood*

THE EVOLVING EARTH

Editor: *L. R. M. Cocks*

BRITISH MUSEUM (NATURAL HISTORY)

CAMBRIDGE UNIVERSITY PRESS

Published by the British Museum (Natural History), London and the Press Syndicate of the University of Cambridge
The Pitt Building, Trumpington Street, Cambridge CB2 1RP
32 East 57th Street, New York, NY 10022, USA
296 Beaconsfield Parade, Middle Park, Melbourne 3206, Australia

First published 1981

Printed in Great Britain at the University Press, Cambridge

British Library Cataloguing in Publication Data

The evolving earth. – (Chance change and challenge)
I. Evolution
II. Cocks, Leonard Robert Morrison
III. British Museum (Natural History)
551 QE28.3 80-42171

ISBN 0 521 23810 2 hard covers
ISBN 0 521 28229 2 paperback

Cover illustration

Crab Nebula in the constellation Taurus
The Crab Nebula is the remnant of a supernova, some 6000 light years distant, which was witnessed in China in AD 1054. A supernova is a massive, exploding star which can eject more than the amount of matter in the Sun at a velocity of 10 000 km per second. During such explosions 'heavy' elements such as plutonium and uranium are synthesised and distributed throughout interstellar space. If an expanding supernova remnant encounters an interstellar cloud of dust and gas, a shock-wave is produced which may cause the collapse of the cloud. Such compression may have triggered the formation of the Solar System – see Chapter 1.

FOREWORD

Chance, change and challenge

The basis of natural history

When the Museum moved to South Kensington in 1881, controversy over evolution was at its height and although the reference collections were not at first employed directly to advance evolutionary ideas, the work of its staff has always contributed ultimately to evolutionary knowledge and to explaining the principles underlying natural history. It is therefore appropriate that the Museum's centenary year should be marked by the publication of two volumes on evolution. This subject not only links all living organisms, but also the organic to the inorganic world, and the Past to the Present, thereby providing a theme to which all five science departments in the Museum (Botany, Entomology, Mineralogy, Palaeontology and Zoology) can contribute.

The wide range of subjects currently studied in the Museum ensures that there are many areas and individual spheres of interest and expertise within the broad spectrum of theoretical and empirical evolutionary studies. This knowledge is tapped here in a series of very personal essays which reflect the widely differing approaches to the study of evolution.

The title summarises, in simple terms, the essential elements of evolution. CHANCE – the interplay of cosmic forces that brought the Earth into existence and the often fortuitous, random mutations and other genetical changes that collectively contribute to the evolution of life. CHANGE – the formation and destruction of continents and ocean floors as rocks and sediments are generated, eroded, and recycled, and the infinitely variable patterns of life as plants and animals diversify. CHALLENGE – the ever-changing physical and biotic environments to which living organisms respond, and in so responding, evolve.

Although, in its generally accepted sense, the concept of evolution dates back some 200 years, it did not attract great interest among naturalists (and the general public) until after the publication of the famous paper on 'Evolution by natural selection' by Darwin and Wallace in 1858, followed almost immediately by Darwin's monumental *Origin of Species by means of Natural Selection* (1859). The ensuing stream of publications on evolution and evolutionary theory eventually became a flood which as yet shows no sign of diminishing. The present volumes do not set out to review the research output of the last century, neither are they intended to serve as text-books: their aims are more modest, although not perhaps less important. They show how the evolution of life is linked to that of the Earth itself, thus providing a broader perspective than is possible with most text-books. Intelligible accounts of highly specialised and very different work should help to disseminate ideas between specialists in fields between which there is usually little contact. Finally, students should benefit from a set of essays which are not always uniform in philosophical outlook and which demonstrate the disagreements that are part of the continuing challenge of research and of the theory of evolution itself.

The first volume – *The Evolving Earth* – demonstrates the interrelated contributions of Mineralogy, Geology, Oceanography and Palaeobiology to an understanding of the changing physical and chemical backgrounds against which plants and animals arose and evolved. In so doing it also describes the environments which have supported living organisms, and shaped their adaptation and diversity during the last 3·5 thousand million years. The revolutionary changes in cosmological and geological thought during this century are reflected in the sections on the origin of the Earth, and on continental drift and plate tectonics. Controversial hypotheses, such as that of an expanding Earth, which still require further testing and evaluation, are included. Apart from those essays which review the background to continental drift, biogeographers should be interested especially in the later chapters on Mesozoic and Cenozoic palaeogeography which provide a framework for the study of present-day distributions.

The second volume – *The Evolving Biosphere* – is necessarily more selective and less comprehensive than its companion since its scope is potentially greater. It is concerned with the mechanisms and interactions which produce and account for the diversity, coexistence, coevolution and distribution of plants and animals in the world today. There is naturally much emphasis on speciation, the basic process underlying these phenomena, and the one subject not discussed in the *Origin of Species* since nothing was known in Darwin's day of the possible mechanisms involved. The arrangement of these essays is essentially similar to that followed by Darwin in '*The Origin*', thus serving to underline changes in thinking on evolution and evolutionary processes since the mid-nineteenth century.

Major problems, both philosophical and practical, which still hinder our understanding of evolution are not avoided. The contributions show the diversity of interpretation and opinion held by students of evolution, and highlight the dynamic state of modern evolutionary biology. They also show how taxonomists have contributed to the advance and interpretations of evolutionary theory, and how in turn a deeper appreciation of evolutionary processes, and hence of phylogeny, has influenced taxonomic theory and the practice of classification.

These volumes were conceived and planned by a group of colleagues which first met in 1977 to consider the feasibility of a book of this sort. I should like to express special gratitude to the General Editor, Dr P. H. Greenwood, and the volume editors, Drs L. R. M. Cocks and P. L. Forey, for their time-consuming and dedicated work in bringing these publications to fruition. They, I know, recognise the special help they have been given by Dr C. G. Adams and Mr J. F. Peake at various stages in their work. The Museum is grateful to all the contributors, especially those from outside who so willingly filled the gaps in our knowledge. Their help demonstrates the close links forged over the years between the Museum, other government establishments and the universities. Thanks are also due to Mr C. J. Owen (Co-ordinating Editor) and Mr E. Dent (Production Controller).

R. H. Hedley
Director, British Museum (Natural History)
January 1981

Contents

Introduction

The theme of this book is the dynamic and progressive evolution of the planet Earth; its purpose to present a summary of those parts of the Earth Sciences which are especially relevant to biologists and others who wish to know about the processes that have changed and shaped our planet during the last 4500 million years. The book is also of interest to those who wish to have some idea of the Earth's changing geography, particularly during the last 250 million years, a palaeogeography that has constrained the free movement of animals and led to evolutionary centres separated from each other by physical and climatic barriers. The processes of biological evolution are not dwelt upon here, since they are discussed in our companion volume, and in this book animals and plants are mentioned only as they affect the thinking of Earth scientists on contemporary problems, apart from the final chapter, which deals with mammals in more detail. For illustrations of, and information on, the fossil record, the reader should turn to such works as Moore (1953–79) for the invertebrates, with illustrations of their changing ecology through time in McKerrow (1978); to Romer (1966) for the vertebrates, and to Boureau (1964–77) for the plants.

Although the Earth is inanimate, it is a highly energised and very mobile body. Its energy derives from the radioactive elements within the crustal rocks and underlying mantle. The generation of heat, followed by the escape of molten rocks, gases and vapours from the interior, led gradually to the formation of the continents, atmosphere and hydrosphere, which were in turn further heated by the Sun. These energy sources fuelled the forces that have slowly moulded our planet, and changed it from a barren and inhospitable wasteland to the largely fertile and pleasant world we know today. The unimaginably long time-scale involved, and the lack of direct evidence about the Earth's primitive state, mean that ideas concerning its origin and early history can only be speculative. Such ideas can at any time be controverted by new discoveries, and the best that can therefore be said about the ideas in this book is that they reflect current opinion.

From the mid-nineteenth to the mid-twentieth centuries, most geologists accepted Lyell's Principle of Uniformitarianism, which postulated the almost self-evident truth that the present is the key to the past. In other words, if we can understand the forces and processes shaping the Earth today, we should be able to interpret its history as recorded in the rocks. Lyell's famous principle was based on the assumption that these forces have always functioned unchanged, and at approximately the same rates. Although recent research has shown this to be fallacious, it is nevertheless true that, without a clear understanding of present-day processes, the record of the past would be indecipherable.

This book is divided into five sections, and the first short section summarises modern thought on the origin of the Earth and of life upon in it. Section II deals with the principles by which the continents evolved, and the shallow seas which lie around and upon them, and how the various igneous and sedimentary rocks and soils formed and were deposited. The opportunity is taken in the Introduction to Section II to present a very short review of some elementary facts for those readers with little or no geological background knowledge. Section III discusses the oceans, the evolution of sea water and the varied deep ocean sediments; has chapters on the atmosphere and climates which discuss their original formation; shows how the original reducing atmosphere changed to an oxidising one at about 2200 million years ago, and charts the subsequent modifications which both oceans and atmosphere have undergone.

Section IV reflects the revolution which has transformed geological opinion within the last twenty years: a fundamental change in attitudes from the belief that the Earth remained essentially stable and similar for several thousand million years to the acceptance that continental plates have always been continually on the move. The effects of these movements are studied in more detail in Section V, in which palaeogeographical maps are presented for successive periods, with particular emphasis from Mesozoic times to the present day.

The deciphering of Nature's riddles is as old a game as Man himself: though many have been answered, many remain.

References

Boureau, E. (ed.). 1964–77. *Traité de paléobotanique*. Masson, Paris. 9 vols, continuing.

McKerrow, W. S. (ed.). 1978). *The ecology of fossils*. 385 pp. London: Duckworth, and Massachusetts Institute of Technology Press.

Moore, R. C. (ed.) 1953–79. *Treatise on invertebrate paleontology*. Geological Society of America and University of Kansas Press. 25 volumes, continuing.

Romer, A. S. 1966. *Vertebrate paleontology*. 468 pp. University of Chicago Press.

PART I

Origins

Introduction

In the single chapter of this section the origin of the Earth and the origin of life are discussed, and this forms a prelude to the ensuing sections which go on to describe the subsequent evolution of the planet.

Astronomers and other scientists will no doubt speculate for many more years on the history and future of the Universe, but the age of our own planet is now known with reasonable certainty, and the round figure of 4500 million years seems unlikely to be wrong by more than a few per cent. Scientists are also becoming more agreed than previously on the origin of the solar system and the method of formation of our planet. This may be partly attributed to the new data which have poured in from other planets and satellites following the space research programme, but the evidence for planetary origins is still inferential. Meteorites have also provided valuable clues to the composition of original planetary matter.

The origin of life is more contentious, but the concept of a single event at approximately 3800 million years ago is favoured by most workers, although this is still a substantially older age than the earliest known fossils, which are bacterial prokaryotes dated at about 3500 million years. Although the exact form and chemical history of this single event is as yet unknown, the reactions which took place can now be guessed at, following the major advances in biochemistry which have occurred during the past thirty years.

CHAPTER 1

The origin of the Earth

R. Hutchison

We still do not know how the Earth formed. Much knowledge has been gained from the study of rocks, but the oldest rocks on our planet are 3800 million years old, and before that time our ideas can only be speculative. However, much relevant information can be derived from the examination of lunar rocks – which are known to have ages of up to at least 4200 million years – and of meteorites, those fragments of rock and, rarely, metal that reach the Earth's surface from space. Some meteorites formed over 4500 million years ago and have since remained effectively unaltered, and these objects preserve a record of some of the earliest processes in the development of the solar system.

One fundamental problem we face is that we do not know whether planetary systems are common in the universe. If they are common, it is probable that the formation of the Sun led to the formation of the planets, but if planetary systems are rare, it may then be probable that our own planetary system formed by a later process completely unrelated to that which formed the central star, the Sun. Astronomical observation has established that a small number of bright stars have motions consistent with the presence of dark companions. Each of these bright stars must therefore have one or more objects in orbit around it. Unfortunately, techniques are not sufficiently advanced to enable astronomers to determine whether any star, other than the Sun, has a number of planets in essentially a single orbital plane, although this problem is now being tackled with the aid of space technology. In the meantime, we cannot give more weight to a theory in which Sun and planets formed together than to one in which the planets originated by a separate process after the formation of the Sun.

The origin of planets

Williams (1975) suggested that theories of the origin of the planets are of three main kinds:

1 Those stating that planetary matter is derived from the interaction of the Sun with a passing star,
2 Those stating that planetary matter is captured by the Sun during its passage through an interstellar gas-cloud, and
3 Those stating that the solar system originated in a single process.

The earliest theory in the first category may be attributed to the French natural philosopher, G. L. L. Buffon who in 1745 postulated a collision between a comet and the Sun to account for the ejection of matter from the Sun. The masses of comets are now known to be insignificant compared with those of the Sun and planets, and so Buffon's theory must be rejected. Earlier this century, J. Jeans suggested that a closely passing star drew out a filament of gas from the Sun; this filament subsequently separated into several masses each of which condensed to form a planet. H. Jeffreys later supported Jeans' theory by showing that stellar interaction was necessary to explain the high angular momentum of the planets relative to the Sun. However, it was later shown that a filament of dense, hot material drawn from the Sun would expand and dissipate before it could cool and condense.

By the 1950s the Jeans–Jeffreys theories had fallen into abeyance, but in 1960, M. M. Woolfson put forward a new theory in which he assumed that the Sun formed in a young stellar cluster, and was approached by a young, less massive, star to within about 40 astronomical units (one A.U. is the mean distance between the Sun and the Earth, $149{\cdot}6 \times 10^6$ km). Because the young star was assumed to have been of low density and cold, matter torn from it would not have expanded and dissipated before condensation began. In fact, gravitational interaction could have held such matter between the Sun and the young star while the condensable elements formed the seeds from which planets might have grown (Fig. 1.1).

Theories in the second category postulate capture of the material which evolved into the planets during the Sun's

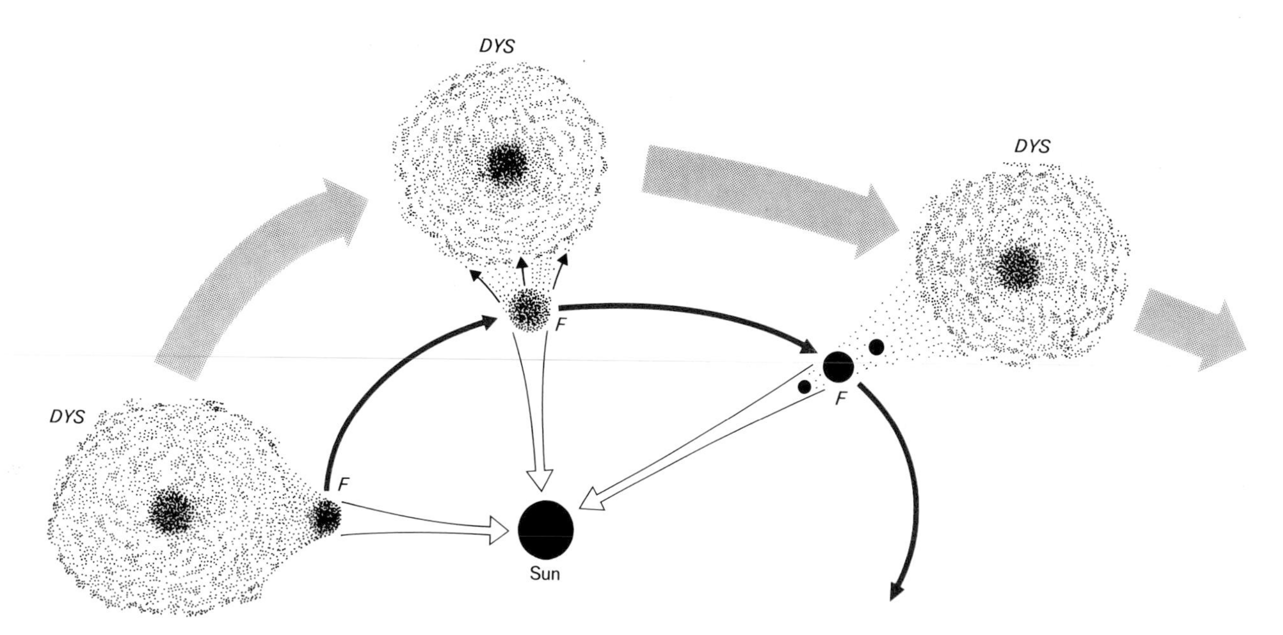

Figure 1.1 Diagrammatic representation of Woolfson's theory of the origin of the planets. A diffuse, young star (DYS) approaches the more massive Sun. A filament (F) is torn from the DYS and is held at the position where the Sun's gravity balances that of the DYS. As the DYS swings around and away from the Sun, angular momentum is imparted to the filament as it breaks up and condenses into planets.

passage through an interstellar gas-cloud; this could have occurred in about one million years. If a second star had been present, its gravity would have increased the rotation of the captured matter to give the planets, when formed, their high angular momentum. Solar radiation would have driven off the gases from the material near the Sun, thus accounting for the small mass and high density of the inner planets. Alternatively, if the captured gas became partly ionised by solar radiation, the Sun's magnetic field would have caused the retention of only dust in the region of the inner planets, and of gases, such as hydrogen, further away. Magnetism could then be invoked as the mechanism by which angular momentum was transferred from the Sun to the planets, although it might also be explained by assuming that the young Sun captured matter from an already rotating gas-cloud.

Thirdly, there is a group of currently fashionable theories proposing that the Sun and planets formed as a result of a single process. R. Descartes propounded the first in 1644, while recently W. H. McCrea suggested that the solar system formed from an interstellar gas-cloud whose mass might have been 100 times that of the Sun, and which would have consisted chiefly of molecular hydrogen, which radiates energy efficiently. Contraction of the cloud under gravity, with supersonic turbulence, would have caused it to break up into a number of fragments, or 'floccules'. After random motion, with collision, disruption, and re-formation, some twenty floccules might have coalesced to form the proto-Sun. Motion and collision of the remaining floccules brought some material into a single plane and some into the Sun; the remainder would have been ejected from the nascent solar system. Floccules rotating around the Sun in a single, nebular, plane became the planets. These, however, would have been of uniform size and composition, and it was therefore necessary to propose a mechanism by which the denser materials could have been segregated from less dense ones. Settling of early condensed dust grains on to the nebular plane, and, ultimately, into planetary cores, was the mechanism selected. However, the most popular theory today is based on that of I. Kant who, in 1755, postulated that a gas-filled universe, with density variations, would break up and that contraction and condensation would occur around the denser centres. Rotation within these primordial condensations led to the angular momentum of stellar systems. Modern work suggests that collapse of a pre-solar gas-cloud could have occurred in less than a million years. Centrifugal force would have prevented material beyond a certain distance from falling into the Sun, and solar radiation would have tended to drive off the gases, leaving gas-free dust in the inner solar system. The aggregation of dust grains which fell to the nebular disc led to the formation of the two groups of planets. As we shall see, the million year time-scale and the mechanism of collapse of a gas–dust cloud are consistent with some observations on meteorites.

Meteorites

The Earth is constantly being bombarded by natural debris from space. Estimates vary, but 10000 tonnes per year may be an acceptable estimate of the rate. Most of this material consists of friable, dust-sized grains which are consumed in the upper atmosphere as meteors, or 'shooting-stars' and have trajectories indicative of a cometary origin. A bright fireball usually accompanies the atmospheric entry of the larger or tougher of such solid objects. Of the many fireballs reported annually throughout the world, only five or six result in the recovery of these extra-terrestrial objects, which, after landing, are known as meteorites. Such meagre information as is available indicates that meteorites come from the asteroid belt, the region between Mars and Jupiter occupied by the minor planets or planetary fragments, in orbit around the Sun. Gravitational interactions between asteroids and Mars or Jupiter cause collisions among the asteroids, which may ultimately send fragments into Earth-crossing orbits (and thence to museums).

Meteorites can be divided into three main types on the basis of their chemical composition. Stony meteorites are those most commonly seen to fall, whereas falls of iron or of stony–iron types are seldom witnessed. In addition to meteorites actually seen to fall (known simply as 'falls') there are also 'finds', meteorites discovered where they have lain for an unknown time. In general, falls and finds are named after their discovery positions on Earth. Over 2200 different meteorites are known; of these about 1400 are stony, 660 iron, and 80 stony–iron meteorites.

Stony meteorites are of two major types, chondrites and achondrites. Chondrites have never been completely molten, and hence are thought to yield the best samples of average, condensable, solar system material. In contrast, achondrites are essentially igneous rocks or planetary 'soils' composed of fragments of igneous rock and their constituent crystals. Only the chondrites are discussed here, but the nature and origin of meteorites are comprehensively treated by Sears (1978).

Some 80 per cent of falls are chondrites, which characteristically contain millimetre-sized spheres of silicate, called chondrules, set in a matrix of silicates, fragments of other chondrules, metallic iron-nickel, and the iron sulphide, troilite (Fig. 1.2). Such a mixture of silicate,

Figure 1.2 Ordinary chondritic meteorite, Parnallee, India, fell 1857. A shows prominent chondrules (C), and troilite (T). B illustrates the distribution of shiny metal, the chondrules being metal-poor. Width of specimen 7 cm.

metal, and sulphide is almost unknown among terrestrial rocks. Chondrites are subdivided into five major groups which are chemically distinct, with no intermediate types:

1 Enstatite chondrites have an atomic magnesium/silicon ratio (Mg/Si) of 0·75. They are highly reduced, contain no oxidised iron and have an unusual suite of sulphide minerals, including the calcium sulphide, oldhamite.

2–4 Ordinary chondrites comprise three groups, all of which have an atomic Mg/Si ratio close to 0·95. Metallic iron-nickel averages about 18 per cent, 8 per cent and 2 per cent by weight in the three groups, with some overlap. Thus from one group to the next there is a gain or loss of metal relative to silicate.

5 Members of the fifth group, the carbonaceous chondrites, have atomic Mg/Si ratios near 1·05. These meteorites are the most highly oxidised among the chondrites and, with few exceptions, contain little or no metal. Five members lack chondrules but are linked by their chemical similarity with those that contain them. Carbonaceous chondrites without chondrules are low-temperature mineral assemblages with up to 20 per cent water (Frontispiece). These meteorites have a chemical composition which best matches the abundances of the condensable elements in the Sun's photosphere – a thin layer of gas at the base of the solar atmosphere – from which emanates most of the Sun's light. Other carbonaceous chondrites have centimetre-sized white inclusions rich in calcium, aluminium, and titanium. These inclusions appear to be high-temperature condensates and may have been the first materials to have condensed from a nebula of solar composition (Fig. 1.3).

Various pieces of evidence indicate that chondrites were formed by the aggregation of their components on the surfaces of planetary bodies only a few hundreds of kilometres in radius, probably the asteroids. However, it is possible that carbonaceous chondrites have cometary origins. Most enstatite and ordinary chondrites formed by the accretion of cold particles and were subsequently heated, or else they were hot at the time of formation and cooled slowly; in either case chondrules and matrix were caused to crystallise and become integrated – compare Figures 1.2 and 1.4 – and a source of heat is thus required.

Whatever their origin, chondrites appear to be the only available samples of primitive solar system material. Isotopic age determinations by several methods show that many chondrites have remained unaltered for over 4500 million years. One of the first attempts to measure the age of formation of meteorites was made by Patterson in 1955. He measured the lead isotope ratios of several stony meteorites, and of the lead in the iron sulphide (troilite) from an iron meteorite. The isotope ^{206}Pb (lead-206) is at the end of the decay series of ^{238}U (uranium-238), and ^{207}Pb is at the end of the ^{235}U decay series. Because the uranium isotopes decay to lead at different rates, if uranium is present the ratio of ^{207}Pb to ^{206}Pb changes with time. In a closed system, the ^{207}Pb/^{206}Pb ratio defines the time since the system became closed, provided that the ^{207}Pb/^{206}Pb ratio at the time of closure is known. Iron sulphide contains negligible amounts of uranium, and so the lead extracted from the troilite present in the iron

Figure 1.3 Allende meteorite, with white inclusions of calcium, aluminium and titanium-rich minerals. FC = fusion crust. I = white inclusion. The matrix is largely composed of iron-rich olivine. Width of specimen 5 cm.

Figure 1.4 Barwell chondrite. One of the fragments of the stony meteorite shower which fell on Christmas Eve, 1965, at Barwell in Leicestershire. Note the black fusion-crust and rare chondrules in comparison with Fig. 1.2. Barwell and similar chondrites were either cooled slowly or reheated, causing crystal growth and merging of chondrules with matrix. Width of specimen 9 cm.

meteorite should have the primordial isotopic composition because lead produced by uranium decay should not be present in the troilite. Lead-204 has no radioactive progenitor. Patterson plotted meteoritic $^{207}Pb/^{204}Pb$ against $^{206}Pb/^{204}Pb$; the points defined a straight line on which lay the ratios obtained from the lead extracted from the troilite. Such a line is called an isochron; its slope is the $^{207}Pb/^{206}Pb$ ratio, which is consistent with an age of 4500 million years (Fig. 1.5).

Patterson went further and plotted on the same diagram the isotopic composition of lead in modern sediment from the Pacific Ocean; it was found to be within the limits of experimental error of the meteorite isochron, thus showing that the age of the Earth is consistent with the age of meteorites. It is to Patterson's credit that 4500 million years is still the generally accepted estimate for the Earth's age, but it should be remembered that we do not yet have a means of determining the age of the Earth independently of that of meteorites.

Meteorites often retain evidence of completely extinct radionuclides. One such example is ^{244}Pu (plutonium-244)

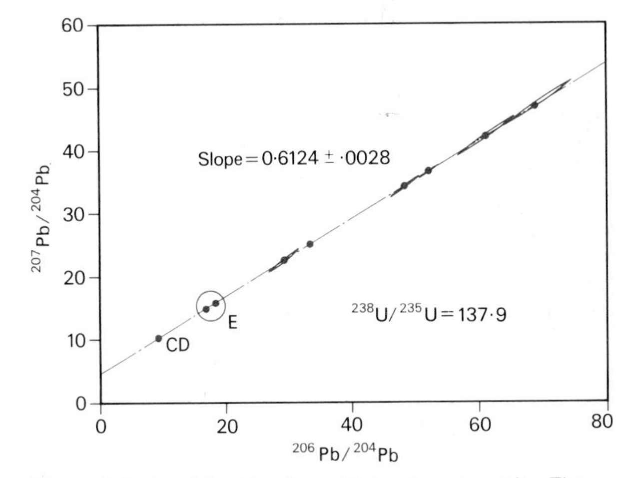

Figure 1.5 Lead–lead isochron, Richardton chondrite. The slope of the line represents an age of 4538 million years. CD = the isotopic composition of lead extracted from the troilite (FeS) of the Cañon Diablo iron meteorite. E = region of modern, terrestrial lead composition. (From data by Unruh, Tatsumoto and Hutchison.)

which decays with a half-life of 82 million years, producing a number of fission-products, notably isotopes of the noble gas, xenon. Some xenon extracted from meteorites and lunar rocks has an isotopic composition identical with that of xenon from artificially produced ^{244}Pu. But from our point of view another short-lived isotope, ^{129}I (iodine-129), is more important. Its half-life of only 17 million years enables us to use the decay of ^{129}I as a chronometer in determining the length of time it took the solar system to form after the event in which the radionuclide was synthesised.

The daugher product of ^{129}I decay is ^{129}Xe. Following the discovery in 1960 of an excess of this isotope in an ordinary chondrite, ^{129}Xe has been observed in a variety of meteorities. Today, meteorite samples for I-Xe determination are first irradiated with neutrons in a reactor. A proportion of the atoms of the only naturally occurring isotope of iodine, ^{127}I, are converted to ^{128}Xe. Gas is then driven from the sample by stepwise heating, and the ^{129}Xe/^{128}Xe ratio is measured in the gas released at each temperature. This allows the direct correlation of ^{129}Xe release with iodine abundance. There can be no doubt that ^{129}Xe occupies sites in crystals previously filled by iodine atoms. By converting the measured ^{129}Xe/^{128}Xe ratios to ^{129}I/^{127}I, the time of formation of each meteorite relative to an arbitrary standard may be calculated. Obviously, a high ^{129}I/^{127}I ratio indicates that a meteorite had formed and cooled to retain xenon before much of the ^{129}I had had a chance to decay. Surprisingly, we find that members of most classes of meteorites had formed and cooled within a 20 million year period.

Furthermore, we can extrapolate the ^{129}I/^{127}I ratios back in time. Because of its instability, it is unlikely that ^{129}I was synthesised in the same abundance as the stable isotope, ^{127}I. Also, some of the stable ^{127}I in meteorites and the solar system in general may have been synthesised in a number of events which predated solar system formation. When the ^{129}I/^{127}I ratio is extrapolated back ten half-lives, or 170 million years before the formation of meteorites, it becomes unacceptably high. Therefore we can safely conclude that the solar system formed within about 200 million years of the nucleosynthesis of some heavy elements (Fig. 1.6). Because some gas wells on Earth emit gases with excess ^{129}Xe it appears that parts of Earth are as old as meteorites, and that the Earth's interior has not yet been completely outgassed. This means that some of the gases incorporated within the Earth during its formation have not been lost to the atmosphere on heating.

Even more exciting than the excess of ^{129}Xe in meteorites was the discovery of excess ^{26}Mg (magnesium-26) in some components of carbonaceous chondrites. The isotope ^{26}Mg is the daughter product of ^{26}Al (aluminium-26), which has a half-life of only 740 000 years. It is possible that the white, calcium-aluminium-titanium-rich inclusions which have ^{26}Mg excesses originated outside the solar system. Alternatively, by analogy with ^{129}I, they might indicate that ^{26}Al was synthesised within about 5 million years of the formation of the solar system.

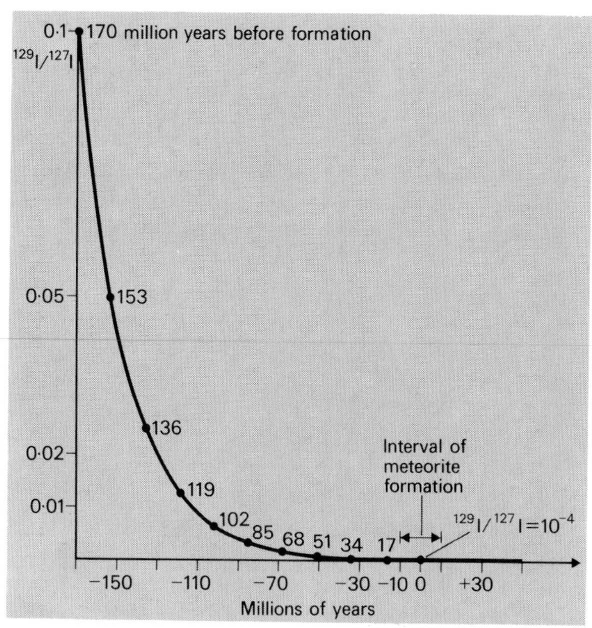

Figure 1.6 Iodine–xenon chronology of the time of formation of meteorites relative to the synthesis of ^{129}I. ^{129}Xe is expressed as the parent-isotope, ^{129}I. Zero time is the time of formation of many meteorites, when the ^{129}I/^{127}I ratio was approximately 10^{-4}. Extrapolated back ten half-lives of ^{129}I (170 million years), the ratio becomes 10^{-1}, and is approaching the theoretical upper limit. The ^{129}I must therefore have been synthesised within about 200 million years of the formation of meteorite parent-bodies.

But what was the process by which some of the heavy elements were synthesised? The most favoured mechanism involves the release of nuclear energy in the cores of dense, massive stars, thus causing them to explode. During these supernovae explosions, rapid nuclear reactions take place, producing heavy atomic nuclei. Such explosions send out shock-waves as matter is radially expelled. It is thought likely that a shock-wave from the supernova in which ^{26}Al was produced passed into a neighbouring gas–dust cloud and caused its collapse, thus triggering off the formation of the Sun and planets. The short half-life of ^{26}Al indicates that the whole process took less than a few million years, which is consistent with the views of astronomers. However, an earlier supernova explosion must then be postulated to account for the production of the longer-lived ^{244}Pu and ^{129}I. Again, this is compatible with observation for, as our galaxy rotates, each region passes through a dense, spiral arm every hundred million years or so, and it is in spiral arms that supernovae form. Thus, during one passage through a spiral arm the pre-solar gas–dust cloud may have been seeded with the debris from one supernova. During the next passage, some one hundred million years later, the shock-wave from a second supernova might have caused the collapse of the cloud and the introduction of newly synthesised ^{26}Al. It is evident from this discussion that the former existence of short-lived radionuclides is entirely consistent with a theory in which the Sun and planets formed together.

Figure 1.7 The Mare Imbrium photographed from the Earth. The mare is surrounded by bright highlands. Note the arcuate margin of the mare basin which is sometimes broken by craters, some of which were partially flooded by the later outpouring of lunar lavas. (Courtesy of Hale Observatories)

The Moon

The Moon, unlike the Earth, retains evidence of events which occurred more than 3800 million years ago. At least the upper few hundred kilometres of the Moon melted about 4400 million years ago and the low density mineral anorthite ($CaAl_2Si_2O_8$) floated on the melt to form a layer about 60 kilometres thick. Anorthite crystals, together with trapped liquid, solidified to form an ancient lunar crust dominated by anorthosite (a nearly pure anorthite rock) and by related rock-types. Today this ancient crust is exposed as the Lunar Highlands (Fig. 1.7).

Very few of the lunar rocks so far studied are older than 4200 million years because the early lunar surface was intensively bombarded, probably with the debris remaining from the earliest phase of planetary formation. Between 4200 and 3900 million years ago, a number of large objects struck the Moon, mostly on the side nearest the Earth. These major impacts formed huge basins, some of which are circular and it is thought that the ancient lunar crust extends beneath them. Outpourings of lava occurred from 3900 to 3200 million years ago in one basin or another, and they were wholly or partially filled with basaltic lavas. After the volcanism ceased, the surface features of the Moon altered little, and during the last 3200 million years the only change has been produced by sporadic impact cratering.

It seems likely that the Moon formed by the accretion of cold solid particles. Calculations indicate that even if it had formed in only 16000 years by the accretion of small bodies, its low gravity field would have resulted in an increase in temperature of only a few hundred degrees centigrade. A source of heat is therefore required to have produced partial melting by 4400 million years ago. Several possibilities have been suggested. Heating could have occurred as a result of the radioactive decay of the short-lived ^{26}Al, or the Sun may have emitted charged particles with a flux thousands of times stronger than the present solar wind. Impingement of charged particles on the Moon would have produced electrical heating. But whatever the heat-source, it is likely that the larger and more massive Earth would have been even more intensely heated than its satellite. Heat for the production of lava on the Moon between 3900 and 3200 million years ago was probably provided by the decay of the long-lived radionuclides – uranium and thorium – resulting, in a body of lunar size and over a long period of time, in a build-up of heat. Maximum temperatures were not attained until some hundreds of millions of years after the formation of the planetary body.

The Earth

Before speculating about the origin of the Earth it is first necessary to try to estimate its overall chemical composition and to establish its physical characteristics. This achieved, it is then possible to use data from meteorites to construct a theory for the Earth's formation.

There is compelling evidence that the Earth has a dense core. Rocks exposed at the surface have an average density somewhat less than 3, and the bulk density of the Earth is 5·52. Therefore there must exist, in the Earth's interior, material with a density greater than 5·52. The behaviour of earthquake waves indicates the existence of a core with a radius of 3500 km (Fig. 1.8). Since the Earth's radius is 6380 km, the upper boundary of the core is some 2900 km beneath the surface. The core has two parts – a solid inner core, with a density greater than 12 and a radius of 1300 km, and a liquid outer core of density about 10 and a thickness of some 2200 km. These high densities are partly accounted for by the high pressure obtaining at such depths, but they are also consistent with an inner core rich

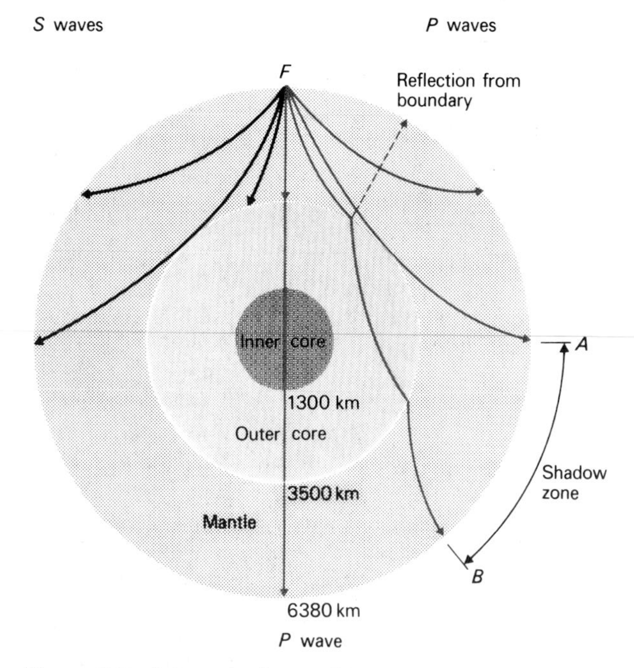

Figure 1.8 Schematic diagram illustrating the paths of S and P waves from an earthquake at F. Transverse (S) waves are not transmitted through the outer core which is, therefore, inferred to be liquid. Compression (P) waves are transmitted through the outer core. P wave FA is refracted upwards because of the increased velocity (increased density) with depth in the mantle. This is the limiting, core-grazing case. P wave FB enters the core but is partially reflected by the core-mantle boundary. The remainder is refracted downwards because of the decrease in velocity from mantle to liquid outer core. AB represents the 'shadow-zone' which receives no direct waves from the earthquake.

in iron-nickel and an outer core formed of metal plus some iron sulphide or some other light element such as silicon, carbon or oxygen. The core as a whole constitutes 32 per cent of the mass of the Earth.

The outer 68 per cent of the Earth's mass is attributable to the mantle: the crust, oceans and atmosphere can be disregarded here because their total mass is insignificant. We have access to material from depths of up to about 200 km. Blocks of a dense rock called garnet-peridotite (Fig. 1.9) are brought near the surface during the explosive emplacement of a rare type of rock called kimberlite (named after Kimberley in South Africa) and which is the primary source of diamond. Other rocks from higher levels in the mantle are transported to the surface during mountain-building or by basaltic volcanism, and there is now evidence for considerable chemical and mineralogical variation in the top 200 km of the mantle. However, for our purposes it is safe to assume that the accessible part of the mantle is dominated by the mineral olivine, $(Mg,Fe)_2SiO_4$: lesser amounts of pyroxene and garnet are present, but at the relatively low pressures obtaining at high levels, garnet is replaced by spinel. The formulae of these minerals are:

Ca-poor pyroxene	$(Mg,Fe)_2Si_2O_6$	(with some Ca and Al)
Ca-rich pyroxene	$Ca(Mg,Fe)Si_2O_6$	(with Al, Cr and Na)
Garnet (pyrope)	$Mg_3Al_2Si_3O_{12}$	(with Cr, Fe and Ca)
Spinel	$MgAl_2O_4$	(with Fe and Cr)

An estimate of the chemical composition of the top 200 km of the mantle can be obtained from these rocks (Table 1), but knowledge of rocks from greater depths is limited. The chemical composition of the top 200 km may apply to the upper mantle as a whole and may even apply to the lower mantle, (see Fig. 1.8). Alternatively, the lower mantle may have the composition of a low-calcium, iron-rich pyroxene. It is clear that we do not yet know how much iron oxide is present in the Earth, but for convenience the content of FeO in the top 200 km is taken as being representative of the mantle as a whole.

The bulk chemical composition of the Earth

In Table 1 the bulk composition of the Earth is first calculated by combining 32 per cent by weight of iron-nickel metal and iron sulphide, with 68 per cent silicate mantle. The second estimate uses various chemical ratios in chondrites and in the Earth to derive a total composition (see Anders, 1977). For example, uranium decay is apparently the main source of the Earth's internal heat, so the uranium content can be calculated from heat flow measurements at the surface. Ratios in chondrites of calcium to uranium, aluminium to uranium, and titanium to uranium are then used to fix the bulk calcium, aluminium and titanium contents of the Earth. The remainder of Anders' estimate of the planetary composition is similarly compiled, but additional assumptions are introduced.

These are only two of a number of estimates of the composition of the Earth. It must be remembered that the constraints on composition are few and all models are based on incomplete evidence. In spite of this, the two differently derived estimates presented in Table 1 are remarkably similar, the main differences being the lower MgO and higher Al_2O_3 and CaO contents in the Anders' model relative to the estimate from upper mantle plus core.

Neither estimate of the bulk composition of the Earth exactly matches that for a single group of meteorites, and this suggests formation either from the mixing of different kinds of meteoritic material, giving the appropriate composition, or from some kind of chondritic material which was subsequently reprocessed. Any processing after formation from chondritic material must have caused the loss of silicate relative to iron, for estimates of the total iron content of the Earth are about 35 per cent by weight and higher than in any group of chondrites. It is difficult to envisage the attainment of the high temperatures required to drive off silicate from a massive body like the Earth, and thus raise the iron to silicate ratio. Instead, a process called 'heterogeneous accretion' is now preferred.

This heterogeneous accretion hypothesis for the Earth's origin may be outlined as follows. The nebula from which the solar system formed somehow collapsed owing to

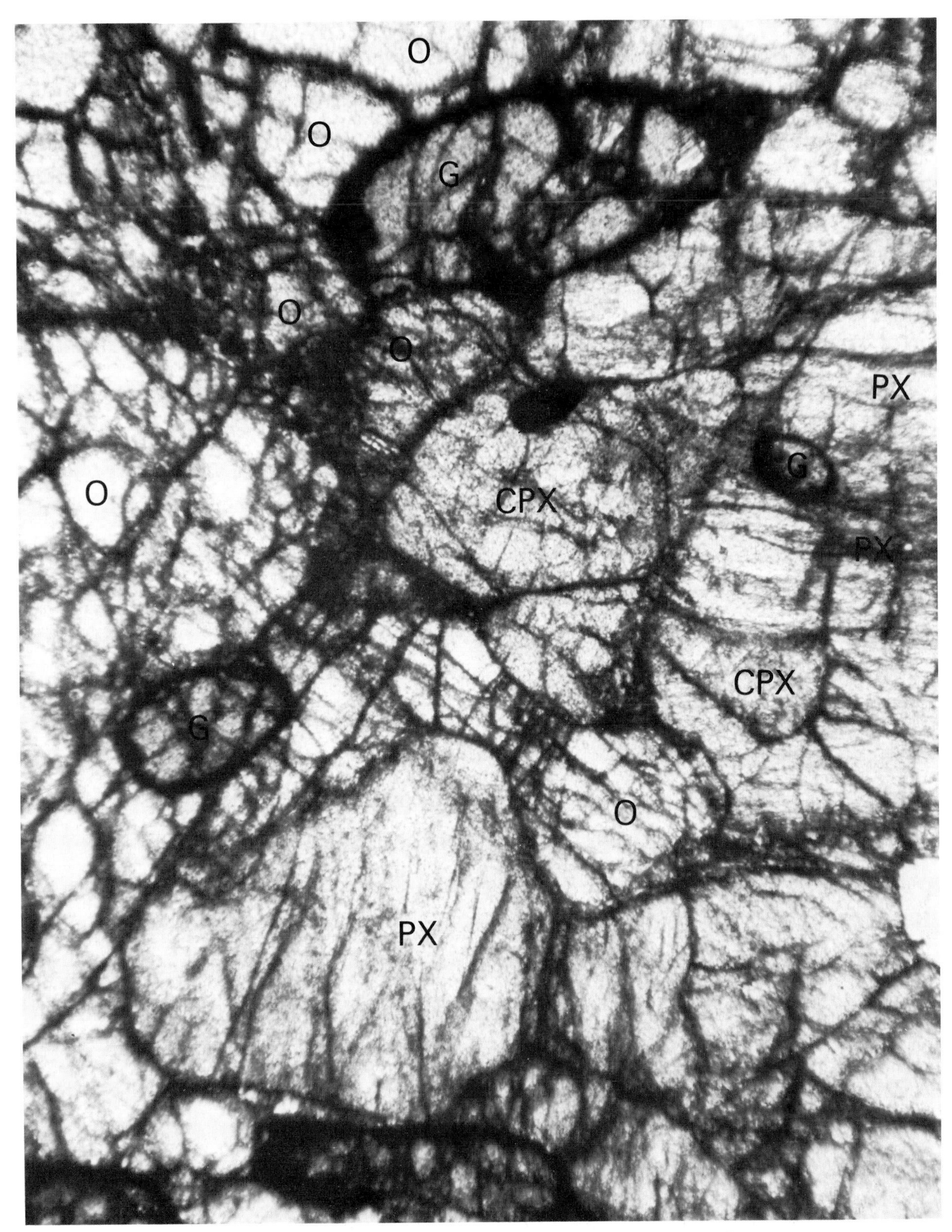

Figure 1.9 Thin-section of garnet-peridotite block carried to the surface in the De Beer's Mine kimberlite pipe, South Africa. Transmitted light. G = garnet; O = olivine; CPX = calcium-rich pyroxene; PX = calcium-poor pyroxene. The specimen has been partially altered by the secondary action of water. Width of field is 5 mm.

compression caused by a nearby supernova. A rise in temperature of the nebula was followed by cooling during which matter began to condense from the gas. The nebula at this stage was probably disc-shaped, with a thickening around the centre. Near the centre, where gas pressure was high, metallic iron-nickel condensed before the silicates, olivine and pyroxene. Early formed metal, together with grains of other refractory, condensed oxide and silicate minerals rich in calcium, aluminium and titanium, tended to become concentrated in the mid-plane of the disc, where they began to aggregate into small bodies orbiting around the denser centre of the nebula. Metal grains would be mutually attracted by magnetic forces if temperatures were sufficiently low whereas non-metals would be unaffected. Alternatively, because at higher temperatures metal is still more malleable than the brittle silicates and oxides, it would stand a better chance of growing into larger grains. Consequently, the first part of the Earth to form may have been the core.

As the metal-rich planetary seed grew, its gravity would have increased until it became sufficiently great to capture the silicates which it encountered in orbit. This process would have proceeded at an ever increasing rate until all the available solids had been accreted, when the Earth would have had a metal-rich core and silicate mantle but, like the present-day Moon, no water.

We know that the Moon partially or completely melted within a few hundred million of years of its formation. We know too that the parent-bodies of meteorites were heated some 4500 million years ago. Therefore it seems certain that the Earth must also have melted soon after its formation, and any metal or sulphide mixed with silicates of the mantle would have sunk to enlarge the core. Moreover, unless melting had been too rapid to allow nickel to become concentrated in metal and sulphide, the mantle would have lost all its nickel to the core. By contrast, the modern upper mantle contains about a quarter of one per cent of nickel oxide (Table 1), which is more than might be expected from a study of meteorites, for those containing both metal and silicate always have nickel strongly concentrated in the metal, with negligible amounts in the silicate. It is argued, therefore, that the upper mantle must have received a nickel-bearing component after core formation had ceased. Because of the high temperatures which prevailed after the Earth's formation, water and other volatiles must also have been added later. For further information the reader is referred to Turekian & Clark (1969), and for an account of the nature and origin of the solar system to Smith (1979).

The addition of both nickel and volatiles (water, CO_2, etc.) to the Earth indicates that one group of carbonaceous chondrites may have acted as carriers. Lunar 'soils' contain one or two per cent of a carbonaceous chondrite component identified by its 'fingerprint' – the ratios of refractory metals such as platinum and iridium. There is, therefore, evidence to support the theory that intense bombardment of the Earth by bodies of carbonaceous chondrite composition occurred before the first known water-laid rocks were deposited 3800 million years ago. Calculations show that the Earth needed to receive carbonaceous chondrite amounting to two or three times the mass of the Moon, if all the nickel in the upper mantle is to be accounted for in this way. Most of the water, together with compounds of carbon and other volatiles must have escaped from the Earth.

The early history of the Earth may be summarised as follows: after a period of early melting, the planet consisted of core and mantle, any primitive atmosphere having been driven off either by internal heating or by intense radiation from the Sun. The surface of the Earth cooled by radiation and solidified within a few hundred million years. Then, from perhaps 4400 to 3800 million years ago, it was intensely bombarded by objects remaining from planetary formation. In the later phases these were dominantly of carbonaceous chondrite composition and possibly cometary in origin. Below its brittle surface, the Earth was still molten and convecting vigorously, and incoming bodies would have blown holes in the thin crust. Under the prevailing high pressure gradient, anorthosite could not have formed so thick a layer as that on the Moon because, below about 30 km on Earth, anorthite transforms to the dense mineral garnet, and rocks rich in garnet would have foundered. The Earth's early crust must therefore have remained thin and easily breachable. The impact structures on Earth would have been as large, or larger, than the

Table 1 Chemical composition of the whole Earth

	1	2	3
SiO_2	45·0	30·6	30·7
TiO_2	0·09	0·06	0·17
Al_2O_3	3·50	2·38	3·34
Cr_2O_3	0·41	0·28	0·68
FeO	7·87	5·35	5·03
NiO	0·25	0·17	–
MnO	0·11	0·08	0·08
MgO	39·0	26·52	21·91
CaO	3·25	2·21	2·70
Na_2O	0·28	0·19	0·22
K_2O	0·035	0·02	0·02
Sil.	99·80	67·86	64·85
FeS		4·0	5·05
Fe metal		26·0	28·75
Ni metal		2·0	2·04
Total		99·86	100·69

1 Silicate upper mantle, from table IV, no. 8, in Hutchison, *et al.* 1975.
2 Bulk Earth, 68 per cent analysis 1, plus 32 per cent core. This model neglects chromium and titanium which may be in the sulphide of the core.
3 Bulk Earth composition of Anders, 1977.

circular basins on the Moon, that is, thousands of kilometres in diameter. Impact, with convection, would have ensured the mixing of newly added material over at least the top few hundred kilometres of the Earth. The Earth's gravity field was sufficiently strong to retain some volatiles, so that depletion in potassium, lead and water was not so extreme as on the Moon. Gradually the bombardment declined, an atmosphere was formed, and parts of the crust stabilised to become proto-continents.

Finally, a word of warning is required. Although the heterogeneous accretion hypothesis is currently favoured, new evidence may at any time lead to its rejection and the consequent re-examination of those difficulties which now render other theories less acceptable. For example, one detailed hypothesis (Ringwood, 1966) demands a primitive Earth with a composition akin to that of carbonaceous chondrites. Heating, accompanied by reduction of most of the oxidised iron and nickel, is postulated to have caused core formation and the present shelled structure. However, up to 20 per cent of the original mass of the Earth must subsequently have been lost, mainly as water and CO_2. Furthermore, in the upper mantle the atomic magnesium/silicon ratio is considerably higher than in carbonaceous chondrites, as is the total iron content. For these and other reasons the carbonaceous chondrite hypothesis has fallen into disfavour, but it might have to be reconsidered should heterogeneous accretion be discarded.

There remains the problem of the Moon. It has recently been suggested that the Moon's tungsten is now in the Earth's core, indicating that our satellite was derived from the Earth's mantle after core formation. If this were so, the loss of the tungsten and iron-poor Moon, together with other material dissipated in space, might account for the apparent iron-enrichment (relative to chondrites) of the residual Earth. The further one looks, the more diverse are the possibilities.

The origin of life

It is now appropriate to consider how life might have originated on Earth. There are two chief possibilities. Life could either have been introduced from elsewhere in the universe, a view currently unpopular, but recently supported by Hoyle & Wickramasinghe (1978); or it could have arisen on Earth directly from abiotic precursors, which will be discussed first. In their basic form the following ideas were originally put forward independently in the 1920s by the Russian, A. I. Oparin, and by J. B. S. Haldane in England. At some time before about 3800 million years ago, surface and atmospheric temperatures on Earth fell below 100 °C. The first precipitation must have been highly corrosive and rich in dissolved halogens or their compounds, such as hydrogen chloride. These would have been derived from a carbonaceous chondrite component; such gases still emanate from volcanoes today. These early acid rains presumably attacked surface rocks and dissolved sodium and potassium, giving rise to hot, primitive oceans. Carbonaceous chondrites today contain carbon in a variety of compounds with hydrogen, oxygen and nitrogen. Much of this carbon is in the form of 'tarry' hydrocarbons which are extremely difficult to extract and identify; collectively they are known as 'intractable polymer'. However, there occur smaller additional amounts of other carbon compounds, including amino acids. Thus a carbonaceous chondrite component would have supplied the young Earth with hot brines and the building blocks required for protein.

At this time the primitive atmosphere would have contained no free oxygen, so no ozone layer could have existed. Energetic ultra-violet radiation would have reached the Earth's surface unchecked, causing dissociation of surface water. Hydrogen would have been released to the atmosphere, while water tended to become oxygenated and a chemical potential would therefore have developed at the water–atmosphere interface. With a source of energy, such as ultra-violet light, and water, carbon, hydrogen and nitrogen available, chemical reactions were inevitable. Under sterile conditions in the laboratory and using the forms of energy and the materials available on the primitive Earth, it is relatively easy to synthesise a range of carbon compounds, including amino acids.

Amino acids are characterised by the presence of amino (NH_2) and carboxyl (COOH) groups. Two amino acid molecules may be joined together by the peptide bond (–CO–NH–) caused by the elimination of a water molecule, the H being derived from the amino, and the OH from the carboxyl groups. Polymerisation of amino acids then probably produced protein-like compounds called 'protenoids'. In brines, protenoids tend to concentrate in microspheres, globules about 1 μm in diameter and which, although abiotic, nevertheless possess certain properties resembling those of cells. But there is still an important distinction between microspheres and simple cells: the former neither replicate nor produce energy by feeding or photosynthesis as do true, biotic cells.

However, we may conclude that once complex carbon compounds had been synthesised, it would have become possible for some molecules to act as templates on which new ones could be built. This may be envisaged as being similar to the 'seeding' effect of a single crystal (for example, of salt) broken into fragments and dropped into a saturated solution. Each fragment acts as a nucleus on which crystal growth occurs. Furthermore, under certain conditions, microspheres become surrounded by a double layered wall which acts like a cell membrane. In such a primeval 'soup' some of the abiotically synthesised carbon compounds might have become 'food' for primitive cells in that they could pass through cell membranes and release energy by chemical reactions within the cell. (Such a possibility was recently put forward for sustaining Martian life, if it exists.) Given some hundreds of millions of years, viruses and bacteria may have arisen on Earth from abiotic carbon compounds.

The alternative hypothesis is that life on Earth originated

elsewhere. This theory is often known as *panspermia*, meaning seeds of life scattered everywhere in the universe. The question then arises as to how life arrived on Earth. An obvious answer is that it came with carbonaceous chondrites during the early bombardment of our planet. This idea has been taken further by Hoyle & Wickramasinghe (1978), who suggest that comets acted, and still act, as the couriers of life. There are good grounds for linking carbonaceous chondrites with comets.

In the 1960s, carbonaceous chondrites were examined for evidence of extraterrestrial life. Although there have been various reports of micrometre-sized 'organised elements', most have been discounted either as results of terrestrial contamination or as inorganic minerals. Carbon compounds were also investigated and it was found that laboratory mixtures of water, ammonia, and carbon monoxide could yield compounds in which the carbon isotopic ratios matched those of compounds in carbonaceous chondrites. In these experiments, heat, ultra-violet light or spark-discharge were used as sources of energy, and iron-nickel metal or serpentine minerals catalysed the reactions. However, the compounds extracted from both laboratory experiments and meteorites were found to have roughly equal amounts of both D and L optical isomers; in contrast, terrestrial organic compounds have a predominance of L isomers. Thus it appears that organisms on Earth yield compounds of carbon which can be distinguished from those produced under sterile conditions in the laboratory or extracted by solvents from carbonaceous chondrites.

Hoyle and Wickramasinghe argue that, in the cores of comets, reactions between substances such as water, ammonia, and carbon monoxide would cause heating. Hence the cometary cores could have contained water in which the synthesis of complex molecules may have proceeded. In some carbonaceous chondrites, there is indeed evidence for the former presence of liquid water. Both laboratory syntheses and the Hoyle and Wickramasinghe model require elevated temperatures; in contrast, recent measurements of deuterium/hydrogen ratios in the carbon compounds of some carbonaceous chondrites indicate that deuterium was fractionated from hydrogen below 0 °C. If correct, this indicates that the laboratory experiments and the Hoyle and Wickramasinghe theory may not be appropriate to the formation of carbon compounds in carbonaceous chondrites.

There is no overwhelming evidence favouring the Hoyle and Wickramasinghe hypothesis rather than that of a spontaneous origin of life on Earth. In favour of the latter, Sagan & Khare (1979) argue that viruses could not be transmitted to Earth by comets, as Hoyle and Wickramasinghe propose, because viruses are specific to the cells of terrestrial organisms which they attack. Such organisms evolved in the conditions specially obtaining on our planet, so the viruses must have evolved together with the higher organisms.

References

Anders, E. 1977. Chemical composition of the Moon, Earth, and eucrite parent body. *Philosophical Transactions of the Royal Society of London*, Series A, **285**: 23–40.

Bernal, J. D. 1967. *The origin of life.* 345 pp. London: Weidenfeld & Nicholson.

Fox, S. W. & Dose, K. 1972. *Molecular evolution and the origin of life.* 359 pp. San Francisco: Freeman.

Hoyle, F. & Wickramasinghe, N. C. 1978. *Lifecloud.* 189 pp. London: Dent.

Hutchison, R., Chambers, A. L., Paul, D. K. & Harris, P. G. 1975. Chemical variation among French ultramafic xenoliths – evidence for a heterogeneous upper mantle. *Mineralogical Magazine.* **40**: 153–70.

Ringwood, A. E. 1966. Chemical evolution of the terrestrial planets. *Geochimica et Cosmochimica Acta* **30**: 41–104.

Sagan, C. & Khare, B. N. 1979. Tholins: organic chemistry of interstellar grains and gas. *Nature*, London, **277**: 102–7.

Sears, D. W. 1978. *The nature and origin of meteorites.* 187 pp. Bristol: Adam Hilger.

Smith, J. V. 1979. Mineralogy of the planets: a voyage in space and time. *Mineralogical Magazine* **43**: 1–89.

Turekian, K. K. & Clark, S. P. 1969. Inhomogeneous accumulation of the Earth from the primitive solar nebula. *Earth and Planetary Science Letters* **6**: 346–8.

Williams, I. P. 1975. *The origin of the planets.* 108 pp. Bristol: Adam Hilger.

Wood, J. A. 1979. *The solar system.* 196 pp. New Jersey: Prentice Hall.

R. Hutchison
Department of Mineralogy
British Museum (Natural History)

PART II

The evolution of continents

Introduction

page

This section of the book has enormous range. It starts with the largely molten world of the Pre-Archaean, more than 3800 million years ago, and finishes with the soils round Hemel Hempstead in Hertfordshire, England (Fig. 5.3). Its emphasis is historical, or on the use of geochemistry and sediments as historical documents. The main exception is the chapter describing soils, for these have overwhelming practical importance. History is interesting, but practical needs come first.

Curiously enough the historical picture that now emerges from geological and geochemical data is somehow like the Book of Genesis. This described a similar passage from the fantastic to the commonplace. At the beginning 'the earth was without form and void; and darkness was upon the face of the deep.' As the Book of Genesis goes on, however, the features of the real world became more and more recognisable, and more prosaic. Adam begat a son at 130 years and lived for another 800. At the end of the Book, however, Joseph dies at the almost believable age of 110. In a like way modern science passes gradually from the strange Pre-Archaean ocean; through the origin of geosynclines and life; into the Phanerozoic in which all the modern phyla of animals are from the start represented; and so finishes in the Outer Suburbs of London.

The main difference between the two accounts is a millionfold increase in time range and greatly improved documentation. Our earliest scientific documents are greenstones from Greenland almost 4000 million years old. It is no wonder that they show the world as a thoroughly different place.

The fall of the Book of Genesis was a gradual one, and views about later events mostly changed before those about earlier ones. One of the first changes came with the late eighteenth-century discovery (to Europeans) of Sanskrit which, by establishing the Indo-European family of languages, threw doubt on the myth of the Tower of Babel. Later, as the nineteenth century passed, the Flood was replaced by glaciers, extinct fossil species became known,

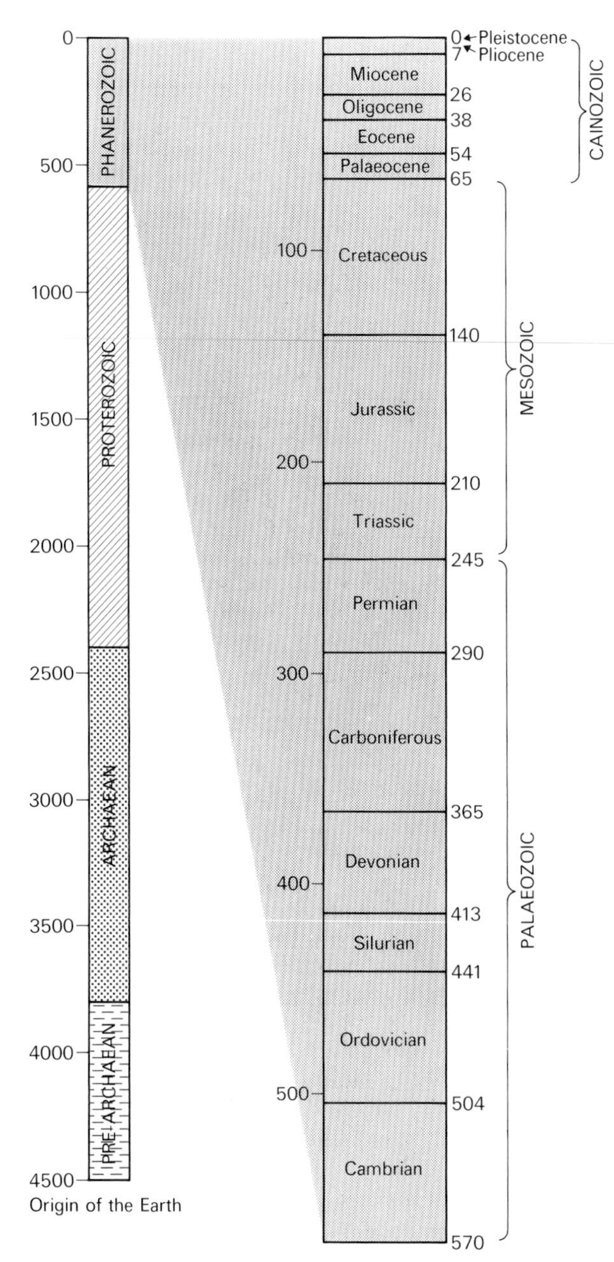

Figure II.1 The geological time scale

and the fixity of species was overthrown by Lamarck, Wallace and Darwin. Distinctive twentieth-century contributions were radioactive dating, the origin of plate tectonic theory, and the elucidation of the atomic structure of minerals. Without doubt the new history is still incomplete and partly wrong. By the year 2081 it will be greatly altered but it is coherent and complex and therefore its main features will stand.

The opportunity is taken in this Introduction to present some simple geological facts to amplify the succeeding chapters. The foundation of geologists communication is the geological time scale (Fig. II.1), which shows the time since the formation of the Earth divided into a succession of eras, the most recent of which is the Phanerozoic, starting at about 570 million years ago. This era includes all strata yielding fossils with hard parts, and has itself been divided into a number of geological systems (commencing with the Cambrian), largely based on the successive faunas and floras which the rocks of that system contain. The correlation of rocks of each system, both over local areas and round the world, is effected by the comparison and analysis of their contained fossils. The attachment of ages in years to these subdivisions has been made possible by the study of radioactive decay in some long-lived isotopes formed at the same time as igneous rocks crystallised. When such igneous rocks are interbedded with fossiliferous sedimentary rocks, then the real age in years of the latter can be deduced. However, for detailed relative ages and correlation work, fossils provide finer subdivision than isotope ages, since successive fossil faunas may evolve over no more than half a million years, whereas isotopic dates are rarely accurate to within less than 2 or 3 percent – a considerable margin of error when rocks older than the Mesozoic are being considered.

Rocks are divided by geologists into three main classes: firstly, sedimentary rocks, laid down by gravity under air or water, or precipitated from solution at normal surface temperatures and pressures; secondly, igneous rocks, which have cooled from liquid melts from within the Earth's surface; and thirdly, metamorphic rocks, which are pre-existing rocks of any origin which have been dragged down so that they have become wholly or partly recrystallised by the increased heat and pressure at depth. Sedimentary rocks include the familiar limestone, coal, sandstone and mudstone; igneous rocks granite and lavas; and metamorphic rocks marble and slate. Each rock is usually made up of several distinct minerals. Chapter 4 deals with the various environments in which the different sedimentary rocks were laid down, and with the evolution of those environments through geological time. Igneous rocks are perhaps less familiar to the reader, but they form the vast bulk of the rocks present in the Earth's crust, and Figure II.2 shows some of their commoner types of occurrence. If the molten magma reaches the Earth's surface as a volcano or submarine lava flow (usually within the temperature range 850 to 1200 °C) then the subsequent crystallisation is usually rapid and the resulting rocks fine-grained. The most abundant fine-grained lava type is a dark-coloured rock termed basalt. If, however, the magma crystallises slowly at depth, then the resulting rocks will have larger crystals and be coarse-grained. As can be seen from Figure II.2, sometimes fingers and sheets of magma can penetrate from a larger magma reservoir up into the so-called country rock. If these sheets are essentially vertical then they are termed 'dykes', and if parallel to the grain of the strata then they are called 'sills'. Since different rock types have different rates of weathering, then dykes and sills weather differentially at the surface; a good example is the Whin Sill of Cumbria upon which some

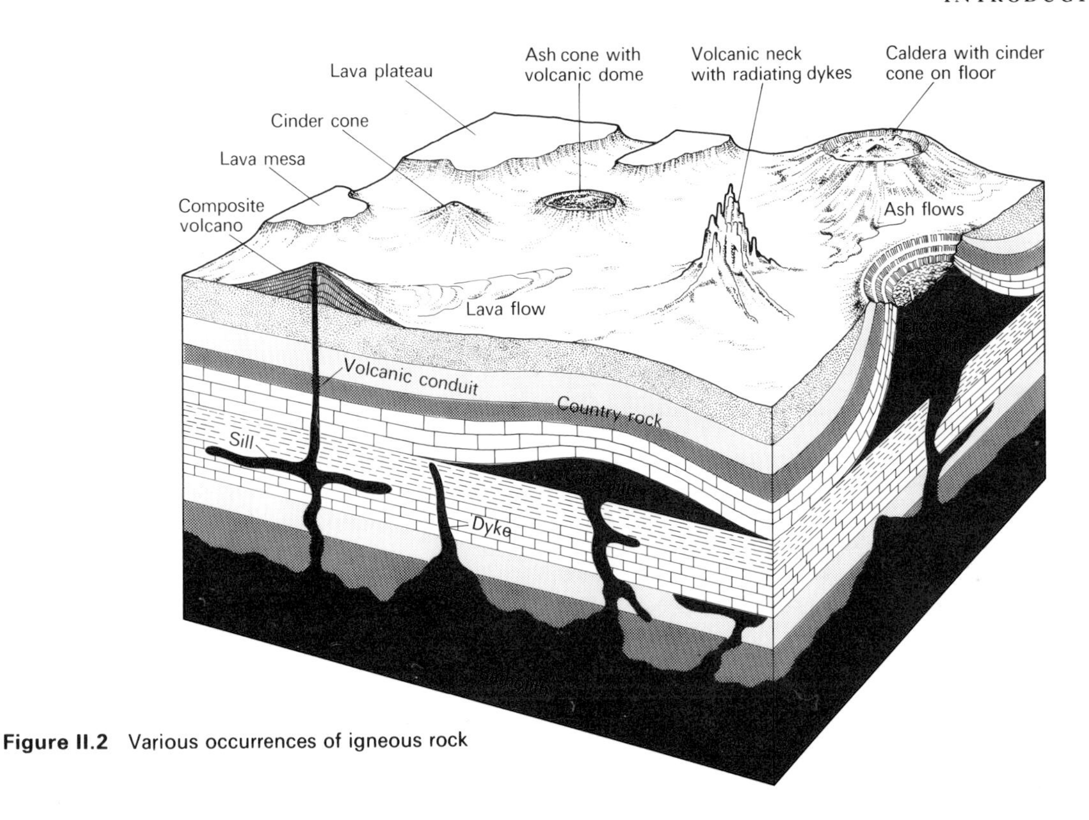

Figure II.2 Various occurrences of igneous rock

of Hadrian's Wall is built and which stands out from the surrounding limestone into which it was intruded. Chapter 3 goes into this subject in more detail, and also indicates how and where some ore deposits of economic value are found.

Metamorphic rocks are formed locally by the heating of 'country rock' (see Fig. II.2) round an igneous intrusion, a process known as 'contact metamorphism'. However contact metamorphism is a small-scale phenomenon by comparison with 'regional' metamorphism, when whole areas of pre-existing rocks are dragged down deep into the Earth's crust. The actual type of metamorphic rocks which form are partly dependent on the amount of excess temperature and pressure to which they are subjected, and partly on the minerals already present in their sedimentary or igneous precursor.

Sediments are usually deposited in horizontal layers, and yet most people have seen sedimentary rocks in cliff faces that are tilted or even distorted into zigzag or rounded folds. This folding or bending of rocks can be found on all scales, and the folds are often accompanied by cracks in the rocks. Where no displacement has occurred along these cracks then they are termed, 'joints', but where relative movement has taken place then they are called 'faults'. Progressive stresses over long periods can give displacements along faults which vary from a few centimetres to thousands of kilometres, and the larger faults can represent fundamental fracture zones in the Earth's crust. Movement along faults is one of the chief causes of earthquakes today.

Geologists chiefly work on rocks that have been brought up by earth movements (so-called 'tectonic activity') and subsequently exposed at the surface by weathering. Because some such rocks were buried quite deeply in former geological times, therefore we can gain direct knowledge of the types or rocks which occur within about 40 km of the Earth's surface. Deep boreholes have not extended this information, since the deepest is less than 10 km. It therefore follows that all our knowledge of the Earth below 40 km depth has been gleaned by indirect methods. These are the speciality of the geophysicists, who analyse the reflections of earthquakes, atomic explosions and other tremors, and see how these shock waves have been deflected by the differing layers within and beneath the crust. Other geophysical information has come from measurements of density, heat flow, electrical resistivity, gravity and magnetism, all combining to give us the understanding of the Earth which we possess today.

R. P. S. Jefferies
L. R. M. Cocks
British Museum (Natural History)

CHAPTER 2

Growth and development of the continents

A. R. Woolley

Knowledge of the Earth's evolution is comparable to that of modern history in that the most recent periods are known in great detail, as are the forces and processes that helped shape them. However, the further back one goes in time the larger become the gaps in our knowledge and the less certain are our conclusions. In this chapter an attempt is made to sketch the history of the continents, but so rapid is the pace of current research that this account would have been very different had it been written only twenty years ago – a clear indication that we still have much to learn.

The Pre-Archaean Earth

The oldest and longest period of Earth history is the Precambrian, which has been sub-divided into an older Archaean Era and a younger Proterozoic Era. Since the oldest-known Archaean rocks were formed some 3800 million years ago, there was a period of about 700 million years between the time of the Earth's formation and the crystallisation of these ancient rocks; this is called the Pre-Archaean. The earliest events in the Earth's history were outlined in the previous chapter, but a synopsis is given here for the sake of continuity. Again it must be stressed that any account of this very early phase of Earth history must necessarily be speculative in the absence of direct evidence. Rocks slightly older than 3800 million years may yet be discovered but it is unlikely that they will be much older than 4000 million years, because the instability and high temperatures in the crust prior to that time would have prevented the radiogenic isotopes used to determine the ages of ancient rocks from becoming stabilised in minerals.

After initial accretion, the Earth's temperature rose and a molten silicate mantle formed above an iron-nickel core. Subsequent cooling, partial crystallisation, and differentiation of the mantle eventually led to the formation of an upper liquid layer of basaltic composition. As this basaltic magma began to freeze, three principal minerals, olivine, pyroxene and calcium-rich plagioclase feldspar (anorthite) separated out. The first two minerals sank through the basaltic liquid, but the lighter plagioclase would have collected at the surface and cooled to form anorthosite rock. This anorthosite layer gradually thickened and strengthened, eventually covering the whole of the globe (Fig. 2.1A). However, contraction, and convection in the

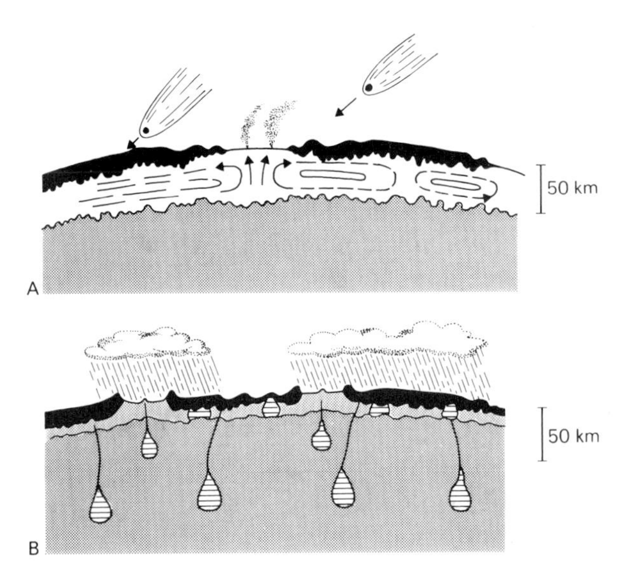

Figure 2.1 Sections through the surface of the Earth towards the close of the Pre-Archaean Era (after Shaw, 1976). (A) Anorthosite (solid black) floats on a still molten, convecting and degassing layer of basalt, beneath which is a layer of solid gabbro. There was frequent bombardment of the surface by meteorites.
(B) Fractional crystallisation of the basalt produces a granite layer 10–15 km thick (pale blue) beneath the anorthosite crust. Precipitation of olivine and pyroxene from the basalt, together with trapped plagioclase feldspar, thickens the underlying gabbro. Heat convected from below, together with pressure release along contraction fractures, yields molten rock reservoirs (horizontal shading). Hot acid rains are falling but the surface is still too hot for water accumulation.

underlying liquid basaltic magma, probably broke the anorthosite into irregular rafts.

Separation of olivine, pyroxene, and anorthite from basaltic liquid gradually changes the composition of the liquid until the final fraction is chemically close to granite. Such changes must have happened in Pre-Archaean times with the result that final consolidation produced a worldwide granitic layer beneath the anorthosite (Fig. 2.1B). This 'granite' was probably about 10–15 km thick, and was underlain by a layer of gabbro (the coarse-grained equivalent of basalt) produced by the accumulation of sinking olivine and pyroxene crystals, together with some plagioclase trapped during the settling process.

The upper part of the Earth may thus have comprised three layers (Fig. 2.1B), but a number of processes tended to disrupt this simple pattern. The effects of contraction and convection have already been mentioned, but there would also have been a considerable bombardment by meteorites, many of which were probably very large: their impact would have disrupted and overturned the pristine 'proto-crust' and caused local remelting.

As the crust continued to cool it would fracture, and volcanoes formed along the fracture zones (Fig. 2.2A). Radioactive heating was probably significant in initiating this volcanism. During the separation of the core and mantle, uranium, thorium, and potassium would move upwards to accumulate in higher levels of the mantle, there to generate considerable heat (see Fig. 2.3). The greater part of the heat still being lost from the Earth probably comes from the decay of these elements. It is thought that the early crust was probably unstable because the early-formed anorthosite layer overlaid a lighter granitic layer produced somewhat later by differentiation of the basaltic mantle. Greater isostatic stability was achieved as less dense granitic magma erupted at the surface or intruded into the anorthosite (Fig. 2.2B). Gradually, less dense crust must have become elevated relative to the denser surrounding material and it would have developed deeper roots in compensation (Fig. 2.2C).

No hydrosphere could exist at this time because surface temperatures were greater than 100 °C, but continued cooling eventually led to the condensation of water as rain. This, as it drained off the more elevated areas, caused weathering, erosion, the transport of weathering products, and finally the accumulation of sediments (see Chapter 4). Weathering must have been very rapid, because the rain water would have contained large amounts of hydrochloric and carbonic acids. It would have leached soluble carbonates, sulphates, halides, borates and other salts from the surface rocks and thus rapidly became saline as it accumulated in the lower-lying areas between the unthickened parts of the crust (Fig. 2.2D). In this way the differentiation of oceanic from continental crust was initiated. However, because the crust was still relatively thin in Pre-Archaean times, the oceans were not so deep nor the continents so elevated as they are today. Much of the Earth would have been covered by a shallow ocean, perhaps about two kilometres deep, standing clear of which were volcanoes surrounded by aprons of radially accumulating volcanic sediments.

Until this time the atmosphere was denser than at present and dominated by H_2O, CO_2, CO, N, SO_2, HCl and a few other trace gases (see Chapter 8); little solar radiation could have reached the Earth's surface. However, as the hydrosphere developed, solar radiation, particularly

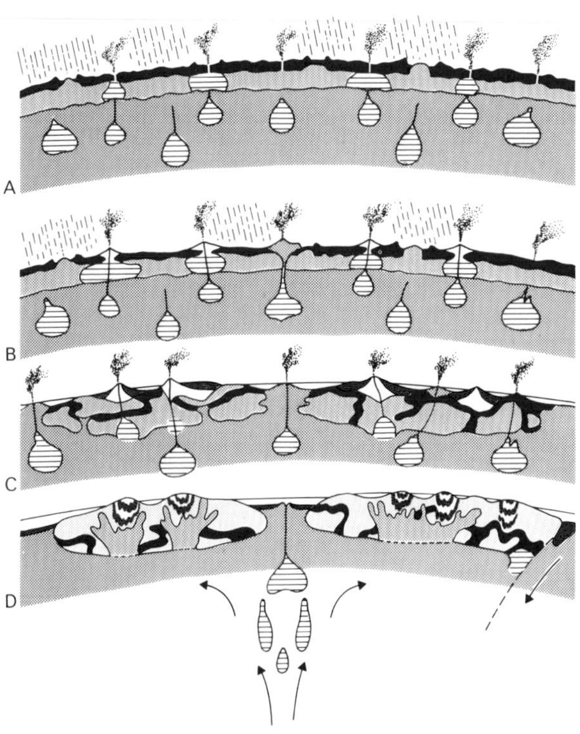

Figure 2.2 Sections through the surface of the Earth at the close of the Pre-Archaean and the early Archaean Eras (after Shaw, 1976).
(A) Contraction fracturing of the surface allows gases to escape and volcanism to begin. An early atmosphere is developed and rain falls.
(B) Volcanism continues with the extrusion of basaltic and granitic magmas, forming large volcanoes and starting to cover the anorthosites.
(C) The upper part of the crust is now a mixture of extrusive volcanic rocks, sediments, and igneous intrusions. Where these rocks locally thicken the crust, its elevation is increased as a result of isostatic compensation. The crust is cool enough for water to condense so that seas now cover low-lying areas.
(D) Thickening of the crust leads to the melting of the granitic rocks at its base which intrude to higher levels. The crust is now more coherent, but starts to break up into blocks carried above convection cells that are developing in the mantle. Separation of the crust exposes the underlying gabbroic layer on to which the water drains from the more elevated continental areas. The leading margin of the right hand continent begins to over-ride oceanic crust and so initiates a subduction zone (see Chapter 11).
Granitic layer, pale blue; sediments, solid blue; anorthosites, solid black.

in the ultra-violet band would, in the absence of oxygen, penetrate to the surface. Atmospheric temperatures must still have been high because, although the Sun then emitted less heat than it does now, the high concentration of CO_2 in the atmosphere would create a 'greenhouse' effect.

This then was probably the situation towards the end of the Pre-Archaean era, perhaps 3900 million years ago. The differentiation of oceanic from continental crust had been initiated, although the crust was rather thinner than today. Sediments had begun to accumulate around elevated areas and in shallow basins, but were subordinate in amount to igneous rocks. These immature sediments were clastic rocks derived with little sorting or transport from volcanic rocks, together with some carbonate, halide, and sulphate precipitated from the expansive oceans. The lack of extensive upland areas and deep ocean basins, together with the reducing atmosphere and the absence of living organisms, inhibited the production of the extensive, reworked and well-sorted sediments characteristic of later eras.

The Archaean Era

The oldest reliably dated rocks occur near Godthaab, west Greenland, and are about 3800 million years old (Moorbath, 1977). Since the age of the Earth is estimated to be about 4500 million years, this leaves only 700 million years for its evolution to the stage at which the rocks in the Godthaab area could have formed, and for the initiation of geological processes which continue to the present day.

The oldest Godthaab rocks are the Isua Supracrustals: 'supracrustal' is a term used for rocks formed at the Earth's surface. The Isua Supracrustals consist principally of a sequence of subsequently metamorphosed volcanic rocks and banded ironstones, although a little marble and a conglomerate also occur. These rocks are distributed around a dome-shaped area of gneiss, known as the Amitsoq Gneiss, which seems originally to have been a series of granitic intrusions that were later metamorphosed. There is a complex sequence of other igneous, metamorphic and sedimentary rocks in the Godthaab area, all of which are younger than the Isua Supracrustals and the Amitsoq Gneiss, and range from 1600 to 3700 million years old.

The relationship between the Isua Supracrustals and the Amitsoq Gneiss is not clear because, although contacts are exposed, they are strongly deformed. However, the occurrence of supracrustals as inclusions in the gneiss indicates that they are the older, and this is confirmed by isotopic dating. It also presents a problem, because the supracrustals were deposited on a surface of which no trace now remains. Solid crust was therefore exposed in the Godthaab region before the Isua Supracrustals were deposited, but its nature and fate are unknown.

Archaean rocks older than 3000 million years are known from all the present-day continents; they consistently comprise two principal suites known as 'granitic-gneiss terrains' and 'greenstone belts' which correspond respectively to the Amitsoq Gneiss and Isua Supracrustals of west Greenland. Granite-gneiss terrains are formed of banded granitic or granodiorite gneisses that have been strongly deformed and heated. The greenstone belts, on the other hand, comprise basic volcanic rocks and sediments that have often been but mildly deformed and heated, so that they contain minerals (like chlorite and hornblende) characteristic of relatively low degrees of metamorphism. The colour of these minerals tends to give the rocks a greenish tinge, hence the name 'greenstone'. The low grade of metamorphism usually allows the nature of the original rocks to be seen.

About 80 per cent of exposed Archaean rocks are of the granite-gneiss terrain type. They are usually strongly deformed and intensely metamorphosed to a degree known as 'granulite' facies and most known granulite facies rocks are of Archaean age. Granulite facies metamorphism is caused by high temperatures but relatively low pressures, and thus indicates that the granite-gneisses were never deeply buried. There is no doubt that geothermal gradients were steeper in Archaean times than at present (Fig. 2.4), probably because the crust was thinner and radiogenic heat production was much higher (Fig. 2.3). The thermal gradient in early Archaean times must have been in excess of 30 °C km^{-1}, and possibly as high as 60 or 70 °C km^{-1}, in contrast to the present gradient of about 10 °C km^{-1}.

A further important rock-type of granite-gneiss terrains is calcic anorthosite. Archaean anorthosites are much more voluminous than those formed during later periods and appear to be characteristic of the early Earth. They usually occur as large layered intrusions up to 500 km in length. It is not clear how they were formed, but it is probably significant that anorthosites are important components of

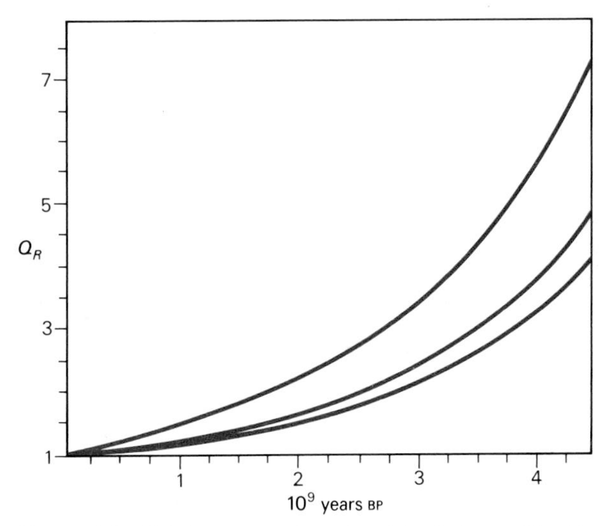

Figure 2.3 Relative production of radiogenic heat within the Earth according to three different hypotheses. Q_R, ratio of heat production in the past to current heat production; BP, years before present (after Dickinson and Luth, 1971).

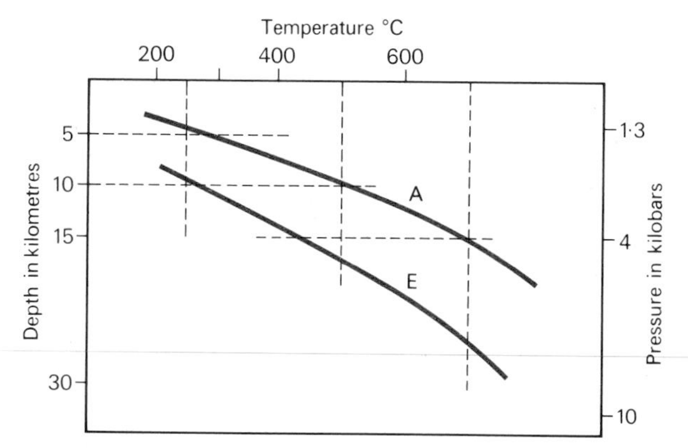

Figure 2.4 Geothermal gradients in the Archaean (A) and for the present day (E) (after Ray, 1970).

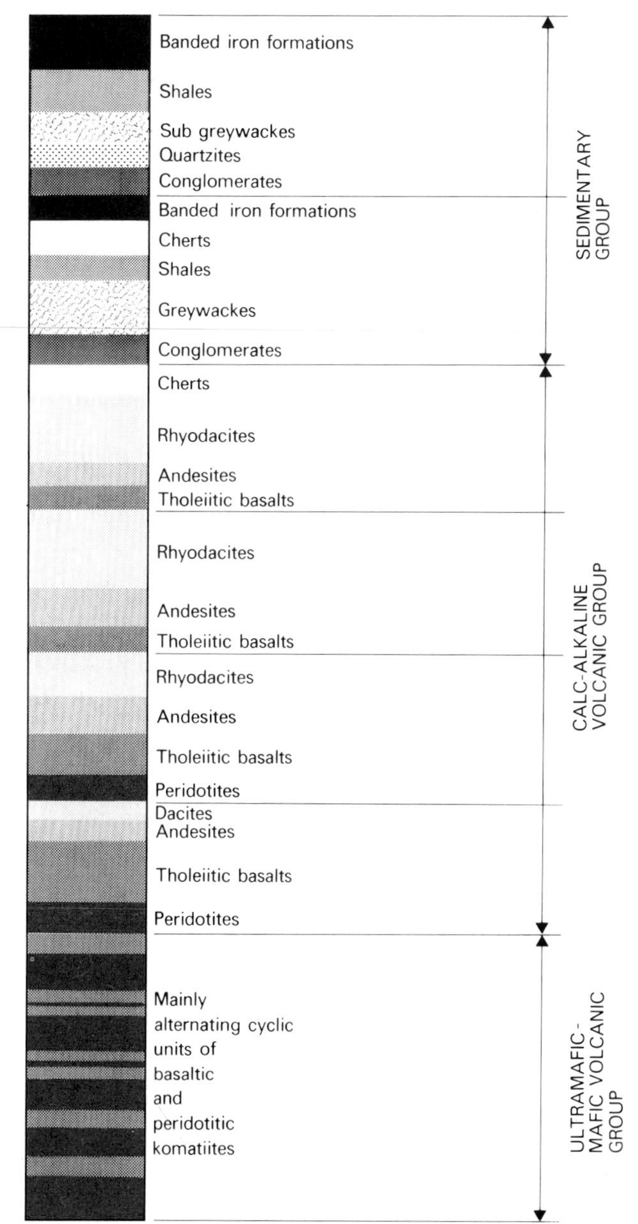

Figure 2.5 An hypothetical stratigraphical succession for an Archaean greenstone belt (based on the data of Anhaeusser, 1971a for the Barberton Mountain Land, Swaziland, but modified by Windley, 1977). Note the relative abundance of basalts and komatiites in the lower part of the succession. Increasing volumes of sediments, including banded ironstones, occur towards the top. Dacites and rhyodacites are extrusive rocks (lavas) with chemical compositions similar to those of some granites.

the highland areas of the Moon, and that the early crust has been postulated as being anorthositic.

Whereas the granite-gneiss terrains were probably formed by the emplacement of large granitic intrusions, the greenstone belts contain a high proportion of basaltic, andesitic and rhyolitic lavas. Some of the basalts show pillow structures characteristic of lavas that have erupted under water. Some of the lavas, known as komatiites (see Fig. 2.5) are exceptionally rich in magnesium: these are apparently restricted to the Archaean. The high magnesium content implies higher melting temperatures than those of normal basalts and this in turn indicates both a thinner crust and higher temperature gradients than at present.

The occurrence of immature sedimentary rocks, particularly in the upper parts of greenstone belt assemblages, suggests that land was beginning to rise from the seas in which the sediments could have been deposited. (Figs. 2.5 and 2.6). The overall thickness of these sequences may reach 20,000 m, as for example in the Barberton Mountain Land of Swaziland, the stratigraphy of which is shown in Figure 2.5.

The form of greenstone belts and granite-gneiss terrains is distinctive and characteristic of the Archaean. The granite-gneisses form oval or circular areas separated by relatively thin bands of greenstones (Figs. 2.7 and 2.8) which form synclines, often of very complex structure, between the ovals (Fig. 2.9). It is probable that the rocks of the greenstone belts were once much more extensive than they are at present, but were substantially reduced by erosion so that only the deeper parts of the troughs have survived. Although contacts between greenstones and granite-gneisses are often exposed, their relative ages can rarely be determined because of shearing. However, some greenstones rest unconformably on granite-gneisses with conglomerates above the contact. This, together with an increasing number of very old isotopic dates derived from granite-gneiss terrains, suggests that the latter are generally the older. There is, however, controversy between those who maintain that the greenstones were deposited in a primitive ocean, with the granite-gneisses representing continental blocks, and those who contend that the greenstones themselves represent early crust that was subsequently intruded by granites generated in the mantle. In recent years several workers have pointed out the chemical and lithological similarities between the igneous and sedimentary rocks of greenstone belts and those associated

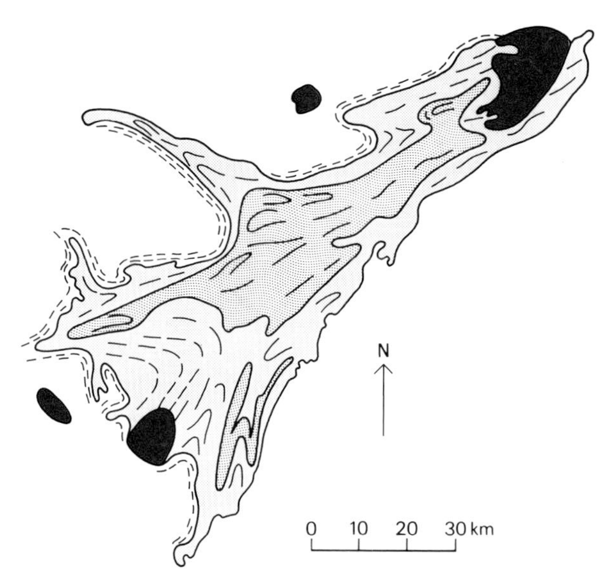

Figure 2.6 Simplified map of the Barberton greenstone belt, Swaziland (modified after Anhaeusser, 1971). Late intrusive granites, solid blue; lower part of succession, dominantly of volcanic rocks, blue shading; upper part of succession, dominantly of sedimentary rocks, light stipple.

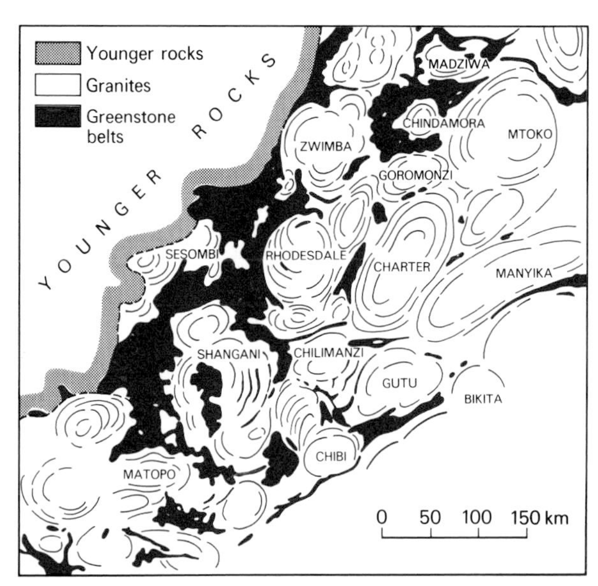

Figure 2.7 Part of the Archaean of Zimbabwe (Rhodesia) showing the circular to elliptical forms of the granite-gneiss terrains, with the greenstone belts between them (after MacGregor, 1951).

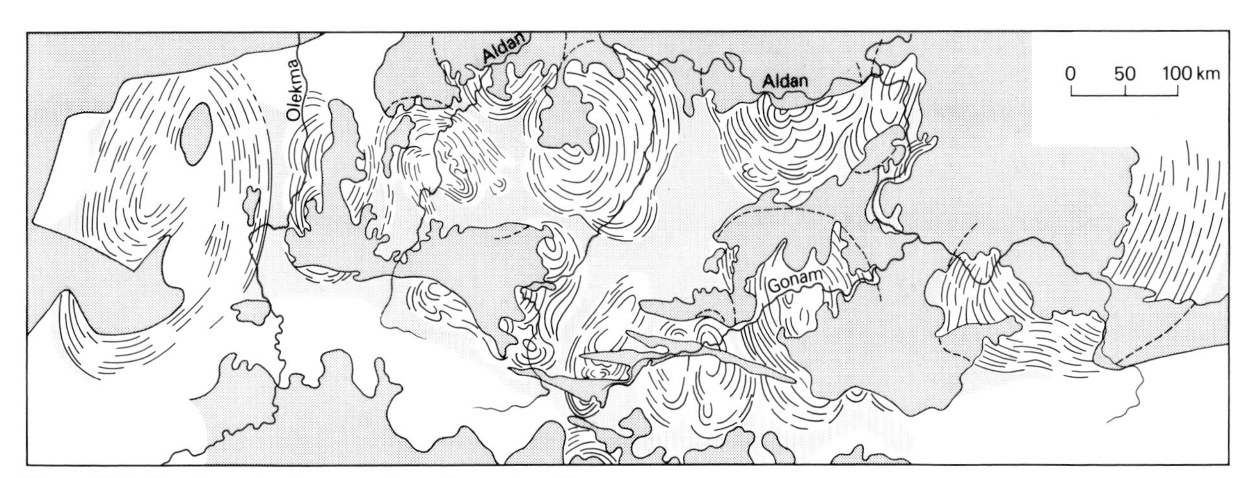

Figure 2.8 Very large circular areas of granite-gneiss with greenstone belts (pale blue) between them, in the Archaean of the Aldan Shield, USSR (modified after Salop, 1977). Grey stippled ornament indicates younger rocks.

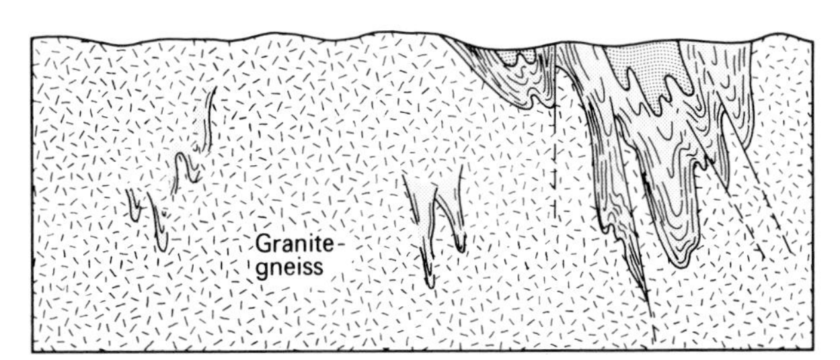

Figure 2.9 Cross-section of a greenstone belt (modified after Engel, 1968). The lower part of the succession (pale blue) is dominantly volcanic, the upper part (stippled ornament) dominantly sedimentary. Remnants of greenstone belt rocks are indicated, engulfed within granite-gneiss.

with modern island arcs. They have concluded that Archaean geology can be interpreted in terms of plate tectonics (see Chapter 11 of this volume), with the greenstone belts representing the sites of ancient subduction zones. However, discussion of the mechanisms whereby the continents developed through time is deferred until the end of this chapter.

Archaean sedimentary rock suites differ considerably from their more recent counterparts. For example, well-sorted mature sediments, such as pure sandstones and quartzites, are almost invariably absent. The few quartzites that occur are probably recrystallised cherts which were originally chemically precipitated. Conglomerates are relatively scarce, particularly in the lower Archaean, and limestones, principally dolomites, gradually increase in importance through the Archaean, but are always comparatively minor in amount. They are undoubtedly chemical precipitates whose formation must have resulted in the gradual removal of carbon dioxide from the atmosphere. Some types of rock appear to be restricted to the Archaean and lower Proterozoic; banded ironstone formations occur in the Archaean and are abundant in the Lower Proterozoic, but are absent from the Upper Proterozoic and the Phanerozoic. The iron in these rocks is usually present as hematite and/or magnetite, and the rocks are typically banded (bedded), with the layers of magnetite being intercalated with layers of chert and silicate minerals. Banded ironstone formations were probably produced by chemical precipitation from water; the chemistry of the hydrosphere and atmosphere being critical to their formation. There has been much debate about the source of the iron, the two most prevalent hypotheses invoking either volcanic exhalations or deep chemical weathering of a continental source area. As the Earth's atmosphere became oxygenated so banded ironstones ceased to be precipitated. (For further discussion of banded ironstones see Chapters 3 & 8).

The deposition of banded ironstones reached its acme in the lowermost Proterozoic and was probably caused by the deposition of iron in shallow waters through the agency of blue-green algae. The total biomass seems gradually to have increased through the Proterozoic in spite of the fact that until the upper Proterozoic, the ozone layer, which absorbs ultra-violet radiation, was still not developed in the Earth's atmosphere.

Graphite occurs in lower Archaean marbles and gneisses, sometimes in quantities sufficiently large to warrant commercial exploitation. It has been suggested that the carbon was deposited from ammonia-rich water and carbon dioxide, as a result of radiogenic synthesis through cosmic radiation. However, on the basis of the carbon's isotopic composition, a biogenic origin seems probable and indicates the existence of biological systems as long ago as 3500 million years. Organic remains occur in the upper part of the Archaean, though they are rare. These are stromatolites, which were probably produced by blue-green algae and bacteria, and they indicate that the atmosphere had become sufficiently transparent for sunlight to reach the surface and thus promote photosynthesis.

The Proterozoic Era

There is no universal agreement on the temporal position of the boundary between the Archaean and Proterozoic Eras; here it is taken at 2500 million years. The principal difference between the two eras is one of tectonic style. Archaean terrains lack a regional structural trend, the granite-gneiss regions and greenstone belts simply defining circular to elliptical areas. The so-called 'permobile' regime refers to the general mobility of the crust at this time. The Proterozoic by contrast has linear mobile belts, and the mechanical properties, thickness and strength of the crust more nearly resemble those of today. In the Proterozoic, thick sediments and extrusive volcanic rocks seem to have accumulated in linear troughs for the first time. These rocks were subsequently metamorphosed, folded, intruded by a variety of magmas, and elevated. Similar geological events have continued to the present day and, although many linear mobile belts have formed in the last 2500 million years, none is known from the Archaean. The onset of this change in tectonic style seems to have occurred first in Africa, whereas elsewhere it developed later – in some places hundreds of millions of years later.

It can be argued that the Archaean greenstone belts are themselves linear sedimentary basins, similar to those of later times; but, as we shall see, there are important differences. The 'geosyncline' concept developed during the nineteenth century, principally as a result of work in the Appalachian region of the United States. A geosyncline starts as an asymmetric, linear, subsiding trough (Fig. 2.10A), bounded on one side by a stable continental block called a 'craton', and on the other by a tectonically active land-mass. Sediments accumulating on the craton side of a geosyncline generally include a relatively high proportion of limestones and 'mature' sediments such as pure quartz sands, and are called miogeosynclinal sediments. On the other hand, sediments accumulating in that part of the geosyncline adjacent to the tectonically active land-mass (the eugeosyncline) are thicker than those of miogeosynclinal type and are predominantly immature sediments, notably greywackes with thick shales, but they also include abundant basic volcanic rocks (basalts) with pillow lavas. The miogeosyncline and eugeosyncline are often separated by a tectonically active median ridge that may be a source for some of the sediment.

If the concept and terminology of geosynclines are applicable to ancient rocks then Archaean sediments appear to be exclusively of eugeosynclinal type, whereas both eugeosynclinal and miogeosynclinal sediments occur in Proterozoic rocks. In recent years, particularly since the advent of the theory of plate tectonics (see Chapters 10 and 11), the geosynclinal concept has fallen from favour, principally because geosynclines do not seem to be forming today. However, Dietz (1963) has suggested that the sediments of modern continental margins can be interpreted in geosynclinal terms as illustrated in Figure 2.10B – miogeosynclinal sediments being correlated with deposits on the continental shelf, and eugeosynclinal sediments with the very thick sedimentary prisms of the continental rises. (Dietz' model accords better with modern ideas of continental tectonics, but the distinction of miogeosynclinal from eugeosynclinal rock sequences still holds, and the important point for the interpretation of continental growth is that miogeosynclinal sequences are absent from Archaean terrains.).

Cratonisation is an important mechanism of continental growth and development and is illustrated in Figures 2.10A and B. Linear mobile belts generally develop at the margins of stable crustal blocks, although they may also

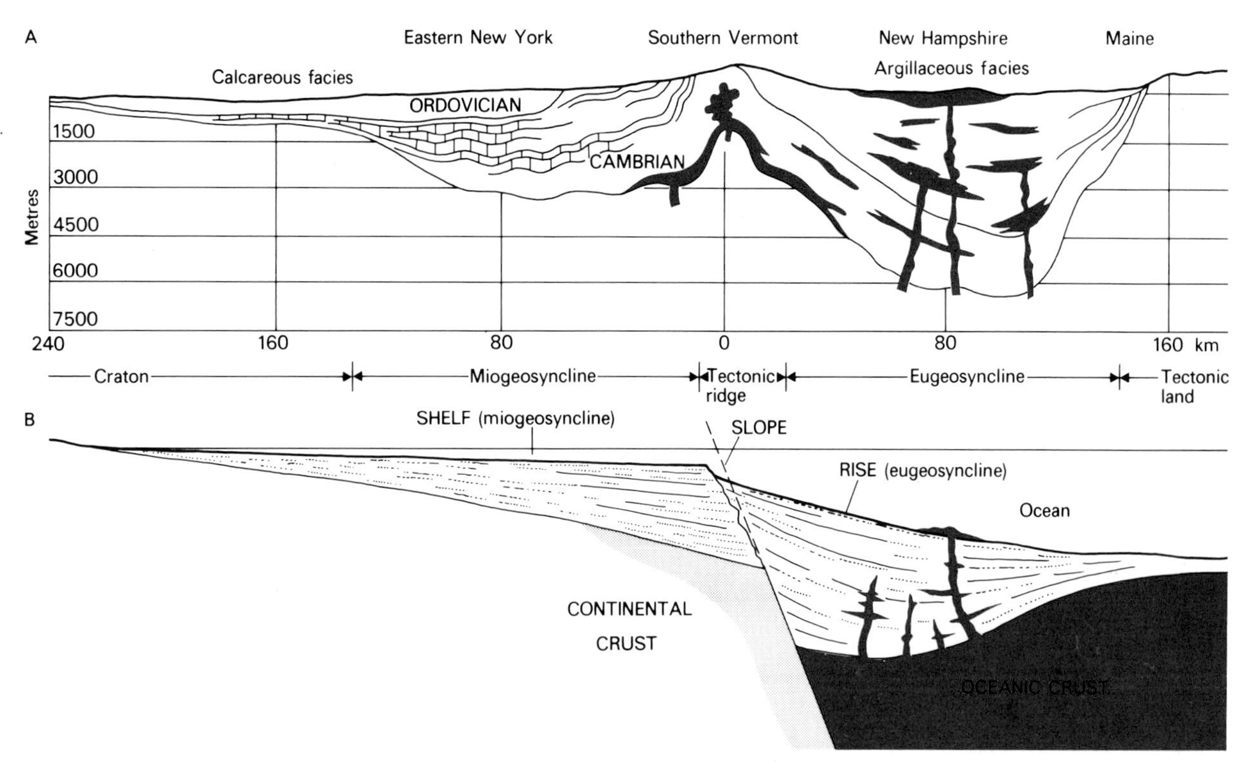

Figure 2.10 (A) Reconstruction of the Appalachian Geosyncline according to Kay (1951). (B) The same section as (A) but according to the interpretation of Dietz (1963). This reconstruction shows the eugeosyncline as a prism of sediments deposited at the base of the continental slope, and piled against the fractured edge of the continent, the eastern continuation of which has drifted away (see Chapter 11). Igneous intrusives, lava flows and oceanic crust shown in blue.

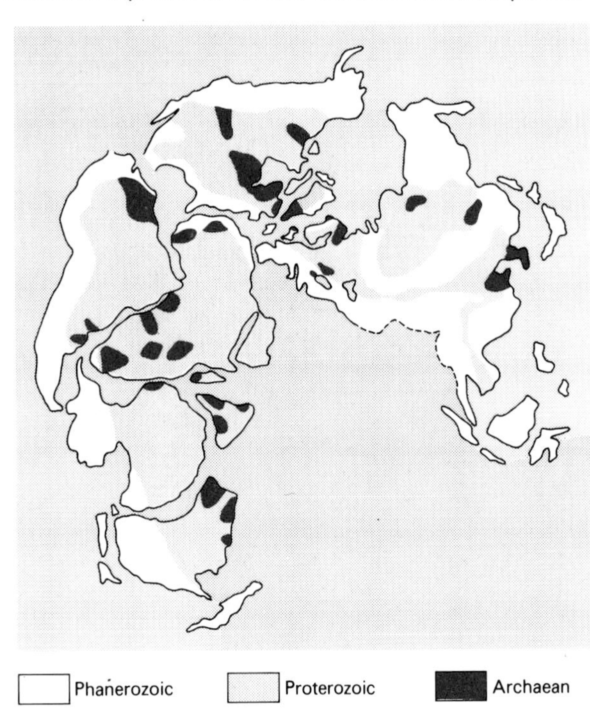

Figure 2.11 A reconstruction of the continents in Permian times showing the Proterozoic cratons with Archaean regions within them (after Windley, 1977).

cut across them. The stable crust, or craton, provides part of the sediment that fills the trough. This material is eventually deformed and may be partially melted as it is 'welded' on to the craton. Any part of the original craton intimately involved in this process is said to have been reworked.

In early Proterozoic times the cratons were Archaean blocks and there seems to have been a relatively large number of them (Fig. 2.11). In terms of Figures 2.10A and B the eugeosynclinal sequence thus became an integral part of the continent. This process of cratonisation appears to have continued through to the Phanerozoic so that the cratons became larger and fewer (Fig. 2.11).

Whether cratonisation increased the volume of the crust, or simply resulted in the crust forming fewer but larger units with the same overall volume, is still a matter of debate; this is discussed further later in this chapter. The evidence is not conclusive but by the end of the Proterozoic there may have been only two continents, one comparable with the late Palaeozoic Laurasia, comprising the present northern cratonic areas of Canada, Greenland, the Baltic area, and Siberia, and a second – the precursor of Gondwanaland – encompassing the present cratonic areas of Africa, peninsular India, South America, Australia, and Antarctica. However, some geologists contend that a single super-continent existed some 2700 million years ago.

Although the volume of continental crust may have increased with time, its area seems to have decreased through the Proterozoic as a result of the onset of sea-floor spreading and associated subduction (see Chapter 11). These processes carry crustal material to the subsiding basins – geosynclines – associated with subduction zones, where it is welded on to the craton, so thickening the crust and gradually increasing the area of oceanic crust. Through Proterozoic time it has been estimated (Hargraves, 1976) that the crust increased in average thickness from about 20 km to about 31 km, while the fraction of the Earth's surface underlain by oceanic crust increased from 40 to 50 per cent. Thickening of the crust would, because of the isostatic buoyancy, have increased the elevation of the continental areas; this in turn would have led to increased rates of erosion and sediment transport.

As cratonisation continued the stable areas became larger, with the formation of extensive platform deposits such as arkoses (coarse, feldspar-rich sandstones), lithic sandstones (containing rock fragments) and limestones. These grade into texturally and mineralogically mature sediments, such as quartzites, typical of the miogeosynclinal facies. Eugeosynclinal facies rocks are also abundant, together with suites of volcanic and plutonic igneous rocks indistinguishable from those associated with more modern geosynclines.

A suite of uranium- and gold-bearing conglomerates and sandstones of platform type and of lowermost Proterozoic age (2500 million years) is found in North America, South Africa, Asia and Europe. There has been considerable controversy as to the origin of the gold and uranium but they were probably derived from Archaean greenstone belts. A unique feature of these rocks is the presence of well-rounded clastic grains of uraninite and pyrite, minerals that are not preserved in modern sedimentary deposits because of atmospheric oxidation.

The gradual build-up of free oxygen in the atmosphere had two marked effects on sedimentation – banded iron formations ceased to be deposited, and 'red beds' appeared for the first time. This change took place about 2200 million years ago. Red beds, such as the well-known Devonian and Triassic sandstones of England, are generally indicative of an arid climate, but they also prove that there was enough oxygen in the atmosphere for the oxidation of the available iron. Highly oxidised iron-bearing minerals such as glauconite and chamosite first occur in the earliest Proterozoic red beds. Whether these indicate world-wide aridity is not yet known, but the oldest recorded evaporite deposits are also of this age. The occurrence of red beds and evaporites, and the abundance of platform and miogeosynclinal deposits, suggest that elevated continental areas were becoming more extensive.

As limestone deposition proceeded, so the amount of CO_2 in the atmosphere decreased and there was a concomitant decrease in the importance of the 'greenhouse' effect. Palaeoclimatological studies by Sinitsyn (1967) suggest that average surface temperatures in early Palaeozoic times were about 30–32 °C and falling; Salop (1977), extrapolating these data, concludes that late Proterozoic surface temperatures were about 40 °C. However, the lowering of temperature through the Proterozoic must have been accompanied by the establishment of latitudinal and altitudinal climatic zonation. Although Proterozoic surface temperatures seem to have been generally higher than those of the present day, it is nevertheless from this period that the first glaciations are recorded.

The Phanerozoic Era

The Phanerozoic is taken as beginning with the Cambrian period when there was a great increase in the numbers and varieties of living things. It may also mark the time of the initial break up of the continents and of the beginning of the plate tectonic regime. Although the processes of sea-floor spreading and subduction appear to have been active during Proterozoic and possibly Archaean times, it is only for Phanerozoic times that the evidence is strong enough to support the theory unequivocally. An account of plate tectonics is given in Chapters 10 and 11 of this volume, but reference needs to be made to it here because it was probably the main mechanism controlling continental development in Phanerozoic times.

As in the Proterozoic, linear troughs continued to develop along continental margins. These received sediments derived both from the continents and the oceanic basins, together with igneous rocks originally generated at mid-ocean ridges and carried towards the continent by movement of the oceanic crust (see Chapter 11). Subsequent burial and metamorphism would weld the marginal troughs to the edge of the continent, resulting in further continental accretion. In this way progressively younger rocks were added to the margins of cratons.

In the Phanerozoic there seems to have been a tendency for linear mobile belts to become longer than their earlier counterparts (see Fig. 2.12) though this could simply be due to their better preservation. The extensive Caledonian–Appalachian belt, for instance, stretching from northern Europe and Greenland through eastern Canada and the United States, is small in comparison with the Cenozoic circum-Pacific belt, and the Alpine–Himalayan system which extends to the East Indies (Fig. 2.12). These mountain belts may be the result of a few large convecting mantle cells; the absence of such belts in the Proterozoic may indicate that mantle convection was then organised into more numerous but smaller cells, operating beneath a greater number of lithospheric plates than in the Phanerozoic.

Some Phanerozoic mobile belts were not confined to continental margins but cut across the continents, so that older continental crust as well as younger sediments were metamorphosed during the ensuing mountain-building movements. In this way old continental crust is reworked

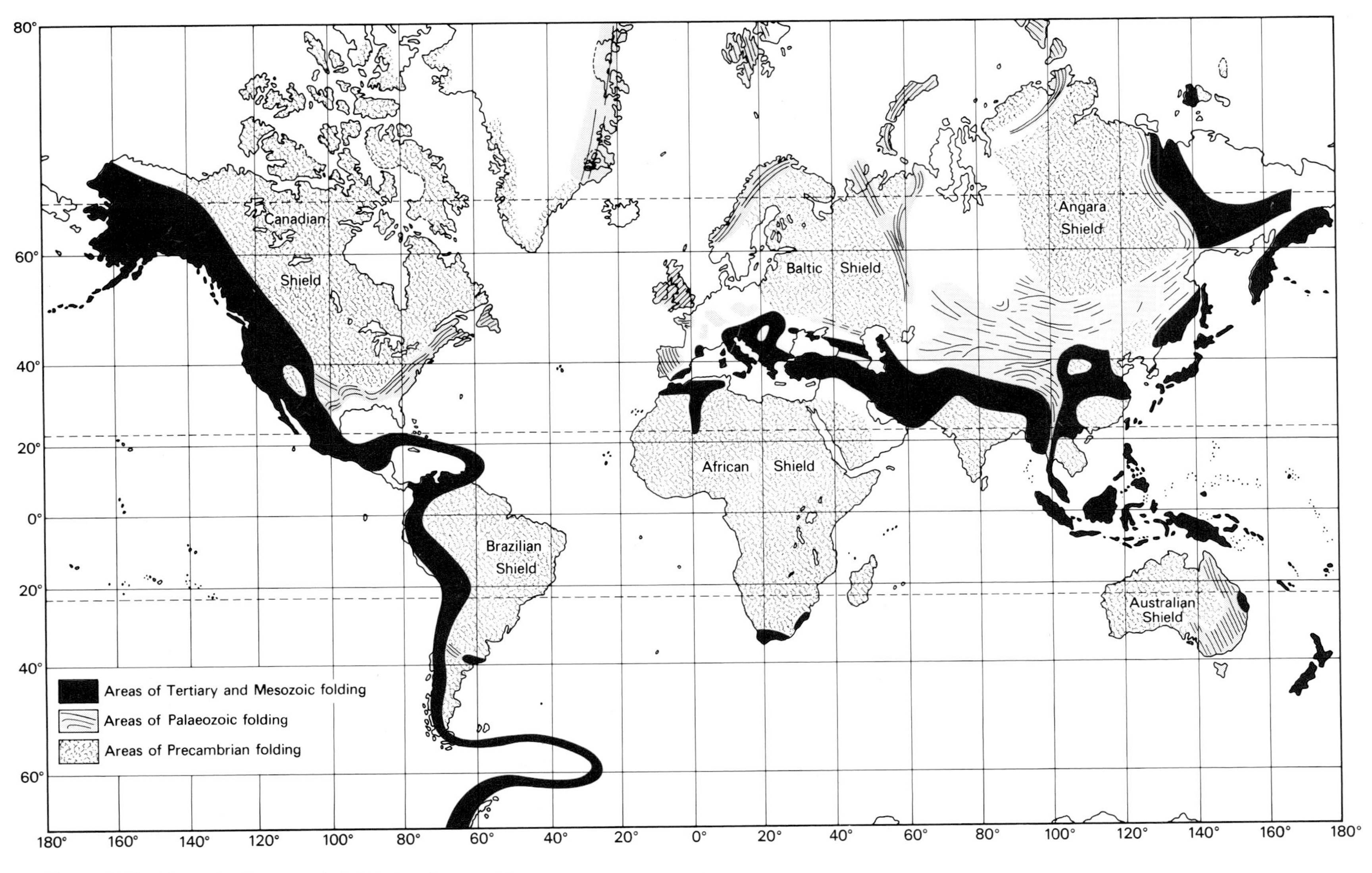

Figure 2.12 The major Phanerozoic fold belts of the world (after Umbgrove, 1947).

and does not constitute a net addition to the continent. Indeed some geologists maintain that there has been little or no addition to the continental crust since Archaean time, but that all the material added as a result of sea-floor spreading and associated processes is simply recycled material of continental origin.

There is strong evidence that the nature of regional metamorphism has changed with time. The grade of metamorphism to which a rock has been subjected, essentially a combination of increasing temperature and pressure, is revealed by its mineralogy – certain minerals or mineral assemblages being formed only at particular temperatures and pressures. Regionally metamorphosed rocks can be divided into three groups (Miyashiro, 1972 and 1973)

Type	*Characteristic minerals*	*Average geothermal gradient*
Low pressure	andalusite	> 25 °C km^{-1}
Medium pressure	kyanite (without glaucophane)	about 20 °C km^{-1}
High pressure	glaucophane and jadeite	about 10 °C km^{-1}

Rocks that have been regionally metamorphosed at high temperatures but relatively low pressures are known as granulites, and these are only of Archaean and Proterozoic age – particularly the former. They reflect the general cooling of the crust with time. In contrast rocks metamorphosed at high pressures and low temperatures, corre-

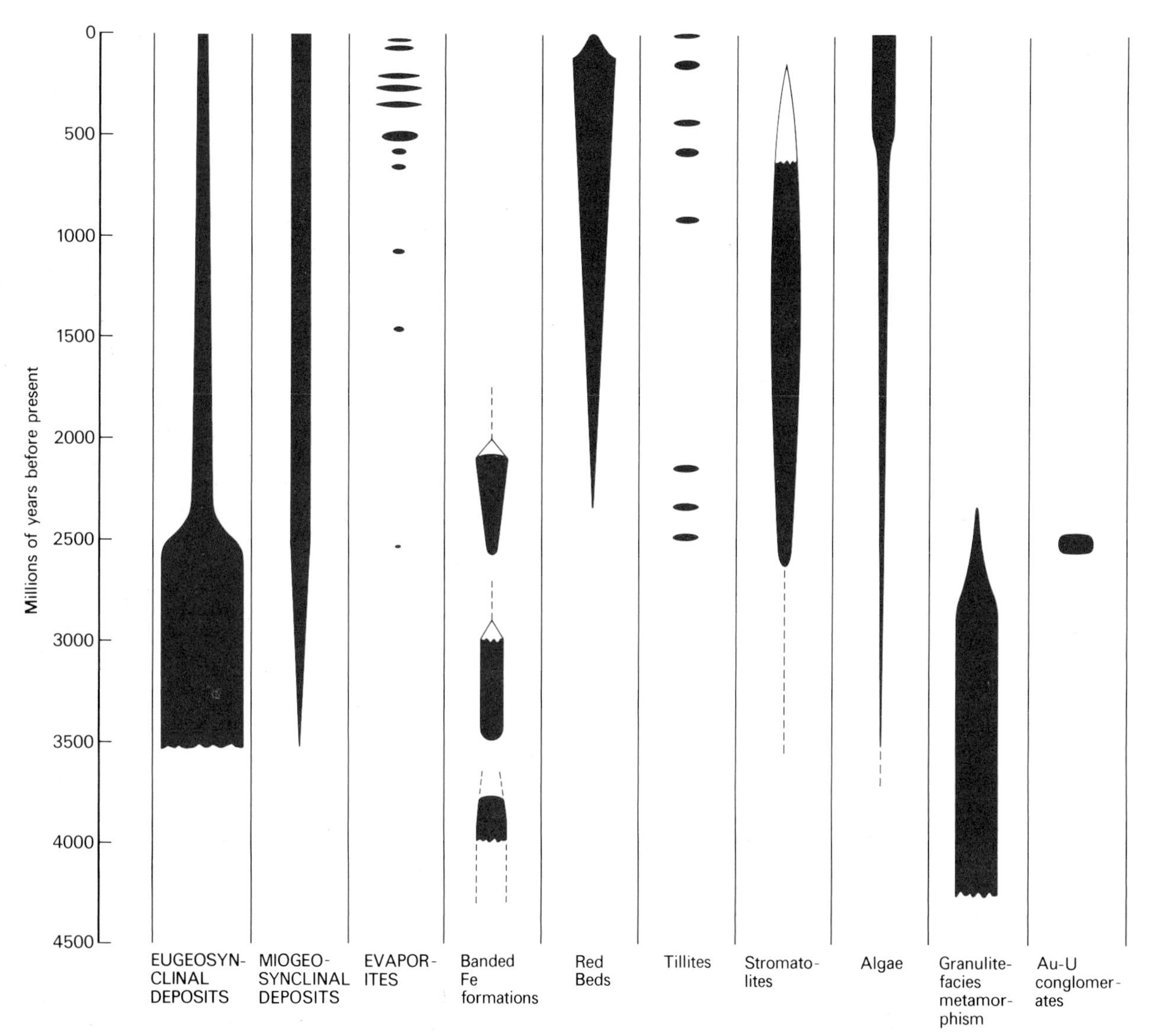

Figure 2.13 A synopsis of some significant geological changes with time.

sponding to geothermal gradients of only 10 °C km^{-1}, are confined to the Phanerozoic, most of them to the last 200 million years. These rocks are sometimes called 'blue schists', because they contain blue amphiboles, notably glaucophane.

Occurrence of blue schists in late Phanerozoic mountain belts suggests that rocks were deeply buried, perhaps beneath a greater thickness of sediments and associated volcanic rocks than those of earlier times. These belts seem to have become longer, narrower, and deeper with time, probably reflecting the general cooling and thickening of the crust, and perhaps also fundamental changes in the lithosphere and underlying mantle.

Some of the most significant geological changes that have taken place on the Earth since the beginning of the Archaean, and which were outlined in the preceding pages, are summarised in Figure 2.13.

The timing of continental growth

Although there is now general agreement as to the main mechanisms of continental growth, some dispute still remains about the time of formation of the sialic (silica and aluminium-rich) crust. One view (e.g. Fyfe, 1976) is that a world-wide primordial sialic crust formed in Archaean times, and that partial remelting and re-cycling has produced all subsequent granites and associated rocks. A variant of this hypothesis is preferred by those who favour an early initiation of plate movements (see Chapter 11). Again, early separation of sialic crust is envisaged, followed by recycling of sialic material through the mantle.

A second view is that the volume of sialic crust has increased continuously with time by the addition of material from the mantle, although it is generally conceded that crustal growth was most rapid early in the Earth's history. Moorbath (e.g. 1977), one of the main protagonists, suggests that 3500 to 3800 million years ago the crust occupied about 5–10 per cent of its present area. Then, after a great continent-forming event at the end of the Archaean (2900 to 2600 million years ago), some 50–60 per cent of the present crust had formed, and this had about the same thickness as now. He suggests that there were further episodes of crustal formation in the periods 1900 to 1700, 1100 to 900, and in the last 600 million years, coinciding with the break-up of the continents. He calls these episodes 'accretion–differentiation superevents' (Moorbath, 1978) and suggests that their periodic nature could be caused by the accumulation and release of heat generated by the decay of radioactive elements, and by periodic changes in the convection systems within the mantle. It is suggested that as the area of crust increased the reworking and remelting processes became increasingly important. Evidence in support of these ideas is obtained from the study of naturally occurring isotopes, particularly those of strontium, rubidium and lead.

Measurement of the amount of the isotopes, rubidium 87(^{87}Rb), strontium 87(^{87}Sr), and strontium 86(^{86}Sr) in

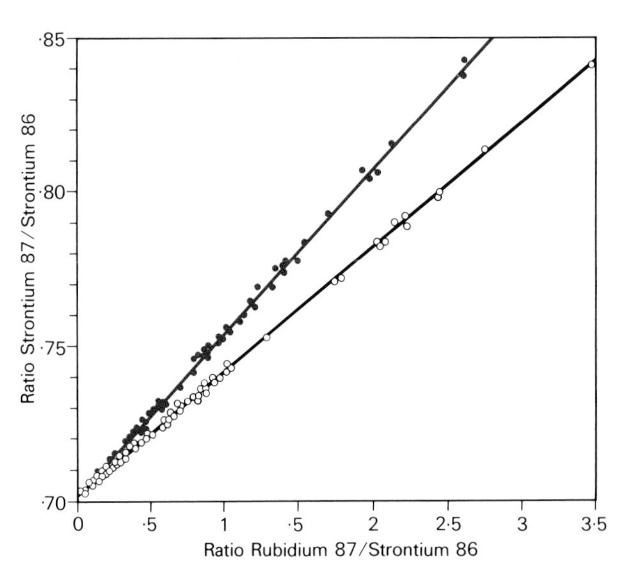

Figure 2.14 Rubidium–strontium isochron diagram for two groups of Archaean gneisses from Greenland (after Moorbath, 1977). The Amitsoq Gneiss (blue) is 3750 million years old, and the Nuk Gneiss (black) is 2850 million years old.

samples from a rock suite such as the Amitsoq Gneiss of west Greenland, should indicate whether they were either derived directly from the mantle or represent older reworked continental crust. The isotope ^{87}Sr is produced by the radioactive decay of ^{87}Rb and therefore the amount present in a rock increases with time, whereas the quantity of ^{86}Sr, the stable isotope of strontium, remains constant. The increase in the ratio of $^{87}Sr/^{86}Sr$ is thus directly proportional to the ratio of rubidium to strontium in the rock (Fig. 2.14). The $^{87}Sr/^{86}Sr$ ratio for the upper mantle is known from measurements on modern lavas derived from the upper mantle and erupted from oceanic volcanoes, and the ratio at the time of the formation of the Earth is known from meteorites and lunar rocks (see Fig. 2.15). Because there is considerably more rubidium in the crust than in the mantle, the temporal increase in $^{87}Sr/^{86}Sr$ crustal ratio is much greater than in the mantle. Although different $^{87}Sr/^{86}Sr$ ratios result when a large number of rocks containing different amounts of rubidium is analysed from a suite of rocks, the data can then be extrapolated back to the time of their formation when all the ratios would have been the same. If the rocks were derived directly from the mantle, the ratios will converge at a value representing the $^{87}Sr/^{86}Sr$ ratio in the mantle at the time they were formed (Fig. 2.15). If, on the other hand, the rock suite represents recycled older continental crust, then the value at which all the ratios are the same will not correspond to a mantle ratio, but to some higher value (Fig. 2.16). This technique has been applied by Moorbath (1977) and others to rock suites from the Archaean rocks of west Greenland, including the Amitsoq Gneiss, and it has been found that they have $^{87}Sr/^{86}Sr$ ratios equivalent to that of the mantle at the time of their formation (Fig.

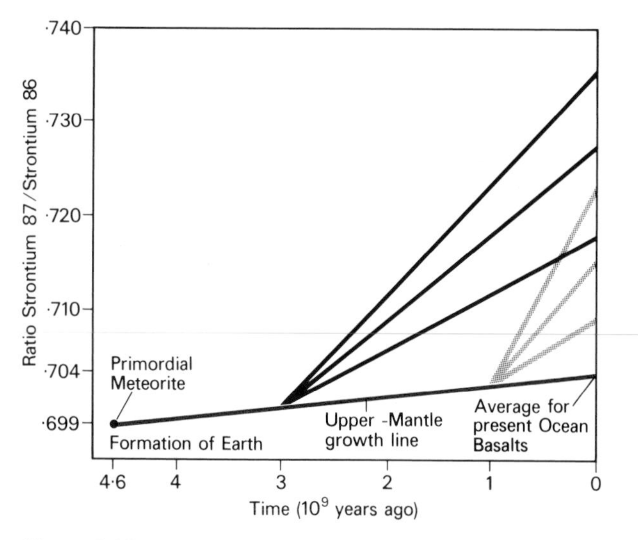

Figure 2.15 The ratio of Strontium 87 to Strontium 86 in the upper mantle has increased from 0·699 at the time of formation of the Earth to 0·704 at present, as indicated by the blue line. By plotting $^{87}Sr/^{86}Sr$ ratios for a varied suite of rocks and extrapolating back in time, it is possible to determine the time at which the rock suite had a common $^{87}Sr/^{86}Sr$ ratio, and its value at that time (the initial ratio). If the initial ratio falls on the upper mantle growth line it is concluded that the material was derived from the upper mantle and not from remelted continental crust. Examples are shown of two suites of rocks that derived from the mantle 3000 million and 1000 million years ago (after Moorbath, 1977).

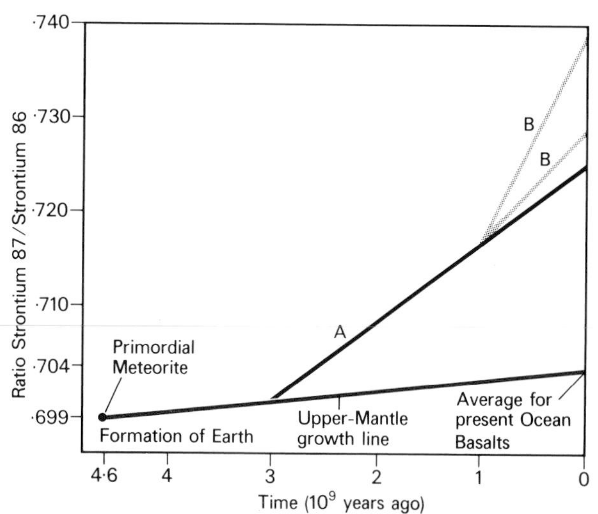

Figure 2.16 A plot similar to that of Figure 2.15 but showing a suite of rocks (lines B) with $^{87}Sr/^{86}Sr$ ratios that converge at a value much higher than that of the upper mantle. It is inferred that continental crust was derived from mantle material 3000 million years ago (line A) and that it was subsequently reworked 1000 million years ago (lines B) (after Moorbath, 1977).

2.17), indicating that these rocks were derived from the mantle and not from recycled continental crust. For a fuller discussion of the application of isotopes to this problem see Moorbath (1977).

The strontium isotope technique has revealed that, although old crust is reworked and regenerated from time to time by later orogenic events, crustal rocks generally seem to have been derived from the mantle. Thus it can be concluded that the continental crust has increased in volume with time, although there is evidence that the process is slowing down, and that there is more reworking of old crust in younger than in older orogenic belts.

Although strontium isotopes seem to solve the problem of timing crustal growth, difficulties remain. For example, if crustal materials can be recycled through the mantle then their isotopic compositions might be expected to be homogenised with those of the mantle, the isotopic ratios of which would be virtually unaffected owing to its immense volume. Recycled materials could not, therefore, be distinguished. Similarly it is possible that homogenisation to mantle ratios might take place under the high temperatures and pressures prevailing at the base of the crust where melting and mobilisation might engender intrusions having isotope ratios close to those of the mantle from which they arose. However, experimental evidence

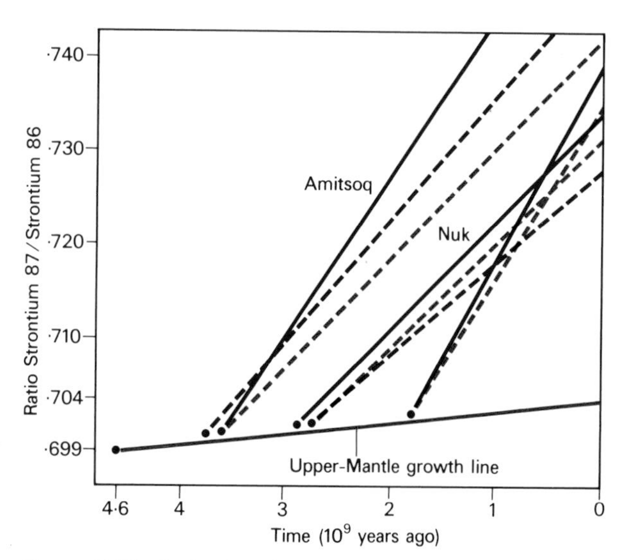

Figure 2.17 A plot showing $^{87}Sr/^{86}Sr$ growth lines for a range of Archaean granite-gneiss terrains from Greenland (solid black lines), North and South America (dashed blue lines), and the shield of southern Africa (dashed black line). Because they all converge close to the upper mantle growth line it is inferred that the crustal rocks in these areas were not produced by reworking of older crust, but were derived from the mantle (after Moorbath, 1977).

suggests, but does not prove, that this homogenisation is unlikely to have taken place on a large scale. Future workers will probably have to solve this problem through fresh and radical approaches.

References

Anhaeusser, C. R. 1971. The Barberton Mountain Land, South Africa – a guide to the understanding of the Archaean geology of western Australia. *Geological Society of Australia Special Publication* **3**: 103–120.

Anhaeusser, C. R. 1971a. Cyclic volcanicity and sedimentation in the evolutionary development of Archaean greenstone belts of shield areas. *Geological Society of Australia Special Publication* **3**: 57–70.

Dickinson, W. R. & Luth, W. C. 1971. A model for plate tectonic evolution of mantle layers. *Science*, New York **174**: 400–404.

Dietz, R. S. 1963. Collapsing continental rises: an actualistic concept of geosynclines and mountain building. *Journal of Geology* **71**: 314–333.

Engel, A. E. J. 1968. The Barberton Mountain Land: clues to the differentiation of the Earth. *Transactions of the Geological Society of South Africa* **71** (annex): 255–270.

Fyfe, W. S. 1976. Heat flow and magmatic activity in the Proterozoic. *Philosophical Transactions of the Royal Society of London* Series **A280**: 655–660.

Hargraves, R. B. 1976. Precambrian geologic history. *Science*, New York **193**: 363–371.

Kay, M. 1951. North American geosynclines. *Geological Society of America Memoir* **48**, 143 pp.

Macgregor, A. M. 1951. Some milestones in the Precambrian of Southern Rhodesia. *Proceedings of the Geological Society of South Africa* **54**: 27–71.

Miyashiro, A. 1972. Metamorphism and related magmatism in plate tectonics. *American Journal of Science* **272**: 629–656.

Miyashiro, A. 1973. *Metamorphism and metamorphic belts*. 492 pp. London: Allen and Unwin.

Moorbath, S. 1977. The oldest rocks and the growth of continents. *Scientific American* **236**: 92–104.

Moorbath, S. 1978. Age and isotope evidence for the evolution of continental crust. *Philosophical Transactions of the Royal Society of London* Series **A288**: 401–413.

Ray, S. 1970. The plutonic concept. *Journal of the Geological Society of India* **11**: 54–60.

Salop, L. J. 1977. *Developments in palaeontology and stratigraphy, 3: Precambrian of the Northern Hemisphere*. 378 pp. Amsterdam: Elsevier.

Shaw, D. M. 1976. Development of the early continental crust Part 2: Prearchean, Protoarchean and later eras, pp. 35–53 *in* Windley, B. F. (Ed.) *The early history of the Earth*. London: John Wiley.

Sinitsyn, V. M. 1967. *Introduction to palaeoclimatology*. 232 pp. Leningrad: Nedra.

Umbgrove, J. H. F. 1947. *The pulse of the Earth*. 2nd ed. The Hague: Nijhoff.

Windley, B. F. 1977. *The evolving continents*. 385 pp. London: John Wiley.

A. R. Woolley
Department of Mineralogy
British Museum (Natural History)

CHAPTER 3

Element distribution and the formation of rocks

P. Henderson

The division of the Earth into core, mantle, and crust, established early in the Earth's history, sets the broad scene for our inquiry into the processes that govern element distribution within our planet. The wide diversity of the Earth's materials and of the processes acting on them give rise to an intriguing set of problems for investigation. Over 2000 different minerals have been identified and described; numerous distinct combinations of minerals into rock types and ore deposits exist; and chemical and mechanical processes act over a wide range of temperature and pressure conditions. However, the inaccessibility of the core and mantle means that much of our knowledge of the Earth's materials and the factors which affect them has had to come from observations on the crust, despite the fact that it represents slightly less than a half a per cent of the total mass of the Earth.

The mean thickness of the crust is only 17 km, averaging about 33 km under the surface of the continents and about 5–6 km under the oceans. Throughout the mass of continental crust, metamorphic rocks are the most abundant, while the ocean crust is mostly of basaltic igneous rock. The continental crust consists of a few large, geologically stable areas (cratons), composed of Precambrian rocks, which are surrounded by more tectonically active parts consisting of mountain chains or volcanic zones. The Canadian Shield, with its ancient metamorphic terrain surrounded by the younger fold belts of the Rocky and Appalachian Mountains on its western and south-eastern margins, represents one of the best examples of this aspect of crustal structure. With depth the continental crust becomes richer in certain elements, particularly magnesium, calcium and iron, but slightly poorer in aluminium and the alkali metals, sodium and potassium. The oceanic crust is more uniform, with a thin veneer of sediments lying on two layers – an upper one of a mixture of sediment and basalt and a lower one of basalt or its coarse grained equivalent, gabbro.

In general, all the various main types of sedimentary, metamorphic or igneous rocks may be found throughout the stratigraphical column. Thus, except for their fossil content or their structural disposition there may be little to distinguish the various rocks of, say, Ordovician age from those of other ages occurring throughout the whole crust, although a particular combination of conditions may give rise to some unusual, localised variants (e.g. chalk as a widespread type of limestone in the Cretaceous period). In the Precambrian, however, certain rock types were formed – massive anorthosites and banded ironstones – which are not found in other parts of the stratigraphical column (see also Chapter 2). We will discuss later whether or not these occurrences may reflect some evolution of the Earth's crust.

Composition of the crust

A knowledge of the average composition of the crust aids our understanding of the overall composition and zonal structure of the Earth and of the geochemical processes operating within it. It provides a useful reference frame for assessing the extent of element fractionation or mixing processes which have given specific rock types or mineral occurrences. However, the heterogeneous nature of the crust, both laterally and vertically, makes the task of estimating its average composition a difficult one. It is helpful to start by distinguishing ocean crust from continental crust and then estimating the average composition of each.

A number of methods for estimating the composition of the continental crust have been tried, including that based on a simple average of available rock analyses. For more accurate estimates it is necessary to take account of the proportions of the occurrences of the different rock types; this in turn requires us to know or to make assumptions about the nature of the deep crust. We may obtain information about the deeper layers from seismic studies, which can be helpful in establishing the proportions of the mafic and silicic rocks.* Such a study was carried out by Pakiser & Robinson (1967) for the crust of the

* Mafic rocks are those rich in magnesium and iron; silicic rocks contain greater than 66 per cent SiO_2 by weight, see also Figure 3.4.

United States. They were able to distinguish two distinct crustal layers – an upper one of slower seismic velocity, correlating with rocks relatively rich in silica, and a lower layer of faster wave velocity, with rocks of basic (or mafic) composition. Their data, together with average rock compositions, allowed them to compute an estimate for continental crust composition (see Table 1). Earlier, Poldervaart (1955) had calculated the average composition of four structural regions of the Earth: the deep oceanic region; the sub-oceanic region (i.e. continental shelf and geosynclines); the continental shield region; and the younger folded belts region. His estimate for the composition of the Earth's crust based on a model of crustal structure (in which he made a number of assumptions, such as that the underlying rock of the ocean floor is olivine basalt) is also given in Table 1.

Alternative methods of estimation may be adopted. One is to make use of the marked difference in some trace element abundances in the various principal rock types. For example, Taylor (1964) used the very different abundances of the rare earth elements in granitic and basaltic rocks to show that a mix of 1:1 mafic to silicic igneous rock would produce the observed rare earth element distribution in 'average sediment'. This igneous rock mix might thereby represent the composition of average continental crust. With the given proportions of the mafic and silicic components it is then possible to compute, from average rock compositions, the concentration of major and trace elements. Taylor's estimate for the major elements is given in Table 1 and for a selection of trace elements in Table 2. The method is subject to some criticism in that it relies upon no differential loss of the rare earth elements into solution during the derivation of the sediment, and we still cannot be sure that such a requirement is satisfied. However, the methods discussed here, as well as a number of other independent ones, all yield estimates for the continental crust that are in close agreement. This is perhaps sufficient reason to suggest that the average, major element, composition of the crust has now been ascertained.

The estimated composition shows that only nine elements: O, Si, Al, Fe, Mg, Ca, Na, K and Ti (on a water-free basis) make up more than 99 per cent by weight of the crust. Of these, silicon and oxygen are the most abundant, so it is not surprising to know that the commonest rock-forming minerals are silicates, compounds in which each silicon atom is bonded to four tetrahedrally-disposed oxygen atoms to give an $(SiO_4)^{4-}$ unit. The SiO_4 tetrahedra are bonded to other elements which may be Mg, Fe, Ca, Na, etc. dependent upon the mineral group. The simple oxide SiO_2 is also grouped with the silicates. These minerals constitute approximately 95 per cent of the Earth's crust.

Table 1 Average percentage chemical composition of the Earth's crust

	Poldervaart 1955			**Pakiser & Robinson 1967**	**Taylor 1964**
Oxide	*Oceanic crust*	*Continental crust*	*Entire crust*	*Continental crust*	*Continental crust*
SiO_2	46·6	59·4	55·2	57·9	60·3
TiO_2	2·9	1·2	1·6	1·2	1·0
Al_2O_3	15·0	15·6	15·3	15·2	15·6
Fe_2O_3	3·8	2·3	2·8	2·3	–
FeO	8·0	5·0	5·8	5·5	7·2*
MnO	0·2	0·1	0·2	0·2	0·1
MgO	7·8	4·2	5·2	5·3	3·9
CaO	11·9	6·6	8·8	7·1	5·8
Na_2O	2·5	3·1	2·9	3·0	3·2
K_2O	1·0	2·3	1·9	2·1	2·5
H_2O	–	–	–	–	–
P_2O_5	0·3	0·2	0·3	0·3	0·2

All estimates on a water-free basis
* *All Fe expressed as FeO*

Table 2 Abundances of some minor and trace elements in the continental crust (from Taylor, 1964)

Atomic number	*Parts per million*	*Atomic number*	*Parts per million*
3 Li	20	40 Zr	165
4 Be	2·8	47 Ag	0·07
5 B	10	50 Sn	2
9 F	625	55 Cs	3
16 S	260	56 Ba	425
17 Cl	130	57 La	30
21 Sc	22	58 Ce	60
23 V	135	63 Eu	1·2
24 Cr	100	70 Yb	3
27 Co	25	71 Lu	0·5
28 Ni	75	72 Hf	3
29 Cu	55	74 W	1·5
30 Zn	70	79 Au	0·004
31 Ga	15	80 Hg	0·08
32 Ge	1·5	82 Pb	12·5
33 As	1·8	83 Bi	0·17
37 Rb	90	90 Th	9·6
38 Sr	375	92 U	2·7
39 Y	33		

Rock-forming minerals

The rock-forming silicates may be conveniently classified according to the type of linking of the silicon-oxygen tetrahedra to each other (Fig. 3.1). Minerals in which the tetrahedra are isolated are termed orthosilicates, an example being the olivine group $(Mg,Fe)_2SiO_4$. In this group there is a continuous range of possible compositions from a

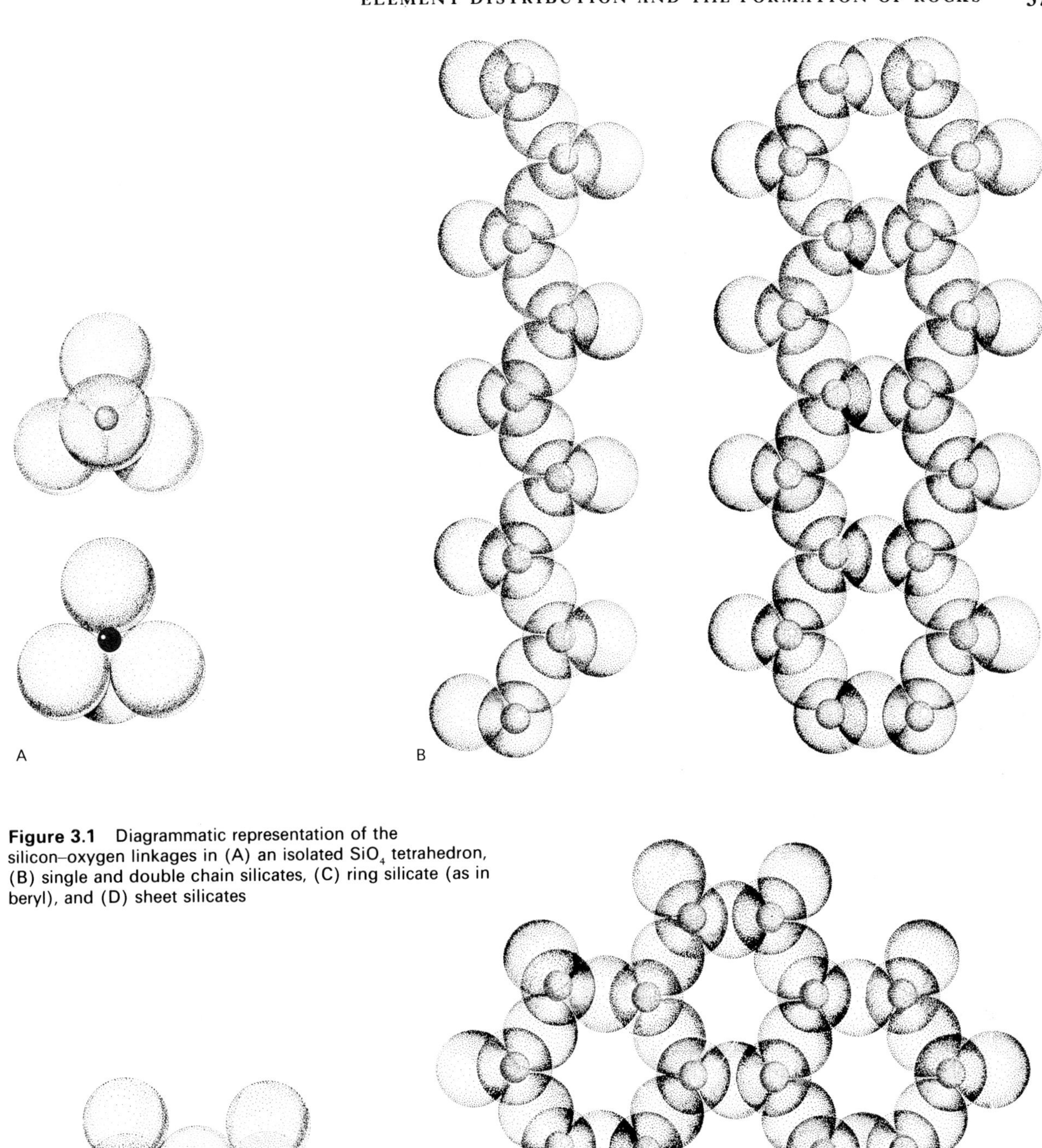

Figure 3.1 Diagrammatic representation of the silicon–oxygen linkages in (A) an isolated SiO_4 tetrahedron, (B) single and double chain silicates, (C) ring silicate (as in beryl), and (D) sheet silicates

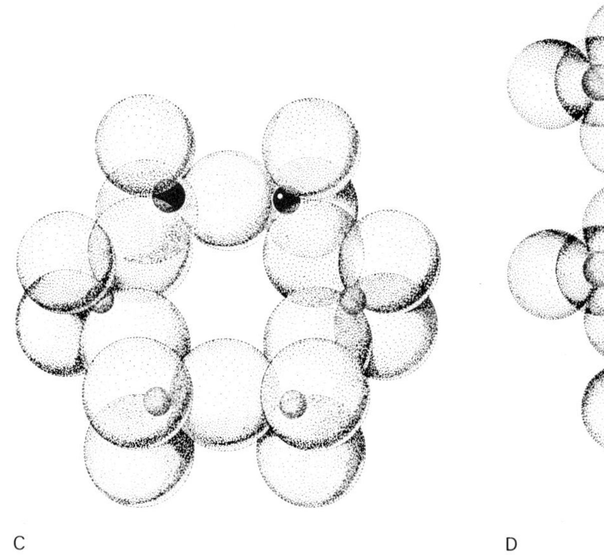

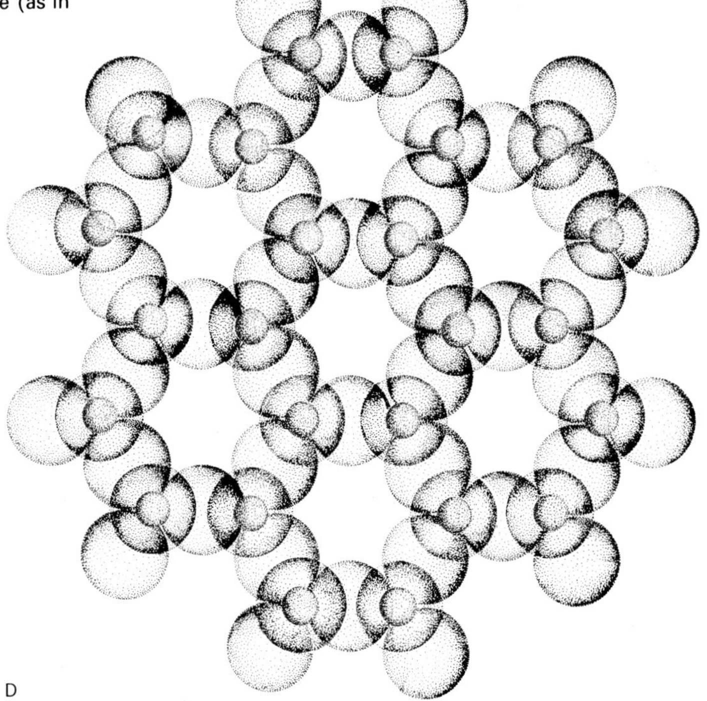

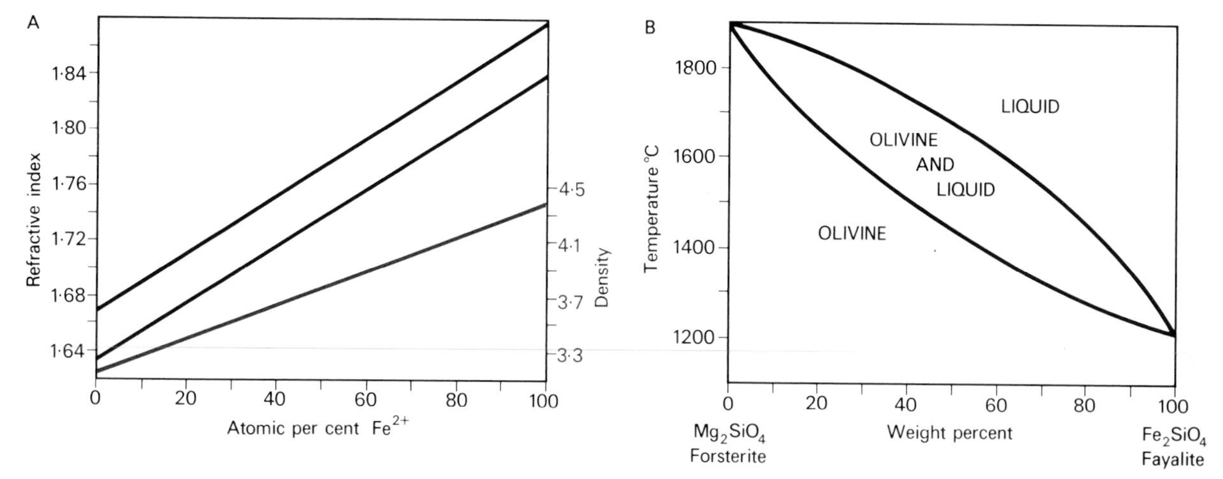

Figure 3.2 (A) Variation in density and refractive indices (least and greatest) in the (Mg,Fe)–olivines; (B) Melting relationships in the olivine system Mg_2SiO_4 – Fe_2SiO_4.

magnesium member – forsterite (Mg_2SiO_4), to an iron member – fayalite (Fe_2SiO_4). This phenomenon, in which one element (e.g. Fe) may substitute for another (e.g. Mg) to give a series of compositions without change in the crystal structure, is called solid solution. It is shown by many of the silicates. Figure 3.2 shows the variation with composition in some of the properties of the olivine solid solution series.

The sharing of some or all of the oxygen atoms by adjacent SiO_4 tetrahedra leads to the possible development of ring, chain, sheet, or framework structures (Fig. 3.1). In a chain silicate two oxygen atoms of each tetrahedron are shared by two other tetrahedra to form $(SiO_3)^{2-}$ anionic groups, which in turn are bonded to other elements. One of the most important groups of chain silicates is that of the pyroxenes. Examples are: enstatite (Mg_2SiO_3); diopside ($CaMgSi_2O_6$); jadeite ($NaAlSi_2O_6$). The sharing of two oxygens of each tetrahedral unit between two adjacent units could also lead to development of a ring structure if the two ends of the chain were linked together. Ring silicates occur with three, four, or six-linked tetrahedra; the beryllium mineral – beryl ($Be_3Al_2Si_6O_{18}$) – is a ring silicate containing groups of six-linked tetrahedra (Fig. 3.1C). Chains of tetrahedra may link by further sharing of the oxygen atoms to form double chains (Fig. 3.1B). The large group of silicates the amphiboles, which includes the common mineral hornblende, $NaCa_2(Mg,Fe,Al)_5(Si,Al)_8O_{22}(OH,F)_2$, are double-chain compounds. In hornblende (and in some other silicates) some of the silicon atoms are replaced by aluminium atoms but the tetrahedral arrangement is preserved.

When three oxygens of each tetrahedra are shared by others, a sheet structure is developed (Fig. 3.1D). The micas, with their perfect basal cleavage, e.g. muscovite $KAl_2(Si_3AlO_{10})(OH)_2$, and the clay minerals, e.g. kaolinite $Al_2Si_2O_5(OH)_4$, are sheet silicates. Finally, when all the oxygens are shared by adjacent tetrahedra, a three dimensional framework is produced, as in the common mineral quartz (SiO_2) or in the feldspar group which has calcium, sodium, or potassium ions occupying some of the interstices of a $(Si,Al)O_4$ framework so as to balance the electrical charges created by the replacement of some silicon (Si^{4+}) by aluminium (Al^{3+}). Common feldspars are albite ($NaAlSi_3O_8$), anorthite ($CaAl_2Si_2O_8$) and orthoclase ($KAlSi_3O_8$).

The silicates present a varied and diverse range of structures not only in terms of the silicon–oxygen linking but also for the types of polyhedra available for the co-ordination of other elements and for the symmetry of their molecular arrangement. This is also true for the non-silicate groups, which include many types of compounds such as carbonates (e.g. calcite $CaCO_3$), sulphates (e.g. baryte $BaSO_4$), phosphates (e.g. apatite $Ca_5(PO_4)_3(OH)$), oxides (e.g. magnetite Fe_3O_4), halides (e.g. fluorite CaF_2), and sulphides (e.g. galena PbS).

All the various members of the rock-forming silicates and non-silicates can be found in rocks and mineral deposits of all ages. The broad nature of mineral occurrences, governed by the overall composition of the crust, was established early in the history of the Earth and does not appear to have changed since.

Changes in crustal composition

Unfortunately we cannot readily apply the methods for estimating the composition of the continental crust to previous geological times, but much of the stratigraphical record lends support to the constancy of crustal composition over a considerable period, although the time-restricted occurrences of the massive anorthosites (gabbroic rocks, very rich in feldspar) and the banded ironstones could be indicative of an early overall change. The oldest known anorthosites (nearly 4000 million years old) might be the remnants of a primaeval crust produced from the melting and subsequent cooling of the outer parts of the Earth, and so be akin to the 4000 million year old, and older, anorthosites dated from the highland regions of the Moon. Banded ironstones are sedimentary rocks consisting of alternating layers or iron oxide, iron silicates, iron carbonates or sulphides, and silica in the form of chert, and with ages ranging from about 1700 million years to more than 3500 million years. They are believed (e.g. see Garrels *et al.* 1973) to reflect the particular conditions of sediment formation at that time, in which there was virtually no free atmospheric oxygen, it being stored in carbonate rocks, silicate rocks and water. Thus sediments were produced in a reducing environment, so that iron was present only in the ferrous oxidation state and could form sedimentary iron carbonates and silicates on a relatively large scale. With the advent of photosynthetic organisms and the generation of free atmospheric oxygen, a more oxidising environment was established and the weathering of pre-existing rocks would have led to the sedimentary transport of iron in the ferric state, and the cessation of formation of banded ironstones. Sedimentary iron formations are found in the Phanerozoic, but with different mineral associations, and showing evidence of a genesis involving an oxidising atmosphere.

The occurrence of the banded ironstones can be explained in terms of the evolution of the Earth's atmosphere without resorting to a change in its crustal composition, but the presence of massive anorthosites (and some other Precambrian rock types not discussed here) could indicate the existence of a primitive crust distinct in composition from that of today, but which was quite soon modified to give an almost constant average composition throughout the Phanerozoic or longer. It should also be remembered that Archaean rocks constitute much of the shield areas of the present-day continents.

There is one process that has led to a steady but extremely slight change in crustal composition with time. This is the decay of the radioactive elements, which include uranium and thorium, present in the Earth. Figure 3.3 shows the relative changes in abundance of the isotopes of four elements, and highlights the dramatic reduction in the amount of ^{235}U since the formation of the Earth; there has been virtually no change in ^{87}Rb because of its very slow decay rate. Uranium, by its radioactive decay, is the most important heat-producing element (about 3 $J g^{-1} year^{-1}$) and so has played a significant part in determining the thermal history of the Earth, and more especially of the crust and upper mantle. It must be emphasised that the decay of radioisotopes has caused little change in crustal composition. This is more readily appreciated when the relative abundances of the radioisotopes are known: ^{40}K constitutes only 0·01 per cent of total potassium (the other isotopes being stable), and the present-day $^{238}U/^{235}U$ ratio is close to 140, so the more rapid decay of ^{235}U in relation to that of ^{238}U has had little effect.

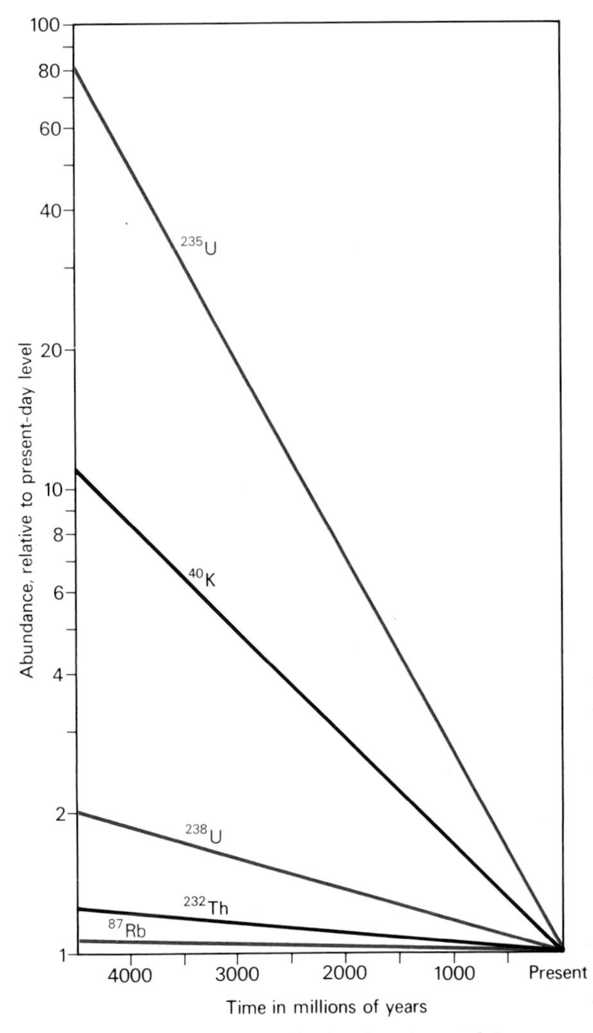

Figure 3.3 Relative change in the abundance of the radioactive isotopes in the Earth over 4500 million years

Element distribution and rock formation

Chemical and mechanical processes acting on the surface of, and within, the crust cause distinct chemical fractionation or mixing of its components. The rates of these processes vary considerably from region to region and are dependent upon a number of factors including tectonic activity, climate, and local crustal structure and composition. It is principally those interactions involving solid and liquid, or solid and solid that govern element distribution in the crust. Solid-liquid interactions are seen in the chemical weathering of rocks, in the precipitation of salt deposits, or in the crystallisation of a magma. Solid-solid transformations occur typically during the genesis of metamorphic rocks. Solid–gas or liquid–gas reactions do occur, as is shown for example by the existence of volcanic sublimate deposits or by reactions at the atmosphere–hydrosphere interface, but their overall effect on the pattern of element distribution in the solid crust is relatively small.

All these interactions give rise to regional and local changes in crustal composition but to a large extent they proceed in differing directions so that their sum effect would yield no discernible, persistent changing of the crust's average composition with geological time. Since at least the beginning of the Phanerozoic the overall composition has been close to, or at, a steady state condition.

What processes have occurred to give this steady state composition? One answer could be simply that during or after accretion of the Earth the heaviest elements sank under the gravitational field towards the centre, and a residual scum of the lightest elements was left to form a crust (see also Chapter 1). A closer examination of the crust's composition shows that while this explanation may account adequately for the distribution of many elements amongst the core, mantle, and crust, there are some serious anomalies. Undoubtedly the high density of metallic iron and nickel would have been important in the accumulation of these elements in the core, but geochemical and geophysical evidence shows that there has been an extreme outward migration of uranium in the Earth, despite it being one of the densest elements.

The early melting of the Earth (see Chapter 1) means that the crust and mantle were derived from a once fluid, silicate shell. This being so, it is reasonable to apply our observations on element distribution during the fractional crystallisation of a magma to help elucidate the chemical factors governing crustal composition.

Igneous Rocks

A basic magma (i.e., one with SiO_2 less than about 53 per cent) undergoing slow cooling will first crystallise mafic minerals such as pyroxene and magnesium-rich olivine, together with some minor oxide phases such as chromium-rich spinel, $(Mg,Fe)(Cr,Al)_2O_4$. These minerals, of a density greater than that of the melt, will tend to sink towards the floor of the magma chamber, so leaving a residual liquid relatively enriched in those elements that are not incorporated into the early crystallising phases. Continued cooling leads to the formation of lower-temperature minerals and successive changes in the residual magma composition. During this fractional crystallisation process, silicon, phosphorus, sodium, potassium and titanium are the major elements that rise in concentration in the residual liquid phase, while magnesium, calcium and aluminium tend to fall. Iron may rise or fall depending upon the oxidising conditions of the melt.

Of the trace elements, scandium, chromium, cobalt and nickel are amongst those incorporated into the early crystallising phases. Lithium, rubidium, caesium, barium, zirconium, the rare earth elements, and uranium are examples of those elements which become enriched in the final magma fractions. To a large extent this pattern of behaviour may be explained by the differences in the sizes and charges of the ions. The alkali metal ions with a charge of +1 do not readily substitute for those ions of +2 or +3 charge, such as Ca^{2+}, Mg^{2+}, Fe^{2+}, Fe^{3+} and Al^{2+}, which constitute the early crystallised minerals, whereas ions similar to them in charge and size, including Cr^{3+}, Co^{2+} and Ni^{2+}, will substitute easily. Ions such as Ba^{2+} and La^{2+}, although of appropriate charge, are too large to be accommodated stably in the available co-ordination sites. Similarly, highly charged ions such as Zr^{4+}, U^{4+} or U^{6+}, and Th^{4+} are not readily incorporated.

The partition of trace elements into the crystallising minerals depends upon the available co-ordination polyhedra in both the solid and liquid phases. For example, strontium, as the relatively large ion Sr^{2+}, does not substitute for Mg^{2+} or Fe^{2+} ions occupying six-fold co-ordination sites in olivine but will replace cations in the irregular and large nine-fold co-ordination sites of calcium-rich feldspars, or in the irregular eight-fold co-ordination sites of calcium-rich pyroxenes. Unfortunately our knowledge about the availability of cation sites in silicate melts is very limited but there is a significant increase in the number of sites of low co-ordination number (particularly four-fold) at the expense of those of higher number, compared with the crystalline solid of the same composition. The retention by the magma of the small cations (e.g. Be^{2+}) during fractional crystallisation is a result of these ions having greater relative stability when occupying the smaller co-ordination sites.

Other factors, in addition to the structural aspects discussed above, affect element distribution. Temperature and pressure changes lead in certain cases to changes in the partition of elements between minerals or between a mineral and its parent melt. Thus it is possible to use selected pairs of minerals as geothermometers or geobarometers provided that an experimental calibration has been made. The geochemical behaviour of some transition metal ions is affected by their particular electronic configurations, which are distinct from those in non-transition metal ions (Burns, 1970). Also, the role of covalent bonding, especially in the development of magmatic sulphide minerals, can be important. Many of these additional factors are well understood, so that we can predict with reasonable certainty the compositional changes that solid and liquid undergo during the fractional crystallisation of a magma of known composition.

The solid products from the differentiation of a basic magma range from rocks such as peridotite (consisting of olivine, pyroxene, and a spinel) and gabbro (Ca-rich feldspar, pyroxene, magnetite (Fe_3O_4), and sometimes olivine), which are relatively rich in magnesium, calcium and the trace elements chromium and nickel, to granite, rich in the alkali metals (Fig. 3.4) and such traces as uranium, barium and the rare earth elements. The typical mineralogy of a granite is quartz, alkali-feldspar, and biotite or hornblende, with accessories such as rutile (TiO_2), apatite, zircon ($ZrSiO_4$), and magnetite.

The fractional crystallisation of a basaltic magma can lead to the development of layered igneous rocks in which

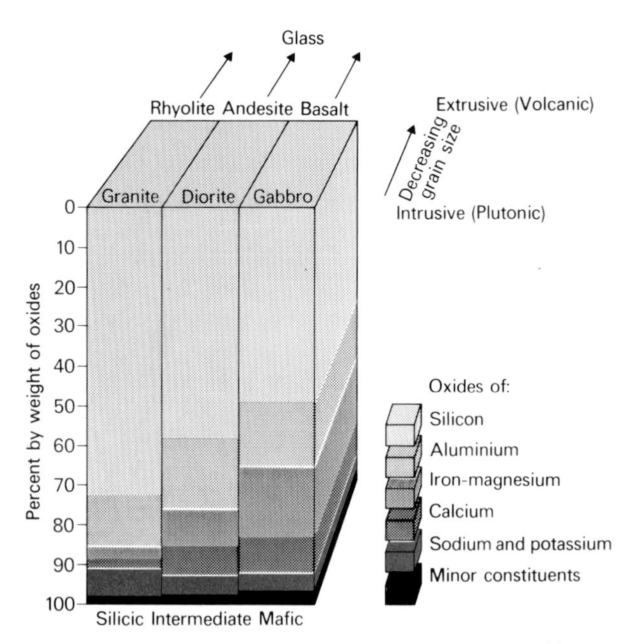

Figure 3.4 Average composition of the major types of lava and their intrusive counterparts

layers rich in a particular mineral or mineral assemblage may alternate with, or be succeeded by, layers rich in other minerals (e.g. feldspar). Individual layers can often be traced for considerable lateral distances. They arise from the sorting of crystals of different sizes and densities as they sink (or rise) through the magma and also from the different times of crystal nucleation of the different minerals during the magma cooling. One such layered intrusion, made famous by the detailed study of Wager & Deer (1939), is the Skaergaard plutonic complex, Kangerdlugssuaq, east Greenland, where about 500 km³ of basic magma was emplaced rapidly into the upper crust. The layers range in rock type from olivine-rich gabbro at the exposed base to a granitic rock at the top of the layered part of the intrusion. Throughout the layered sequence there is a steady systematic change in the compositions of the minerals that show solid solution. For example, olivine at the exposed base is magnesium rich, but is a pure iron-olivine (fayalite) at the top (Wager & Brown, 1968).

The Muskox layered intrusion, near Coppermine, North West Territories, Canada, is particularly interesting in that the feeder, through which the magma passed, has been exposed. The form of this intrustion is shown in Figure 3.5. Unlike the Skaergaard intrusion, which was derived from one mass of magma, the Muskox one developed from successive pulses of magma into the chamber. Other examples of layered igneous intrusions include those of Rhum, Inner Hebrides, Scotland; the Stillwater Complex, Montana, USA, and the Bushveld Complex, near Pretoria, South Africa. The latter is important as a source of platinum metals (see below).

Many layered intrusions show the development of a granitic rock as the final product of the differentiation of the basaltic (or basic) magma. However, the amount of granite produced by this process is too small to account for the large masses of granite that occur in parts of the continental crust. These batholiths are more probably

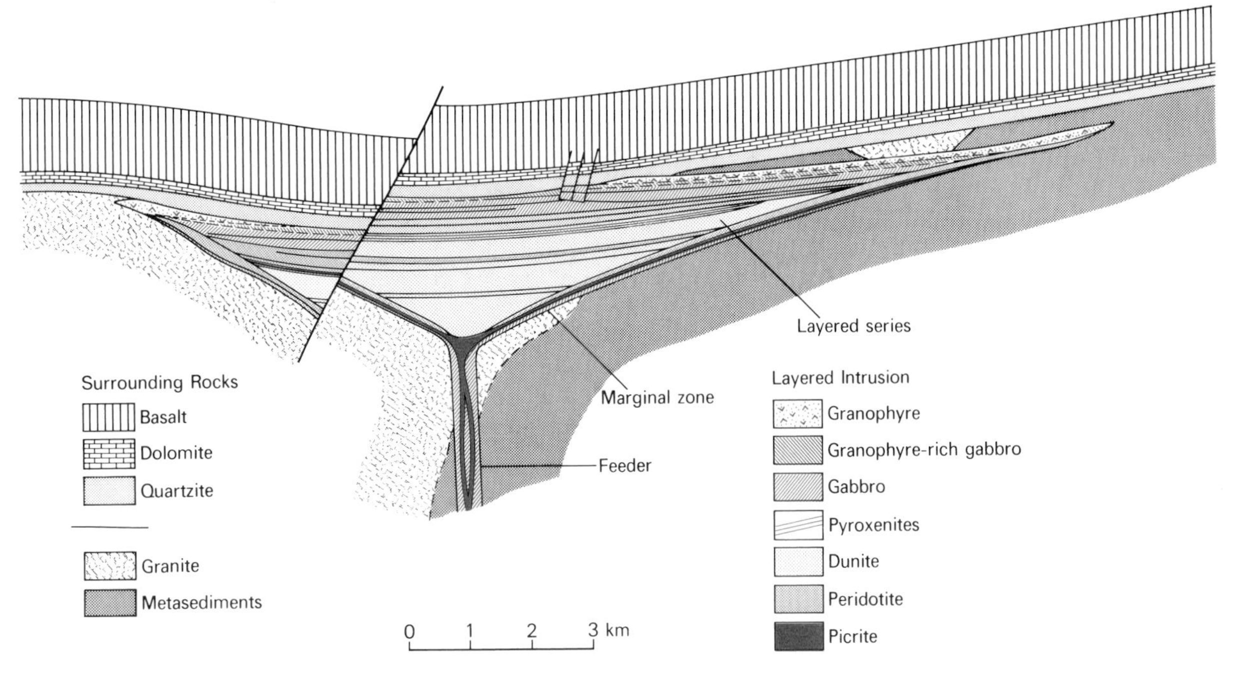

Figure 3.5 Cross-section (restored) of the exposed part of the Muskox intrusion, North West Territories, Canada (After Irvine and Smith 1967)

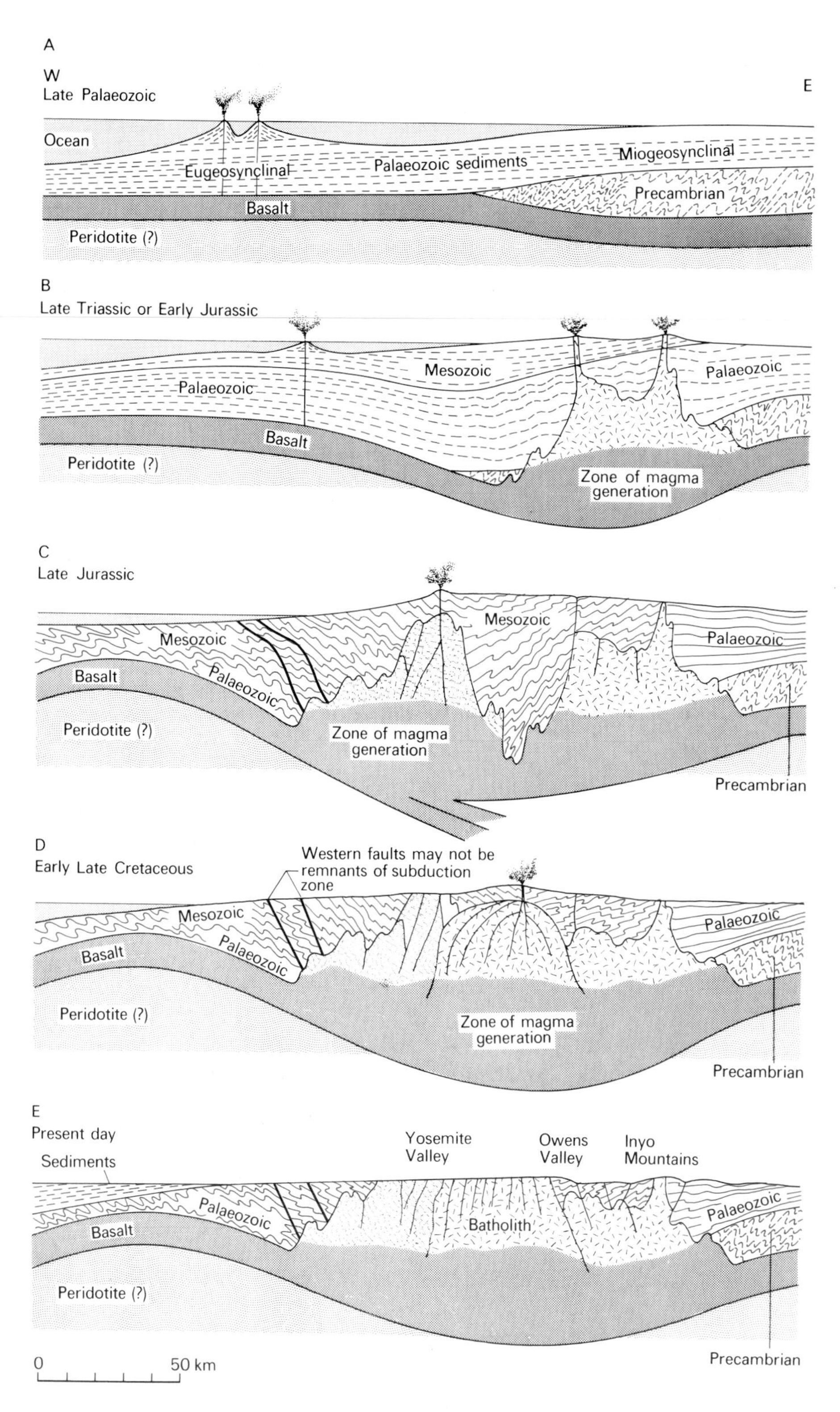

Figure 3.6 Schematic cross-sections of the Sierra Nevada, USA, showing its possible evolution (A) to (D) and the existing structure (E) (After Bateman and Eaton 1967)

derived by the partial or complete melting, at depth, of pre-existing sediments which are rich in silica (as in the case of the Sierra Nevada batholith, Figure 3.6). Indeed it is often possible to distinguish, by their trace element and isotopic compositions, granites derived by fractional crystallisation of basic magma from those produced by the melting of pre-existing rocks.

One way this may be done is to make use of the radioactive isotope of rubidium, ^{87}Rb. This isotope undergoes slow decay, by emission of beta particles, to a stable isotope of strontium, ^{87}Sr. Therefore, the total number of ^{87}Sr atoms ($^{87}Sr_p$) in a rock will depend upon the number originally incorporated ($^{87}Sr_o$) as well as the number of ^{87}Rb atoms remaining (^{87}Rb) and the age (t) of the rock. The relationship between these parameters is obtained using the radioactive decay law, approximated for the slow decay rate of ^{87}Rb, to give an expression that holds true for all but the oldest rocks:

$$^{87}Sr_p = {}^{87}Sr_o + {}^{87}Rb \,.\, \lambda t$$

where λ is the decay constant of ^{87}Rb. It is more convenient to modify this equation by dividing throughout by the number of atoms of ^{86}Sr in the rock, (this stable isotope does not come from the decay of any known natural radioisotope so its amount in the Earth is constant). Thus:

$$\left(\frac{^{87}Sr}{^{86}Sr}\right)_p = \left(\frac{^{87}Sr}{^{86}Sr}\right)_o + \frac{^{87}Rb}{^{86}Sr} \cdot \lambda t$$

The $(^{87}Sr/^{86}Sr)_p$ and $^{87}Rb/^{86}Sr$ ratio values of a rock specimen are determinable by mass-spectrometric analysis but as there are two unknowns (i.e. $(^{87}Sr/^{86}Sr)_o$ and t) the equation is not soluble unless two or more samples from the same rock mass (i.e. same t) are analysed. Provided that the different samples, or analysed minerals, have sufficiently different isotopic ratios they will, when plotted as in Figure 3.7, define a straight line which joins the points of same age (isochron). The slope of the line is λt and so it gives the age of the samples; the intersection of the line with the ordinate axis gives the $(^{87}Sr/^{86}Sr)_o$ value.

Now the rate at which the $^{87}Sr/^{86}Sr$ ratio increases with time in the Earth's crust is greater than in the upper mantle (see Chapter 2). This is because rubidium, as a result of its ionic charge (+1) and large ionic radius, has been preferentially concentrated in the crust during the differentiation of the Earth. Therefore a granite derived from the fractional crystallisation of a basaltic magma produced in the upper mantle will have a lower $(^{87}Sr/^{86}Sr)_o$ ratio, on blue line in Figure 3.7, than a granite of the same age but derived by fusion of pre-existing crustal rocks (black line in Fig. 3.7). Hence, the determination of the $^{87}Sr/^{86}Sr$ ratio of a rock at the time of its formation helps to establish the provenance of the magma (Faure, 1977). Granitic rocks derived from a mantle source have $(^{87}Sr/^{86}Sr)_o$ values of about 0·700, while those derived from a continental source have values of around 0·720 or higher. This method has been applied with success to a number of granite bodies in Africa and elsewhere, but results for some batholiths in North America seem to require more complex interpretations (Cox *et al.*, 1979).

The partial melting of upper mantle material is known to be the principal process for the production of basaltic magma. Only a small fraction (a few per cent) of the material need melt as a result of local thermal variations before the liquid will be able to migrate upwards and away from the refractory and denser residue. The melt may then collect in a chamber, still at some depth below the Earth's surface, where it undergoes cooling and crystal fractionation (as described above) or it may reach the surface as lava and other volcanic products. Under surface conditions the lava tends to cool quickly, there is insufficient time for crystal fractionation, and a uniform, fine-grained rock results. Sometimes this extruded lava is not entirely liquid but may contain crystals, often with well developed shapes, of the early forming minerals. The resultant rock will have these crystals (called *phenocrysts*) set in a fine-grained or glassy matrix. The upwelling magma may also penetrate fissures in the country rocks, or migrate along bedding planes of strata, to form dykes and sills. Unless these intrusions are very thick, they will also cool quite quickly, but the rock is usually more coarsely crystalline than its extrusive equivalent. At some subsequent time, earth movements and erosion will cause the intruded rocks and deep seated masses of cooled and crystallised magma to be exposed at the surface.

The variety of magma compositions leads to a wide range in lava viscosity which, in turn, affects the nature of the accompanying volcanism. Basaltic magmas with their low viscosities are associated with volcanism involving extensive lava flows, such as are seen in Iceland or in the Tertiary igneous province of Scotland. On the other hand, silicic lavas such as rhyolite (Fig. 3.4) are often so viscous that they barely flow and so give rise to volcanic domes (e.g., Mt Pelée, Island of Martinique) or, where gaseous emanations are important, to explosive volcanic activity and the production of cinder cones (see Introduction to this section of the book). Macdonald (1972) gives a comprehensive account of volcanic activity and structure.

It has been shown experimentally (see Yoder, 1976) that a basaltic melt is produced first during melting at high pressure of garnet-bearing peridotite, a rock that is probably abundant in the upper mantle. Indeed basalt is the most copious magma type to come from the mantle, but, despite this, the continental crust does not have a basaltic composition. The large masses of existent granitic rock make the continents the most easily fusible and least dense fraction of the whole Earth.

Another magma type that erupts onto the surface of the continental crust gives the volcanic rock andesite. This contains feldspar, pyroxene and/or an amphibole and is chemically intermediate between basalt and rhyolite (Fig. 3.4); diorite is its coarse-grained intrusive counterpart. Andesite is a common constituent of mountain chains that

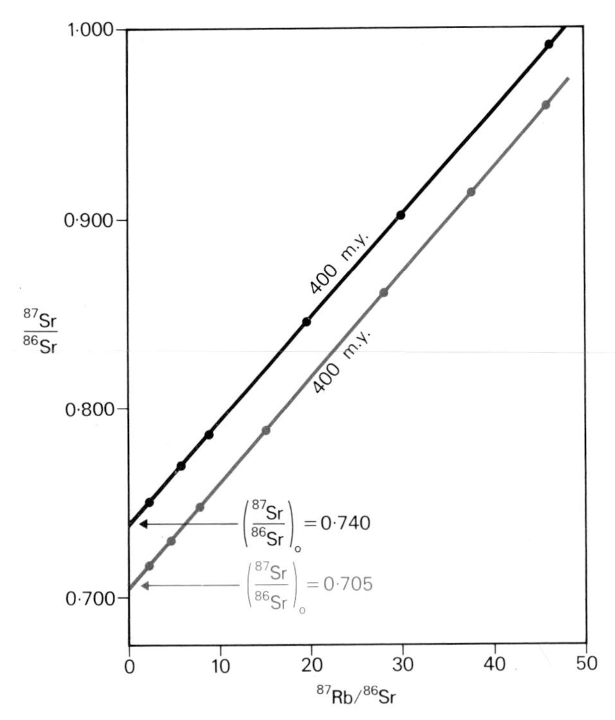

Figure 3.7 Rb–Sr isochron diagram of two granites of the same age but of different derivations. The blue isochron is for a granite mass derived by fractional crystallisation of a basaltic magma from the upper mantle; the black isochron is for a granite derived from the melting of crustal rocks

are associated with crustal deformation near subduction zones (e.g. the Andes). It is considered to come, at least in part, from the partial melting of the oceanic crust and wet sediment of a descending lithospheric plate (see Chapter 11).

Ore deposits

Many hydrothermal ore deposits are formed during the last stages of magmatic differentiation, when metal-rich fluids and volatiles are given off and pervade the surrounding country rock. There the metals are precipitated through reaction of the fluids with constituents of the country rock, or with sulphide-bearing ground waters to give the common ore minerals – pyrite (FeS_2), galena (PbS), sphalerite (ZnS), cinnabar (HgS), chalcopyrite ($CuFeS_2$) and others. For the ore to be economic the metals must usually be enriched by a large factor relative to their average crustal abundance (Tables 1 and 2). For example, the 'concentration factor' for copper in a workable deposit will be at least 60.

Our knowledge of the nature of hydrothermal solutions has come, in part, from detailed studies of fluid inclusions within the ore minerals, as the fluids probably represent the ore-forming medium. The inclusions are usually very small (a few μm across); often consist of a saline fluid, sometimes with 'daughter' minerals and a gas phase; and have different compositions at different stages of ore development. On the assumption that the original material trapped in the ore mineral was a homogeneous fluid, the temperature of mineral formation may be established from a relatively simple experiment. Furthermore, determination of the relative abundances of the stable isotopes of oxygen in the fluid can indicate whether the water was from a magmatic or a meteoric (i.e. atmospheric) source, as the $^{18}O/^{16}O$ ratios in the two sources are usually distinct.

It has been suggested (Sillitoe, 1974) that the ore-bearing provinces of the Andes were generated by the action of hydrothermal solutions produced from the magmatic activity of subduction. There the intrusive and extrusive rocks form long, narrow belts roughly parallel to the continental margin and which are often younger towards the east. The belts contain a parallel sequence of ore deposits, consisting of a coastal belt rich in iron and successive inland belts rich in copper, then copper–lead–zinc–silver, and finally one rich in tin, although this generalised pattern is disrupted in places by tectonic boundaries.

Fluid inclusion studies of the important economic copper deposits of the Andes have yielded some interesting results. In one area the primary copper mineralisation was formed from hot (350–600 °C) saline fluids of magmatic origin. The precipitation of the ore minerals was probably caused by changes in the composition of the hydrothermal solution through its interaction with the host rock. Later stage mineralisation (mainly pyrite) has produced by solutions at a relatively low temperature ($<$ 350 °C) and of low salinity. The oxygen isotope studies show that these later solutions were predominantly of meteoric origin.

It is possible that changes in the temperature and composition of the hydrothermal solutions as they permeate, and interact with, the surrounding rocks can lead to the development of zoned ore deposits. For example, in south-west England four general zones of mineralisation associated with the granite intrusions can be recognised, although the pattern was complicated by the structural controls and the multi-stage nature of the depositional process. The zone furthest from the granite is rich in iron and manganese, and includes minerals such as limonite (hydrated iron oxide), haematite (Fe_2O_3), and siderite ($FeCO_3$). Next is a lead–zinc–silver zone with galena, sphalerite and argentite (Ag_2S), which is adjacent to a copper–arsenic–tungsten zone of arsenopyrite (FeAsS), wolframite ($(Fe,Mn)WO_4$), and chalcopyrite. Finally a tin zone of cassiterite (SnO_2) is closest to, and extends into, the granite (Edmonds *et al.*, 1975).

Where a granite intrusion is surrounded by limestone, a specific mineral assemblage develops by the replacement of the calcite and dolomite ($CaMg(CO_3)_2$) with silicates of calcium, magnesium and iron at the contact of the two rock types. This association, known as skarn, may be a source of tin, tungsten and copper minerals.

The solfataric activity associated with volcanoes can be

a significant contributor to ore development, especially by deposition of native sulphur or by the production of sulphur-bearing waters capable of causing the precipitation of metal sulphides from solution. The Matsuo deposit in central Honshu, Japan, contains more than 50 million tonnes of native sulphur, some of which was erupted in the molten state. Also in Japan, at Kuroko, there are stratabound sulphide deposits produced, in the sandy and shaly sediments of previous isolated basins, by subaqueous volcanic activity involving sulphur-rich solutions.

The production of ore deposits associated with igneous activity need not involve the late-stage, volatile products of a differentiated magma. Many ores are produced by direct crystallisation from a magma as, for example, in layered intrusions, which are an important source for many metals, such as chromium, nickel, titanium and platinum. The Bushveld layered intrusion of South Africa contains a platinum-bearing layer of great lateral extent. The layer varies between one and five metres and has, besides the silicates, a remarkable variety of minerals, including sulphides of iron, nickel and copper. The platinum occurs partly in the native state, as ferroplatinum alloy, and partly as sulphide and arsenide. Gold and other platinum metals also occur in the deposit, which is considered to have been precipitated from a liquid phase that was immiscible with the main body of magma.

It needs to be said that igneous activity is not a pre-requisite for ore formation. Saline ground waters, at temperatures elevated by the geothermal gradient take up heavy metals by interaction with the surrounding rocks, and so produce potential ore-forming solutions, and some ores are laid down in marine sediments.

Metamorphic rocks

Where a magma is intruded near the Earth's surface into sedimentary rocks it frequently has a visible effect (contact metamorphism) on the surrounding rocks. The heat from the magma can produce significant changes to the texture and mineralogy of the host rock, but rarely produces a change in the chemical composition except where ore minerals are formed. The mineralogical changes involve solid–solid transformations, the low temperature phases of the sedimentary rock being made over by heat into new minerals formed only at higher temperatures. A shale some-way from the intrusive contact will show a loss of its bedded structure, and micas will have developed within it at the expense of the clay minerals. Nearer the contact the metamorphosed rock will be more crystalline and other minerals such as andalusite (Al_2SiO_5) or pyroxene may have grown. If the metamorphism is rapid, a rock at the contact will develop a hornfelsic texture with a fine grain-size and no foliation.

High pressures and temperatures produced on a regional scale, as in the formation of mountain fold belts, lead to the development of regionally metamorphosed rocks in which characteristic minerals occur. Garnets (e.g. almandine, $Fe_3Al_2Si_3O_{12}$) nucleate and grow by replacement of the micas and quartz of the original rock through diffusion and chemical exchange of components in the solid state. Other minerals characteristic of metamorphism form by the same process; examples are staurolite and chloritoid, both of which are iron–magnesium–aluminium silicates, although the mineral content is dependent upon the chemical composition of the original rock.

The fact that some minerals crystallise only above a particular temperature and/or pressure permits their use as indicators of the grade of metamorphism. The mineral andalusite will, at high pressures, convert to another crystal structure to give the mineral kyanite, or at high pressure and temperature to give sillimanite, without any change in composition. Relatively slight metamorphism of shales produces muscovite, then biotite, garnet and staurolite, and at higher grades, kyanite and finally sillimanite. The sequence of mineral changes in basaltic rocks starts with the development of zeolites and chlorite, followed at higher grades with epidote and amphibole, then garnet and pyroxene.

Tectonically affected terrains often show zones of progressive metamorphism over large areas. For example, the Dalradian rocks to the south-east of Aberdeen, Scotland, display a mappable sequence of zones, with approximately parallel boundaries, ranging from mica and quartz bearing rocks of low metamorphic grade to mica–garnet–sillimanite–quartz bearing rocks of higher grade.

In the process of mineral replacement elements are redistributed only on a local scale. The chemical composition of a hand specimen of the metamorphic rock will be the same as that of the original, but the texture and mineral content may be very different. The resultant distribution of the elements is again governed by controls of crystal structure, temperature and pressure as already described for igneous rocks. However only the processes of igneous and sedimentary rock formation lead to the large-scale fractionation of elements in the Earth's crust. Subsequent chapters describe the nature of sedimentary rocks and of soils, which are the products of further mechanical and low-temperature chemical redistribution of these elements.

References

Bateman, P. C. & Eaton, J. P. 1967. Sierra Nevada Batholith. *Science*, New York **158**: 1407–17.

Burns, R. G. 1970. *Mineralogical applications of crystal field theory*, 224 pp. Cambridge: University Press.

Cox, K. C., Bell, J. D. & Pankhurst, R. J. 1979. *The interpretation of igneous rocks.* 450 pp. London: George Allen & Unwin.

Edmonds, E. A., McKeown, M. C. & Williams, M. 1975. *British Regional Geology: South-West England* (4th ed.). 136 pp. London: Institute of Geological Sciences.

Faure, G. 1977. *Principles of isotope geology.* 464 pp. J. Wiley & Sons.

Garrels, R. M., Perry, E. A., Jr. & Mackenzie, F. T. 1973. Genesis of Precambrian iron-formations and the development of atmospheric oxygen. *Economic Geology* **68**: 1173–1179.

necessary high currents and wave velocities could be achieved. Thus whenever conglomerates were found in ancient rocks, geologists tended to interpret them automatically as indicating original shallow water. However the science of sedimentology underwent a revolution soon after the Second World War, when P. H. Kuenen (e.g. in Kuenen & Migliorini, 1950) showed that fast dense undercurrents flowing along the floor of a basin can exist independently of movement in the main mass of water above them, and can also transport sediment in great amounts. Knowledge of such turbidity currents, as they have come to be called, changed the thinking of sedimentologists since the turbidity currents provide a mechanism whereby particles of almost any size can be transported along the ocean floor well below wave base. This realisation has been the spur for a revaluation of all kinds of sedimentary rocks during the past 30 years, and this, together with much new work on the chemically-deposited rocks, has resulted in a much better understanding of the different environments in which the various rock types were formed.

One of the best-documented large-scale natural turbidity currents occurred off Canada in 1929. An earthquake near the Grand Banks, off Newfoundland, triggered slope failure in a large pile of soft sediments, and an enormous slump, later estimated to be in the order of 100 cubic kilometres in volume, was carried by a turbidity current down the continental slope and on to the abyssal plain, eventually spreading over an area of some 100000 square kilometres. As the turbidity current moved downslope it broke various submarine cables at different times, and these broken cables served as stopwatches, enabling accurate estimates to be made later of its speed and travelling distance (Figs 4.1, 4.2). Not all turbidity currents are of this scale; they can even be generated artificially in a handbasin. The sediments which are eventually deposited by a turbidity current usually show grading (sorting the various constituent grains from larger sizes to smaller sizes) both horizontally, that is the farther from the sediment source the finer the sediment that is present, and also vertically, that is a turbidite layer at one locality is coarser at its base than at its top.

However, other factors also determine particle size. Winnowing, both by winds on land and by currents under water, can sift a sediment clean of its finer particles, leaving a higher proportion of coarser particles than the sediment which was first deposited. Mechanical abrasion can change the particles in two separate ways, firstly by causing them to break into smaller units by mechanical collision, and secondly by smoothing the jagged projections and corners from the particles. For example a sandstone originally formed in the desert will often have rounded 'millet-seed' grains, whilst a marine sandstone of similar overall grain size may have all of its sandgrains with angular corners after relatively speedy transport and burial. This rounding occurs at different scales, as witnessed by the rounded cobbles of storm beaches, which contrast

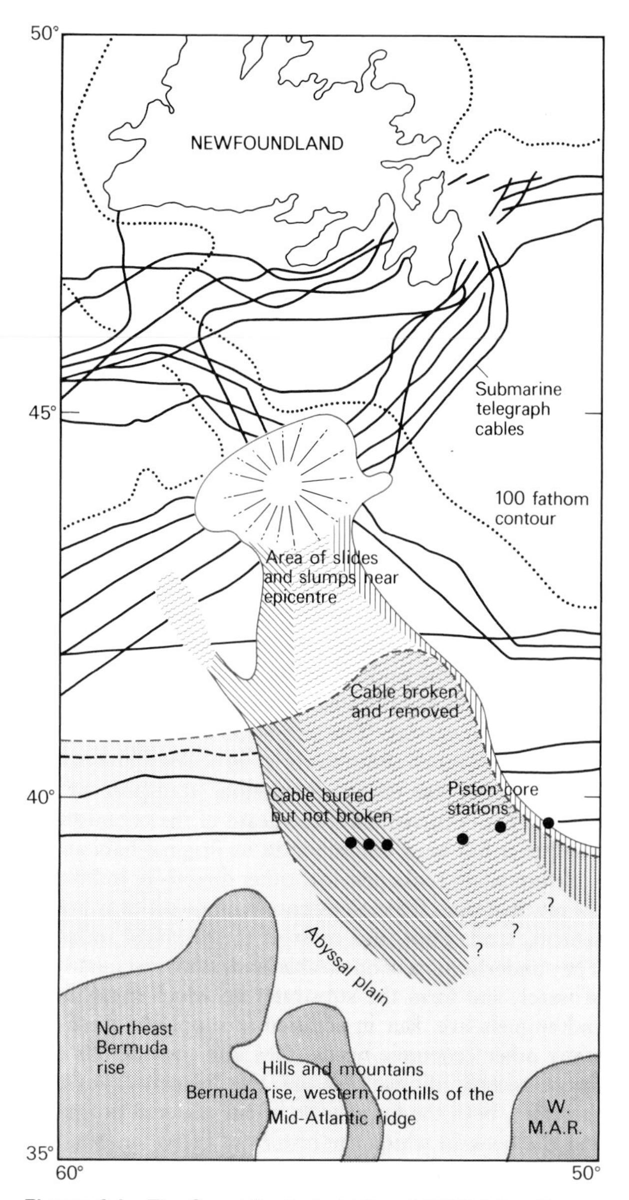

Figure 4.1 The Grand Banks turbidite of 1929 (after Heezen and Hollister, 1971).

with angular rock fragments of a similar size which may be present on the same beaches as the result of landslides. Erosion can also occur through processes other than direct water action. Bombardment by wind-borne particles can play a significant role, particularly in dry climates. The actions of freezing and thawing, as well as heating and cooling, can also help to break down different rocks in different ways under different circumstances.

Another approach in recent years that has also played a considerable part in the better understanding of sediments is the simulation of their deposition under various artificial circumstances, for example in flume tanks. From such experiments hard data have been gained on the actual

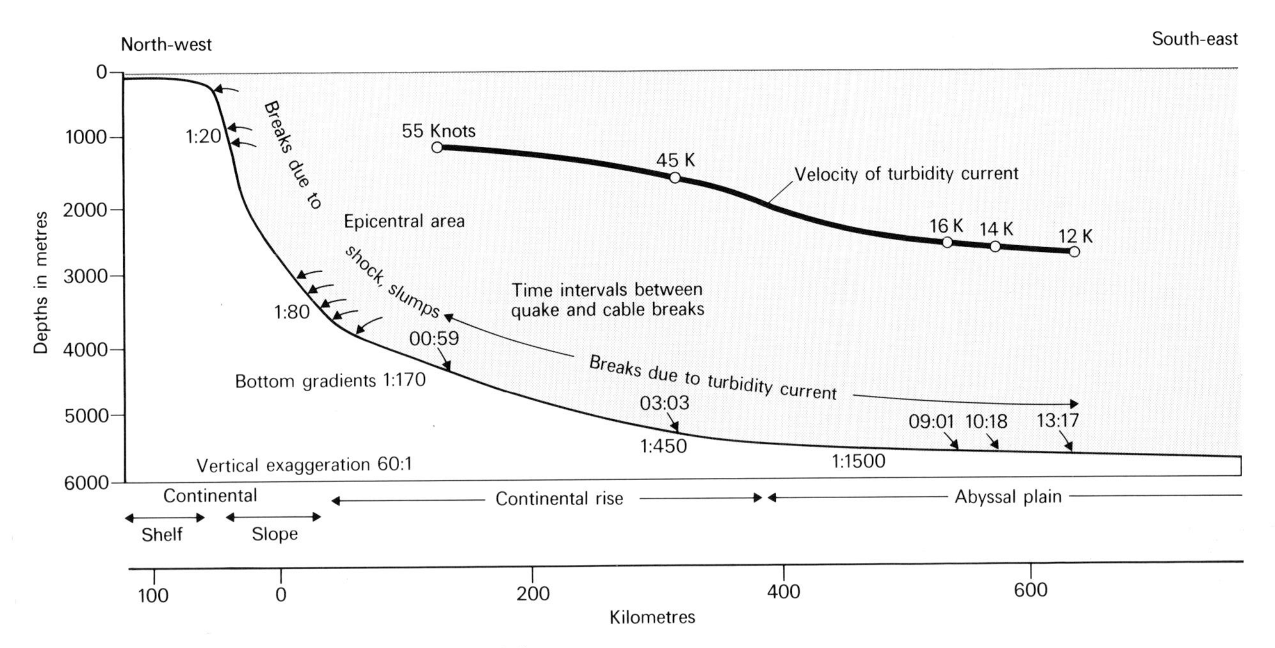

Figure 4.2 The velocity of the Grand Banks turbidity current, calculated from the times of successive cable breakages which it caused (after Heezen and Hollister, 1971).

behaviour of sediments of different grain sizes, shapes and composition under different regimes of slope, current velocity and other factors. Oil companies have been in the lead in using such data, for example in the analysis of whole ancient sedimentary basins in their search for petroleum products. Such studies have been complemented by work on recent sediments using sophisticated arrays of oceanographic and geophysical techniques to present a three-dimensional picture of great detail. Many data have also come from smaller-scale experiments designed by engineers wishing to minimise silting in harbours and reservoirs, and to help them assess the relative effectiveness of various dam types and breakwaters under different circumstances.

Chemical processes can and do affect many types of sediment, both in their primary deposition and also in their diagenesis, and some of these processes are reviewed later in this chapter. Finally the modifying acts of animals and plants are often an important sedimentary process. Such acts include the stabilisation of sediment surfaces, for example by algal mats, preventing erosion by all but the worst storms, and the destructive effects of burrowing by many types of animals and by the roots of plants. In addition the debris of the animals and plants themselves can be an extremely significant part of the final sediment, obvious examples being coal and many limestones.

Depositional environments

The main sedimentary environments will now be reviewed in turn; for expanded treatments the reader is referred to the excellent recent summaries of Reading (1978) and Walker (1979).

1 *Deserts*

Despite the popular misconception that deserts consist chiefly of sand dunes, these are not usually the major element of the total desert area; the largest surfaces consist of bare rock and alluvial outwash from it, as well as chemical and evaporite deposits such as sabkhas or playa lakes. However, the bare rocks will only be represented in a future geological outcrop by an unconformity; they will not be dealt with further here, and the desert alluvial deposits, although formed only during occasional periods of flash floods, do not differ significantly from the more normal alluvial deposits described below, except in their penecontemporary reddening due to weathering of their contained iron compounds. The inland sabkha and playa lake deposits are chiefly made up of chemicals which have evaporated from flood or ground water, and are comparable to those evaporite deposits found near the edges of warm seas and discussed further below. Figure 4.3 (after Glennie, 1970) shows the distribution of the various terrains in the desert of the Oman peninsula today, and gives a good indication of which areas are undergoing net erosion, in contrast to those which are accumulating sediments which might be preserved in the future geological record.

Dune sediments can be very different from each other, and their composition and grain size are influenced chiefly by such factors as the variability of older rocks in the source area, the prevalence and strength of the chief winds, and the frequency of the occasional storms. The nature of

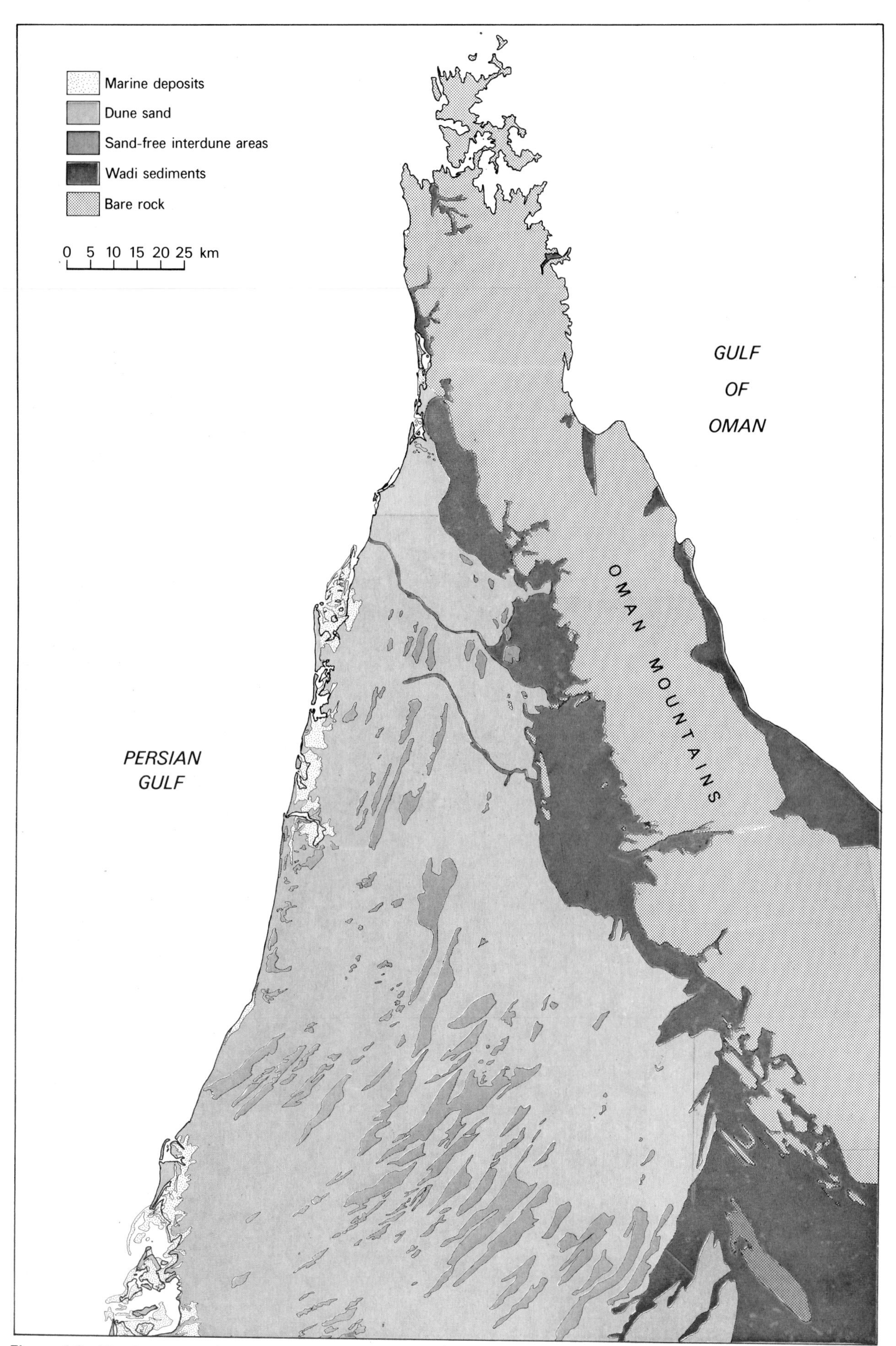

Figure 4.3 The desert area of the Oman peninsula, Arabian Gulf, showing the areas of bare rock, and the various sediment types (after Glennie, 1970).

the bedding within the dunes themselves is determined both by the thickness of the accumulation of the loose sand cover and also by the variation in the grain sizes present. In thick accumulations, such as those of the Sahara, there are three scales of beds, from very large ones (called draas), on which are superimposed normal dunes, which in turn have superimposed upon them a system of ripples. In areas of thinner sand accumulation, over most of Australia for example, only dunes and ripples are developed. All of these dune types have been recognised in the geological record, for example in the Permo-Triassic New Red Sandstone of central England.

Considerable volumes of silt-grade and finer material can be moved as dust storms, and these settle to form a deposit known as loess. Loess deposition is also associated with colder climates, and particularly periglacial ones, where the grinding action of the ice has produced an abundance of fine material. Because loess is very often subsequently redeposited by water action, it is sometimes difficult to identify in the geological record, but is important to engineers since its shear strength is low: the effects of its subsidence can be disastrous on building foundations, particularly in earthquake areas such as north China, where loess is widespread.

2 *Alluvial and freshwater environments*

Although it is tempting to think that most sedimentary particles are carried by water from the mountains to the sea, in fact a large proportion of the particles never reach marine environments but are deposited permanently quite close to their source area. Some of these alluvial sediments are fan deposits, where water, often in the form of flash floods, lays down the particles in sheets which resemble the segments of a cone in shape. Such sediments are often poorly sorted near areas of high relief. Between the major periods of sediment influx and spreading, the surface of the fan usually becomes dissected by channels of variable size, each of which also carries and deposits some sediment, and whose courses often change after a major event. The scale of these fans varies from a few metres to hundreds of kilometres, like the enormous fans seen today over the north Indian and Nepalese plains at the base of the Himalayas. Mudflows also occur, and these can carry with them much poorly-sorted larger material. In some deposits the mud is subsequently winnowed away to leave only the larger particles in the form of conglomerates.

Other alluvial deposits are the products of rivers, particularly in more temperate climates. These areas seldom experience flash floods, but their rivers may carry a large quantity of sediment over much of the year. When the gradient becomes sufficiently low, the sediment will drop out progressively in relation to its grain size, and braided rivers will form over alluvial plains whose areas can be substantial. The sedimentology of these deposits is often very complex, but has been intensively studied over many years, and the various environments of deposition can be deduced from the rocks formed when the alluvial deposits have been preserved. The river flow itself can form the sediments into dunes, sandwaves, point-bars, and a variety of smaller structures such as ripples, and from their structure in ancient rocks can be deduced the dominant current directions and other parameters. Ancient river terraces can also be recognised, although the assessment of their relative heights above sea level becomes progressively more difficult in older rocks.

Freshwater deposits also include those laid down in lakes, which today form about one per cent of the Earth's continental surface and can be important sites of sediment accumulation. Lacustrine sediments are closely comparable with those formed in shallow marine environments, but differ from them not only in their dissimilar animal and plant (and subsequently fossil) content, but also in their generally quieter settling conditions and in their greater variation due to their increased dependence on the local climate. The depth of the lake is also important, for example shallow and impersistent lakes in warm climates can accumulate chemical deposits at their margins. In temperate and cold climates, lake deposits often reflect marked seasonal variations.

3 *Deltas*

When a river meets the sea its current velocity drops, and its contained sediment load is deposited. At some river mouths there are destructive sedimentary environments, where the wave and marine current action is so great that the river sediment is swept away laterally or outwards on to the continental shelf. However, where the marine waves and currents are not so intense, then the sediments are deposited locally to form 'constructive' deltas, varying in size from a few metres to the substantial structures seen today near the mouth of the Mississippi and some other great rivers. Such deposits are of great importance in the geological record because they seem to be commonly preserved: the volume of rocks of deltaic origin is immense. Figure 4.4 is a block diagram showing the principal sedimentary facies laid down around a typical delta. The areas to the side of the river and its distributary channels can be occupied by subaerial levee deposits, sometimes flanked by alluvial and marsh sediments, but the main bulk of the particles is deposited to seaward near the distributary mouths and on the delta front slope. The coarser sediments will be deposited nearest to the river mouth, in the form of sheet sands and distributary mouth bars, later to be preserved as cross-bedded sandstones. The deposits become progressively finer off-shore, forming distal bars of silts. The finest particles are carried further out to sea to form prodeltaic deposits which on consolidation will form fine siltstones and mudstones. As the delta builds out to sea, earlier prodeltaic deposits will be covered by delta-front deposits, and delta-front deposits by distributary mouth deposits and so on. Thus deltaic sequences are commonly cyclic in vertical section, a cycle ceasing when a main channel changes its position laterally. When the river discharge is steady, and generally with a high

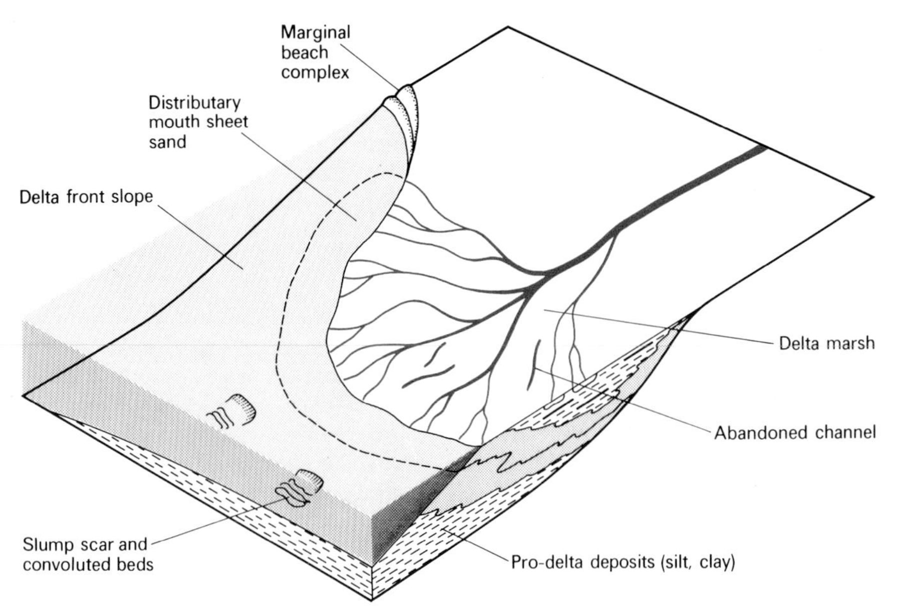

Figure 4.4 Block diagram of the sedimentary environments near a lobate delta (after Miall *in* Walker, 1979).

suspension load, then the delta will tend to be in the form of a birds-foot, such as that of the Mississippi. However when the river has a fluctuating or seasonal discharge, with a higher proportion of bed-load in the sediments, then the typical delta form will be more lobate in outline, such as that of the Nile. Salinity varies enormously round a delta mouth: when there are strong ocean currents and waves nearby then marine faunas and floras will occur quite near the shore; in other situations reduced salinity assemblages may be found many kilometres out to sea.

In addition to the primary cause of cyclicity in deltaic deposits outlined above, sequences can also be affected by changes in sediment input (owing to primary changes in the paths of distributary channels), eustatic changes in sea level, changes in climate (and therefore weathering rates) in the source area, and more fundamentally by the decreases in gradient as the deltaic sediments themselves fill the receiving basin topography. When the sediments are sufficiently shallow to enable subaerial plants to grow on them, then the amount of vegetation can permit the formation of coal deposits at the top of a cycle, particularly laterally to the distributaries flowing at that time. During Upper Carboniferous times, large quantities of sediment were supplied to the margins of many basins in Europe and the United States, and major deltas were formed: these are the source of most of our coal today.

4 Shorelines and estuaries

The most critical parameter affecting sedimentary environments at a shoreline is the average height of the tide. Figure 4.5 shows the dramatic effect that the tide has on the distribution of some of the various types of depositional coastal morphology – estuaries, salt marshes, and tidal flats and sand ridges can only develop in areas of substantial tides, whilst barrier islands, and tidal deltas and inlets are confined to micro- and mesotidal coasts. Apart from this, the sediments are very dependent on the protected nature or otherwise of the shoreline; those areas on an exposed oceanic coast with a strong onshore prevailing wind naturally tend to be areas of erosion, at least above wave base, although sedimentation may occur offshore. If the prevailing wind is consistently oblique to the shoreline

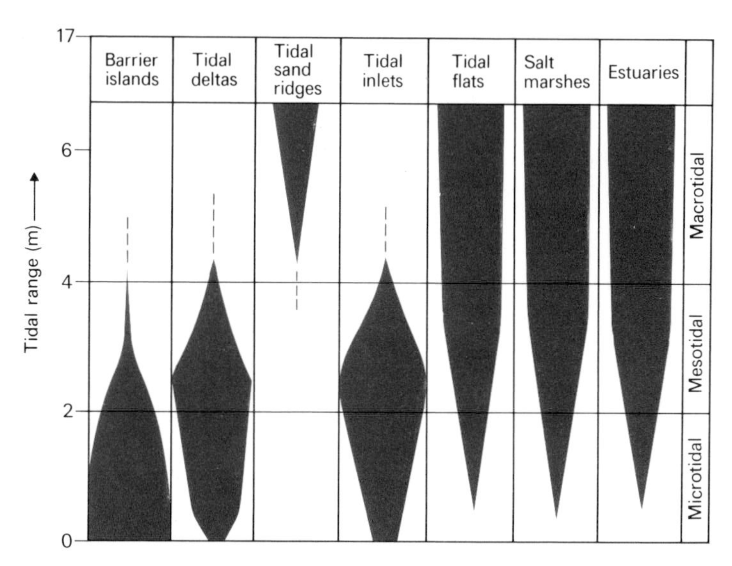

Figure 4.5 The relationship between various tidal ranges and the morphology of the adjacent coastlines (modified after Hayes, 1976).

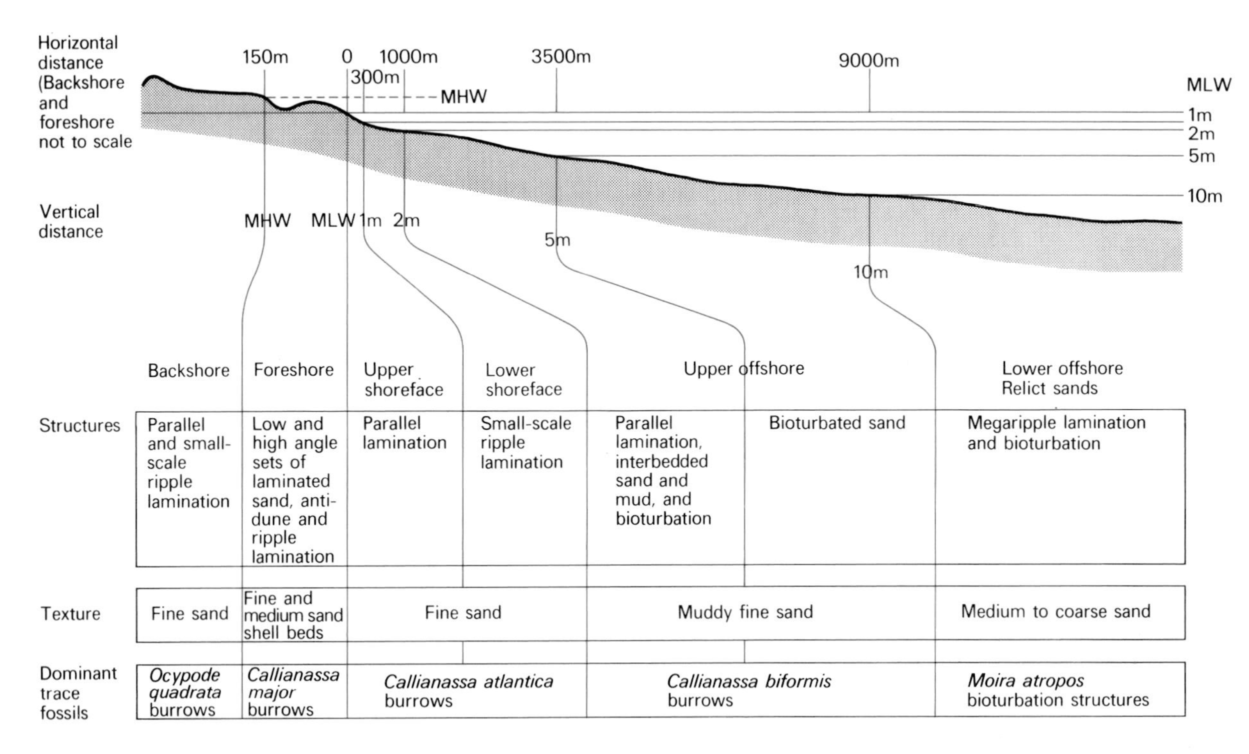

Figure 4.6 The different sedimentary assemblages, textures and dominant trace fossils from backshore to offshore in Sapelo Island, Gulf of Mexico (modified after Howard and Reineck, 1972).

then longshore drift will occur. The profile across a typical beach consists of: aeolian dunes, backshore, foreshore, shoreface (between mean low water level and fair weather wave base), and offshore; on each of these sediments can accumulate, and their structures can be very variable. The migration of features such as foreshore ridges and shoals can give patterns of sediments with variable vertical sequences, often with high-angle laminations if the particles were deposited at steeply inclined margins to the features. The sediments can be derived both from the land by direct shore erosion, and also from the longshore drift of deposits originating from river mouths in destructive situations (see above under deltas). Whether or not these sediments will eventually be preserved as rocks is dependent on the whole coastline undergoing subsidence due to earth movements, or the sea undergoing an eustatic rise in level.

The structures found within the rocks are dependent not only on their primary lamination on deposition, but also upon the amount of bioturbation that occurred prior to lithification of the sediments. One of the chief differences between younger and older rocks is the relatively bioturbation-free deposits of the Precambrian (though some bioturbation is over 1000 million years old); there was a steady increase in the numbers and variety of trace fossils during the Palaeozoic, and the post-Palaeozoic to Recent sediments are often intensively burrowed. Figure 4.6 shows the facies found on a mesotidal beach today, and offshore from it, and depicts how the dominant bioturbating agent changes with increase in depth and variation in sediment type. It also shows the variations in the structures found within the sediments, which have different ripple scales and laminations; these differences are due to the interaction of the sediment types with the amount of fair weather wave influence, and the strength and depth of the local currents. Coarser beach deposits, such as the commonly-occurring shingle beaches of the south coast of England, are usually confined to the foreshore and backshore; only occasionally are these eventually preserved.

Estuaries are important features of some coastlines today, but the largest are confined to macrotidal environments. Although in the nineteenth century many rocks were interpreted as having been formed under estuarine condition, most of these, such as the so-called Great Estuarine Series of the British Jurassic, are now considered to have been deltaic deposits. At present there are relatively few examples of ancient estuarine rocks known.

5 *Shallow clastic seas*

Shallow seas are defined as those between 10 and 200 m in depth and can be divided into two types, a marginal or pericontinental type which covers the modern continental shelf, and an epeiric or epicontinental type which lies upon a continental area. Thus the first type simply occupies the area between the land and the deep ocean, and usually has no special geographical name, whilst examples of epeiric seas include the Baltic, North Sea and Hudson Bay. Today most shallow seas are not yet in equilibrium after the post-glacial transgressions of the Pleistocene, and so direct analogies with ancient sediments are often difficult to make. However, an important part of the geological record consists of clastic rocks originally deposited on continental margins or in epeiric seas; they are often loosely termed 'shelly facies' by geologists. The fossil shells so often found in them are rarely in their original position of growth, except when a living assemblage has been quickly and permanently overwhelmed by a storm deposit; more usually the shells are disarticulated and broken through the winnowing action of deep currents and substantial waves.

The rocks are chiefly sandstones, siltstones and mudstones, and their degree of maturity (i.e. their angularity and sorting) varies considerably, depending on how many times the sediments have been reworked prior to final deposition. Modern sediments often form complex patterns of dunes, sandwaves and migrating megaripples. In epeiric seas strong currents are often caused by tides, for example the pattern and direction of the sand banks found on the floors of the North Sea and the English Channel today are controlled by tidally-induced currents. As with the shoreline environments discussed above, the bioturbation in shelf sediments can be intense. Finer sediment can remain in suspension for long periods, although suspension feeders can significantly increase sedimentation rates by extracting the fine particles (which might otherwise have been swept away by currents) and combining them into faecal pellets. It has been estimated (Verwey, 1952) that mussels deposit approximately 150000 metric tonnes of sediment per year in the Dutch Wadden Sea.

6 *Environments of chemical and biological deposition*

The composition of sea water is considered further in Chapter 6, but its various salts are deposited as sediments in many ways, both directly by evaporation and precipitation and indirectly through the biological mechanisms of animals and plants. Evaporite deposits are found forming today in areas of low clastic input and high evaporation, notably the Arabian Gulf, although they are also known at the margins of inland lakes, for example Salt Lake, Utah. In the Arabian Gulf the salt flats, or sabkhas, are flooded twice daily by the tide, and the evaporite deposits are slowly built up, cemented by algal mats. In the various subdivisions of the sabkha environment, the minerals anhydrite, halite, gypsum, dolomite and celestite are differentially deposited. In the Gulf of California, Mexico, where the climate is slightly more arid than in the Arabian Gulf, a concentration of halite deposits occurs. Ancient sabkha deposits have been recognised in several places, for example in the late Jurassic Purbeck Beds of southern England. However, more important to the geologist are large bodies of evaporites which were clearly not deposited in sabkha environments. These so-called 'saline giant' deposits, such as the Devonian Elk Point Basin of Saskatchewan and Alberta and the Permian Zechstein Basin of northern Europe and the North Sea, are the largest evaporite deposits known, and their origin is a matter of dispute among geologists, who argue both shallow- and deep-water origins for them. For example, some anhydrite beds have been found redeposited as turbidites and mass-flow breccias and slumps in the German Zechstein (Schlager & Bolz, 1977), which appears to be strong evidence favouring a deep-water origin for at least part of that formation. These saline deposits are relatively weak in comparison with their surrounding rocks, and subsequent tectonic pressure often causes mobilisation and the formation of salt domes.

On the continental shelf and in epeiric seas, certain authigenic minerals can be deposited. Phosphate deposits occur in areas of slow clastic sedimentation and are the result of direct precipitation of calcium phosphate as nodules or laminae from sea water enriched through concentrations of phosphorus-absorbing phytoplankton. Glauconite is formed by several different processes such as direct precipitation or the alteration of some detrital silicate minerals, but mainly by the alteration of organic matter, particularly faecal pellets. The iron-rich mineral chamosite is the warmer-water equivalent of glauconite and usually forms between 10 and 170 m deep. Colloidal silica is sometimes deposited; this aggregates during diagenesis and subsequently hardens to form chert or flint rock.

However, the main deposits in this category are those of calcium or magnesium carbonate, i.e. limestone and dolostone. Because the precipitation of carbonate is easiest in warm shallow sea water, most carbonate sediments are formed on the shallower parts of the continental shelf in the tropics, in areas like the Bahama Banks. Some of these deposits remain to form *in situ* limestones, others are transported shorewards to beaches and others are transported offshore, sometimes by turbidity currents and other mass-flow movements, to deeper-water environments; the latter join the carbonates produced by the fall-out of calcareous zooplankton and phytoplankton. Table 1 (from James *in* Walker, 1979) shows the main organisms that produce and bind the carbonate which forms in addition to that precipitated chemically.

True reefs form only a minority of carbonate sediments in the geological record; most limestones were originally relatively level-bedded deposits, and reef structures, or bioherms, are only occasionally preserved. The bioherms usually display a vertical succession consisting firstly of a pioneer stage, which provides a base upon which the second, colonisation, stage of reef-building metazoans (such as corals today) can consolidate. The bulk of the bioherm mass is usually built up subsequently in a third, diversification, stage, in which the primary reef metazoan group usually includes a variety of genera and growth forms. After this there is often a final, domination, stage in

Table 1 Modern carbonate producing and binding organisms, their counterparts in the fossil record, and their effect on sedimentation

Modern organism	*Ancient counterpart*	*Sedimentary aspect*
Corals	Archaeocyathids, corals, stromatoporoids, bryozoans, rudistid bivalves, hydrozoans.	The large components, often in place of reefs and mounds.
Bivalves	Bivalves, brachiopods, cephalopods, trilobites and other arthropods.	Remain whole or break apart into several pieces to form sand and gravel-size particles.
Gastropods, benthic foraminifers	Gastropods, tintinids. tentaculitids, salterellids. Benthic foraminifers, brachiopods.	Whole skeletons that form sand and gravel-size particles.
Codiacean algae – *Halimeda*, sponges	Crinoids and other pelmatozoans, sponges.	Spontaneously disintegrate upon death into many sand-size particles
Planktonic foraminifers	Planktonic foraminifers, coccoliths(post-Jurassic).	Medium sand-size and smaller particles in basinal deposits.
Encrusting foraminifers and coralline algae	Coralline algae, phylloid algae, renalcids, encrusting foraminifers.	Encrust on or inside hard substrates, build up thick deposits or fall off upon death to form lime sand particles.
Codiacean algae – *Penicillus*	Codiacean algae – *Penicillus*-like forms.	Spontaneouslydisintegrate upon death to form lime mud.
Blue-green algae	Blue-green algae (especially in Pre-Ordovician).	Trap and bind fine-grained sediments to form mats and stromatolites.

which only a relatively small number of taxa occur; and subsequently the top of the bioherm is reached, often with sedimentary indications of the effects of shallowing water and surf action. Patch reefs are more common than barrier reefs in the geological record. Slope deposits, either offshore from a simple carbonate shelf, or one fringed by a reef, are of widespread occurrence, and are often interbedded with clastic material derived from other sources.

7 *Deeper-water environments*

Deep ocean sediments are dealt with in detail in Chapter 7, but between the edge of the continental shelves and the abyssal plains there is often a varied topography, including large basinal areas of sediment accumulation. These are mainly filled with turbidite deposits, like those from the Grand Banks earthquake described earlier. However, other types of mass flow deposits also occur, sometimes consisting of very badly sorted sediments, ranging in grain size from fine muds to massive blocks all

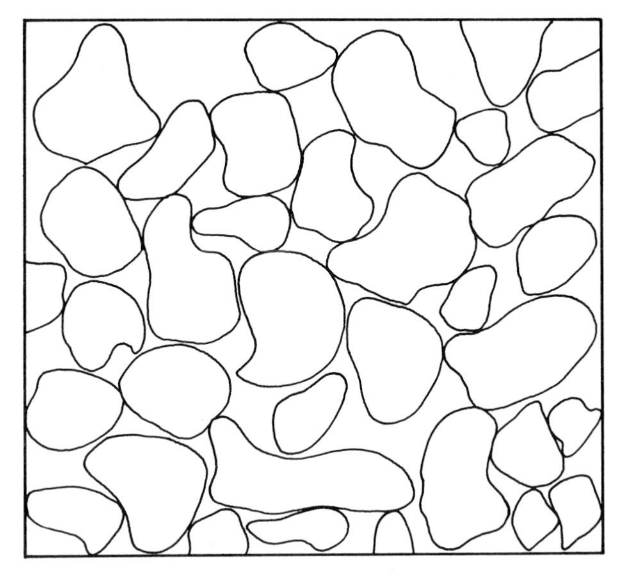

Figure 4.7 Unconsolidated sand.

jumbled together. These mélanges are termed 'olistostromes' and can contain blocks up to several hundred metres in size, many of which would have been derived from submarine fault scarps. Such deposits are also triggered by earthquakes, or can even be the by-product of a chain reaction resulting from a freak storm. Turbidites and olistostromes are composed chiefly of clastic rocks, but the particles can also consist of igneous, metamorphic or carbonate fragments, and organic remains are often found broken up and redeposited within them. When turbidites have lithified, the observer is struck by the very parallel-sided beds that are formed: their scale can range from a few millimetres to about 15 m in thickness, and each bed can be traced laterally across the face of the outcrop, and often for many kilometres beyond it.

Diagenesis

So far we have considered the routes and processes by which sedimentary material is formed, transported and deposited. We are now going to consider what happens to sediments after deposition, and how they are transformed into rocks. The subsequent stage in the rock cycle, when increasing temperature and pressure lead to substantial recrystallisation of primary minerals, is metamorphism, but the boundary between metamorphism and diagenesis is ill-defined. The process by which sediments become rock is known as consolidation or lithification, which comprises both physical and chemical changes, the most important of which are compaction (i.e. loss of volume) and cementation (usually involving the growth of new minerals). Both of these lead to a decrease in porosity, an effect which can have wide implications, as we shall see later.

Arenaceous sediments

Sands are generally the sediments least altered by diagenesis, and particularly by compaction. On deposition they may have a porosity of about 35–40 per cent, and this is altered only marginally by rearrangement of particles leading to loss of volume. A typical sand (Fig. 4.7) is composed largely of inelastic, approximately equant quartz grains, and, even with considerable overburden pressure due to later sedimentation, there is very little scope for compaction to take place since the rearrangement of particles is unlikely. In this case the major process of lithification is through the growth of cements derived from the sediments themselves, and known as authigenic.

Autochthonous cements in sandstones are formed from three main sources: the dissolution and reprecipitation of quartz, the breaking down of feldspars, and the alteration or recrystallisation of detrital clays and micas. Minor components include carbonates (from shells) and iron minerals. Variations in pressure, temperature and pH with depth control the formation of these cements, of which the most important is probably quartz.

Pressure solution of quartz, usually leading to local reprecipitation as overgrowths (known as syntaxial), appears to be fairly common in sandstones (Fig. 4.8), and can lead to loss of porosity both by reduction in volume and by the growth of cement (Fig. 4.9A).

Degradation of feldspars and the alteration of micas both lead to the formation of a clay mineral cement, which is usually kaolinite (Fig. 4.9B) in the case of feldspars, and illite or mixed-layer clay (Fig. 4.9C) in the case of micas. These cements have the effect of reducing not only the porosity but the overal permeability to a degree out of proportion to the volume of the cement. This is because the cement causes a constriction of the flow by precipitation in the pore throats (connecting passages).

Allochthonous cements in sandstones can replace, displace or fill the pores, and they are similar to the autochthonous ones: quartz, clay minerals, and occasional carbonates. The composition and passage of pore waters are of course the controlling factors, but the controls on these themselves are complex and depend on a combination of lithological, stratigraphical, structural, and tectonic factors.

A typical sandstone will have a combination of allochthonous and autochthonous cements, and it is often possible to determine the sequence of deposition of these by careful examination in thin-section and with the electron microscope. Cementation can begin at a few metres depth of burial and continue until the onset of metamorphism, reducing porosity to near zero. A representative sequence might be:

Increasing depth of burial ↓

1 Syntaxial quartz cement precipitated from pore water.
2 Breakdown of feldspar to form kaolinite.
3 Alteration of mica to give illite.
4 Pressure solution of quartz and further syntaxial overgrowth.
5 Etching of quartz by carbonate precipitated from pore water.
6 Precipitation of amorphous silica from pore water.

Such a sequence might be interrupted in appropriate cases by the injection of hydrocarbons by tectonic activity from neighbouring strata, and it is the importance of diagenetic changes in sandstones in relation to the effect on their reservoir properties that has led to the recent upsurge in research in this field.

Argillaceous sediments

The study of diagenesis in clays was relatively neglected until about 15 years ago. Apart from general observations, such as the apparent domination of Palaeozoic clays by the minerals illite and chlorite, there was little systematic understanding up to then of late postdepositional changes in clay sediments. Rapid advances were made, however, after studies of thick series of clays of similar lithology unaffected by weathering, and the term 'burial diagenesis' came to be used for the changes found (Dunoyer de Segonzac, 1970; Perry & Hower, 1970).

However, we must first consider the state of a clay at the time of deposition. This is very different from that of a sand: the initial porosity is much higher, about 80–90

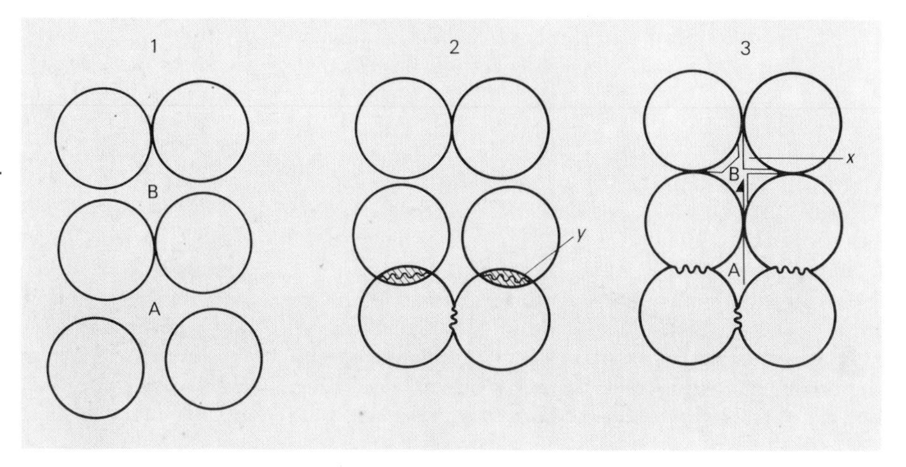

Figure 4.8 The sequence of stages in the development of sand grains cemented by pressure solution. In Stage 1 the sand is uncompacted; in Stage 2, pressure solution has begun (e.g. at point *y*). In Stage 3 some fluid from pore space A escapes to space B, reduces pressure and precipitates quartz; point *x* is an area of authigenic quartz. (modified after Sibley and Blatt, 1976).

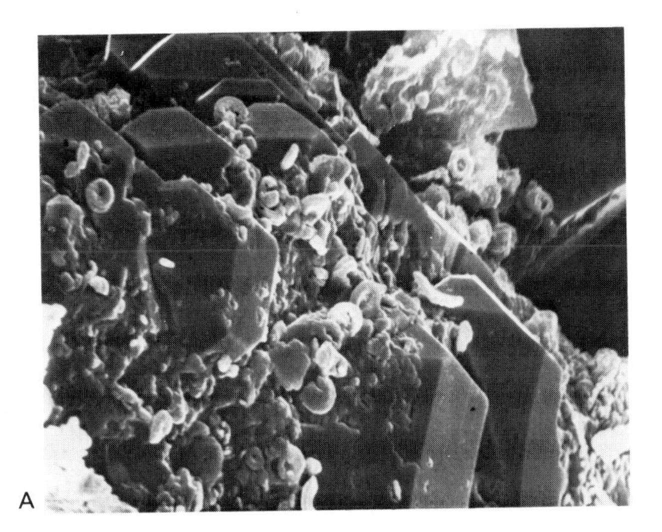

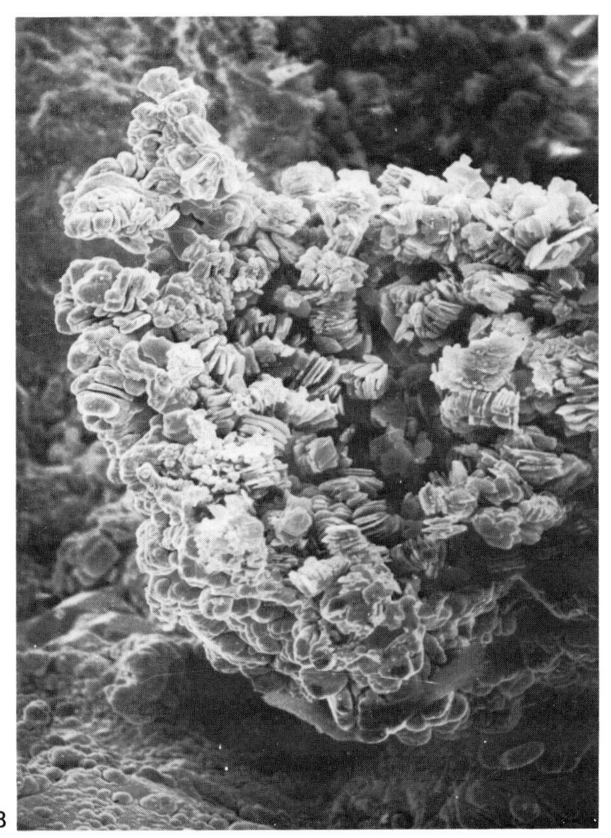

Figure 4.9 A. Scanning electron micrograph (SEM) of syntaxial quartz overgrowths in a coccolithic sandstone, illustrating the reduction in porosity.
B. SEM of authigenic kaolinite forming from degrading feldspar in sandstone.
C. SEM of 'fibrous illite' formed at the degrading edges of detrital mica.

per cent, and grain-to-grain contacts are thus rare. This situation changes very rapidly during the first few metres of burial: as water is expelled the clay particles become rearranged and the strength increases considerably (Figure 4.10).

During this initial period there may be precipitation of gels and dissolved chemicals as clay minerals, and metastable minerals such as soil iron oxides may be transformed into early cements (leading for example to the production of clay ironstones). Porosity and permeability are reduced dramatically, and further mineralogical changes tend to be caused by the removal or alteration of existing material; the introduction of new material through pore water can be extremely slow.

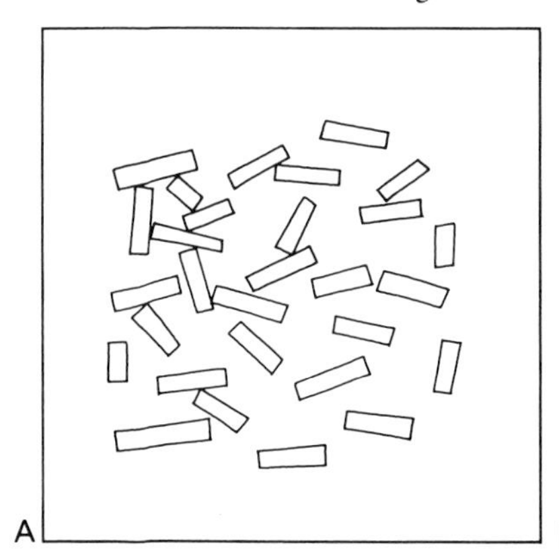

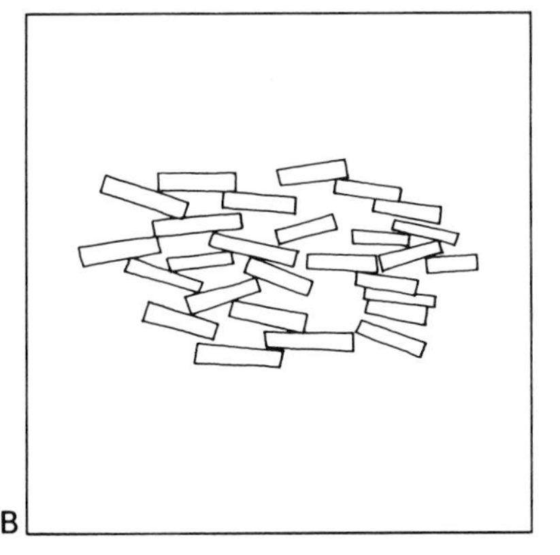

Figure 4.10 Idealised clay particles, A, at deposition, B, after some compaction.

Clays are the major original deposition rocks of hydrocarbons, so one must not forget the effect of organic matter during diagenesis. Curtis (1977) has summarised the various zones of diagenesis as affected by organic matter and activity (Figure 4.11).

The results of the studies of deep burial mentioned above showed in particular how an initial mineralogy of kaolinite and mixed-layer illite-smectite, with a preponderance of smectite layers, changed gradually with burial into an assemblage of chlorite and illitic mixed-layer clay. The earlier observation that older sediments contained mostly illite and chlorite was thus explained.

More recent work has related clay mineral facies to burial conditions in more detail (e.g. Velde, 1977), and

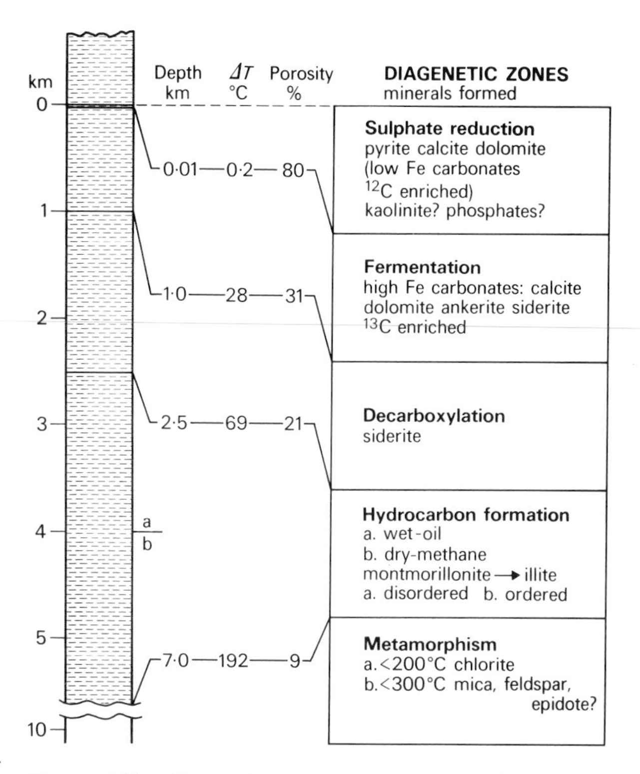

Figure 4.11 Diagenetic zones. T °C is the increase in temperature with depth, at an average gradient of 27·5 °C per km (after Curtis, 1977).

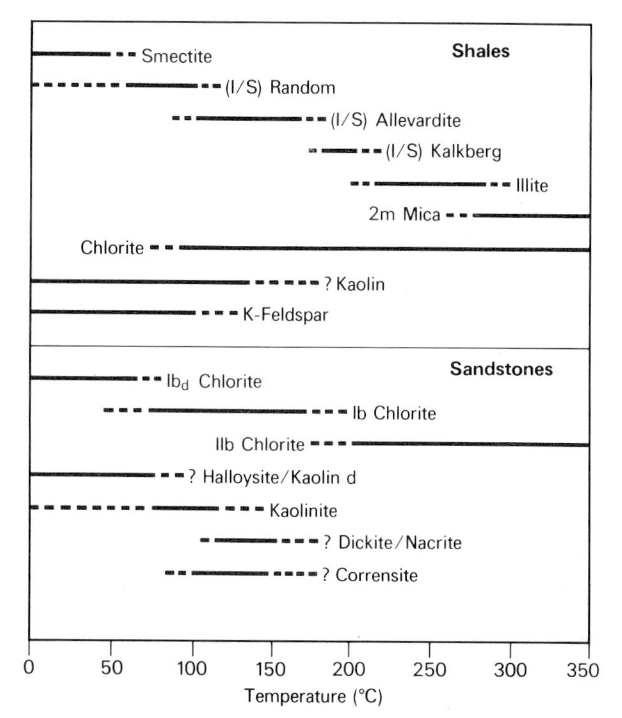

Figure 4.12 Temperature-dependent mineral assemblages in shales and sandstones (after Hoffman and Hower, 1979).

Hoffman & Hower (1979) have shown how clay mineral assemblages may be used as a geothermometer (Figure 4.12). As diagenesis grades into metamorphism, illite becomes recrystallised as mica, but chlorite persists.

Carbonate sediments

As with the two previous groups of sediments, we can broadly divide the diagenetic processes in carbonates into two phases: those occurring shortly after deposition and those which result from burial, where increases in temperature and pressure are significant. The main processes in the earlier phase are micritisation and cementation. Micritisation is the transformation of the original carbonate fabric into an aggregate of randomly oriented needles, 2–5 μm in length, of cryptocrystalline calcite. In carbonate grains (e.g. ooids) this change begins at the perimeter and proceeds inwards towards the nucleus. It is promoted by microbiological activities, such as boring by algae and fungi. The susceptibility of various particles to micritisation varies both with structure and mineralogy:

Increasing susceptibility (↓):
- Coralline algae (high-Mg calcite)
- Benthic foraminifera (high-Mg calcite)
- Gastropods and bivalves (aragonite and low-Mg calcite)
- Corals (aragonite)

The time-scale of micritisation is of the order of hundreds of years.

Early cementation is by either aragonite (cryptocrystalline or radial fibrous) or high-Mg calcite (equigranular or radial fibrous): it generally takes place some way below the sediment–water interface and is achieved by the formation of cemented lumps which later coalesce (as in the hardgrounds found at intervals in the Chalk). Favourable conditions for cementation are found in a warm sea saturated with $CaCO_3$ where there is freedom from bottom traction and a suitable substrate. However, cementation can occur, even in the more agitated environment where ooids form, provided precipitation is fast enough.

A different form of early diagenesis in carbonates occurs where they are exposed subaerially. In rainwater (with pH of 4–6), high-Mg calcite is more soluble than aragonite which in turn is more soluble than low-Mg calcite. This leads to differential solution (Fig. 4.13) which may be followed by reprecipitation of the dissolved material as low-Mg calcite. Dolomitisation can also occur subaerially, in sabkhas. An increased Mg/Ca ratio, resulting from the fixation of Ca as gypsum and anhydrite, leads to the replacement of pre-existing carbonate (mostly aragonite) by dolomite (Fig. 4.14).

Later diagenesis can reduce porosities of carbonates to below 5 per cent (from an initial 40–70 per cent), but this is rarely due to compaction since the early cementation stabilises the fabric. The main processes after burial are continued precipitation, solution, and recrystallisation or transformation. As with clays and sands, pore-water chemistry and effective circulation are the controlling

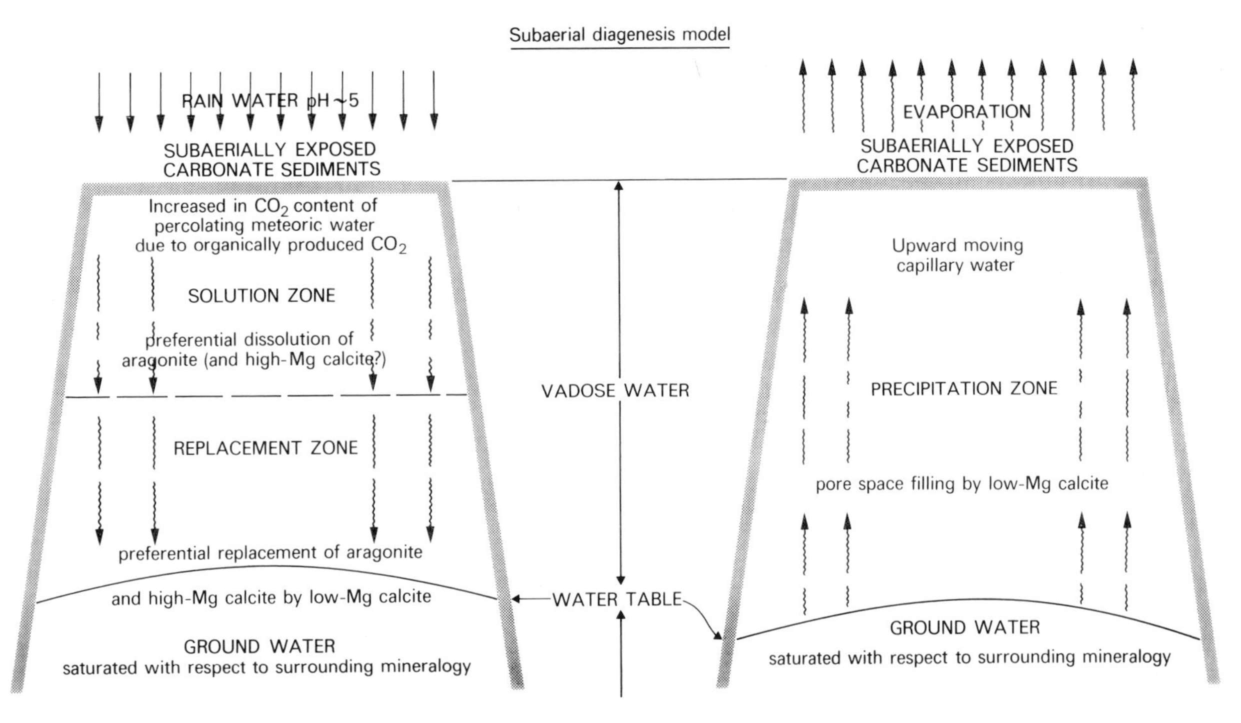

Figure 4.13 Subaerial diagenesis on an oceanic island as a consequence of downward-percolating meteoric water and upward-moving capillary water (after Purdy, 1968).

factors in precipitation, which becomes less important a process as permeability decreases.

Pressure solution in carbonates is more common than in sandstones; the force necessary, and hence depth of burial, is less. The undulose sutures called stylolites that result are easily recognisable.

Lastly, there is an overall trend in carbonate sediments (similar in a way to the increasing chlorite and illite in older clays) of a change from an original mineralogy of aragonite and high-Mg calcite to low-Mg calcite and dolomite. For a detailed treatment of all these processes the reader is referred to Bathurst (1975).

The final outcome of long-term diagenesis in carbonates is an interlocking mosaic of crystals giving very low porosity. Cementation sequences can be extremely complex and difficult to interpret by conventional microscopic techniques. The recent application of cathodoluminescence petrography to this problem has shown how many phases of cement can be involved in filling a single pore (Fig. 4.15).

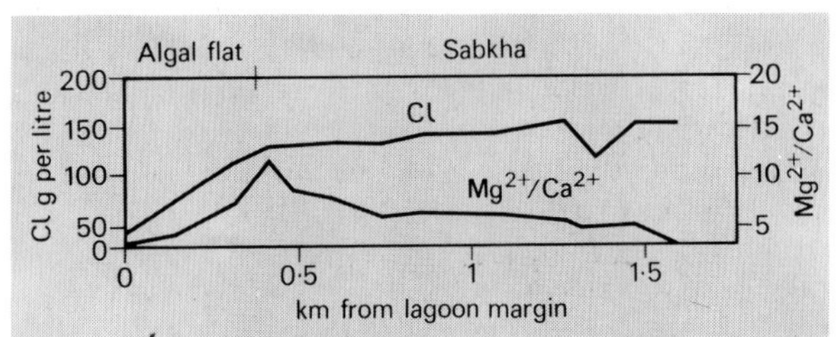

Figure 4.14 Variation in brine concentration and Mg^{2+}/Ca^{2+} with distance from the outer edge of an algal flat (after Illing *et al.* 1965).

The history of sedimentation

The first chapter of this book deals with the origin of the Earth and its early history, and from its suspected time of formation at about 4500 million years ago until the time of the oldest rocks preserved today (which have been isotopically dated at 3800 million years), there is no direct evidence of sedimentary activity. However, as described in Chapter 2, these old rocks are themselves sediments, albeit metamorphosed ones, and it is clear that whenever the primitive oceans (of whatever composition) first formed under a primitive atmosphere, the first sediments were laid down soon after. The chief principles of sedimentation must have been unchanged then as now: the physical and chemical weathering of pre-existing rock; the mechanical transport of these fragments by fluids or gases, or their chemical transport in solution; and the final deposition of the sediments under gravitational settling or chemical precipitation. However, some parameters have changed over the successive geological eras, and modern sediments are very different from those in the early Precambrian. A review of sediments through time now follows.

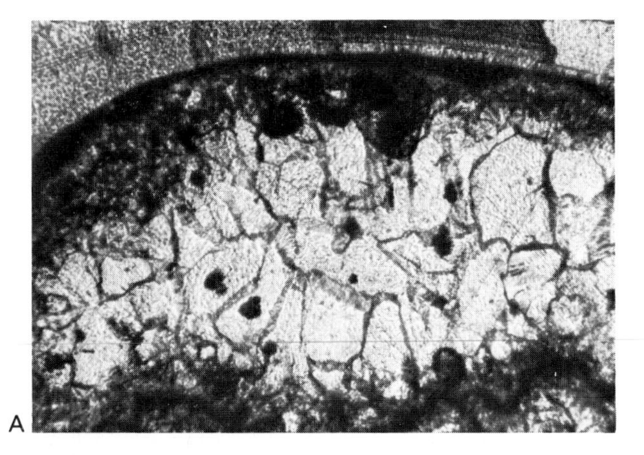

A

B

Figure 4.15 An infilling carbonate cement viewed in transmitted light (A) and cathodoluminescent light (B) (photographs provided by M. Mayall).

1 *The Archaean*

The raw material for the earliest sediments must have consisted chiefly of volcanic rocks, which broke down to give particles of various size, shape and chemical composition. As discussed further in Chapter 8, the Earth's atmosphere in pre-Archaean and Archaean times was a reducing one, consisting largely of carbon dioxide, nitrogen, water vapour and inert gases, with subsidiary amounts of hydrogen, methane and ammonia. There was no free oxygen. With this harsher chemical environment, weathering rates would probably have been faster than today, and the lack of vegetation cover, and the slow development of only the most primitive types of soil, would have made the erosion quite rapid, particularly in areas of high precipitation. However, in contrast to later periods, there would have been no frost action. The first main class of sedimentary rock types to form would have been greywackes, chiefly deposited by turbidity currents, and the oldest greywackes known are found in the Archaean greenstone belts (see Chapter 3); although their subsequent metamorphism precludes an accurate description of their original sedimentary particles, they were probably chiefly basic or intermediate igneous rocks. However, the dynamics of the ocean would probably have been comparable with those of today. As wave action persisted, and as the individual sedimentary particles became involved in more than one cycle of erosion, transport and sedimentation, so the proportion of well-rounded particles increased during geological time. Similarly the physically weaker or chemically less stable particles became relatively smaller, or were ground to dust, whilst the stronger and more stable fragments remained. Thus the sediments gradually became more mature in successive eras, with, for example, a higher proportion of rounded quartz grain as time went by. From an assessment of sediment quantities deposited during Tertiary time, it has been calculated that each sedimentary particle has been involved in an average of five complete sedimentary cycles since the early Precambrian.

The original landscape would presumably have been jagged and angular, and thus the earliest streams and rivers would have had relatively steep gradients: such rivers would have had considerable erosive power. Only after long periods of erosion would broad alluvial plains have developed, across which meandering river systems and their associated fluviatile deposits could have built up. Some workers have postulated that sea water covered much, if not all, of the Earth's surface during Archaean and early Proterozoic times, but their arguments have been convincingly dismissed in a review by Windley (1980) of the various Archaean and Proterozoic sediments which could only have been formed on or near dry land. These include palaeosols, shallow-water conglomerates, fluvial deposits, red beds, subaerial volcanic deposits and evaporites, all of which have substantial Precambrian sedimentary records.

Besides volcanogenics, greywackes, feldspathic sandstones (arkoses), quartz sandstones, siltstones and mudstones, an important group of sediments, whose deposition was virtually confined to Archaean and early Proterozoic times, were banded ironstones; these are discussed in Chapters 3 and 8 (see also Fig. 8.8). The changing proportions of the various sedimentary rock types found at successive periods of geological times are shown in Figure 4.16.

2 *The Proterozoic*

A key period in the history of sedimentation was about 2200 million years ago. Prior to this time the oceans and atmosphere had been weakly reducing in chemical composition, with banded ironstones and other distinctive rocks being deposited; after this time free oxygen became available, and the predominant iron oxide in sediments included the red-coloured ferric ion. Some iron-rich sequences in Canada range through this critical period, and show black and green beds beneath later red beds containing haematite. The sheer bulk of iron deposited since then has never been as great, although sandstones and other rocks with a smaller proportion of red iron compounds have often since been deposited, particularly in desert environments.

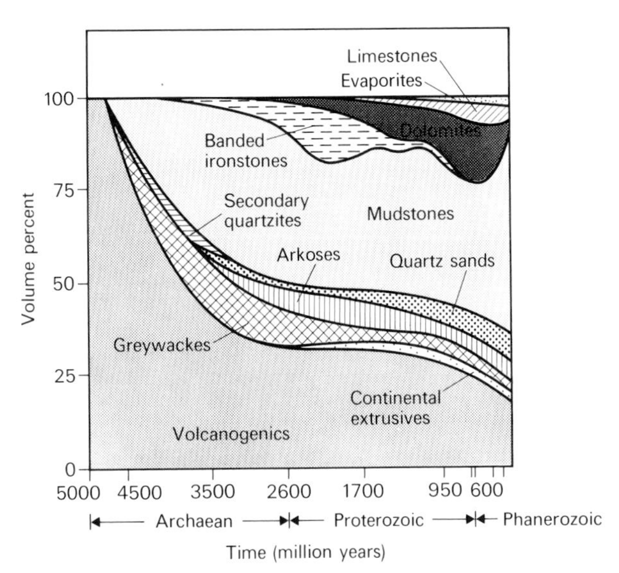

Figure 4.16 The relative proportions of the different major sediment types over geological time (after Ronov, 1964).

Limestones became common for the first time in the Proterozoic, and they frequently contain algal stromatolites (see Fig. 8.7), which have been used for relative dating of younger rocks of the era. These Precambrian limestones have in general a higher magnesium content than modern limestones, that is a higher proportion of dolostone. A form of coal (called 'schungite') is found for the first time in rocks about 2000 million years old, and, although it is now metamorphosed, it probably formed initially from local accumulations of blue-green algae.

Chert is known from the middle Proterozoic, in rocks up to 1600 million years old, and occurs in the form of nodules within dolomites, which also include algal stromatolites and oncolites. However these cherts are almost certainly of secondary origin (Sathyanarayan & Müller, 1980) and primary cherts, formed directly from siliceous organisms such as radiolaria, are not known until latest Precambrian times.

The variability in the climate, with periodical glacial episodes and other changes, would certainly have affected local sedimentary regimes; however, the broad impression is of continuity in the various types of sedimentary environment in the later Precambrian as a whole, some of which persist to the present day, although affected by the important biological changes which have occurred since, particularly during the Phanerozoic.

3 The Phanerozoic

There is good evidence that, by the beginning of the Phanerozoic, the general compositions of both sea water and atmosphere were broadly comparable with those of today (see Chapters 6 and 8). However, the amount of oxygen dissolved in the sea water was probably less, as witnessed by the abundance of euxinic black shales in the early Palaeozoic. Progressive ventilation through Palaeozoic time resulted in a gradual decline in the proportion of anoxic sediments, and by about 400 million years ago oxygen levels were probably similar to those of today. Thus, although the proportion of land above sea level has varied at different times during the Phanerozoic, the general environments of sedimentation have been broadly similar during the last 570 million years. However, although sedimentary processes have remained largely unaltered, the sediments themselves have changed very considerably due to the effect on the environments of evolving organisms. These have acted in many ways, but their effects may be grouped into three main headings – firstly, the physical accumulation of skeletal remains as rocks; secondly, the biochemical stabilisation of chemical compounds from solution in sea water; and thirdly, the biological binding and redistribution of soft sediments. Each heading will be dealt with in turn.

The physical accumulation of skeletal remains in different rocks during successive geological periods within the Phanerozoic has depended on the evolutionary history of the individual groups of animals and plants. For example, although rare schungite coal is known from the Precambrian, the deposition of substantial coal-bearing formations did not occur until after land plants had become well established and diversified in late Palaeozoic (Carboniferous) times. Inorganic limestone deposition has been recognised in the late Archaean, and sabkha deposits are known from rocks as early as 1900 million years old, but the biological content of limestones has differed greatly with time. The widespread stromatolites of the Precambrian became virtually extinct, and confined to specialised restricted environments, after the appearance of herbivorous metazoans, particularly gastropods, in the early Palaeozoic. Table 1 has already shown the various carbonate organisms that have preceded the common reef-building biota of today, and to illustrate the consequent variability of limestone types through time one may cite the widespread chalk of the Cretaceous age, whose characteristic texture is due to its contained coccoliths.

The various plankton that eventually die to produce organic-rich sediments, which may later be concentrated into commercial deposits of oil, tar and natural gas, only became common during the late Precambrian, and the major oil-producing rocks are all of Phanerozoic age. Some such organic-rich sediments are of immense size; for example, the Athabasca Tar Sand of eastern Canada is up to 60 m thick and extends over more than 50000 square kilometres.

The second effect of organisms is in the direct contribution to the sediments through the biological activities of animals and plants whilst they are alive. The concentration of minute sediment particles into larger, and therefore quicker-settling, groups of particles by the digestive processes of plankton is a powerful agent for sedimentation before the minute particles get swept away by waves and currents. Faeces of all sizes can form important constituents of some sediments, particularly carbonates. Silica, phosphorus and iron are concentrated in some animal skeletons and so chert, phosphorite and glauconite deposition have all increased in abundance during the Phanerozoic through organic agencies and subsequent diagenesis. Oolites have also become more abundant due to the increasing quantities of organic nuclei that have become available.

The third group is the physical effect on the sediments through the activities of animals and plants. The stabilising action of a vegetation cover, which may vary in scale from algal mats to mangrove roots, can prevent soft sediments from being winnowed and eroded. On the land too, the vegetation cover which has steadily increased in the last 400 million years has had the important effect of reducing erosion rates and hence the sediment supply. Primitive plants first invaded the land in late Silurian and early Devonian times, and by the Carboniferous large fringing swamps were able to provide all the vegetable matter for the coal deposits. With successive layers of humus, land soils became more diversified, leading in turn to more ecological niches and a more varied flora and fauna. However, flowering plants (angiosperms) did not evolve until Mesozoic times, and widespread grasses only became abundant during the Tertiary period: each of these developments profoundly affected terrestrial sediments.

In contrast to the stabilising effect, there is also a destructive effect, that of bioturbation in soft sediments. Mud-eating and bottom-stirring animals, and plants with roots, did not exist in Precambrian times, and, as a result, delicate laminations are preserved in Precambrian sedimentary rocks which are usually destroyed in the Phanerozoic. A recent survey of the Firth of Clyde failed to find any sediment which had not been affected by bioturbation. There are a few traces of bioturbation in the late Precambrian, and it slowly increased in abundance during the Palaeozoic – it is not until the Mesozoic and subsequently that virtually all shelf sediments have been reworked by organisms.

This brief review of changes in sedimentary environments through time has only been able to chart some of the most significant events and progressive sequences. The soils which lie upon those sediments on land are discussed in the next chapter, and deep ocean sediments are reviewed in Chapter 7. However, the proper appreciation, use and conservation of the Earth's sedimentary resources as a whole are of immense importance, not only to geologists but to all of mankind.

References

Bathurst, R. G. C. 1975. *Carbonate sediments and their diagenesis.* (2nd ed.) 658 pp. Amsterdam: Elsevier.

Curtis, C. D. 1977. Sedimentary geochemistry: environments and processes dominated by involvement of an aqueous phase. *Philosophical Transactions of the Royal Society of London*, Series A, **286**: 353–372.

Dunoyer de Segonzac, G. 1970. The transformation of clay minerals during diagenesis and low-grade metamorphism: a review. *Sedimentology* **15**: 281–346.

Glennie, K. W. 1970. Desert sedimentary environments. *Developments in Sedimentology* **14**: 1–222.

Hayes, M. O. 1976. Morphology of sand accumulation in estuaries: an introduction to the symposium, pp. 3–22 *in* Gronin, L. E. (Ed.) *Estuarine Research II, Geology and Engineering*. London: Academic Press.

Heezen, B. C. & Hollister, C. D. 1971. *The face of the deep.* 659 pp. New York: Oxford.

Hoffman, J. & Hower, J. 1979. Clay mineral assemblages as low grade metamorphic geothermometers: application to the thrust-faulted disturbed belt of Montana, U.S.A. *Society of Economic Paleontologists and Mineralogists Special Publication* **26**: 55–79.

Howard, J. D. & Reineck, H. E. 1972. Physical and biogenic sedimentary structures of the nearshore shelf. *Senckenbergiana Maritima* **4**: 217–223.

Illing, L. V., Wells, A. J. & Taylor, J. C. M. 1965. Penecontemporary dolomite in the Persian Gulf. *Society of Economic Paleontologists and Mineralogists Special Publication* **13**: 89–111.

Kuenen, P. H. & Migliorini, C. I. 1950. Turbidity currents as a source of graded bedding. *Journal of Geology* **58**: 91–127.

Perry, E. A. & Hower, J. 1970. Burial diagenesis in Gulf Coast pelitic sediments. *Clays and Clay Minerals* **18**: 165–177.

Purdy, E. G. 1968. Carbonate diagenesis: an environmental survey. *Geologica Romanica* **7**: 183–228.

Reading, H. G. ed. 1978. *Sedimentary environments and facies.* 557 pp. Oxford: Blackwell.

Ronov, A. B. 1964. Common tendencies in the chemical evolution of the Earth's crust, ocean and atmosphere. *Geochemistry*: 713–737.

Sathyanarayan, S. & Müller, G. 1980. Origin of nodular chert in the carbonate rocks of the Kaladgi Group (younger Precambrian), Karnataka, India. *Sedimentary Geology* **25**: 209–221.

Schlager, W. & Bolz, H. 1977. Clastic accumulation of sulphate evaporites in deep water. *Journal of Sedimentary Petrology* **47**, 600–609.

Sibley, D. F. & Blatt, H. 1976. Intergranular pressure solution and cementation of the Tuscarora orthoquartzite. *Journal of Sedimentary Petrology* **46**: 881–896.

Velde, B. 1977. *Clays and clay minerals in natural and synthetic systems.* 217 pp. Amsterdam: Elsevier.

Verwey, J. 1952. On the ecology of distribution of cockles and mussels in the Dutch Wadden Sea, their role in sedimentation and the source of their food supply, with a short review of the feeding behaviour of bivalve molluscs. *Archives Néerlandaises de Zoologie* **10**: 171–239.

Walker, R. G. (Ed.) 1979. Facies models. *Geoscience Canada Reprint Series* **1**: 1–211.

Windley, B. F. 1980. Evidence for land emergence in the early to middle Precambrian. *Proceedings of the Geologists' Association* **91**: 13–23.

L. R. M. Cocks
Department of Palaeontology
British Museum (Natural History)

A. Parker
Department of Geology
The University
Reading RG6 2AB

CHAPTER 5

Soils

J. A. Catt and A. H. Weir

The word 'soil' means different things to different people. To farmers and gardeners it is the topmost 20–30 cm of the Earth's surface, darkened by decomposing organic matter and explored by the roots of smaller plants. To botanists it is the medium in which plants grow, and in this sense could include any material to which water and the necessary nutrients have been added. To civil engineers it corresponds to any unconsolidated rock material – this includes most Quaternary deposits and some pre-Quaternary, and is roughly equivalent to the term 'regolith'. To geologists and soil scientists (pedologists) it is a layer of variable depth on the Earth's surface, which is affected by various processes of physical reorganisation and chemical alteration dependent upon the presence of the atmosphere above. This sense is followed in this chapter. The altered zone is divisible into horizons roughly parallel to the land–atmosphere interface and these are often unconformable with the underlying structure of the rocks.

Early soil scientists, often with a geological background, considered soil to be fragmented rock mixed with organic compounds formed by decomposition of plants, but this oversimplified view ignores the living components of soil, their complex relationships with the non-living constituents, and the large variations resulting from climatic differences, time and human influences. These factors combine in various ways to give many clearly different physico-chemical systems. However, even today many of the factors determining soil type or controlling plant growth are poorly defined and understood, and a universally acceptable system of soil classification has yet to be devised. The preparation and interpretation of soil maps are consequently rather subjective sciences, and land use planning based on the capabilities of different soil types is in its infancy.

Soil-forming factors

Dokuchaev, Jenny and other early pedologists regarded five factors as important in soil formation – climate, organisms, parent material, relief and time. More recently, it has been realised that hydrological and human factors (previously included with relief and organisms respectively) are at least as important as the other five. However, all seven factors are interdependent to some extent, and most are multiple variables (e.g. organisms include the separate effects of fauna and flora). Quantification of their individual or combined effects is often difficult.

Climate

The two most significant climatic variables are temperature and rainfall, but the usual atmospheric measurements of these are only indirectly related to soil temperature and moisture, which are the important factors determining rates of chemical weathering and leaching. For example, vegetational changes can affect soil moisture independent of rainfall. Soil temperatures below about 1 m depth are fairly constant (approximating to mean annual air temperature) but in higher soil horizons they are more variable, and the rates of many processes (e.g. decomposition of organic matter) are more closely related to mean maximum air temperature. In tropical areas, soil clay, organic matter and nitrogen contents tend to increase with increasing rainfall, whereas pH and base saturation decrease. However, some of these effects are often overridden by differences in relief, vegetation, hydrology or parent material, and by human interference. In arid regions, rapid evaporation from the soil surface encourages the crystallisation of carbonates and sulphates on or just below the surface, and the maintenance of neutral or alkaline conditions. In colder regions, the frequency of freeze–thaw cycles largely controls the rate of physical weathering of rocks, the mixing of soil materials by cryoturbation, and sometimes the rough size-sorting of material to produce patterned ground, and low temperatures inhibit chemical weathering and leaching. In both warm and cold arid regions, the forces exerted by crystallisation of salts are now recognised as important in the mechanical disintegration of rocks (salt weathering).

The types of layer silicate clay-minerals or hydrated iron and aluminium oxides formed by weathering, which often influence soil structure (state of aggregation) and thus the ease with which air and water can circulate, are determined by climatic, hydrological and parent material factors. In the wet tropics, intense leaching gives soils rich in kaolinite or hydrated oxides (gibbsite, haematite, goethite), but long dry seasons, during which the soil dries out to wilting point, favour the formation of smectite clays. In temperate and cold regions, minerals undergo less drastic chemical weathering changes, and the composition of soil clays is determined much more by the nature of the parent material.

Parent material

The unaltered rock, or unconsolidated sediment from which the soil was formed, is often difficult to determine, especially in warmer regions where soil-forming processes frequently destroy many of the parent material's original characteristics. As deposition of new material on the land surface by glacial, fluvial, eolian and slope processes is extremely widespread, the assumption that the soil at any one place is formed entirely by alteration of the rock encountered at depth beneath the profile is often wrong. Thin superficial deposits are especially common at the surface in temperate regions, which generally have undergone more environmental changes during the Quaternary than other parts of the Earth. They give rise to layers of different particle size distribution, stone and organic matter content, colour and chemical or mineralogical composition, and the horizons produced by subsequent pedological processes are superimposed on these, though their boundaries are often predetermined by the original lithological discontinuities between the deposits.

The widespread Quaternary sediments are generally unconsolidated, and soil development in them can proceed without the need for preliminary comminution by mechanical weathering. In arid and many cold or cool-temperate regions these younger unconsolidated deposits often give rise to the only soil deep enough for trees or arable crops; the older consolidated sediments and crystalline rocks devoid of superficial sediments bear either no soil cover or merely a thin accumulation of partly decomposed organic matter. Elsewhere the *in situ* alteration of hard rock to soil is initiated either by fracturing and fragmentation through thermal expansion/contraction and root penetration, or by softening through chemical dissolution of cementing agents or alteration of less resistant mineral components.

Probably the most important characteristic of unconsolidated parent materials in determining soil properties is particle size distribution, because it influences soil strength, structural stability, permeability, water retaining capacity and cation exchange capacity. In general sandy soils have little bearing strength, small water-retaining and exchange capacities, weak structure and high permeability; as a result they are prone to drought, rapid loss of nutrients by leaching, and locally to erosion. Silty soils also have weak structure and are easily eroded, but their available water capacities are greater than either sandy or clayey soils, and their exchange properties and permeability often prevent such rapid leaching as occurs in sandy soils. Clay soils usually have low permeability and consequently suffer from waterlogging and poor air circulation, especially in wetter climates. Loams contain approximately equal amounts of sand, silt and clay, and tend to possess most of the properties favourable to plant growth with few of the various disadvantages of the three chief constituents.

Chemical and mineralogical composition of parent material is also important in determining some soil characteristics. It affects the supply of many major and minor plant nutrients to the soil solution, and determines the nature of the residues formed by weathering processes. Granites, granite-gneisses and coarse sediments are rich in quartz, which is resistant to most weathering, and they usually form leached, acid soils with small exchange capacities. These are inherently less fertile than soils formed from basic igneous rocks, limestones and clay-rich sediments, which contain larger amounts of weatherable minerals rich in Ca, K, Fe, Mg and many trace elements important to plant growth, and usually have large exchange capacities. However, some ultrabasic igneous rocks are so rich in heavy metals (mainly Co, Cr, Cu and Ni) that the soils on them are toxic to many plants.

Relief

Relief has direct effects on soil formation processes, and also indirect effects arising from its dependence upon geology and its influence on parent material, climate and hydrological factors. One of the direct effects is to control the balance between the weathering and soil formation *in situ* and the loss of weathered material downslope. On flat or gently sloping sites a large proportion of the rainfall penetrates the soil and a deep altered and leached mantle is formed; on steeper slopes weathering and leaching are probably slower in absolute terms, and much of the altered mantle is also removed by increased runoff, so that the soil is shallower, less weathered and often more stony.

Many features of the topography result from geological factors such as the variable resistance to erosion of different rock types. As a result, the topography is associated with changes in soil type caused fundamentally by changes in parent material, a relationship used extensively in soil mapping. However, other changes in parent material result from (rather than cause) topographic features. For example, material eroded from slopes by runoff is often sorted during transportation – sand and coarse silt being deposited first near the slope foot, and finer silt and clay being carried some distance beyond. Consequently footslopes and the floors of minor dry valleys often have coarse soils in hillwash deposits, whereas the alluvial sediments of rivers contain mainly clay and fine silt, and the soils on them are correspondingly finer.

Relief has a strong indirect influence on soil formation,

especially in tropical areas, through its effect on climate. Altitudinal soil zonation often results from the decrease in temperature with height, which averages about 1 °C for every 170 m, and slows down the rates of mineral weathering and decomposition of organic matter. Increased precipitation on higher ground also affects soil type by intensifying leaching and erosion.

Relief also indirectly affects soil development by its control of the water table. Soil waterlogging due to a high groundwater table is more common on valley floors than on valley sides, hill slopes or plateaux. However, on flat upland sites subject to high rainfall, waterlogging often occurs well above the regional water table because precipitation in excess of the soil's infiltration capacity is not removed quickly across the ground surface.

Hydrology

In addition to waterlogging caused by a high groundwater table or inadequate surface drainage (i.e. poor site drainage), the higher horizons of some soils may also become waterlogged because of drainage impedance within the profile (i.e. poor profile drainage). Impedance may result from a layer of originally impermeable (e.g. clayey) parent material, or from a thin impermeable horizon formed by pedogenetic processes (e.g. an iron-pan) or by soil cultivation (e.g. a plough pan).

The main visible effect of waterlogging, however caused, is the occurrence of greyish colours due to reduction of ferric oxides and hydrated oxides in the presence of organic matter; this is termed 'gleying'. Temporarily waterlogged horizons are mottled grey and brown, but those permanently beneath the water table are often uniformly grey, sometimes with a greenish or bluish tinge. In the wet–dry tropics, the upper horizons of gleyed soils may dry out seasonally, and the iron mobilised in the reduced form is then reoxidised and reprecipitated to form ferruginous concretions. Manganese and some other trace elements are also made soluble by anaerobic conditions, and amounts of them in the soil solution may then reach toxic levels. However, the main limitation to plant growth caused by waterlogging is the decreased aeration of the soil. Oxygen is largely replaced by carbon dioxide, methane, hydrogen and ethylene, thus inhibiting root growth. In many species an acceleration of glycolysis causes ethanol accumulation and partial membrane dissolution, but flood-tolerant species prevent this by transferring oxygen from leaves to roots. Anaerobic soil conditions also accelerate microbiological processes converting nitrate to nitrogen and nitrogen oxides (denitrification); less nitrate is therefore available for plant uptake.

Organisms

Although botanists emphasise the influence soil type has on vegetation, it is also true that vegetation affects many soil properties. The rates of supply and decay of dead plant material both above and below ground largely control the amounts of soil organic matter, though turnover rate is also influenced by soil temperature, particle size distribution and drainage, and dilution may occur locally through deposition of fresh inorganic sediment on the soil surface. Nitrogen fixation by leguminous plants locally improves soil fertility, whereas some savanna grasses inhibit the release of organically bound nitrogen as ammonia or nitrate (nitrogen mineralisation). Many plant decomposition products, such as polyphenols and organic acids, accelerate leaching and processes of mineral weathering, while living roots assist mineral breakdown by removing some elements from the soil solution. Forest vegetation tends to conserve soil moisture by maintaining a stable microclimate at the soil surface and encouraging infiltration rather than runoff, whereas a more open cover allows the soil surface to be damaged by rain impact, causing capping (impedance to infiltration) with increased runoff and erosion.

Macrofauna affects soil mainly by burrowing, disturbance and mixing. The effect of most animals is only locally important, but earthworms in temperate regions and termites in the tropics have a wider influence, often building up considerable depths of well-structured, stone-free soil over large areas. The soil mesofauna, mainly arthropods, plays an important role in breaking down plant remains, which are then converted by microbiological activity to humus, hydrocarbons or carbon dioxide. Various microbes (bacteria, viruses, actinomycetes, fungi, protozoa and algae) are also important in maintaining the cycle of nitrogen between soil, plant and atmosphere, in assisting root uptake of nutrients (such as phosphorus), and in the oxidation of some soil minerals such as pyrite (FeS_2).

Time

Time differs from other soil-forming factors in usually affecting stage rather than process, though some processes cannot begin until others are well advanced, and in this sense are controlled by time. However, comparatively little is known about rates of soil formation. Some processes, such as the breakdown and incorporation of organic matter, leaching of soluble salts from newly exposed marine sediments, and gleying, are fairly rapid, often affecting the top 20–100 cm below the surface in a matter of months, years or decades. Others, such as the weathering of silicate minerals, are much slower, especially in temperate or cold regions, and probably need a hundred to a million years before achieving effects detectable by field observation. Mechanical disintegration of consolidated rock proceeds at very variable rates, depending on climate, but world average rates of total surface denudation, ranging from 50 mm per 1000 years for gentle slopes to 500 mm per 1000 years for steep slopes and mountainous areas, give an indirect approximation of physical weathering rates.

Immature soils are those in which profile development has not proceeded as far as it could in the climatic, topographic and biological environment concerned, because insufficient time has elapsed since initial exposure of the parent material to soil formation. They occur typically on recent deposits, such as river alluvium or lavas and ashes of active volcanoes, and are easily identified in their extreme form. However, the mature state for any particular combination of environmental factors is less easy to recognise. In temperate and colder regions, extensive glacial and periglacial erosion during the Quaternary removed existing Tertiary or interglacial soils from all areas except broad flat plateaux, and many new deposits were also formed. In the 10000 years since the last cold period it is unlikely that fully mature profiles have developed very extensively in these fresh deposits or on the eroded surfaces of older parent materials. Also, during this period, man has considerably altered the vegetation of large areas, so at least one environmental factor in soil development has remained constant for much less than 10000 years. In the tropics, less erosion and deposition occurred during the Quaternary, and man has probably affected the environment less, so that there has been enough time for mature profiles, closer to equilibrium with their environment, to develop. However, even the most mature, base-deficient soils may be chemically rejuvenated, at least for short periods, by small additions of fresh material, such as volcanic ash or loess.

The sequence of changes a soil profile undergoes during progressive development towards maturity, either in fairly constant environmental conditions or as a result of changes with time, is often referred to as soil evolution. Stages of evolution can be reconstructed from internal soil features, or less directly by comparison of profiles thought to form a chronosequence, such as might be found on successively older river terraces or buried by successive volcanic, eolian or glacial deposits. The word palaeosol is generally used for soils formed in past environmental conditions sufficiently different from those of the present day to cause the appearance of different profile morphological features. Such soils may be buried beneath more recent deposits, or remain at or close to the surface within the zone affected by current pedological processes; in the latter case (relict soil), recent pedological features are often superimposed on older palaeosol features. A younger age limit for palaeosols is usually set in temperate regions at the last major change of climate (approximately 10000 years ago). However some workers restrict the term palaeosols to profiles formed on past landscapes, and thus include only buried (and possibly also exhumed) soils of any age, even those buried quite recently (e.g. by mechanical earth-moving).

A more logical stratigraphical approach to buried (or fossil) soils, and surface soils (relict or exhumed) with features inherited from past periods, is provided by Morrison (1978), who proposed the term 'Geosol' for 'an assemblages of soil profiles that have similar stratigraphic relations...and physical and chemical pedogenic features that permit recognition, lateral tracing, and mapping as a stratigraphic unit, although specific pedogenic features can vary with changes in soil-forming environment (parent material, climate, vegetation, topography/drainage) from place to place'. The various soil profiles comprising a geosol would be approximately contemporaneous, and the age of the geosol defined by stratigraphical principles (i.e. it is younger than the youngest stratigraphical unit on which it is developed, but older than the oldest deposit overlying buried profiles).

Man

Man has had a profound effect on soils since at least the Neolithic period. Some changes are beneficial, allowing crop yields to be increased; others are detrimental, leading to pollution or soil degradation. Many are two-edged swords – for example, artificial drainage of waterlogged recent marine sediments can improve aeration and allow crops to be grown where otherwise unproductive swamps exist, yet if pyrite is present it will be oxidised to produce sulphuric acid, so increasing acidity and limiting crop yields.

Deforestation by slash and burn, followed by soil cultivation and growth of arable crops, has often resulted in the gradual loss of soil organic matter and nutrient reserves built up during the forested phase. The subsequent weakening of soil structure leads to increased susceptibility to erosion by runoff and wind, and an overall reduction in fertility. This is especially serious in strongly weathered and leached tropical soils, where some nutrients, such as potassium, are almost entirely associated with organic matter in the topsoil. Cultivation also homogenises the top 15–30 cm of the soil profile, thus modifying the characteristic surface horizons formed under forest.

Irrigation in arid areas usually has a beneficial effect on fertility and may increase soil organic matter. The accelerated mineral weathering that also often results from irrigation can increase available nutrients, but may also limit fertility by increasing soil salinity or alkalinity. Other human influences on soil include decreasing natural acidity by liming, increasing organic matter by spreading farmyard manure, peat or seaweed, improving the particle size distribution of sandy soils by marling, improving soil aeration by artificial drainage, and overcoming the limitations of steep slopes (erosion and cultivation difficulties) by terracing.

Soil-forming processes

Some processes of soil formation are simple (e.g. oxidation, hydrolysis, decalcification), whereas others are composite, involving several simple processes – for example, gleying involves the reduction, dissolution and mobilisation, and often oxidation and reprecipitation, of various compounds of iron and other metals. Most of the simple processes result either from the passage of water through the soil or from biological activity. Rainwater is slightly acid because of dissolved carbon dioxide from the atmosphere, and in the soil acidity is often initially increased by organic decomposition products; however, further reactions with the soil may subsequently decrease acidity. Because of higher temperatures the various chemical reactions between water and soil compounds are approximately three times faster in tropical areas than in temperate regions, and at least six times faster than in polar regions.

Weathering

Weathering is a composite soil-forming process, which is conventionally divided into physical (or mechanical) and chemical effects, though the two are often interdependent. As much physical weathering depends upon temperature fluctuations, including ice formation, it predominates in colder regions of the earth, but is also important in arid areas, where there is little vegetation to insulate rocks from temperature extremes.

The effect of chemical weathering on rocks depends partly on the type and intensity of the processes involved and partly on the susceptibility of individual minerals. Some primary minerals are completely dissolved, but others, especially silicates, lose only some of their constituent elements and leave behind fine-grained residual minerals. In the silicates, resistance to weathering depends upon the linking of SiO_4 tetrahedra (isolated tetrahedra, chain and sheet structures being more easily altered than three-dimensionally linked framework structure, see Chapter 3), and also upon the abundance of cleavage planes, which increase the surface area open to chemical attack. Ionic dimensions may also be important, as isomorphous substitution in feldspars and other mineral groups seem to affect weatherability. However, the significance of crystal structure and composition in determining resistance to weathering is still poorly understood.

Chemical weathering processes involve three chief types of reaction, hydrolysis, carbonation and oxidation–reduction; these may act singly or (more often) in combination. Simple hydrolysis reactions include:

$$SiO_2 + 2H_2O \rightleftharpoons \underset{\text{(silicic acid, removed in solution)}}{Si(OH)_4}$$

$$\underset{\text{(kyanite)}}{Al_2SiO_5} + 5H_2O \rightleftharpoons \underset{\text{(gibbsite residue)}}{2Al(OH)_3} + \underset{\text{(in solution)}}{Si(OH)_4}$$

$$\underset{\text{(kaolinite)}}{Al_2Si_2O_5(OH)_4} + 5H_2O \rightleftharpoons \underset{\text{(gibbsite residue)}}{2Al(OH)_3} + \underset{\text{(in solution)}}{2Si(OH)_4}$$

Hydrolysis and carbonation combined are important in reactions such as:

$$\underset{\text{(calcite)}}{CaCO_3} + H_2O + CO_2 \rightleftharpoons \underset{\text{(in solution)}}{Ca^{2+} + 2HCO_3^-}$$

$$\underset{\text{(anorthite)}}{CaAl_2Si_2O_8} + 3H_2O + 2CO_2 \rightarrow \underset{\text{(kaolinite residue)}}{Al_2Si_2O_5(OH)_4} + \underset{\text{(in solution)}}{Ca^{2+} + 2HCO_3^-}$$

$$\underset{\text{(albite)}}{2NaAlSi_3O_8} + 11H_2O + 2CO_2 \rightarrow \underset{\text{(kaolinite residue)}}{Al_2Si_2O_5(OH)_4} + \underset{\text{(in solution)}}{2Na^+ + 2HCO_3^- + 4Si(OH)_4}$$

$$\underset{\text{(orthoclase)}}{3KAlSi_3O_8} + 4H_2O + 2CO_2 \rightarrow \underset{\text{(illite residue)}}{KAl_2(Si_3Al)O_{10}(OH)_2} + \underset{\text{(in solution)}}{2K^+ + 2HCO_3^- + 6Si(OH)_4}$$

Hydrolysis and oxidation combine in reactions such as:

$$\underset{\text{(pyrite)}}{4FeS_2} + 15O_2 + 8H_2O \rightarrow \underset{\text{(haematite residue)}}{2Fe_2O_3} + \underset{\text{(in solution)}}{16H^+ + 8SO_4^{2-}}$$

The important process of hydrolysis seems to occur through the attraction of water dipoles to exposed ions with unsaturated valencies on the crystal surface; dissociation into H^+ and OH^- ions results, the OH^- bonding to exposed cations and the H^+ to oxygens and other negative ions. H^+ ions also replace cations at the mineral surface. In minerals such as feldspars, where aluminium is in tetrahedral co-ordinations with oxygen, the removal of cations and conversion of oxygens to hydroxyls allows aluminium to assume its preferred octahedral co-ordination. As a result the surface layers become unstable and polyhedra are released to form amorphous colloids. However, the transition to some layer silicates, such as illite, may not go through this amorphous phase, as part of the aluminium in this mineral persists in tetrahedral co-ordination, and mobile cations (K^+) are partially retained. The residual weathering product in such cases inherits part of the structure of the original mineral.

Oxidation of ferrous iron to ferric is an important disruptive process in some silicates (biotite, glauconite, pyroxenes, amphiboles), because of the change in ion size. Alternate reducing and oxidising conditions also promote disintegration of alumino-silicates containing ferrous iron, as the iron removed in solution under reducing conditions is replaced during succeeding oxidative phases by aluminium from the lattice, thus weakening the structure. More widely, any replacement of surface cations by H^+ causes lattice disintegration by removing aluminium, which is then hydrolysed to produce more H^+. The hydroxy-aluminium may form gibbsite, but is often deposited between the layers of expanding layer silicates (vermiculite and smectite) to form chlorite-like clay minerals. The expanding clay minerals may be derived from primary micas by removal of interlayer potassium and other cations, or may nucleate and grow from solution using dissolved silica and cations such as Ca^{2+} or Mg^{2+}. The first weathering products of volcanic ash or glass are often tiny polygons of allophane or threads of quasi-crystalline imogolite, both formed from silica and alumina, which are at least partly in solution.

Leaching

The ions extracted from primary minerals by various weathering processes are removed by leaching where downward percolation or lateral movement of water is maintained. Some may be redeposited as secondary minerals elsewhere in the profile, but large amounts are completely removed from the soil in the groundwater. If water movement is slow, salinity increases and the weathering processes slow down or stop as the reactions approach a state of equilibrium. Leaching is therefore important in maintaining the direction of many of the reversible processes involved in weathering.

The solubility of different elements in soil water is very variable, and is influenced by pH, redox potential (Eh) and the types and amounts of organic compounds. Metals that form strongly ionic bonds with oxygen, such as potassium and sodium, are much more soluble in polar liquids like water than those forming hybrid bonds (e.g. Ti, Al, Si), or non-metals which generally form covalent bonds (e.g. C, S, P). However, the anions formed by covalent bonds between non-metals and oxygen (e.g. carbonate, sulphate, phosphate) are negatively charged and therefore soluble in polar liquids, especially those in which an excess of anions over metal cations is balanced by H^+ (i.e. acid solutions). The solubility of calcium carbonate, for example, is effectively zero at pH 9 and above, but increases rapidly below pH 8. Alumina is insoluble in the pH range 5 to 8, but highly soluble at pH 4 and 10. In contrast, the solubility of silica is small and fairly constant in neutral and acid conditions, but increases rapidly above pH 9. In very acid soil ($pH < 4$), there is consequently rapid removal of alumina but a relative accumulation of silica. Such low pH values are generally achieved only by the products of partial organic decomposition, or locally by oxidation of pyrite to form sulphuric acid, as carbonic acid is only weakly ionised, giving pH values scarcely below 6 at atmospheric pressure.

The effect of Eh on solubility is of greatest significance with elements such as iron, manganese and possibly titanium, that exist in more than one valence state when combined with anions, and have different solubilities in each state. Ferrous iron forms strongly ionic bonds and is consequently soluble at higher pH values than ferric iron, which forms less ionic hydrid bonds; for example ferrous and ferric hydroxides are soluble only below pH 9 and pH 3 respectively. As pH 3 is rarely attained in soils, this explains why the 'free' iron (i.e. not combined in the rather less reactive silicate minerals) is mobilised mainly in reducing conditions, as in gleying.

In soil, Eh is controlled by the accessibility of atmospheric oxygen and the abundance of organic matter, which is a strong reducing agent because of the ease with which it is partly or completely oxidised to carbon dioxide. The Eh for any given reaction is also affected by pH, such that at lower pH values less reducing conditions are needed to convert ferric compounds to their ferrous equivalents. Mobilisation of iron as ferrous ions in gleying is therefore assisted by soil acidity as well as by waterlogging and an abundance of organic matter to eliminate oxygen.

Apart from their effects in controlling pH and Eh, organic compounds can also mobilise many cations in soil by forming complexes analogous to the ring structures that chelating agents (e.g. ethylenediaminetetraacetic acid or EDTA) form with metal ions by replacement of hydrogen ions. Laboratory leaching experiments have shown that EDTA can effectively decalcify rock material and then remove iron and aluminium from it, redepositing these two elements as hydrated oxides in the still calcareous 'parent material' beneath. Some soil organic matter possesses similar chelating properties, and could be responsible for much leaching, the dissolution of otherwise immobile secondary minerals, and even the breakdown of some primary minerals by direct removal of cations, but all these effects have yet to be proved.

Although some elements are fairly easily mobilised by weathering and leaching processes, they do not necessarily enter the groundwater in appreciable quantities because of various 'fixation' processes in the soil. Potassium released from primary micas and other minerals, or applied to the soil in fertilisers, is commonly fixed in layer silicate clay minerals, in interlayer sites on the edges of crystals that are transitional between expanding minerals, such as smectites and vermiculites, and non-expanding micas. Sodium is fixed much less, thus accounting for the small amounts relative to potassium that are retained in soils, but the large amounts in rivers and sea water. Phosphate is also fixed in soils, though often rather less firmly or permanently, by association with iron and aluminium hydroxides and hydrated oxides. Initially the fixation results from surface adsorption of phosphate anions – these remain available to plants and thus constitute a useful reserve of phosphorus which is released when needed. However, more permanent fixation can occur if the adsorbed phosphate is occluded by deposition of another layer of iron or aluminium hydroxide, or if less soluble iron and aluminium phosphates form. Vivianite (iron phosphate) forms in reducing conditions, and extremely insoluble minerals of the plumbogummite group, principally crandallite ($CaAl_3(PO_4)_2(OH)_5 \cdot H_2O$), occur where calcium is available, for example on limestones.

Podzolisation

This is a composite soil-forming process leading to the development of a characteristic horizon sequence, consisting of a bleached grey or white layer, beneath a thin organic surface horizon, and resting on darker horizons usually enriched in iron, aluminium and organic matter. The bleached horizon usually forms by acid ($pH < 4$) leaching of iron and aluminium, leaving a chiefly siliceous residue; this is accomplished by organic decomposition products, which may act partly by chelation. Some of the iron and aluminium is redeposited lower in the profile, often forming a weakly cemented horizon, and organic matter may also be redeposited either in the same horizon or slightly above it. However, in lowland tropical areas horizons of visible iron accumulation are often absent or very deep, because the mobilised iron is moved further by the continuous leaching.

A similar sequence of soil horizons can also form locally by waterlogging and complete reduction of ferric compounds to ferrous above a pre-existing impermeable layer. This gleyed horizon often underlies a wet peaty layer, which ensures exclusion of oxygen, and the impermeable layer beneath often results from compaction by repeated ice-lens formation (fragipan horizon). The gley podzols originating in this way are thus characteristic of cooler and wetter regions which have experienced recent or Pleistocene permafrost. A thin iron-pan often occurs at the base of the strongly gleyed horizon, thereby worsening the already poor profile drainage.

Clay illuviation

Subsurface horizons of some soils are enriched in clay which has been washed down from overlying horizons by water percolating through pores and channels. The clay forms coatings (cutans or argillans) on aggregates, stones and pore walls, which are visible in the field as shiny surfaces (compared with the roughness of aggregate interiors), or in thin section under the polarising microscope as layers of oriented, birefringent and often laminated clay. Large accumulations can impair profile drainage and result in weak gleying of the horizons affected.

As the clay is moved particle by particle, it must first be dispersed in upper soil horizons, and as di- or tri-valent cations on clay surfaces result in strong electrostatic attraction between particles (flocculation), dispersion cannot occur until the upper soil horizons have been leached of calcite or other minerals capable of supplying an abundance of such cations. Nor can it occur where conditions are acid enough to mobilise Al^{3+} (e.g. in podzols). However, small amounts of iron are often translocated with clay, in solution as ferrous ions, complexed with organic compounds, or as very fine particles of ferric hydroxide or hydrated oxides, and are deposited in the clay coatings (or argillans) as ferruginous laminae.

Subsoil clay enrichment can also result from inhomogeneity of parent material or clay formation by weathering, so the field or microscopic recognition of argillans is critical to confirmation of clay illuviation. Unfortunately field identification is difficult in soils which are clay-rich in any case, and clay particles can become oriented and therefore birefringent in thin section because of stresses in soils, such as those caused by shrinking and swelling of expansible clay minerals (e.g. smectite). In addition illuvial clay does not always form recognisable argillans in soils subject to frequent shrink–swell, probably because individual conducting channels are too short-lived to develop them. Stress-oriented clay can often be distinguished in thin section from argillans, because it occurs in domains that are not necessarily associated with pores, have poorly defined margins, and are less well oriented and therefore less birefringent. However, argillans themselves may be dissociated from the voids in which they formed; disturbances by burrowing animals, deep cultivation, tree-fall or cryoturbation can break them and embed the fragments in the soil matrix. Recognition of clay illuviation may therefore present many difficulties, especially in clay-rich soils.

Ferrallitisation

In soils not subject to gleying or extremely acid conditions, the iron and aluminium released by weathering of primary minerals is scarcely mobile, and gradually accumulates in the soil at the expense of silica, alkalis and alkaline earths, which are leached away at varying rates. Small amounts of iron released by early oxidative weathering give the soil an overall uniform brown colour, due mainly to the mineral goethite ($\alpha FeO \cdot OH$), a process known as brunification. Further release coupled with dehydration in warmer conditions leads to red colours (rubification) caused by formation of haematite (αFe_2O_3). Moderate clay translocation can occur at this stage, and some of the iron may form small hard concretions. More intense chemical weathering and leaching in hot climates over long periods leads to further concentration of iron and aluminium relative to silica. The only common layer silicates are then 1:1 lattice minerals, such as kaolinite, and increasing amounts of gibbsite accompany the iron oxides and hydrated oxides. Hard nodules or massive subsoil accumulations of mixed iron and aluminium oxides (laterite) are formed, and in tropical areas, where these may reach a total thickness of 30 m or more, they behave geomorphologically as a resistant rock stratum, often giving a characteristic landscape of plateau remnants.

The development of such thick laterite crusts simply by residual accumulation would probably take many millions of years, and alternative origins, usually involving precipitation of dissolved iron within a zone of fluctuating groundwater levels, have been proposed. The source of the iron could then be some distance laterally or vertically from the site of accumulation. Many massive laterites are underlain by deep pallid zones of white or mottled kaolinitic clay, which might have provided much of the iron, but upward translocation on such a large scale could only occur locally where groundwater rises to the surface. Alternatively, the pallid zone could result either from deposition of silica removed from the lateritic layers or from further deep weathering of bedrock after formation of the laterite. Lateral translocation of iron would also give a rather patchy distribution of laterite, so vertical translocation from overlying horizons subsequently removed by erosion seems most likely. Thus, although some mobilisation of iron was probably important in the development of laterites, pedogenetic concentration of iron and aluminium by selective removal of silica (ferrallitisation) must be regarded as the main formative process.

Soil classification

Many different systems of soil classification have been devised for either national or international use, but at present none is used or accepted universally. Some are based on field or laboratory determined profile characteristics, irrespective of their origin; others are related to processes of soil formation inferred from profile characteristics or measurable soil-forming factors (parent material, climate, vegetation). Some differentiate between classes using a few selected properties as independent variables; others use all known properties as naturally inter-related factors.

As space permits the detailed presentation of only one classification system, the most appropriate is that produced in 1974 for the United Nations Food and Agriculture Organisation's Soil Map of the World, because it is simple and quite widely used on an international basis. Also it forms the legend to the most detailed and complete world-wide soil map yet published. The system is based on genetic processes, but also emphasises the capability of the soils for crop production. Certain 'key properties' are used to differentiate the soil units, and groups of properties are used to define 'diagnostic horizons' helpful in recognising units. The names of the units generally consist of an adjective and a noun. Many of the nouns are derived from traditional Russian nomenclature (e.g. Chernozem), which has long been used in many parts of the world. These have been supplemented by new terms, often of Latin or Greek root but sometimes of mixed origin (e.g. Greyzem, English and Russian), to give 26 nouns in all, each of which is represented on the maps by a different capital letter. Subdivision of these major units by the use of adjectives gives a total of 106 soil units. The number of subdivisions in each major unit ranges from none in Lithosols (I), Rendzinas (E) and Rankers (U) to nine in Cambisols (B), namely: Eutric (Be), Dystric (Bd), Humic (Bh), Gleyic (Bg), Gelic (Bx), Calcic (Bk), Chromic (Bc), Vertic (Bv) and Ferralic (Bf) Cambisols. Some of the adjectives are derived from major units (e.g. Gleyic Cambisols from Gleysols), implying subordinate properties related to those of the major unit. Others, such as gelic (presence of permafrost) refer to special properties not used as characteristics of the 26 major units.

It is not possible here to describe all 106 units, nor to show their distribution on a small-scale map of the world. Brief descriptions of the 26 major units are given, and the meanings of some adjectives are given subsequently in the text as necessary. Figure 5.1 shows in greatly simplified form the world distribution of the 26 major units.

A Acrisols (Latin *acris* = acid): soils with an argillic diagnostic horizon (i.e. one enriched in layer silicate clays washed from overlying horizons) and a base saturation (i.e. exchangeable Ca+Mg+Na+K as percentage of total cation exchange capacity) < 50 per cent in at least part of the profile above 125 cm depth. Iron enrichment, giving coarse reddish mottles, nodules up to 2 cm across, or soft accumulations which harden on drying (plinthite), is common within 125 cm of the surface.

B Cambisols (Latin *cambiare* = change): soils with evidence of a little alteration *in situ*, such as increased clay content (not due to illuviation), brunification or decalcification, to at least 25 cm below the surface.

C Chernozems (Russian *chern* = black, *zemlja* = earth): soils with a black mollic surface horizon (i.e. one darkened by > 1 per cent organic matter, with a base saturation > 50 per cent, and a well-developed structure, but containing less organic matter than a histic diagnostic horizon) at least 15 cm thick, which overlies an horizon with redeposited calcium carbonate or sulphate within 125 cm of the surface, sometimes with an intervening argillic horizon.

D Podzoluvisols (Russian *pod* = under and *zola* = ash, indicating the ashen colour of the bleached horizon, and Latin *luvere* = wash, indicating the downwashing of clay): soils with an argillic horizon showing an irregular or broken upper boundary, caused either by deep tonguing of a pale-coloured clay- and iron-depleted (i.e. albic) horizon into the argillic horizon, or by the formation of nodules 2–30 cm across, the exteriors of which are reddened and cemented with iron.

E Rendzinas (Polish *rzedzic* = noise, indicating the sound made by a plough in shallow soil over rock): soils with a mollic horizon immediately overlying calcareous sediment with > 40 per cent $CaCO_3$.

F Ferralsols (Latin *ferrum* = iron, and aluminium, indicating large contents of these elements): with an oxic diagnostic horizon (i.e. one with little or no primary alumino-silicate minerals, such as feldspar, mica, amphibole or pyroxene, > 15 per cent clay, but low base saturation and small cation exchange capacity because of a predominance of iron and aluminium oxides and hydrated oxides) more than 30 cm thick. Red and yellow colours are common, and plinthite may occur within 125 cm of the surface.

G Gleysols (Russian *gley* = wet, mucky soil): soils formed from unconsolidated sediments (other than recent alluvial deposits) that show the effects of waterlogging within 50 cm of the surface. Grey or bluish colours predominate, and there is often evidence of segregation of iron (rusty mottling, iron pans).

H Phaeozems (Greek *phaios* = dusky): soils with mollic surface horizons, but lacking redeposited calcium carbonate or sulphate in subjacent horizons. However, primary carbonate may occur below 20 cm depth, and an intervening argillic horizon is sometimes present.

I Lithosols (Greek *lithos* = stone): shallow soils with severe restriction of root development caused by hard consolidated rock within 10 cm of the surface.

J Fluvisols (Latin *fluvius* = river): soils developed in recent alluvial (marine or fluvial), lacustrine or colluvial deposits, and in which organic matter content decreases irregularly with depth. Sulphides often occur, and their oxidation following artificial drainage may give very acid conditions (pH < 3·5) and yellow mottles due to the mineral jarosite. Carbonate content and base saturation are very variable.

K Kastanozems (Latin *castaneo* = chestnut): soils with 'chestnut brown' mollic surface horizons at least 15 cm thick, and lower horizons containing redeposited calcium carbonate or sulphate, sometimes with an intervening argillic horizon.

L Luvisols (Latin *luvere* = wash): soils with an argillic diagnostic horizon and a high base saturation (> 50 per cent) at least between 20 and 50 cm depth, but lacking the distinctive characteristics of Podzoluvisols, Planosols and Nitosols.

M Greyzems (English *grey*, Russian *zemlja* = earth): soils with black mollic surface horizons at least 15 cm thick, which have grey or white bleached coatings on aggregate surfaces.

N Nitosols (Latin *nitidus* = shiny, indicating shiny aggregate surfaces from redeposited clay): soils with thick argillic horizons, extending to depths of more than 150 cm below the surface. The clay content remains within 20 per cent of its maximum to at least 150 cm, and unaltered parent material is not encountered within this depth. The term was originally restricted to soils formed by deep tropical weathering of basic igneous rocks, but is now applied to profiles on a wider range of parent materials.

O Histosols (Greek *histos* = tissue): soils with a wet organic (e.g. peat) horizon at least 40 cm thick at the surface or within the top 80 cm of the profile. This histic horizon should contain > 30 per cent organic matter if the remainder is clay, > 20 per cent if the remainder contains no clay.

P Podzols (Russian *pod* = under, and *zola* = ash): sandy soils with an horizon below 12·5 cm depth which is > 2·5 cm thick and continuously cemented with redeposited organic matter and often also iron and aluminium oxides and hydrated oxides (i.e. a spodic diagnostic horizon).

Q Arenosols (Latin *arena* = sand): soils formed from sandy parent materials (other than recent alluvial deposits), often with little or no horizon development. The colour of at least the upper 50 cm of the profile is usually pale, because it is determined by the primary sand particles rather than by pedogenetic coatings on them or by incorporation of organic matter.

R Regosols (Greek *rhegos* = blanket): very immature soils with little or no profile development on unconsolidated sediments.

S Solonetz (Russian *sol* = salt): soils with an argillic subsoil horizon in which sodium accounts for > 15 per cent of the exchange capacity over at least its upper 40 cm. Columnar structural aggregates are common in this horizon.

T Andosols (Japanese *an* = dark, *do* = soil): soils with dark-coloured mollic (> 50 per cent base saturation) or umbric (< 50 per cent) surface horizons formed from volcanic deposits. They often contain much volcanic glass in fractions > 2 μm, but when a large proportion of this is weathered, amorphous or poorly crystalline clay (allophane) is formed, and the soil has a low bulk density (< 0·85 g cm^{-3} at 0·3 bar water retention) and smeary consistency.

U Rankers (Austrian *rank* = steep slope): soils with umbric surface horizons < 25 cm thick, usually resting on hard, non-calcareous rock. Soils on recent alluvial deposits are excluded.

V Vertisols (Latin *vertere* = turn, indicating the natural turnover of the soil in the upper part of the profile): soils with > 50 per cent clay in at least the top 50 cm, which develop vertical cracks at least 1 cm wide to this depth during dry seasons. The cracks allow surface soil to become mixed with subsoil (self-mulching) and encourage the formation of surface knolls and basins with an amplitude of 5–200 cm (gilgai microrelief) and wedge-shaped structural aggregates below 25 cm depth.

W Planosols (Latin *planus* = flat, indicating level topography with poor drainage): soils with an albic horizon, often strongly weathered, overlying a slowly permeable horizon (e.g. an argillic or other clay-enriched horizon, or fragipan, but not a spodic horizon) within 125 cm of the surface, and consequently subject to gleying in part at least.

X Xerosols (Greek *xeros* = dry): semi-arid soils with just enough organic matter to darken the surface horizon, but not enough to constitute a mollic horizon. Argillic horizons are sometimes present, and deposition of calcium carbonate or sulphate may occur within 125 cm of the surface, but very saline soils (electrical conductivity > 15 mmhos cm^{-1} at 25 °C) are excluded.

Y Yermosols (Spanish *yermo* = desert): desert soils with similar properties to Xerosols, but less organic matter (< 1 per cent in top 40 cm if clayey, < 0·5 per cent if sandy).

Z Solonchaks (Russian *sol* = salt): strongly saline soils, with electrical conductivity > 15 mmhos cm^{-1} at 25 °C.

Further details of these units are given in Volume 1 (Legend) of the Soil Map of the World (FAO, 1974). Subsequent volumes are explanatory memoirs to the separate map sheets, and provide detailed descriptions of the morphological, chemical and physical properties of profiles representing the map units.

Larger scale maps than the FAO world soil map need a different type of soil classification, often one placing greater emphasis on geological, hydrological and human factors. In detailed soil mapping the smallest unit of classification is the pedon, a three-dimensional body of soil large enough to evaluate the diagnostically important features of all the soil horizons at a given site. Its areal dimensions are of the order of 0·1 to 15 m. Pedons having the same particle size classes, colours, bulk mineralogical composition and geological origin between specified depth limits (successive horizons) are grouped as soil series, which have lateral dimensions up to several kilometres. Series are the most useful unit of soil classification for mapping at 1:50000 or larger scales, though two or more phases are sometimes distinguished within a single series on the basis of minor internal or external characteristics likely to be of agricultural or other human significance (e.g. depth to bedrock, stone content, gradient).

Categories higher than the soil series vary widely between systems of classification. In the current English system (Avery, 1973) Subgroups, Groups and Major groups are defined mainly on the bases of composition and origin of soil material within specified depths and the presence or absence of diagnostic horizons. Like those of the FAO classification, these principally reflect type and degree of pedological alteration. The Major groups are approximately equivalent in importance to the FAO world map units, though the differentiating criteria are never exactly the same.

The most detailed and meticulously defined system of classification is that used in north America (Soil Survey Staff, 1975), and in time this may supplant other systems, including the FAO classification. However, it has an even more difficult terminology than the FAO system, and at present does not have world-wide applicability.

Present soil distribution

World-wide soil distribution

In reducing the FAO world soil map to the scale of Figure 5.1, much generalisation was necessary and some of the 26 major units were grouped together in areas such as some of the mountain chains. Also some quite widespread and agriculturally important soils, such as the Fluvisols, are greatly under-represented, because they occur in areas generally too small to separate. Another limitation of the map results from the variable quality of the soil information from different parts of the world. Where knowledge of the soils themselves is sparse, the map is based largely on inferences from climatic and natural vegetation data, though this is less true than with earlier world soil maps.

Despite these limitations, several features of the map are worthy of comment. First, the world's soils are dominated by comparatively few types, principally Regosols, Podzols, Luvisols, Lithosols, Kastanozems, and Cambisols in polar and temperate regions, and Ferralsols, Lithosols, Yermosols, Arenosols and Luvisols in equatorial areas. Four of these nine types, the Podzols, Lithosols, Yermosols and Arenosols, are among the least fertile. Of the remaining 17 types, some occupy a few quite large areas, for example the Chernozems in the centre of the Euro-Asian landmass, and the Vertisols in central Africa, India and Australia, but most occur only in small areas.

Second, the map shows that Ferralsols are rather less important in many equatorial regions than previous world soil maps, such as that of Bridges (1978, p. 102), have suggested. Although Ferralsols dominate equatorial parts of South America, they are accompanied by extensive areas of Nitosols, Arenosols and Gleysols in equatorial West Africa, and are replaced by Cambisols and Luvisols in East Africa and India, and largely by Acrisols in south-east Asia.

Third, in many parts of the northern hemisphere Podzols, Podzoluvisols, Luvisols, Greyzems, Chernozems, Kastanozems, Xerosols and Yermosols form broad east–west zones. However, in southern parts of North America, where the continent narrows and the climate becomes more oceanic, the zones of Podzols, Luvisols and Chernozems, are drawn southwards, and the Kastanozems form a relatively narrow zone extending north–south. These large-scale patterns of soil distribution demonstrate genuine effects of macro-climate, as the map in such areas is based largely or entirely on primary soil information and not inferred from climatic data. Other patterns shown by the map, such as the north–south zones of Phaeozems, Kastanozems and Yermosols in southern parts of South America, also correlate with macro-climatic zones, but probably more by design than in fact, as primary soil mapping in such areas is sparse.

Fourth, although it is not yet possible to calculate the areas of each map unit for every continent, the figures for North America, South America and Africa (Table 1), taken from the relevant explanatory memoirs, do show some interesting differences. Regosols, many of them Gelic Regosols of the northern permafrost areas, are the most abundant soil type in North America; together with Podzols, Lithosols and Gleysols, they occupy approximately 40 per cent of the northern sub-continent and support mainly tundra vegetation and unproductive forest. Leached soils with argillic subsurface horizons (Luvisols and Acrisols) together cover a further 17·4 per cent.

In contrast, 35 per cent of South America is occupied by Ferralsols. Xanthic (yellow) and Orthic Ferralsols are most abundant, together forming 29 per cent of the sub-continent, and a further 4·3 per cent is occupied by the Cerrado phases of the Rhodic (red) and Acric (small cation exchange capacity) Ferralsols. Many of these soils occur on old plateau surfaces that have escaped dissection and erosion, and are amongst the oldest soils in the world. Ferralsols that are most impoverished of plant nutrients by lengthy weathering and leaching are grouped as the Cerrado phase, because they can only support the cerrado vegetation of tall grasses and stunted trees. The younger soils formed in valleys dissecting the old plateaux are more fertile, and support tropical forest vegetation. Other important soils in South America are the Acrisols, Lithosols and Luvisols, which together occupy a further 25 per cent of the sub-continent.

The most common African soils are the desert Yermosols which, together with the Xerosols of the northern desert peripheries and the seasonally arid south, form 22 per cent by area. Arenosols and Lithosols (27 per cent together) are

Figure 5.1 World distribution of soils, based on FAO Soil Map of the World.

Table 1 Relative areas occupied by FAO soil types in North America, South America and Africa

Symbol	*Soil type*	*North America* km² × 10³	%	*South America* km² × 10³	%	*North & South America* km² × 10³	%	*Africa* km² × 10³	%
A	Acrisol	1381	6·8	1972	11·2	3353	8·8	862	3·1
B	Cambisol	1644	8·1	352	2·0	1996	5·3	1042	3·7
C	Chernozem	405	2·0	0	0·0	405	1·1	0	0·0
D	Podzoluvisol	55	0·3	0	0·0	55	0·1	0	0·0
E	Rendzina	149	0·7	0	0·0	149	0·4	26	0·1
F	Ferralsol	7	0·0	6192	35·2	6199	16·3	4200	15·0
G	Gleysol	1368	6·7	597	3·4	1965	5·2	618	2·2
H	Phaeozem	714	3·5	502	2·9	1216	3·2	7	0·0
I	Lithosol	1934	9·5	2038	11·6	3972	10·5	3460	12·4
J	Fluvisol	135	0·7	622	3·5	757	2·0	721	2·6
K	Kastanozem	1899	9·4	506	2·9	2405	6·3	26	0·1
L	Luvisol	2151	10·6	1231	7·0	3382	8·9	2523	9·0
M	Greyzem	53	0·3	0	0·0	53	0·1	0	0·0
N	Nitosol	108	0·5	257	1·5	365	1·0	1155	4·1
O	Histosol	943	4·6	36	0·2	979	2·6	17	0·1
P	Podzol	2182	10·8	15	0·1	2197	5·8	23	0·1
Q	Arenosol	0	0·0	741	4·2	741	2·0	4053	14·5
R	Regosol	2909	14·4	280	1·6	3189	8·4	1639	5·9
S	Solonetz	107	0·5	148	0·8	255	0·7	94	0·3
T	Andosol	371	1·8	218	1·2	589	1·6	48	0·2
U	Ranker	0	0·0	18	0·1	18	0·0	0	0·0
V	Vertisol	252	1·2	153	0·9	405	1·1	981	3·5
W	Planosol	80	0·4	448	2·5	528	1·4	176	0·6
X	Xerosol	377	1·9	444	2·5	821	2·2	1318	4·7
Y	Yermosol	1070	5·3	722	4·1	1792	4·7	4820	17·2
Z	Solonchak	3	0·0	105	0·6	108	0·3	163	0·6
Totals		20297	100	17597	100	37894	100	27972	100

also important in the south, and Ferralsols (15 per cent) in central Africa. These five soil types account for two-thirds of the continent, and if Luvisols (9 per cent) and Regosols (5·9 per cent) are included, 80 per cent of the total area is occupied by only seven types.

The main differences between Africa, North America and South America in relative importance of different soil types clearly result from major climatic and vegetational differences, such as the presence or absence of large areas of desert, tundra or tropical rain forest. Other factors, such as relief, parent material and time, are less important, though these three are probably more significant than the effects of man, organisms and minor climatic differences. The only soils common in all three continents are the Lithosols, Luvisols and Acrisols, though the last cover only 3·1 per cent of Africa. The distribution of Lithosols depends largely upon relief – shallow soils resulting mainly from erosion on steep slopes associated with many upland areas. However, the widespread occurrence of Luvisols and Acrisols reflects the almost world-wide importance of clay illuviation as a soil-forming process occurring in many different climatic zones.

Soil distribution in Britain

Figure 5.2, based on the FAO soil map of Britain, demonstrates in more detail the relationships between some of the map units and various soil-forming factors, principally parent material, relief and minor climatic differences. The distinction between Eutric (healthy) and Dystric (unhealthy) is founded upon base saturation (< 50 per cent in some part of the soil between 20 and 50 cm depth in Dystric) in the Gleysols and Cambisols, and pH

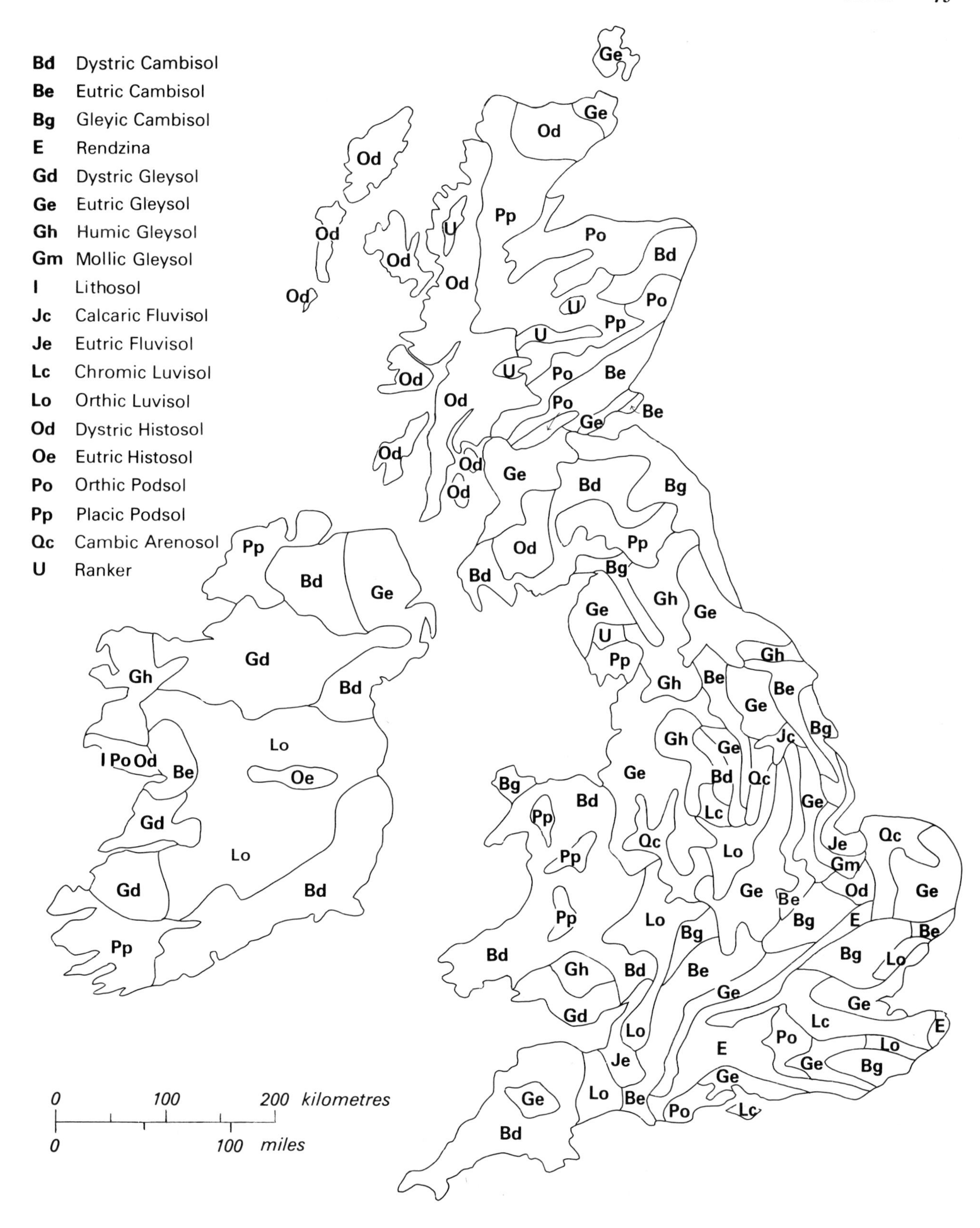

Figure 5.2 Simplified soil map of Britain and Ireland, based on FAO Soil Map of the World.

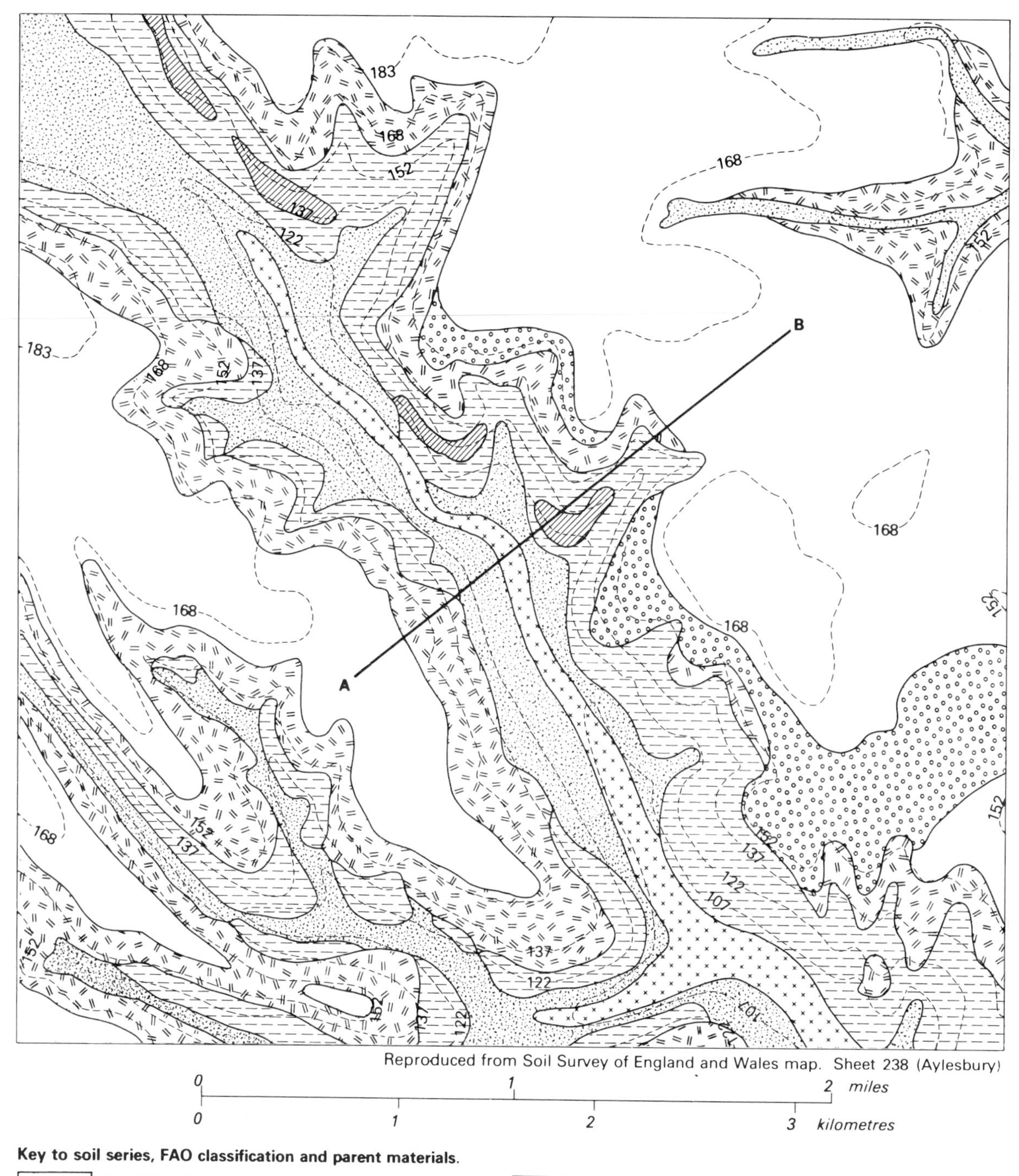

Key to soil series, FAO classification and parent materials.

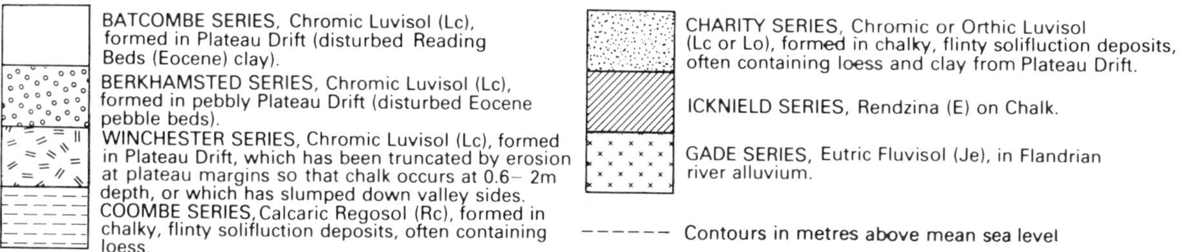

Figure 5.3 Detailed soil map of 25 km² of the Chiltern Hills near Hemel Hempstead, south-east England, showing distribution of soil series in relation to topographic contours.

(< 5·5 in some part of the soil between 20 and 50 cm depth in Dystric) in the Histosols. Gleyic Cambisols are those subject to minor waterlogging, and Humic Gleysols have highly organic surface horizons. Chromic Luvisols have brightly coloured (red or reddish brown) argillic subsurface horizons, whereas Orthic Luvisols are duller in colour throughout. Placic Podzols have a thin iron-pan in or over the spodic horizon, whereas Orthic Podzols do not, and Cambic Arenosols have subsurface horizons which are less calcareous than, and/or different in colour (brunified or rubified) from, the sandy parent material beneath.

The soils of south-east England and the Midlands are mainly Eutric and Gleyic Cambisols (Be and Bg), Eutric Gleysols (Ge) and Luvisols with subsidiary Rendzinas, Orthic Podzols (Po), Cambic Arenosols (Qc), Fluvisols and Histosols. These are all characteristic of a temperate climate with moderate rainfall, and the distribution of each type is determined mainly by parent material, which in turn exerts some control over the subsidiary factors of relief and hydrology. Gleysols occur mainly on low-lying and impermeable Mesozoic, Tertiary and Quaternary clays, Rendzinas on the steeper slopes of the Chalk outcrop, Podzols and Arenosols on sands and sandstones of various ages, Cambisols and Luvisols on sediments of mixed particle size distribution, Histosols on Holocene peats, and Fluvisols in areas of Holocene alluvial deposition such as The Wash.

In contrast, south-western areas of Britain (including Ireland) are characterised by Dystric (i.e. more leached) Cambisols (Bd), Dystric and Humic as well as Eutric Gleysols, and Orthic Luvisols (Lo), with subsidiary Placic Podzols (Pp) and Histosols. Most of the parent materials are similar in composition to those further east, though they are generally older and more consolidated, and the main reason for the slightly different soil groups is the rather wetter climate.

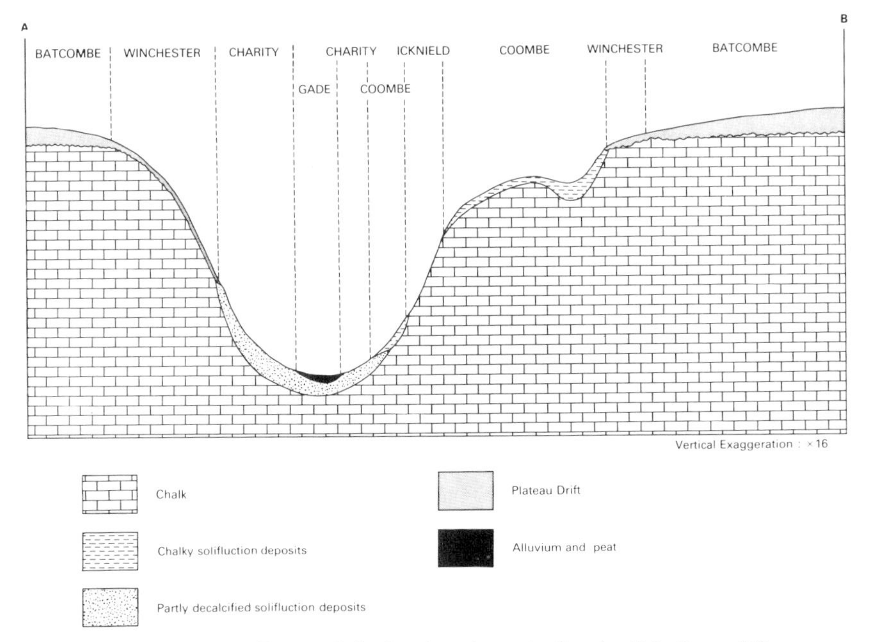

Figure 5.4 Section along the line A – B in figure 5.3, across typical chalk valley (River Gade), demonstrating relationship of soil series to superficial deposits.

In Scotland, Cambisols (mainly Dystric) occur mainly in the drier east and south. Histosols with subsidiary Rankers characterise the wetter western and northern parts, where the hard crystalline bedrock is often close to the surface because it was repeatedly scoured by glaciers during the Pleistocene and has not been much weathered since. Placic Podzols and Rankers typical of the central highlands in northern Scotland give way eastwards and at lower levels to Orthic Podzols developed mainly on Pleistocene glacial deposits. The abundance of Podzols results partly from the abundance of base-deficient parent materials, partly from high rainfall (giving rapid leaching and acid conditions due to the accumulation of peat), and partly from the effect of man in deforesting the area since Neolithic times. In many upland areas of Britain now characterised by Podzols under heath vegetation, the early Holocene soils buried beneath earthworks of Neolithic or Bronze Age are either Dystric Cambisols or Orthic Luvisols, which, as pollen studies show, were formed under deciduous woodland; if the woodland had not been destroyed, these soil types would have persisted to the present day.

Figures 5.3 and 5.4, taken from a detailed soil map originally published at a scale of 1:63 360 (Avery 1964), illustrate the relationships of soil series to parent materials and geomorphological features typical of much of the Chalk landscape in southern England. The part shown in Figure 5.3 is a 5 km × 5 km square north-west of London and within the area delineated as Chromic Luvisols (Lc) in Figure 5.2. Four of the seven series shown correspond to this FAO unit: the Batcombe and Berkhamsted series are formed in clayey and pebbly facies respectively of the Plateau Drift (disturbed and deeply weathered Eocene sediment), which caps remnants of an extensive but dissected sub-Tertiary surface; the Winchester series occurs where the Plateau Drift is < 1 m thick over the Chalk (e.g. at the margins of plateau remnants); and the Charity series is developed on older, partly decalcified solifluction deposits on the floors and lower side slopes of valleys dissecting the plateau. Calcaric Regosols (Rc) of the Coombe series occur in solifluction deposits which have not been appreciably decalcified, because either they are younger or occur beneath steeper slopes (exposing bare Chalk) than the Charity series. The steep valley side slopes cut in Chalk (and still subject to some erosion) are occupied by the Icknield series (Rendzinas), and Eutric Fluvisols (Je) of the Gade series occur in Holocene alluvium overlying solifluction deposits in the lowest part of the valley floor (Fig. 5.4).

The red or reddish brown colours characteristic of subsoil horizons of Chromic Luvisols such as the Batcombe, Berkhamsted, Winchester and Charity series, do not seem to have developed in British soils during either

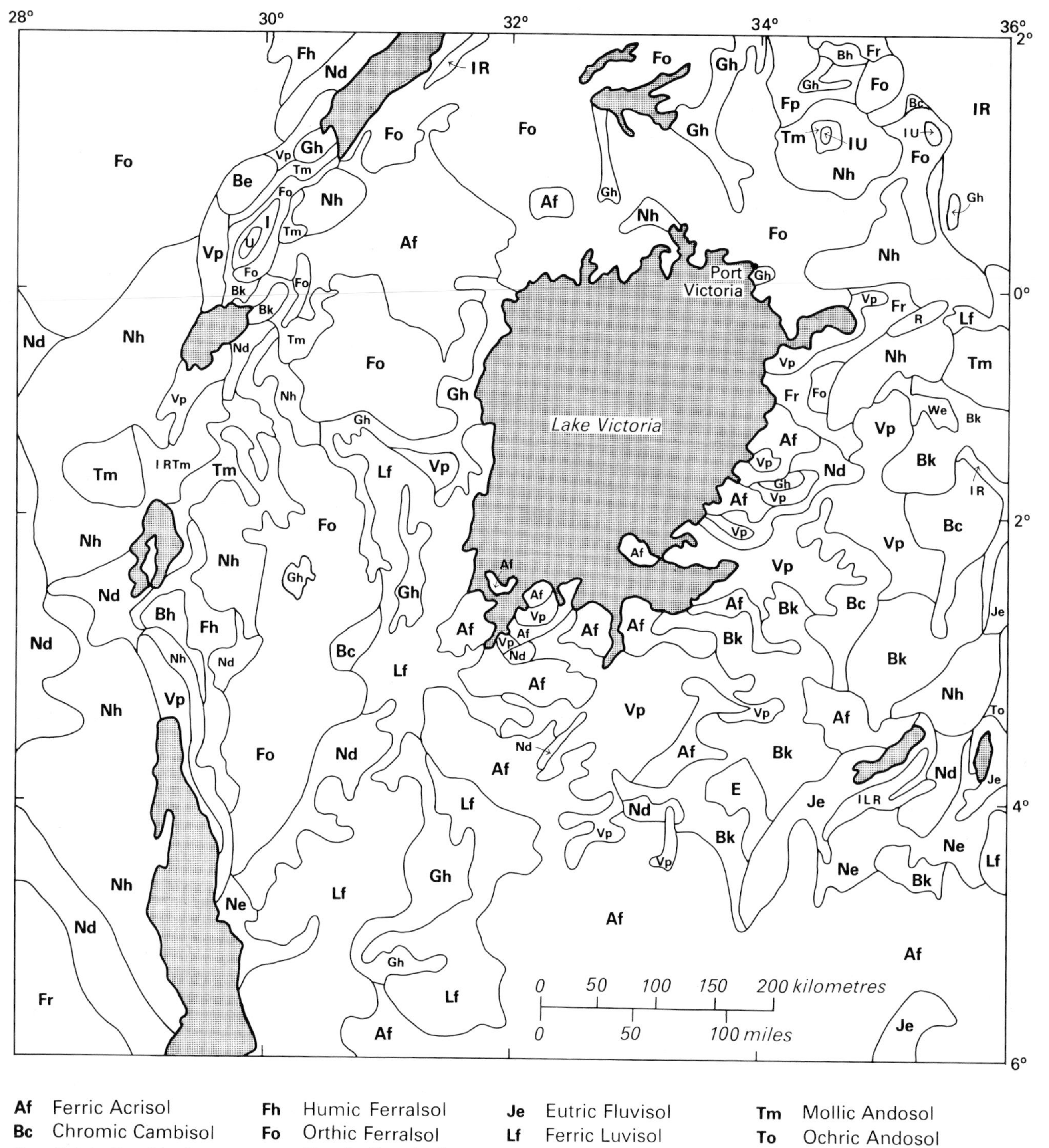

Af Ferric Acrisol
Bc Chromic Cambisol
Be Eutric Cambisol
Bh Humic Cambisol
Bk Calcic Cambisol
E Rendzina
Fh Humic Ferralsol
Fo Orthic Ferralsol
Fr Rhodic Ferralsol
Fp Plinthic Ferralsol
Gh Humic Gleysol
I Lithosol
Je Eutric Fluvisol
Lf Ferric Luvisol
Nd Dystric Nitosol
Ne Eutric Nitosol
Nh Humic Nitosol
R Regosol
Tm Mollic Andosol
To Ochric Andosol
U Ranker
Vp Pellic Vertisol
We Eutric Planosol

Figure 5.5 Soil distribution in parts of East Africa around Lake Victoria, based on FAO Soil Map of the World.

the Holocene or the last Pleistocene cold period (the Devensian Stage), because they are absent from soils on deposits of these periods and occur mainly on older, high-level geomorphological surfaces. The colours result mainly from weak ferrallitisation processes leading to formation of haematite (rubification), which are characteristic of climates rather warmer than Britain has experienced during at least the Holocene. They are consequently regarded as relict (palaeosol) features inherited from either the last interglacial (Ipswichian Stage) or an earlier warm period, possibly even from the late Tertiary in places.

Soil distribution around Lake Victoria, Africa

The soils of this part of central Africa (Fig. 5.5.), including parts of Kenya, Tanzania, Burundi, Rwanda, Zaire and Uganda, totalling 700000 km^2, afford an interesting comparison with those of Britain. The area includes three plateaux cut in the Precambrian Basement Complex at altitudes of 1100–1200 m (the Karamoja Plain, the Lake Victoria Basin, and the Central Plateau), which are bounded to east and west by rift valleys. The eastern part of the region rises from 1500 m on the Serengeti Plains to over 4000 m in Mount Elgon and the Cherengami range, which are composed of Tertiary lavas and ashes. Extrusive igneous rocks are also common in the mountainous area to the west, though peaks such as Mount Ruwenzori are uplifted fault blocks (horsts) free from vulcanism. In the south-west, the land surface declines to 800 m around Lake Tanganyika, which is up to 1400 m deep.

Only six of the 23 soil units in this part of Africa occur also in Britain; these are the Eutric Cambisols (Be), Rendzinas (E), Humic Gleysols (Gh), Lithosols (I), Eutric Fluvisols (Je) and Rankers (U), all of which occupy quite small areas. The two most common soils, together occupying almost half the region, are the Orthic Ferralsols (Fo) and Ferric Acrisols (Af). They have formed mainly on Precambrian Basement rocks, and differ in that the Acrisols have an argillic horizon with minor accumulations of iron oxides on structural surfaces or as small concretions, whereas the Ferralsols are composed predominantly of iron and aluminium oxides and kaolinitic clay. The Ferralsols are more strongly weathered and leached, and have probably developed over a longer period of time than the Acrisols. An earlier stage in the development of these profiles is probably represented by the Ferric Luvisols (Lf), which are common between Lake Victoria and Lake Tanganyika.

On basic lavas associated with the rift valleys there are extensive areas of the more fertile Nitosols. Humic Nitosols (Nh) are the most common, with some Eutric Nitosols (Ne) in the south-east, and Dystric Nitosols (Nd) on high ground in the west bordering the Congo Basin. Humic Nitosols also occur on the footslopes of Mount Elgon, but give way at greater heights to Mollic Andosols and eventually to Lithosols and Rankers on the summit. Strongly weathered Eutric Planosols (We) occur on Tertiary volcanic rocks near the eastern margin of the area.

Pellic Vertisols (Vp), which have dull greyish surface horizons, occur on fine-grained sedimentary rocks or in basins where Pleistocene and more recent clays have accumulated. They are extensive to the south-east of Lake Victoria, where they are associated with Chromic Cambisols (Bc), Calcic Cambisols (Bk), which contain secondary calcium carbonate or sulphate due to rapid evaporation of water from the soil surface, Rendzinas (E) and Eutric Fluvisols (Je).

Figure 5.6 is a large-scale soil map of part of western Kenya superimposed over air photographs which may be viewed stereoscopically to relate relief to soil distribution (Scott *et al.* 1971, p. 164). The area illustrated is near Bumala (at 34° 13′E, 0° 14′N, 26 km NE of Port Victoria) on the eastern edge of the Karamoja Plain, and has the general elevation (about 1200 m) typical of the plateau, with a local relief of 60–80 m. The soils of the region, shown in Figure 5.5 as Orthic Ferralsols, are developed in rocks of the Precambrian Basement Complex, mudstones, grits, lavas and granite. However, Figure 5.6 shows that Ferralsols occur only on the plateau, and that they are Plinthic Ferralsols because they have massive laterite within 30 cm of the surface. The natural vegetation of the plateau is moist savanna, but the Ferralsols are largely cleared and cultivated. On the long upper slopes of the valleys, Ferralic Cambisols occurs. These are at least 1 m deep over weathered rock, and are the most valuable agricultural soils of the area; they are entirely cultivated. Gleyic Cambisols (Bg) occur on the steeper lower slopes of the valleys, and Pellic Vertisols (Vp) along their gently curving footslopes; Humic Gleysols (Gh) and Eutric Histosols (Oe) characterise the valley bottom swamps, and Dystric Gleysols (Gd) small depressions with impeded drainage on the plateau. None of these last four soils is much cultivated.

The age and composition of bedrock in the Bumala district differ from those of the Chiltern area discussed previously, but the later Cenozoic history of the two is not so dissimilar. Both consist of elevated plateaux that have been subject only to subaerial processes since approximately Miocene times, and it is on these surfaces that the most mature soils of either area occur. In both areas a sequence of less mature soils has formed under various hydrological conditions within valleys dissecting the pleateaux; both valley systems could have originated as long ago as the late Tertiary, but have been much modified by Quaternary erosion and deposition, and have similar dimensions.

Some of the most immature valley soils in the Bumala district (Gleyic Cambisols, Dystric and Humic Gleysols and Eutric Histosols) are similar to some common British soils, but the remainder have no counterpart in Britain, either because they contain more iron and aluminium oxides and kaolinite or because they are composed mainly of smectitic clay. As the periods of time available for formation of the respective plateau and valley soils in the Bumala and Chiltern areas are approximately the same, other factors are probably responsible for these differences.

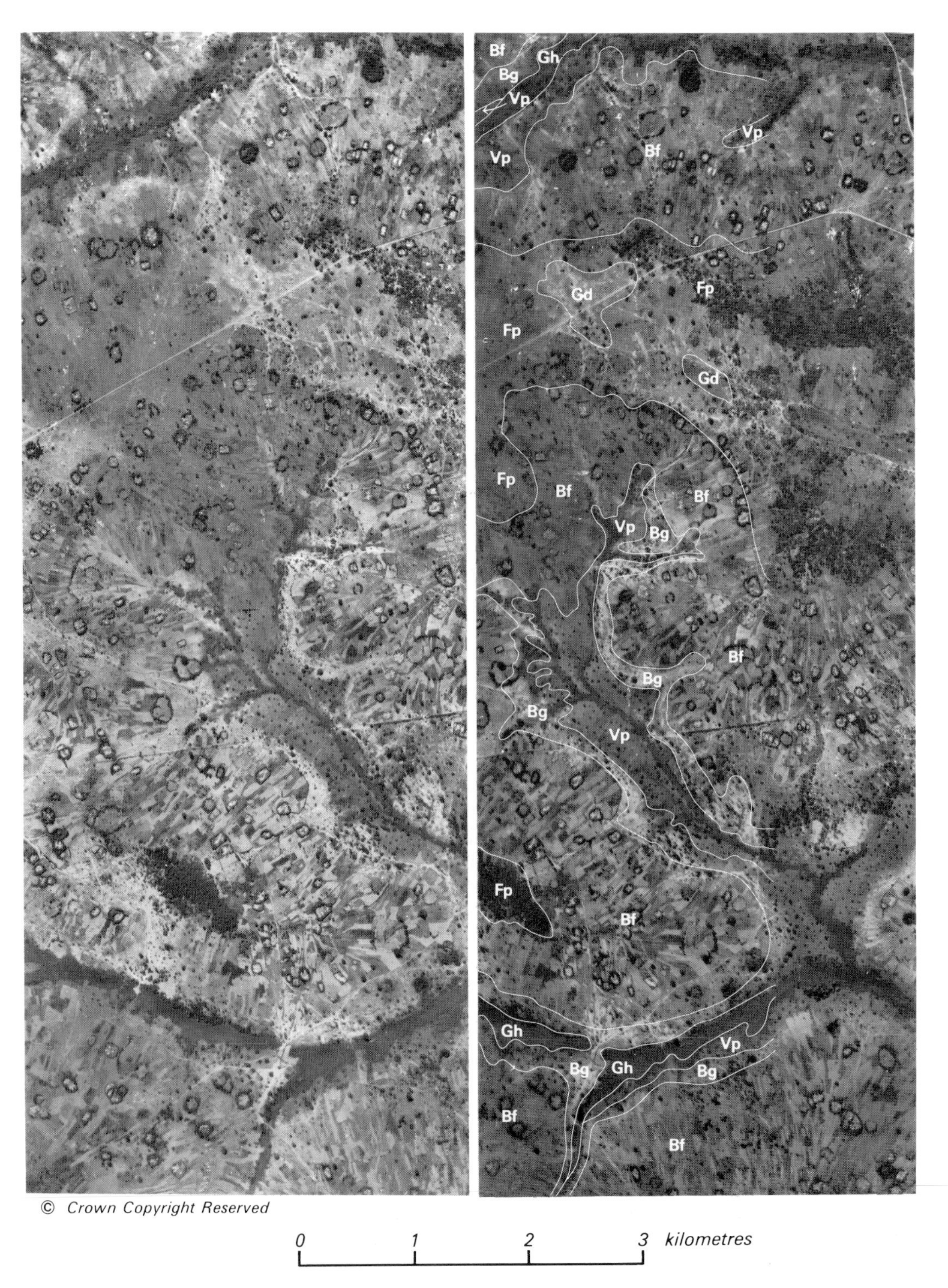

Figure 5.6 Detailed soil map and aerial photograph (stereo-pair) of 32 km^2 of land near Bumala, Kenya, 26 km north-east of Port Victoria, to show relationship of soil types to topographic features (based on Scott *et al.*, 1971).

The strongly calcareous nature of the main parent material (chalk) in the Chilterns is responsible for the occurrence there of Calcaric Regosols and Rendzinas, but in other respects parent material differences probably account for few of the soil differences between the two areas.

However there are significant differences in climate. Present annual rainfall in the Chilterns is 750 mm, spread fairly evenly throughout the year; in Bumala about twice as much rain falls, with a peak of 400 mm in April and May and a dry period in January and February. Average monthly temperatures in the Chilterns range from 3·1 °C (January) to 16·2 °C (July), and frost is possible for six months of the year (November–April); Bumala is frost-free throughout the year, and has average temperatures approximately 15 °C higher than the Chilterns in the winter and 10 °C higher in the summer. Chemical weathering is therefore approximately as fast in the winter at Bumala as in the summer in southern Britain, and in the summer it is probably twice as fast as in the British summer, as a 10 °C rise in temperature at least doubles the rate of many chemical weathering reactions (Young, 1976, p. 64). Leaching is also much more rapid at Bumala, at least during the summer. These present climatic differences are important, but those of cold periods earlier in the Quaternary are probably more so, as the rate of chemical weathering in Britain would then have been much less, whereas in central Africa it would have changed little, but the increased rainfall of pluvial periods would have intensified the rate of leaching in Africa.

The African soils have therefore been subject to much more rapid losses of silica, alkalis and alkaline earths than those of southern Britain. As a result the plateau and older valley soils have been greatly enriched in residual iron oxides, and part of the elements removed in solution have accumulated in poorly drained sites, such as valley bottoms, where they have caused the formation of smectite-rich cracking clays.

Soil formation and distribution in past geological periods

Although it is now realised that many features of present surface soils, especially in temperate regions, are inherited from past periods, either warmer or colder than the present, the exact age of such features is often difficult to determine. Detailed profile studies, particularly those using thin sections, often indicate cyclic changes in the development of temperate soils, such as alternating periods of clay illuviation and disruption of the argillans by cryoturbation, which clearly result from Quaternary climatic oscillations, but the dating of events is usually only relative. More precise dating is possible where buried soils occur within datable sequences of unaltered deposits, such as volcanic rocks, glacial tills, river terrace gravels, or aeolian sediments. This is especially useful if surface soils with relict features can be traced laterally into their buried equivalents. It has been shown in various parts of Europe and north America that clay illuviation and rubification were the main soil-forming processes in well-drained sites during many successive warm (interglacial) periods of the Quaternary. Probably the most complete sequence of Quaternary geosols, at least for the last 800 000 years, is that preserved in the loess deposits of Austria and Czechoslovakia. These correlate well with warm phases over the same period as indicated by isotopic and micropalaeontological studies of deep oceanic cores (Fig. 5.7); the warmest interglacials resulted in formation of Orthic or Chromic Luvisols, and slightly cooler periods are represented by Gleysols, Chernozems and Cambisols, whereas the coldest phases equate with chemically unaltered loess (Calcaric Regosols), which is often frost disturbed (Gelic Regosols).

In tropical regions not subject to the later Quaternary periglacial and glacial erosion that probably erased much of the evidence for earlier pedogenesis from higher latitudes, the oldest soils (mainly Ferralsols) are thought to have originated as long ago as the middle or early Tertiary. However, this dating is only tentative, often being based on the inferred ages of the high level erosion surfaces to which the soils are confined. These could easily have suffered further erosion and lowering over such a long period, making the soils no older than perhaps early Quaternary. Alternatively, in many places old weathered material has moved downslope onto younger erosion surfaces, thereby confusing further the relationship between surface and soil type. Silcretes (e.g. the sarsen stones of the English chalk downland), deep bedrock alteration, and minerals normally resulting from tropical weathering (e.g. gibbsite, haematite and kaolinite) are also often attributed to Tertiary weathering in many temperate parts of the world. However, the pedogenetic origin of many examples is questionable: silica cementation is common as a diagenetic process in deeply buried sediments, most examples of extensive bedrock alteration are in igneous rocks and could result from hydrothermal activity, and local epigenetic mineralisation of soil parent materials often accounts for the supposed tropical weathering products.

For direct evidence of soil development processes and the distribution of soil types in the early Quaternary and older periods, one must turn to the few examples of buried soil in these lower parts of the stratigraphical column. Peaty organic horizons, representing ancient Histosols, occur quite widely (e.g. in deltaic deposits such as the early Tertiary Woolwich and Reading Beds of south-east England), but profiles showing pedogenetic alteration of inorganic materials are much rarer. Those most satisfactorily identified are in essentially terrestrial sequences, such as volcanic deposits, coal measures or redbeds, but opinions are divided even about many of these. For example the pale grey 'fireclays', 'seatearths' or 'underclays', that often underlie coal seams, commonly contain fossilised roots in growth position, and show a fairly consistent sequence of horizons, in which illite, chlorite and felspars often give

Figure 5.7 Correlation of geosols in the Quaternary loess succession of Austria and Czechoslovakia (right) with palaeotemperature curve based on oxygen isotope changes (relative to P.D.B., a standard belemnite from the Cretaceous Pee Dee Formation, S. Carolina) in Equatorial Pacific core V28–238 (left). Redrawn from Morrison (1978), and based on Shackleton and Opdyke (1973) and Kukla (1975).

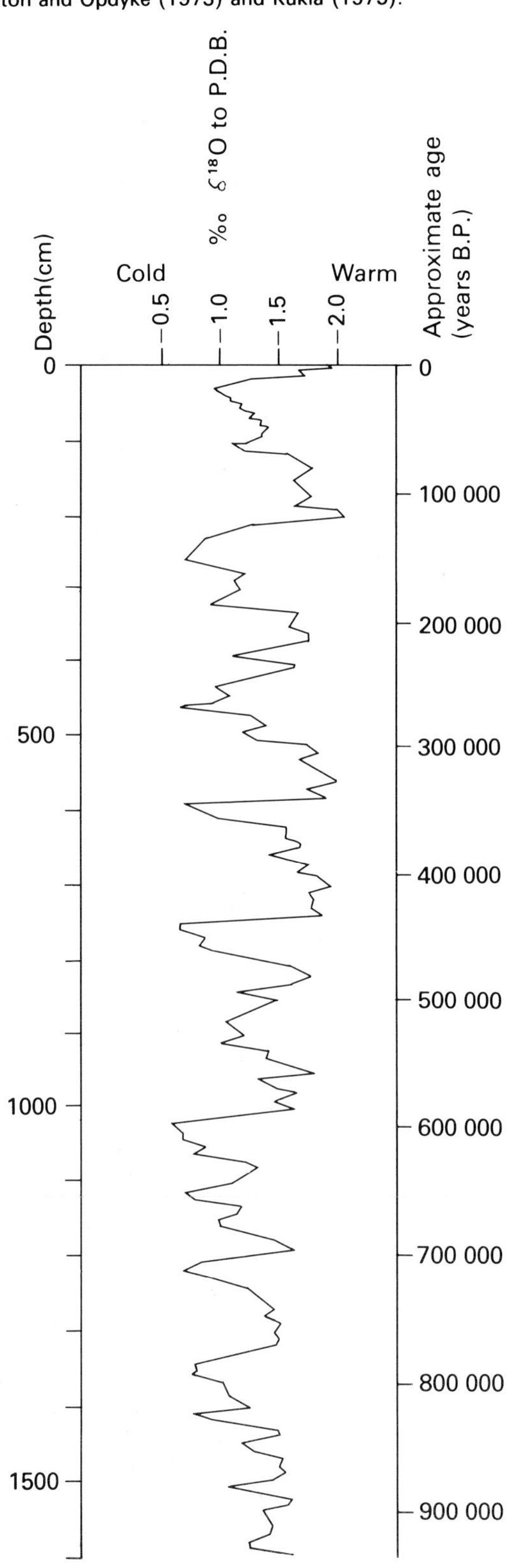

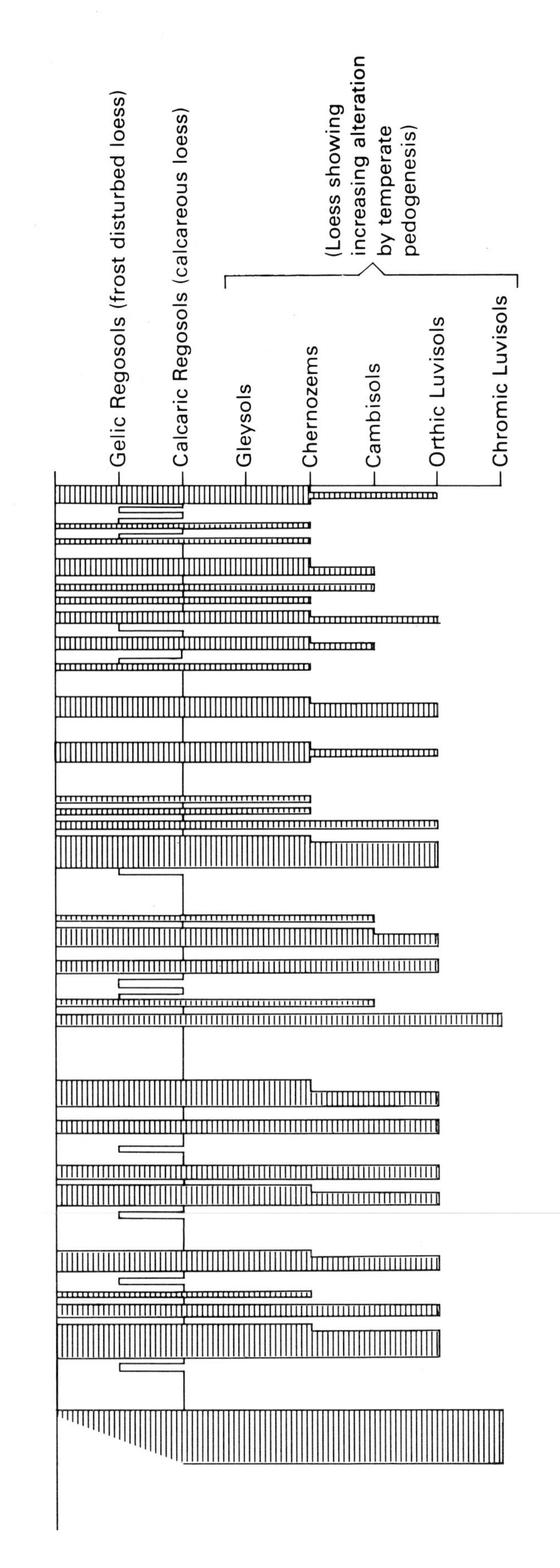

way upwards to increasing amounts of kaolinite. These features are regarded by some workers as indicating development *in situ* of profiles resembling Humic Gleysols, the thick histic (peaty) horizon of which was compressed into coal after burial. However, other scientists relate them to the characteristics of nearby soil that supplied sediment to the coalswamps, and to cyclic changes in the conditions of swamp sedimentation (Wilson, 1965). Less disputed examples of *in situ* older buried soils occur in volcanic successions, where their relationship to temporary land surfaces, such as the top of a lava flow, can often be demonstrated. The early Tertiary 'bauxites' and 'inter-basaltic laterites' of northern Ireland and 'red boles' of the Inner Hebrides and other areas all probably originated as subsoil horizons of Nitosols or Ferralsols. The full soil profile is rarely preserved, but associated terrestrial plant remains often indicate a sub-tropical climate at the time of formation. An example of possible buried palaeosols in the Devonian redbeds of Wales is provided by the 'Cornstones', which resemble recent arid and semi-arid soils enriched in redeposited carbonate (the Calcic Yermosols and Xerosols); they occur in red clays often showing deformations similar to the patterned ground (gilgai) of Vertisols in regions that suffer pronounced dry seasons (Allen, 1973).

The main problems of reconstructing soil conditions in the remote past result from the ease with which soft and friable soil profiles are eroded or changed by diagenesis or metamorphism after burial. Also successions resembling buried soil profiles can be formed either by original lithological differences between deposits, or by deep subsurface geological processes (e.g. intrastratal translocation of iron). As a result there is very little reliable direct evidence of soil types in older geological periods.

Less direct evidence for the nature of soils in older geological periods is obtainable from many sediments because these are often derived more from soils than from unweathered rocks. However, many of the features used to classify soils seldom survive transportation in water and other media, and even if they do, are subject to subsequent diagenetic changes. Consequently, this type of indirect evidence of bulk soil composition usually provides at best an approximate indication of climate (e.g. the seasonally humid, warm–temperate to tropical conditions indicated by the abundance of iron oxides, presumably from Ferralsols, in redbed formations) and generalised information of parent materials, as indicated by the composition of the more resistant detrital components, such as heavy minerals. Sediments deposited close to the source of their detrital components, and transported by mass movement rather than by disaggregation in fluid media, obviously provide the best information of soil characteristics, but these are unfortunately rare components of the stratigraphical column.

The clay minerals probably provide the most useful evidence from sediments of soil conditions in the provenance areas, because their bulk composition in soils is often determined by climatic and hydrological factors. However, changes during and after deposition must be taken into account. In the marine environment these often result in partial or complete losses of smectite, vermiculite and interstratified clays, and increases in mica, glauconite and chlorite. Chemical reactions with sea water and diagenetic changes reverse some of the weathering processes that occurred earlier in the soils, and differential sedimentation rates of individual minerals may also affect the composition of the clay mineral assemblages. Diagenetic changes in layer silicate clays are greater in limestones and coarser sediments, such as greywackes, than in the purely argillaceous deposits, but strongly reducing conditions in clay-rich sediments often profoundly modify other fine components, such as iron oxides, which could provide important information about the soils of provenance areas. In addition, there is the possibility of crystallisation (neoformation) of new clays, such as sepiolite, smectite and attapulgite, from solutions during and after deposition. Despite these limitations, detailed studies of the vertical and horizontal distribution of clays within basins of sedimentation provide much information about soil types of surrounding land areas. In west Africa a dry tropical climate probably with widespread Xerosols and Yermosols during the late Cretaceous seems to have given way in the early Tertiary to a seasonally humid climate with abundant Ferralsols; this change is reflected in the marine sedimentary clays of the area, which consist of detrital illite, smectite and kaolinite in the upper Cretaceous, but contain more kaolinite, iron minerals and neoformed species such as attapulgite (indicating greater leaching of the soil) in the Palaeocene and Eocene. Further examples are discussed by Millot (1970).

More hypothetical reconstructions of soil-forming conditions in the remote past can be indirectly inferred from various lines of evidence, using knowledge of present-day relationships between climate, relief and soil types. From the end of the Permo-Carboniferous glaciations to the later Tertiary, warm climates affected a much larger proportion of the earth's surface than at present (see also Chapter 9). The somewhat diminished land areas throughout this period would have been dominated by soil types similar to those of tropical and sub-tropical regions today – Ferralsols, Acrisols, Nitosols and Luvisols in the more humid areas, and Yermosols, Xerosols, Vertisols and Arenosols in arid and semi-arid regions. Because of the widespread erosion and deposition of new materials during the Quaternary glaciations, a range of immature soils, such as Lithosols, Regosols, Rankers, Rendzinas, Podzols, Greyzems, Chernozems, Kastanozems, Luvisols, Podzoluvisols and Cambisols, are now more widespread than at any time since the Permian, and probably even before then. Some properties used in soil classification are closely linked to the vegetation the soil supports; these include the organic content, colour and structure (aggregation) of surface horizons, and the extent of acid leaching by organic decomposition products. A few modern soil characters therefore may only have evolved since the spread of flowering plants, grasses and broad-leaved trees in the Cretaceous and early Tertiary periods. They may include the well-structured mollic and umbric surface horizons now typical of soils under the more open vegetation of savanna, steppe and temperate grasslands.

During the Permo-Carboniferous glaciations, the polar and high-latitude regions (of those times) would have been dominated by Regosols and Lithosols, probably with subsidiary Histosols, Gleysols and Podzols, as are comparable areas today. Although Gleysols probably characterised lowland humid tropical areas during the Carboniferous, there is some sedimentary evidence that Ferralsols were widespread on intervening uplands. However, the succeeding Permo-Triassic and preceding Devonian periods probably experienced widespread warm-temperate and only seasonally humid climates, with rapid erosion accompanying mountain-building periods; chemical weathering processes were probably less important and Ferralsols less common than in the Carboniferous. Forests (with Gleysols, Luvisols and Acrisols) were smaller, and extensive areas of scrub or other open vegetation would have been dominated by Arenosols, Regosols, Lithosols and Xerosols, with subsidiary Vertisols, Cambisols, Solonchaks and Solonetz.

Before the appearance of land plants in the late Silurian (approximately 420 million years ago), soils would undoubtedly have been very different from those of any time since. Oxygen levels in the atmosphere were lower, especially in the Precambrian, and weathering processes involving oxidation were much more localised. However, the clay fractions of sediments from these periods suggest that other chemical weathering processes, such as carbonation and hydrolysis (e.g. formation of kaolinite from feldspars), were active, which is consistent with an early atmosphere dominated by juvenile volcanic volatiles, including water vapour, carbon dioxide and carbon monoxide. The small amounts of oxygen produced by primitive photosynthesisers in the hydrosphere or by photodissociation of atmospheric water were rapidly fixed as haematite in soils and surface waters containing abundant iron in the ferrous form. The haematite eroded from these soils or precipitated directly in the sea (possibly by bacterial activity) was probably redeposited as the banded ironstones of the earlier Precambrian (before approximately 2000 million years ago) and as the redbeds of the later Precambrian and early Palaeozoic.

Physical weathering processes would have dominated pre-Devonian soil formation, and without land plants and animals the soils would have been devoid of organic matter, unstructured, often very impermeable, and prone to rapid erosion. Even quite gentle slopes would consequently have been dominated by shallow Lithosols, valley floors and footslopes covered by thick, weakly stratified colluvial accumulations, and poorly drained plateaux covered by grey, heterogeneous regolith, possibly with red haematitic mottles in the surface layer. The nearest modern equivalent of the valley floor soils would be the Gleyic Regosols (a type recognised in the FAO classification, but not actually used on the world map), and the nearest equivalent of the plateau soils would probably be the Gleysols. However, in the absence of both organic matter and a strongly oxidising atmosphere, strict correlation with present-day soils (other than Lithosols) is impossible.

Conclusions

Many people concerned with soil would probably feel that the Precambrian and early Palaeozoic are so remote that materials affected by the different atmospheric and land surface conditions then prevalent may be disregarded as soils. However, they are included in the definition given at the beginning of this chapter, and in the context of Earth history as a whole were not unusual, because they probably persisted for a much longer time than the later soils which have developed under vegetation and a more oxidising atmosphere. It is also probable that the chemical and physical constitution of the pre-Devonian soils played an important role in the early evolution of terrestrial life, so palaeontologists should not ignore the processes likely to have been involved in their development.

Soil conditions in subsequent geological periods might have been important in later evolutionary changes, though cause and effect are difficult to disentangle. Compared with earlier periods, the palaeo-pedologist can offer more evidence, direct and indirect, of soil conditions and distribution from the Devonian onwards, because processes and products were much more like those of the present day. However, the recognition and interpretation of buried soils *in situ* and of transported soil materials in sedimentary rocks is a very under-explored field of research. It is only in the later Quaternary of some intensively studied parts of the world, such as North America and Europe, that the distribution of soil types during past periods can be reconstructed with any certainty. In addition the complex relationship between present soil types and natural vegetational variation is understood only imperfectly, so, even if more were known about earlier soil distributions, it would be difficult to relate them satisfactorily to biological changes in space and time. Palaeontologists wishing to relate evolutionary and other changes in plant and animal communities to pedological factors through geological time therefore face many problems which at present are far from being resolved.

Although soils lie at a multi-disciplinary crossroads, they have not been studied sufficiently for their own sake, and consequently have lacked universally accepted study procedures and terminology, which might have stimulated more profitable work and exchange of ideas. Despite some limitations, the FAO World Soil Map now provides a comprehensive basic terminology, which is recommended as an introduction to the subject for scientists of other disciplines and as the best medium at present for world-wide communication. More detailed general accounts of soil classifications (including the FAO system) and of present-day soil-forming factors and processes are given by FitzPatrick (1971), Buol *et al.* (1973), Young (1976), Bridges (1978) and Greenland & Hayes (1978).

As a medium for crop growth, soil is the most important of the earth's natural resources, but it seems to be the one that is most taken for granted, and the least researched and understood. Unlike most natural resources, soil is not necessarily expendable. Treated correctly, it can yield abundantly once, twice or even more times a year. However, careless treatment or greedy exploitation can rapidly result in long-term degradation by structural deterioration and erosion. The maintenance of humus and nutrient levels, of the living soil fauna, and of good soil structure for aeration and drainage, are all vital to food production and thus to man's future on earth. These factors are variously conditioned by the soil's parent material, natural development, history and past treatment by man, by natural vegetation or cultivated crops, by present (geographical) and past (geological and historical) climatic variation, and by physical site circumstances such as height, slope and drainage characteristics. Aspects of knowledge from many diverse and apparently unrelated disciplines are therefore likely to be as important in ensuring a continuing supply of man's most basic need as the most applied research in plant genetics, crop physiology, fertiliser application, or pest and disease control.

Acknowledgements

We thank the Food and Agriculture Organisation of the United Nations for permission to reproduce maps and other data in Figures 5.1, 5.2 and 5.5, and Table 1, the Director of Surveys, Kenya and the Copyright Section of the Ministry of Defence for permission to reproduce aerial photographs CPE/KEN/163(1) used in Figure 5.6.

References

Allen, J. R. L. 1973. Compressional structures (patterned ground) in Devonian pedogenic limestones. *Nature Physical Science*, London, **243**: 84–86.

Avery, B. W. 1964. The soils and land use of the district around Aylesbury and Hemel Hempstead. *Memoirs of the Soil Survey of Great Britain, England and Wales.* 216 pp.

Avery, B. W. 1973. Soil classification in the Soil Survey of England and Wales. *Journal of Soil Science* **24**: 324–338.

Bridges, E. M. 1978. *World soils* (2nd edition), 128 pp. Cambridge: University Press.

Buol, S. W., Hole, F. D. & McCracken, R. J. 1973. *Soil genesis and classification.* 360 pp. Ames, USA: Iowa State University Press.

F.A.O. 1974. *Soil map of the world 1 : 5 000 000.* 10 volumes. Paris: UNESCO.

FitzPatrick, E. A. 1971. *Pedology. A systematic approach to soil science.* 306 pp. Edinburgh: Oliver and Boyd.

Greenland, D. J. & Hayes, M. H. B. 1978. *The chemistry of soil constituents.* 469 pp. Chichester: J. Wiley and Sons.

Kukla, J. 1975. Loess stratigraphy of central Europe, pp. 99–188 *in* Butzer, K. & Isaac, G. L. (Eds.) *After the Australopithecines.* The Hague: Mouton.

Millot, G. 1970. *Geology of clays.* 429 pp. New York: Springer Verlag.

Morrison, R. B. 1978. Quaternary soil stratigraphy – concepts, methods, and problems, pp. 77–108 *in* Mahaney, W. C. (Ed.) *Quaternary Soils* 508 pp. Norwich: Geo Abstracts.

Scott, R. M., Webster, R. & Lawrance, C. J. 1971. *A land system atlas of western Kenya.* 363 pp. Military Vehicles and Engineering Establishment, Christchurch, England.

Shackleton, N. J. & Opdyke, N. D. 1973. Oxygen isotope and palaeomagnetic stratigraphy of equatorial Pacific core V28–238: oxygen isotope temperatures and ice volumes on a 10^5 and 10^6 year scale. *Quaternary Research* **3**: 39–55.

Soil Survey Staff. 1975. *Soil taxonomy. A basic system of soil classification for making and interpreting soil surveys.* 754 pp. US Department of Agriculture Handbook 436.

Wilson, M. J. 1965. The origin and geological significance of the South Wales underclays. *Journal of Sedimentary Petrology* **35**: 91–99.

Young, A. 1976. *Tropical soils and soil survey.* 468 pp. Cambridge: University Press.

J. A. Catt
A. H. Weir
Soils and Plant Nutrition Department
Rothamsted Experimental Station,
Harpenden
Herts AL5 2JQ

PART III

Water and climate

Introduction

The histories of the oceans and atmosphere are crucial to the story of the evolution of our Earth, and, without them, life as we know it could not have existed. Over two-thirds of the Earth's surface is covered by water, most of which is sea water, and the Earth is enclosed within an atmosphere that provides life-support systems and protection from destructive ultra-violet radiation. Together the seas and atmosphere recycle the Earth's chemicals and act as vast reservoirs for the raw materials of life. In this Section our knowledge of the Earth's oceans and atmosphere is reviewed, as are the ways in which they have changed during geological time; their complex chemical interrelationships are also discussed.

The cyclicity of chemical compounds is a recurring theme. Geological processes of weathering, transport and sedimentation continue at the Earth's surface, largely through the medium of the atmosphere, and add both solid and dissolved matter to the seas. At the same time, gases in the atmosphere dissolve in the upper waters of the oceans and are carried to greater depths by water currents and by the activities of marine life. Similar but reverse processes add gases to the atmosphere, and important sources of elements to the oceans and atmosphere are the submarine volcanoes and gaseous releases from the ocean beds, particularly along the mid-oceanic ridges.

Since the beginning of the Deep Sea Drilling Project in 1968, after the abandonment of the ambitious Mohole project to drill through the oceanic crust to the mantle, we have learned a great deal about ocean sediments and the ocean floors, as well as the mechanisms and history of plate tectonics. One product of this intensified study is the realisation that manganese nodules are widespread on the ocean floors (although they have been recorded sporadically for over a century). The origin of these nodules remains one of many unanswered questions, but, if harvested, these nodules could supply us with important reserves of manganese and other trace elements, and already various collecting methods have been proposed.

Although the atmosphere is only a trivial amount of the

total mass of the Earth, its importance during the 4500 million years of the Earth's existence is, and has been, immense. Not only does the atmosphere control many of the weathering processes at the Earth's surface, but it contains gases vital to life. Our Earth appears unique in its high proportion of free oxygen, upon which most organisms rely to oxidise the raw materials from which they derive their energy. However, it appears that an oxygen-rich atmosphere evolved only at about 2200 million years ago. Before then the atmosphere lacked free oxygen, but the degree to which it was of a reducing nature remains in question. Did it consist of gases such as methane, ammonia and hydrogen, or was it in a more oxidised state containing carbon dioxide, nitrogen and water vapour?

These questions impinge on the problem of the origin of life itself, for it has been demonstrated that the activation of methane and ammonia produces simple compounds capable of being built into the complex molecules found in living organisms. A new approach to the study of our atmosphere has come in the last decade through studies of the Moon and certain planets. The Moon's surface is revealed as stark and devoid of soil, water, life or atmosphere, in contrast to the Earth's moist surface. The gases of the planets differ from one another, and from our own atmosphere, allowing deductions to be made about their origins.

Climate is dependent on parameters other than the composition of the atmosphere. Two important and related indicators of climatic conditions are glaciations and the ratios of oxygen isotopes found in oxygen-bearing compounds in the geological record. Analysis of the latter allows the calculation of actual temperatures at the time the compounds were formed. Although this study has been restricted largely to Cretaceous and later examples, the stability of the isotopes could allow the technique to be used far back into the Phanerozoic, providing the primary nature of the materials studied is assured. Glaciations are evocative and enigmatic. Evidence of their past existence can be found in rocks formed as long as 2200 million years ago, that is at much the same time as the atmosphere first became oxygenated. From then onwards, glaciers have periodically spread over large areas, influencing the climate, depth and extent of oceans, and all forms of life.

How will Man's activities affect our seas and skies? Have we seen the last of the Pleistocene ice ages, or are we now enjoying an interglacial period, so that future generations will have to endure advancing ice sheets? As knowledge increases, we are presented with new vistas, but also with new questions. Mankind has quantified numerous properties of the Earth, and described its motion in space, but we wonder at the possible relationships between phenomena on the Sun, on the one hand, and periodicities in the Earth's activities and in the life inhabiting it, on the other.

C. H. C. Brunton
British Museum (Natural History)

CHAPTER 6

Sea water and its evolution

J. D. Burton and D. Wright

The total volume of ocean waters is $1{\cdot}35 \times 10^{21}$ litres, and the oceans have a mean depth of about 3700 m. Sea water accounts for more than 98 per cent of the mass of the hydrosphere, which is the term used to describe the free water at the Earth's surface, and covers just over 70 per cent of the globe. The ocean has a central role in the major sedimentary cycle and its composition is closely linked to those of other reservoirs, particularly the atmosphere. Because of this role, the stability of the composition of the present-day ocean, and how it is controlled, is of great importance in relation to climate. Much attention has been given recently to the effects of pollution on the ocean. Man's intervention, often amounting to an acceleration of existing geochemical processes, is too recent to have generally affected more than the most marginal estuarine and coastal regions, but in a few instances the consequences may be of global significance. Such considerations have stimulated interest in the mechanisms that control the composition of sea water and have added impetus to our fundamental curiosity as to the ways in which that composition has evolved over geological time.

By sampling and analysing samples of sea water from different parts of the ocean a picture has been built up of its chemical composition. The acquisition of such information was the earliest task of marine chemists and, although many details are still unknown, we know the main features of the composition of the present-day ocean waters. The ocean is a geochemically open system which continuously exchanges material on a wide range of time-scales with the atmosphere, the sediments and crustal material at the sea bed, and the continental crust. Recent work has accordingly placed increasing emphasis on the dynamics of the cycling of material through the ocean reservoir. As there are no preserved records to provide direct evidence about the composition of sea water in earlier geological periods, our knowledge of the ways in which the present-day ocean may have evolved is based on indirect approaches.

The nearest thing to a direct record is provided by the chemical, isotopic and palaeontological characteristics of preserved ocean sediments, enabling some deductions to be made as to the nature of their parent solution, and this approach has the advantage that the material examined can often be absolutely dated. However, the evidence is mostly confined to the last 600 million years. Moreover, the relationships between the characteristics of sea water and features preserved in sediments are often complex, and it is often difficult to attribute changes in sediments to a single factor influencing the water column. Interpretations are complicated by the occurrence of changes in the characteristics of sediments following deposition, for example by exchanges of material between the sediment particles and the pore waters trapped in the accumulating sediments.

Another major approach has been through considerations of chemical equilibria which enable limits to be placed on the composition of an ocean under specific conditions. A readily appreciated qualitative example of this is that a variety of reduced chemical forms, such as iron(II) and hydrogen sulphide, are chemically unstable in the presence of free oxygen, and therefore cannot have played a significant part in the chemistry of open ocean water once the atmosphere contained oxygen in excess of the amounts necessary to react with such reducing substances. Quantitative calculations have provided a broad guide to the chemical conditions in an ocean coexisting with other reservoirs, such as the atmosphere and marine sediments, during the probable stages of evolution of the Earth as a whole.

A third major approach has been through considerations of the geochemical balance between processes of input and removal in the ocean. The identification of the sources and sinks and the rates of turnover of chemical species in the present ocean, and the estimation of the mass balance of material between sources and sinks over geological time, provide a framework to concepts of how the ocean has probably evolved.

This account begins with a brief outline of the present-day composition of sea water and the processes which determine it, and is followed by an account of the possible origin and evolution of the ocean. Finally, brief consideration is given to changes in oceanic circulation and to the influence on the sea of Man's recent activities.

The composition of sea water at the present day

Sea water contains a variety of dissolved substances and suspended particles of diverse origins. While the particulates play an important role in the circulation of material in the ocean, the total dissolved material is at much higher concentrations. As an index of the bulk content of dissolved material, oceanographers use salinity, a defined quantity which approximates to the total content of salts and is expressed in parts per mille (‰). The overall range in salinity from estuaries to tropical basins is < 1 to > 40 parts per mille, but the usual range for ocean waters is 33–37 parts per mille and three-quarters of all ocean waters have values between 34 and 35 parts per mille.

A few dissolved constituents dominate the composition of sea water. The concentrations of these major constituents, usually defined as those which occur at a concentration above 1 mg per litre, and their average ratios to salinity are shown in Table 1. These elements show rather constant relationships to salinity and in some cases variations in the ratio of concentration to salinity are within the precision of measurement. Sea water also contains dissolved silicon at an average concentration exceeding 1 mg per litre, but its distribution is variable. This element, with combined nitrogen (mainly as nitrate) and phosphorus, are the principal micro-nutrients for primary organic production in the ocean. Sea water also contains most other elements at low concentrations, many being in the range of 1 ng to 1 μg per litre, as well as a great variety of dissolved organic substances.

Table 1 Major dissolved constituents of sea water*

Constituent	*Concentration* (g kg^{-1}) at S = 35‰	*Ratio of concentration* (g kg^{-1}) to chlorinity (‰)†
Na^+	10·765	0·5557
Mg^{2+}	1·294	0·0668
Ca^{2+}	0·412	0·0212
K^+	0·399	0·0206
Sr^{2+}	0·0079	0·00041
Cl^-	19·353	0·9989
SO_4^{2-}	2·712	0·1400
HCO_3^-	0·142	0·00735
Br^-	0·0674	0·00348
F^-	0·0013	0·000067
H_3BO_3	0·0256	0·00132

* Values are taken or derived from Millero (1974) and refer to the total concentration of each constituent; those for HCO_3^- include the CO_3^{2-} present.

† Chlorinity is a defined quantity approximating to the chloride content.

The chemical nature of sea water is greatly influenced by the presence of dissolved gases. The non-variable gases in the atmosphere, such as nitrogen, oxygen and argon, attain approximately saturation concentrations in the surface ocean under the prevailing conditions of salinity and temperature, and for chemically unreactive gases these concentrations are conserved during mixing in the ocean. Oxygen, a reactive gas, is particularly important for the chemical and biochemical processes in the ocean. Away from the near-surface zone in which photosynthesis occurs, there is no production of oxygen but consumption of the gas occurs in the decomposition of organic material, mainly by respiratory activity, in particular of micro-organisms. This leads to a reduction in concentration, but the supply is exhausted in only a few atypical environments. The presence of free oxygen gives sea water an oxidising character, so that many other elements occur in their higher oxidation states. Biological activity opposes this tendency, so that a number of elements are continuously cycled from reduced to oxidised forms in the processes of production and decomposition of organic matter. Where oxygen is totally depleted, as in the deeper waters of the Black Sea, microorganisms use nitrate and sulphate as oxygen sources. Such environments are characterised by the presence of sulphide and are populated only by a specialised microbiota; in these circumstances more organic matter tends to accumulate in bottom deposits than in normal marine sediments.

Another biologically important gas, carbon dioxide, is present in sea water as dissolved gas molecules and also reacts to give combined forms, carbonic acid, bicarbonate and carbonate ions, of which bicarbonate is the dominant form. However, the bicarbonate and carbonate in the ocean are derived principally from the solutions formed by continental rock weathering processes and carried to the ocean in rivers. Whereas the ocean contains a relatively small fraction of the Earth's inventory of oxygen, the combined inorganic carbon in the ocean represents the major reservoir of carbon dioxide in the exchangeable carbon system, i.e. the material which exchanges with free carbon dioxide on a relatively short time-scale. The presence of the various forms of inorganic carbon in sea water gives a short-term buffering effect on the pH, although the pH may be determined ultimately by equilibria involving the aluminosilicates in marine sediments. The typical pH of present-day ocean waters is about 8·0 with a characteristic range of 7·8 to 8·4.

The preceding outline of the composition of sea water shows a system in effect frozen at the present instant of geological time. However, the ocean is an open system with continual exchanges of material taking place with other parts of the Earth's surface. The composition which results is a reflection of the nature of the inputs and the dynamic balance between the processes that introduce and remove individual constituents. The major processes

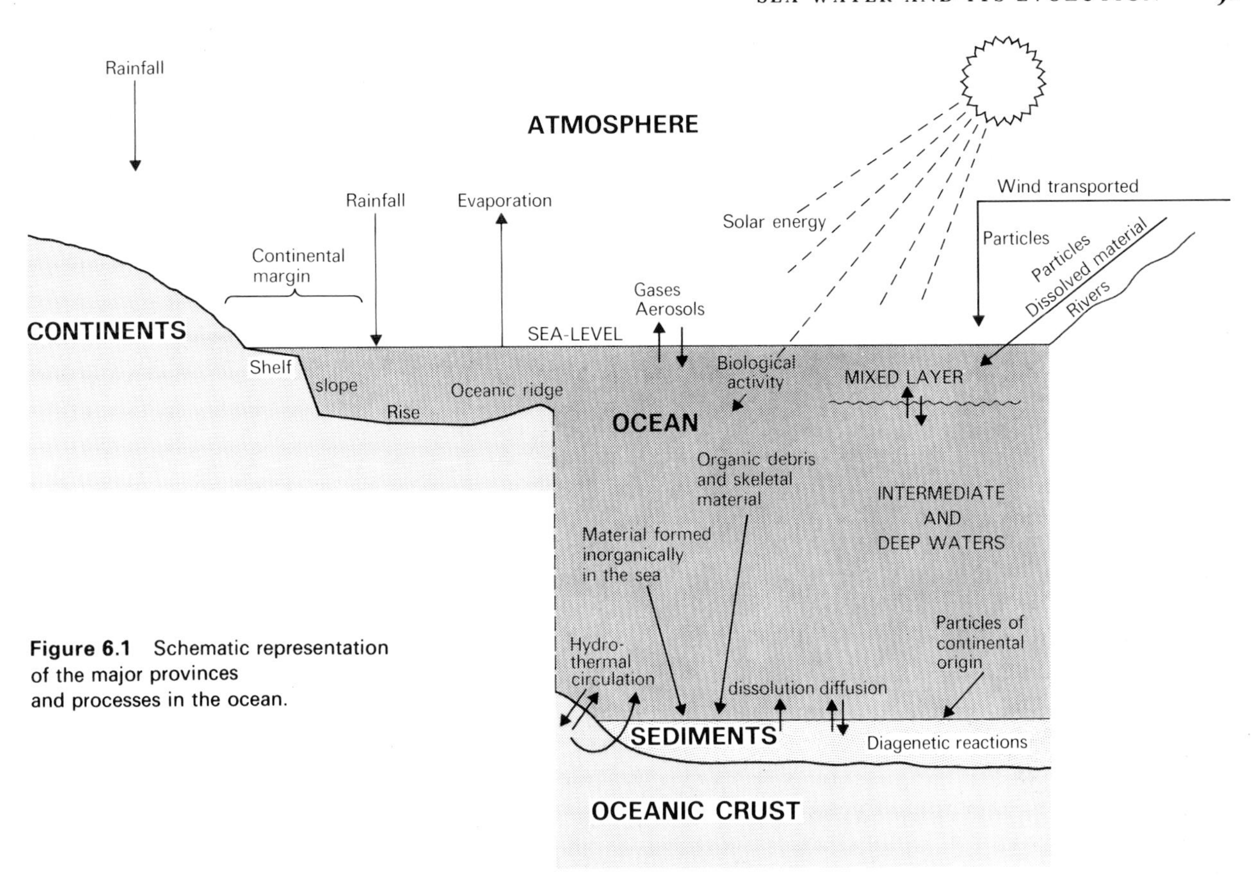

Figure 6.1 Schematic representation of the major provinces and processes in the ocean.

involved in the cycling of material in the ocean are illustrated schematically in Figure 6.1.

The most important processes of input are the weathering and transport of material, mainly by rivers but to some extent by winds, from the continental lithosphere; the transfer of material across the ocean–atmosphere interface, which is the main route of entry for important dissolved gases; and reactions which occur at the sea bed. Of these latter reactions, the most important are the post-depositional changes which occur in the distribution of material between sediment particles and the surrounding pore waters, which can lead to diffusion of dissolved material into the overlying sea water, and the interaction of sea water with oceanic crustal material in tectonically active regions as a result of extensive hydrothermal circulation within the crust.

The removal of material from the ocean occurs mainly by the sedimentation of particles and the incorporation of dissolved substances into solid phases formed in the water column or on the sea bed (see also Chapter 4). Terrestrially-derived particulate material is dominated by the clay minerals, illite, montmorillonite, kaolinite and chlorite, and by quartz. Much of it is deposited in coastal and shelf regions but fine material is transported to the deep-sea basins. Solid phases formed within the oceans are of two main types, those formed by organisms and those produced by inorganic processes. In many deep-sea sediments biogenous material dominates, particularly in the forms of calcium carbonate and silica (opal) originally secreted to form the exoskeletons of surface-living plankton. The most conspicuous inorganic deposits are the ferromanganese nodules which abound on some parts of the sea floor (see also Chapter 7). The interactions between oceanic crustal rocks and sea water, which involve an exchange of material between the two phases, also lead to the net removal of some elements from the ocean, although for other elements they provide a net source.

The geochemical dynamics of the ocean have been examined in most detail with regard to the fate of the river inputs, since we have better quantitative information on these than on the other sources. In the hydrological cycle, water is circulated from the ocean to the atmosphere and a part of this water returns via the continents, bringing with it dissolved and particulate material introduced from the weathering of rock. Present-day total river inputs have been estimated as $1{\cdot}8 \times 10^{16}$ g per year of particulates and 4×10^{15} g per year of dissolved substances (Goldberg, 1971). In Figure 6.2 a comparison is given of the proportions of the major dissolved constituents in sea water and the global average river water entering the ocean. A marked contrast is evident between the dominance of sodium, magnesium, chloride and sulphate ions in sea water and that of calcium and bicarbonate ions in river water: this contrast is enhanced when it is remembered that a

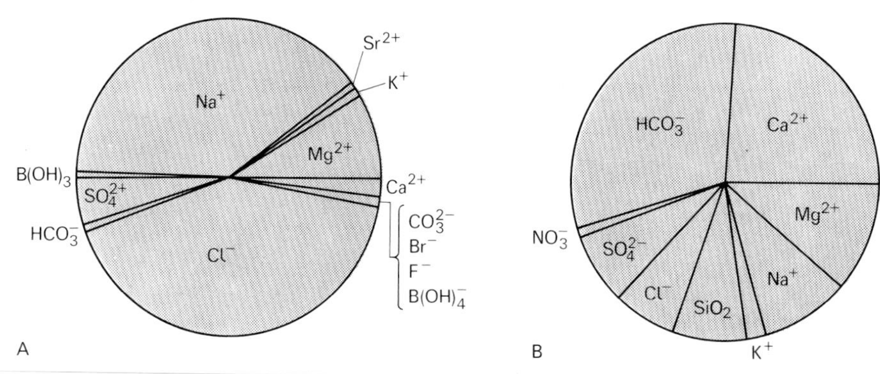

Figure 6.2 The major cations and anions in (A) sea water, (B) the globally averaged river input to the ocean (from Millero, 1975).

proportion of the former ions in river water is derived by the cycling of salt particles carried from the ocean in sea spray to the atmosphere and washed out by rain over land. Sea water is not simply a concentrated version of the river water flowing into the ocean, but has a composition modified by the geochemical reactions occurring in the seas and, as will be shown later, by other input sources.

One way of considering the different geochemical reactivities of the various elements in the ocean is to calculate their mean life-times (residence times) in the oceanic reservoir by dividing the total concentration of the element in the ocean by its rate of input or removal. Relative to river input, these residence times range from high values for elements which tend to form stable ions in solution, such as sodium (10^8 yr), magnesium and potassium (10^7 yr) and calcium (10^6 yr), to values of about 100 years or less for elements such as iron and aluminium, which are rapidly removed in association with particulate material. Elements of long residence time accumulate to concentrations exceeding those in the input solution and because the internal mixing of the ocean is a relatively rapid process they occur in the rather constant proportions already noted. While residence times are to be considered as only approximate indications of turnover times for elements in the ocean, the magnitudes clearly show that most elements have been supplied to the ocean during its history in much greater amounts than those now present.

In considering the geochemical processes in the ocean it is generally assumed, as in the calculation of residence times, that there is a steady state in which the rates of input and removal are approximately equal when considered on an appropriate time-scale. For elements of long residence time such a condition may be achieved despite periods of major imbalance. As will be seen later, the sedimentary deposits of the past 600 million years provide no evidence of a substantial evolutionary change in the ocean during that period, and it is a reasonable assumption that the average gross characteristics of river inputs have been similar to those of present-day rivers over a similar period. On this basis Mackenzie & Garrels (1966) presented a mass balance model in which mechanisms of removal (sinks) were postulated to balance the total input of major dissolved constituents. However it is only in the case of calcium, removed mostly in the deposition of calcium carbonate, that geochemists have been able to demonstrate a fairly close balance between input and removal processes operating at the present time. This may be accounted for by several factors. Some processes, such as deposition of evaporite deposits, occur less extensively now than in previous times. Furthermore, some sinks may not have been identified and the quantitative importance of some removal processes may have been underestimated.

Of particular importance in this last respect is the recent realisation that reactions of sea water with oceanic basalts

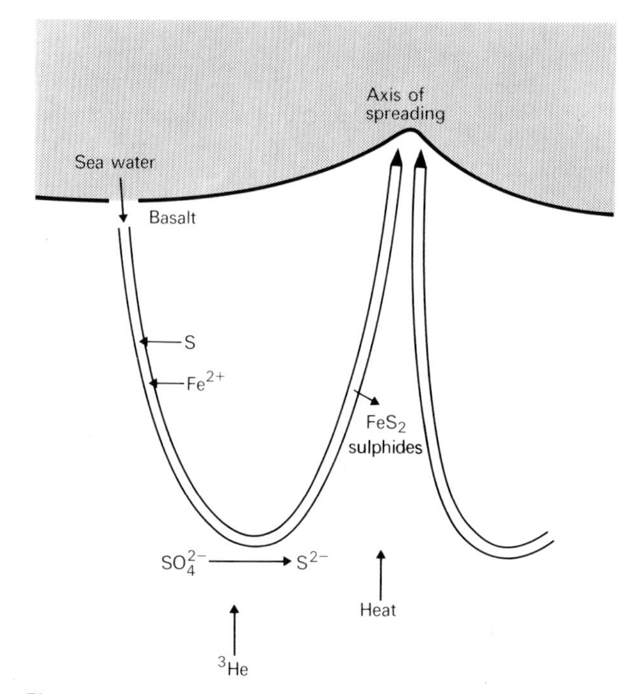

Figure 6.3 Schematic representation of hydrothermal circulation of sea water through the basaltic crust at ridge spreading centres (after Bonatti, 1975).

at ridge spreading centres occur on a sufficiently extensive scale to be important in the budgets of major elements in sea water. Fracturing of fresh crustal material renders it permeable to sea water, which may circulate by convection within the crust, thus leading to reactions under both low and high temperature conditions. This is illustrated schematically in Figure 6.3. Low temperature alteration has been observed in material collected by the Deep Sea Drilling Project, and evidence of substantial high temperature alteration of basalt has been obtained from dredged material (Humphris & Thompson, 1978). On the Galápagos Ridge it has been possible to sample directly deep sea water that has circulated through crustal material and returned to the ocean as a hydrothermal discharge (Corliss *et al.*, 1979). While the scale of interaction remains to be established fully, it seems that the exchanges of material occurring in these processes may be very important in terms of the oceanic budgets of some major dissolved constituents. Such interaction may, for example, remove magnesium and release calcium and silicon on a scale comparable with that of river input. Further work seems likely to alter considerably our perspectives on the detailed processes controlling the composition of sea water, and may significantly alter estimates of the fluxes of elements through the oceanic reservoir. The general arguments presented here regarding the evolution of sea water are not, however, seriously affected by the likely revision of estimates of fluxes and residence times.

This discussion of the characteristics of the ocean has treated the system as essentially a single reservoir. While useful for many purposes, this model does not reflect the existence in the real ocean of various water masses between which mixing occurs over a variety of time-scales. These water masses reflect the major processes of circulation in the ocean. The circulation of surface waters arises initially from the stresses of the major wind systems on the upper water layers and is modified by the Earth's rotation and by topographic barriers. The sub-surface circulation is described as thermohaline since it is driven by the density differences created by surface changes in temperature and salinity. These cause water to sink at high latitudes until it reaches a depth at which it flows horizontally. Of great importance is the stratification over much of the ocean into a near-surface mixed layer comprising the upper 75–100 m, separated by a density boundary – the pycnocline – from the intermediate and deep water masses. Since photosynthesis, and thus the primary production of organic matter, is restricted to the upper layers, the exchanges of material between the mixed layer and the deeper ocean is of major importance in the control of biological activity and the cycle of organic material. This in turn determines many of the non-equilibrium features of the ocean.

The nitrogen cycle provides an excellent example of the importance of such features. A proportion of the nitrogen gas molecules in the atmosphere is converted into combined nitrogen and eventually is transported to the ocean in the form of nitrate. In the ocean, nitrate is taken up by phytoplankton, where it is reduced to ammonia and used to form cellular compounds such as amino-acids and proteins. These reduced forms are largely released back to the water by excretion and by the decomposition of organic remains and they eventually become oxidised to nitrate. If there were no feed-back mechanism in the ocean to convert some of the nitrate to molecular nitrogen, the concentration of nitrate in the ocean would increase at the expense of nitrogen in the atmosphere. Within some 10 million years, the atmospheric reservoir of nitrogen would be completely converted to oceanic nitrate. Recent work (e.g. Codispoti & Richards, 1976) indicates that the major feed-back of nitrogen to the atmosphere is the result of reduction of nitrate to nitrogen (denitrification) in sub-surface zones of limited extent in which the concentrations of oxygen are exceedingly low, but which have not become anoxic in the way that was described above for the Black Sea. Under these sub-oxic conditions, denitrification occurs through the agency of specialised microorganisms which use the nitrate ion as a source of oxygen for the decomposition of organic matter. As a result of its *in situ* production, some nitrogen is eventually released back from the ocean to the atmosphere, thus maintaining its composition. The cycle depends upon the long-term stability of the composition, not of the ocean as a whole, but of major features in its internal structure, that is of the water masses whose identity is maintained by the oceanic circulation processes.

In considering the probable geochemical evolution of the ocean, attention is limited largely to the more abundant of the major dissolved constituents, certain other abundant rock-forming elements such as silicon and iron, and important gases such as oxygen, nitrogen and carbon dioxide. These components play conspicuous roles in marine geochemical cycles, and, by using the approaches already outlined, some reasonable deductions may be drawn as to their concentrations and behaviour under earlier conditions. While the study of less abundant constituents is important for our understanding of the present-day ocean, there is generally little basis on which to speculate about the possible changes in their concentrations over geological time, and it seems probable that little can ever be learned of them.

Origin of the ocean and possible control of its composition by chemical equilibria

The origin and geochemical evolution of the ocean can be considered only within the larger framework of the origin and evolution of the Earth as a whole. It is now generally accepted that the earth was probably formed by the accretion of cold solids, with subsequent differentiation and redistribution of material occurring as a result of the internal heating of the accreted mass (see Chapter 1).

Unlike large, distant planets such as Saturn, which appear to have retained a primary atmosphere rich in light gases, it seems clear that the Earth has lost its primordial atmosphere (see also Chapter 8). One major piece of evidence is the low atmospheric abundance of the inert noble gases krypton and xenon, as compared with their abundance in the solar system. Although there is uncertainty as to how quickly some components of the primary atmosphere may have been lost (Abelson, 1966; Rasool & McGovern, 1966), the atmosphere eventually became dominated by material of secondary origin, and the present atmosphere and hydrosphere have been formed by the release of volatile materials originally trapped in or chemically combined with solid material.

A major line of evidence concerning the origin of volatile materials in the present-day ocean has been obtained from calculations of the total mass difference for individual elements between the igneous rocks weathered and the resulting products in the sedimentary reservoirs, namely the ocean and the sediments and sedimentary rocks (see Figure 6.4). The igneous rocks are an evident major source for elements such as Na, K, Ca and Mg in these reservoirs. If they constitute the sole source, then the weathering over geological time of a certain mass of igneous rocks must have produced the total amount of each element in the sedimentary reservoirs. In other words it should be possible to obtain a consistent solution using any combination of the simultaneous equations which describe the balance between the source of each element and the various products in which it becomes distributed.

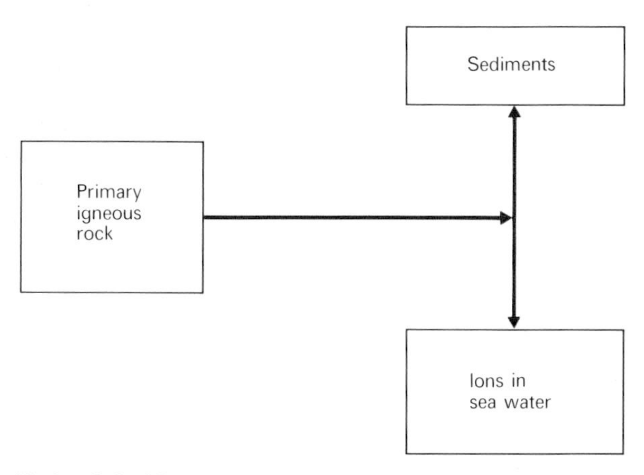

Figure 6.4 Mass balance in the major sedimentary cycle.

Calculations of this kind were first made by Goldschmidt (1933, 1954) who used data on several elements to estimate the amount of igneous rock which must have been weathered in order to supply the observed amounts in the ocean and sedimentary deposits. Horn & Adams (1966) made more comprehensive calculations, using information on 65 elements, and found that for the great majority it was possible to account consistently for the amounts in the oceans and sedimentary reservoirs on the basis that these had been formed by the weathering of 2×10^{21} kg of igneous rock. This represents the minimum quantity weathered, since it is based on the present sediments and does not take into account any sedimentary material which has been recycled, for example by subduction into the mantle. However, the striking feature of such calculations, as was recognised by Goldschmidt and other early workers, is that the amount of weathered material necessary to supply most of the elements is inadequate to supply certain others, including chlorine, sulphur, bromine and boron, the first two of which are the dominant anion-forming elements in the ocean.

Table 2 Estimated amounts of 'excess volatile' constituents supplied to the Earth's surface*

Constituent	*Amount* 10^{20} g
Water	16600
Carbon dioxide	910
Chlorine	560
Nitrogen	42
Sulphur	30
Bromine	1·3
Boron	0·9

* Estimates from Rubey (1951) and Horn & Adams (1966)

The implications of this were discussed fully by Rubey (1951), who showed also that igneous rock weathering could not have provided the amounts of water, carbon and nitrogen present in the atmosphere, hydrosphere and sediments. He concluded that the 'excess volatiles' had been released to the surface earth from magma extruded through volcanoes and hot springs. Estimates of the amounts of excess volatiles released through geological time are given in Table 2. Release by out-gassing of mantle material is now generally accepted as the likely origin of these components of the ocean and atmosphere. Evidence for the input of juvenile material at the sea bed in regions of tectonic activity can be seen in the presence in such areas of an excess of dissolved helium, which from its isotopic composition is clearly of non-atmospheric origin (for example see Clarke *et al.*, 1969). There is less agreement as to whether the major releases occurred early in the history of the ocean or whether the process has led to a more or less continuous build-up.

Rubey (1951) considered that releases have been continuous, arguing mainly from evidence on the relative constancy of the amount of carbon dioxide in the atmosphere and ocean over much of geological time. He thought it probable that the volume of the ocean has increased gradually, the changes in its volume being compensated by isostatic balance with the lithosphere. However, several detailed arguments have been presented by Fanale (1971) in support of the view that, while some crustal degassing has been a continuous process, there was probably a large-scale early degassing of the Earth, giving an early formation of an atmosphere from which there was little loss to space except for the light gases, hydrogen and helium. Increased knowledge of submarine volcanic processes has enabled more direct calculations to be made of the supply of water, chlorine, sulphur and bromine by outgassing in tectonically active regions (Anderson, 1974; Schilling *et al.*, 1978). They support Rubey's view that continuous degassing at approximately the present rates could have supplied the surface reservoirs of these constituents, although higher rates of release may have occurred earlier in the Earth's history. The time-course for the accumulation of the atmosphere and ocean remains uncertain in the light of these studies.

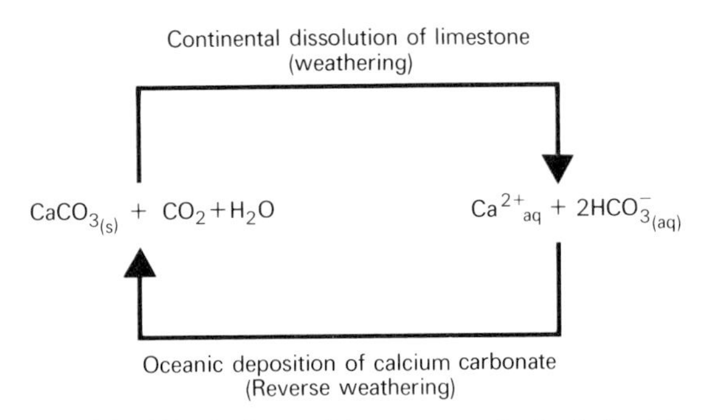

Figure 6.5 Weathering and 'reverse-weathering' of calcium carbonate (limestone).

Whatever this time-course, the work of Goldschmidt, Rubey and others has led to the view that the formation of sea water, together with the atmosphere and the sedimentary deposits, results from the interaction of a solution and suspension derived from the weathering of igneous rocks, as exemplified by present-day river water, with the volatile constituents released from the Earth's interior. This may be represented in the form of an equation:

$$\text{Igneous rock minerals} + \text{Volatiles} \rightarrow \text{Atmosphere} + \text{Ocean} + \text{Sedimentary deposits}$$

Sillén (1961, 1963, 1967) compared the formation of sea water by this means to a large-scale acid-base titration. He derived a model for the resulting composition by considering, on the basis of thermodynamic equilibria, the solid phases and the concentrations of dissolved constituents which could co-exist if the major elements in the igneous rocks and volatiles were added in appropriate proportions to an aqueous solution. One such model involved three volatiles, water itself, hydrochloric acid and carbon dioxide, together with the common rock-forming bases, oxides or hydroxides of silicon, aluminium, sodium, potassium, magnesium and calcium. In a closed system under the conditions at the Earth's surface, nine phases might form in such a system – an aqueous phase, a gas phase (corresponding to the atmosphere), and seven solid phases (nominally calcite, quartz, and the aluminosilicates kaolinite, montmorillonite, illite, chlorite and phillipsite). Sillén showed that the equilibrium concentrations of dissolved ions in such model systems resemble those found in sea water. Moreover the capacity of the aluminosilicates to exchange hydrogen ions and metal cations gives the system the capacity to control pH. Lafon & Mackenzie (1974) extended this approach by a computer simulation of the action of water and volatiles, released by rapid outgassing, on average igneous rock material. When allowance is made for the omission of certain processes from the model, a fair agreement is found between the product of the simulated reaction and present-day sea water.

The importance of equilibria involving aluminosilicates in controlling the composition of sea water has received much emphasis following Sillén's paper (1961). The real ocean cannot of course behave exactly like a closed model system and in some respects, as seen previously for the nitrogen cycle, it is far from equilibrium. Nevertheless, equilibrium models indicate the composition towards which the real system will tend, and suggest underlying mechanisms which may control composition with respect to certain constituents.

Geochemically, the reactions described represent a partial reversal of the continental weathering processes, in that degraded aluminosilicates similar in composition to that of kaolinite react with bicarbonate ions, silica, and a cation such as Na^+, K^+, Mg^{2+} to form a phase which is richer in the cation concerned; carbon dioxide is released in the process. Such reactions parallel the reversal of limestone weathering, which occurs in the ocean when calcium carbonate is deposited (see Figure 6.5), and afford the best explanation for the removal from the ocean of the fraction of bicarbonate which is not removed along with calcium (Garrels, 1965). The hypothesis that reverse weathering reactions of aluminosilicates are important in the geochemical cycles of dissolved silicon, bicarbonate, and the major cations of sea water forms the basis of the mass balance model of Mackenzie & Garrels (1966) already discussed. It has not, however, been possible to verify the hypothesis by identifying the products of these reactions in marine sediments.

Evolution of the ocean over geological time

The conditions for the earliest stages in the formation of the ocean were determined largely by the composition of the early secondary atmosphere formed by volcanic gases. Chemical conditions in this early atmosphere were reducing and the most significant changes in oceanic composition arose with the eventual change to an oxidising atmosphere (see also Chapter 8). Holland (1962, 1964) considered three stages in the evolution of the atmosphere. In the first stage, when metallic iron was still present in the mantle, the volcanic gases were dominated by hydrogen, water and carbon monoxide, with small amounts of nitrogen, hydrogen sulphide and carbon dioxide. At the surface, water condensed to form the primitive hydrosphere and the atmosphere was dominated by methane, formed by the reduction of carbon monoxide and carbon dioxide, and hydrogen. Minor constituents included water vapour, hydrogen sulphide, nitrogen, ammonia and argon.

Major changes in the atmosphere occurred with the loss of hydrogen to space and through alterations in the nature of the input of volcanic gases following the disappearance of metallic iron from the upper mantle, perhaps some 500 million years after the Earth's formation. At this stage the qualitative composition of the volcanic gases probably resembled that of the present-day inputs, dominated by water with minor amounts of carbon monoxide, carbon dioxide, sulphur dioxide and nitrogen. These changes led to the second stage in atmospheric evolution, an atmosphere dominated by nitrogen, with water vapour, carbon dioxide and argon probably being the most abundant minor constituents. The last major change occurred with the accumulation of free oxygen in the atmosphere. This led to the present stage in which nitrogen (partial pressure $= 0{\cdot}78$ atm) and oxygen ($p = 0{\cdot}21$ atm) are dominant constituents and water vapour, argon and carbon dioxide ($p = 3 \times 10^{-4}$ atm) important minor components.

In the first two stages the small amounts of oxygen available were used up in the oxidation of volcanic gases and surface rocks. In the build-up of free oxygen two main mechanisms of production have been considered, namely the photochemical dissociation of water vapour in the atmosphere by ultraviolet radiation, with loss to space of the hydrogen formed, and production by photosynthesis in excess of the amounts used in the oxidation of organic matter. The relative importance of these processes in the early evolution of the atmosphere has been a matter of debate (see for example, Brinkmann, 1969, Van Valen, 1971, and Towe, 1978), but there seems little doubt that photosynthesis has produced most of the total inventory of free oxygen.

As stressed by Garrels & Perry (1974), the major development in the evolution of the ocean, corresponding to these stages of atmospheric evolution, must have been a slow titration of the primitive reduced system by oxygen, leading to an ocean which, on the basis of evidence discussed below, has probably changed but little in its composition over most of the Phanerozoic period. Viewed in this way, the present-day ocean is envisaged as a system energetically dependent upon the continuous input of solar energy, fixed by photosynthesis.

Under the conditions preceding the accumulation of oxygen, the concentration of sulphate in the ocean would have been low, so that the dominant anions were chloride and bicarbonate. A major difference between this and a later ocean would have been the presence of considerable amounts of iron as iron(II). In oxygenated waters this species is readily oxidised to iron(III) which has a very limited solubility. In the precursor of the present-day ocean, the concentration of iron in solution was perhaps about 1000 times higher. Garrels & Perry (1974) estimated the composition of sea water under such conditions, based upon equilibrium with the major solid phases formed by iron(II), calcium, magnesium and silicon; this is shown in Table 3. The concentrations of calcium and magnesium resemble those of present-day sea water and that of

Table 3 Hypothetical composition of ancient sea water*

Constituent	*Concentration* (g kg^{-1})	*Ratio of concentration to that of chloride*
Na^+	10·9	0·55
Mg^{2+}	0·72	0·037
Ca^{2+}	0·44	0·022
K^+	0·39	0·02
Cl^-	19·5	–
SO_4^{2-}	$3{\cdot}82 \times 10^{-9}$	$0{\cdot}19 \times 10^{-9}$
HCO_3^-	0·57	0·03
Fe^{2+}	0·004	$2{\cdot}0 \times 10^{-4}$
Si	0·26	0·013
HS^-†	$1{\cdot}29 \times 10^{-4}$	$0{\cdot}066 \times 10^{-4}$

* From Garrels & Perry (1974).
† The presence of a low concentration of HS^-, the first dissociation product of hydrogen sulphide, reflects the low partial pressure of oxygen (see text).

bicarbonate is about four times higher. The pH of 6·7 is considerably lower, reflecting a high partial pressure of carbon dioxide ($p = 10^{-1{\cdot}2}$ atm); the partial pressure of oxygen is estimated at 10^{-70} atm. Other evidence (Mackenzie, 1975) places limits on the pH at the time of 5·0 to 7·5. The calculations made by Garrels & Perry (1974) assume a concentration of chloride, and of sodium and potassium, similar to that of the present-day ocean. The main mechanism for the removal of chloride is in evaporite deposits and it is uncertain that these were significant at that time. The salinity might therefore have been higher, by up to 45 per cent, than that of present-day sea water (Mackenzie, 1975).

The principal consequences of the development of photosynthetic organisms and the continuous introduction of oxygen into this system may be summarised as follows. The partial pressure of carbon dioxide decreased, as more of it was incorporated into organic material which was not re-oxidised or into carbonate sediments, and the pH rose towards its present value. Iron ceased to be a significant dissolved constituent in the ocean. Sulphate became the main dissolved form of sulphur and a major anion in sea water.

The build-up of free oxygen implies that its rate of production exceeded that of its use in oxidative processes, including the oxidation of the organic matter produced by photosynthesis. If the major mechanism of oxygen production was photosynthesis then this means that some fraction of the organic material produced must be stored in sedimentary deposits. The balance between production and consumption varied over the period of accumulation of oxygen in the atmosphere and is now in a steady state. There is no strong evidence to suggest that this steady state has altered in an important way since early Phanerozoic time. At that time the composition of sea water was essentially like that of today and the major evolution of the ocean up to the present time was already complete.

The view that the evolution of the ocean towards its present-day composition was essentially complete by the beginning of the Phanerozoic is based upon a number of considerations. The bulk composition of sediments deposited over this period indicates that the river input to the ocean has not shown a major systematic change in its character, although the discharge of individual components may have fluctuated considerably on various time-scales. Clear systematic trends do occur in the ratios of certain elements in shales of differing geological ages, corresponding to changes in mineralogy, and carbonate rocks show an increase in the ratio of magnesium to calcium with increasing age. These latter rocks also show systematic changes in isotopic composition of oxygen. Such changes have been discussed by Mackenzie (1975), who concludes that they can be accounted for by post-depositional changes and thus it is not necessary to invoke corresponding changes in the composition of sea water in order to explain them. He considers that this conclusion may be extended to cover the last 1500 million years. These and related studies have reinforced the view of Rubey (1951) that the ocean had evolved to a composition rather similar to that of the present day by a relatively early stage in geological time.

The above considerations do not, however, preclude the possibility of considerable fluctuations in composition within the period concerned. The effect of a change in input or of some other factor influencing the concentration of a constituent in the oceanic reservoir depends upon the length of time the change is sustained and the mean reservoir life-time (residence time) of the constituent in question. Since the major dissolved constituents have long residence times, sustained large changes in input are needed to alter substantially the reservoir concentration. This is illustrated by the calculation by Holland (1965) that if a saturated solution of sodium chloride were added to the ocean, at a rate corresponding to the mean rate of addition of water over geological time, then it would require 500 million years to double the salinity of the ocean. The residence times of most constituents are small, however, in relation to the period during which an ocean with broadly the present-day composition has existed, and so the possibilities for significant fluctuations are considerable. Holland (1972) considered the minerals found in marine evaporite deposits and the sequence of their deposition, and suggested limits for the possible changes in the major composition of sea water during the period of their deposition. He concluded that the concentrations of the major constituents in sea water have probably remained within a factor of two of the present values over the last 700 million years. Broecker (1971) has emphasised that the implications of a geological event for oceanic composition critically depend upon the residence time of the constituents. For example, glacial cycles of some 10000-year duration are too short to alter greatly the concentrations of silicon or total inorganic carbon, still less that of calcium, but could cause a significant change in the concentration of carbonate ions, which would respond on a shorter time-scale.

An attractive approach to the study of changes in conditions in the ocean has been through the investigation of variations in the stable isotope ratios for lighter elements, particularly carbon and oxygen. Water has a low proportion of molecules containing heavy isotopes of hydrogen (^{2}H or deuterium) and oxygen (^{17}O and ^{18}O) in addition to molecules made up of the common isotopes ^{1}H and ^{16}O. When evaporation or condensation take place the heavier molecules are slightly enriched in the liquid phase. There is thus a very slightly higher proportion of ^{18}O, which is the more abundant and readily measured of the two heavy oxygen isotopes, in waters of lower latitudes where evaporation exceeds precipitation, than in waters of higher latitudes where the converse applies. Dilution by continental run-off increases the proportions of the lighter isotopes. In the formation of sea ice the heavier isotopes are enriched in the solid phase. Geological changes can lead to variations in the ratios of stable isotopes in ocean waters, particularly in surface waters, which are more responsive to short-term changes. A major example is the effects of glacial cycles, leading to variations in the distribution of water between the ocean and sea ice, and to changes in freshwater inputs from the continents.

Although there is no direct preserved record of the isotopic variations in sea water, they are reflected in the isotopic ratios in the carbonates formed by planktonic organisms living in surface waters and preserved in bottom sediments. These variations have been studied for oxygen and for carbon, which has two stable isotopes, ^{12}C and ^{13}C, the heavier being of low abundance. A slight degree of discrimination between light and heavy isotopes occurs in the uptake of the elements by organisms from sea water. This discrimination is temperature dependent. Provided that corrections can be made for variations with time in the isotopic composition of the source material, this dependence affords a way of reconstructing the history of surface water temperatures from the oxygen isotopic composition of sedimentary carbonates. Unaltered carbonate material from the Cenozoic and Mesozoic has often been preserved and provides evidence for a significant decrease in Earth surface temperature over the past 150 million years. Examination of sediment cores has provided a detailed reconstruction of the changes in temperature of the surface ocean associated with glacial events (see e.g. Emiliani, 1972). The relative importance of the temperature effect, changes in the isotopic composition of sea water, and post-depositional alteration of composition, is a cause of uncertainty, however, and caution in interpretation is essential.

Much information exists on variations in the isotopic composition of oxygen and carbon in sediments over a period extending back to the Precambrian (Garlick, 1974). The interpretation of these data is very equivocal. There is, however, a notable uniformity in the isotopic composition of carbon in limestones throughout the Phaner-

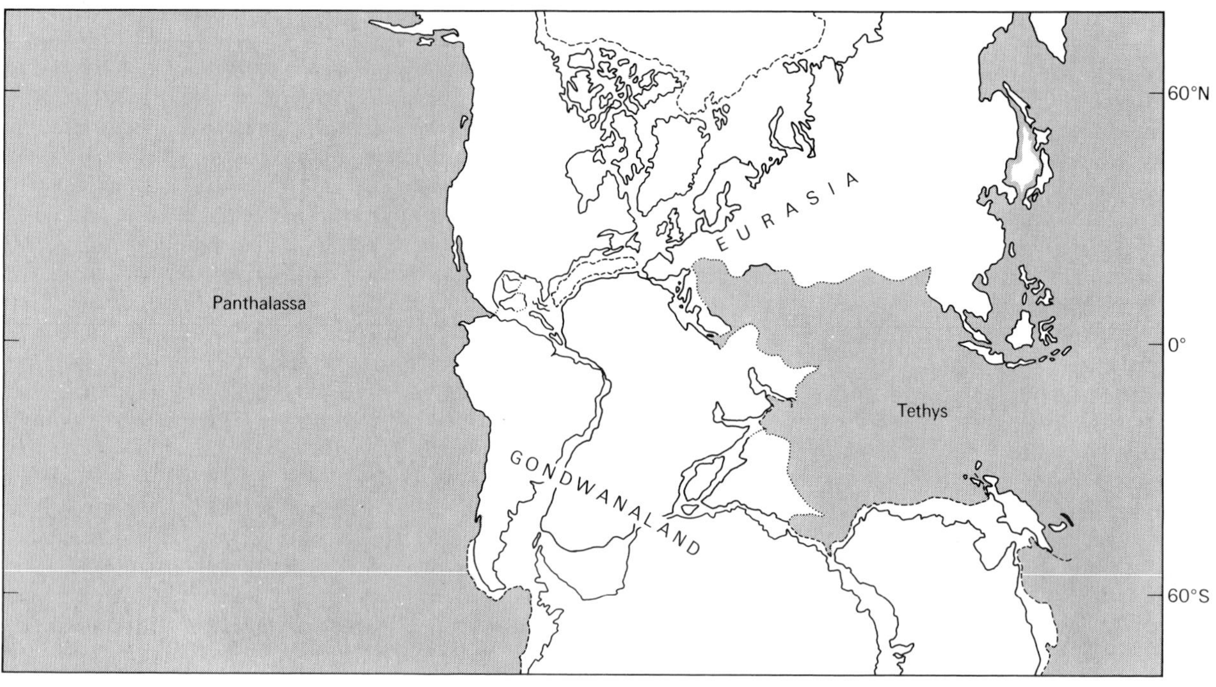

Figure 6.6. Palaeogeographical reconstruction of the distribution of the continents and ocean about 170 million years ago (after Smith *et al.*, 1973). The continental edges correspond to the present-day 1000 m depth contour (---) or an arbitrary separation line (...) between continents.

ozoic, suggesting a corresponding uniformity in the reservoir of organic carbon over that period.

Oceanic circulation and the evolution of the ocean basins

This account of the evolution of the composition of sea water has so far been focused on the general features of the system. The detailed pattern of the circulation of the ocean, which influences many of the detailed features in particular parts of the ocean, must, however, have undergone great changes during the course of the evolution of the ocean basins and the movements of the continents. Modern theories of plate tectonics, together with increased evidence from events preserved in marine sediments, have enabled reconstructions to be made of the most probable major changes. While these must be regarded only as probable models of the sequence of events, they represent a major advance in our understanding of palaeo-oceanography. They are illustrated here by brief summaries relating to the Atlantic Ocean and the Mediterranean Sea.

A reconstruction of the continents and oceans at about the end of the early Jurassic is shown in Figure 6.6. Figure 6.7 illustrates the detailed reconstruction of the circulation in the Atlantic over the past 200 million years, suggested by Berggren & Hollister (1974). This reconstruction assumes the initial rifting of the North Atlantic at about 180 to 200 million years ago. With the separation of Gondwanaland and Eurasia at about 150 million years ago, the Tethys and North Atlantic Oceans became connected. Water exchange between the North and South Atlantic Oceans did not arise, however, until about 90 to 95 million years ago, with the separation of South America and Africa. The development of a strong bottom circulation of cold water in the Atlantic Ocean was a later event arising with the separation of Scandinavia and Greenland some 50 million years ago. Subsequent movements isolated the Atlantic Ocean from exchange with the Indo-Pacific Ocean through the Tethys Ocean and through Panama. Following the early glaciation in the northern hemisphere about 3 million years ago the main changes in the circulation of the North Atlantic Ocean were complete.

The evolution of the Mediterranean Sea has been the subject of much investigation and illustrates the variety of conditions that can arise in marginal seas which are to some degree separated from the primary oceanic circulation. Such waters can be sensitive to changes that are relatively minor in global terms. The Mediterranean Sea is a young feature by comparison with the Atlantic Ocean. The main characteristics of its present-day circulation are an inflow of surface Atlantic Ocean water through the Strait of Gibraltar, which continues as a surface flow with increasing salinity until it sinks in the eastern Mediterranean Sea and then moves westward as a deep current. Water of high

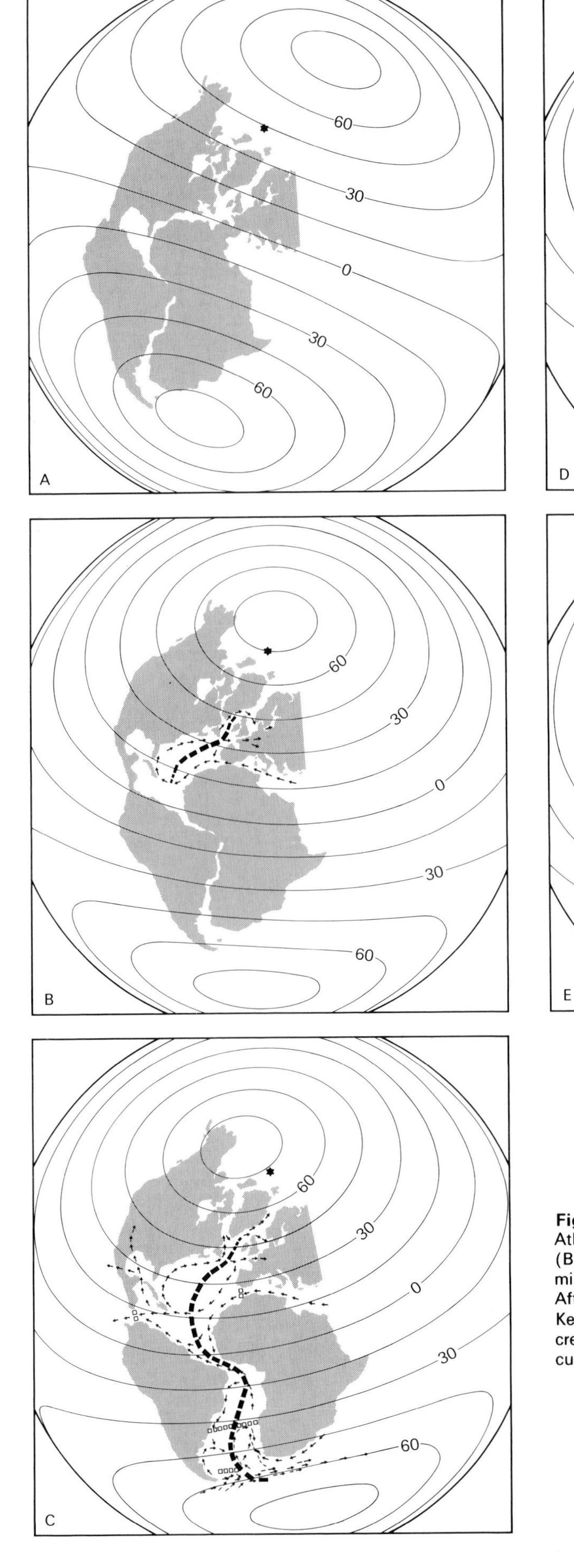

Figure 6.7 Palaeogeographical reconstructions of the Atlantic Ocean for different periods: (A) 200 million years ago, (B) 150 million years ago, (C) 105 million years ago, (D) 35 million years ago. The present-day features are shown by (E). After Berggren and Hollister (1974).
Key: —→ surface currents; --→, deep bottom currents; ■■, crest of Mid-Atlantic Ridge; □ □, barriers to flow of bottom currents.

salinity from the Mediterranean Sea flows over the sill at the Strait of Gibraltar into the Atlantic Ocean, sinking down the continental slope. The Mediterranean circulation is adequate to maintain a supply of dissolved oxygen to the sub-surface waters. An important bottom topographic feature is the sill at the Straits of Sicily, which divides the Mediterranean Sea into two deep basins.

In the evolution of the Mediterranean Sea the first stage was isolation from the Indian Ocean. The extensive evaporite deposits of Messinian age, found in Mediterranean Sea sediments, were probably formed soon afterwards, when it is thought that the basin was isolated also from the Atlantic Ocean (see also Chapter 14). Early in the Pliocene, deep cold Atlantic Ocean water was again able to enter the basin but the circulation in the eastern Mediterranean Sea remained restricted. The restricted circulation of this sector is reflected in the periodic deposition during the late Tertiary and early Quaternary of organic-rich muds deposited under anoxic conditions, such as occur today in the deeper waters of the Black Sea. There is substantial evidence to link these episodes of anoxicity in the eastern Mediterranean Sea with climatic changes as suggested by Olausson (1961). During interglacial periods or other times of high freshwater discharge, outflow of low salinity surface water from the Black Sea may have caused strong stratification in the eastern Mediterranean Sea, leading to stagnation and consequent deoxygenation of deeper waters. These conditions did not extend to the western basin.

Variations in the extent of shallow lagoonal environments have been of evident importance in relation to the amounts of evaporites deposited at different times. Evaporite deposition has been the major sink for oceanic chloride and probably sulphate and has thus had an important effect upon the composition of sea water, as was indicated above. The present period of geological time is one of low deposition of evaporites.

Man's influence on the ocean

The increased mobilisation of materials in the sedimentary cycle by Man's industrial activities, although considerable for a number of elements, is too recent to have had generally significant effects on the oceans, as a consideration of the oceanic residence times of the elements immediately indicates. The problems of marine pollution have been almost entirely confined to restricted marginal environments, immediately influenced by local discharges of wastes. More widespread in its occurrence is the enhancement of lead concentrations in surface ocean waters as a result of its intensive release into the atmosphere over recent decades. The most significant global problem identified so far is the accelerating release of carbon dioxide during the industrial era, a result of the intensive consumption of fossil fuels.

The immediate concern with the effects of carbon dioxide release is that the increased atmospheric concentration may lead to an increase in Earth surface temperatures. This is because carbon dioxide molecules absorb radiation strongly in the infra-red and thus trap some of the radiation of longer wavelength, which is re-radiated from the Earth, following absorption of solar radiation of shorter wavelengths – the so-called 'greenhouse effect'.

The ocean is the major reservoir of the Earth's exchangeable carbon system; most of this carbon exists as bicarbonate in the deeper waters. The increase in carbon dioxide concentration in the atmosphere is thus greatly affected by the rate at which carbon dioxide is exchanged with the surface and deep ocean reservoirs, and the extent to which the ocean can accommodate the increased concentration. These have proved complex questions, but it is clear that a significant part of the excess carbon dioxide has already passed into the ocean. While this represents a mitigation of possible effects on the atmosphere, it raises additional questions as to its effects on the ocean. The effect of an increase in the partial pressure of carbon dioxide in the ocean is to decrease the pH, with a consequent increase in the concentration of bicarbonate ions at the expense of carbonate ions, a change which increases the solubility of calcium carbonate. The initial impact is experienced by the surface layers of the ocean and the eventual effect will be modified by mixing with deeper waters. In the long term, some dissolution of calcium carbonate from sediments may occur. While the implications of such changes have been the subject of considerable controversy it seems probable that if there are any appreciable effects from Man's rapid cycling of fossil fuel carbon, the consequences of the increasing atmospheric concentrations of carbon dioxide, prior to its transfer to the oceans, could be more significant than those arising from changes in the oceanic carbonate system.

References

Abelson, P. H. 1966. Chemical events on the primitive earth. *Proceedings of the National Academy of Sciences of the United States of America* **55**: 1365–1372.

Anderson, A. T. 1974. Chlorine, sulfur and water in magmas and oceans. *Bulletin of the Geological Society of America* **85**: 1485–1492.

Berggren, W. A. & Hollister, C. D. 1974. Paleogeography, paleobiogeography and the history of circulation in the Atlantic Ocean, pp. 126–186 *in* Hay, W. W. (Ed.) *Studies in paleo-oceanography*. Tulsa: Society of Economic Paleontologists and Mineralogists.

Bonatti, E. 1975. Metallogenesis at oceanic spreading centres. *Annual Review of Earth and Planetary Sciences* **3**: 401–431.

Brinkmann, R. T. 1969. Dissociation of water vapor and evolution of oxygen in the terrestrial atmosphere. *Journal of Geophysical Research* **74**: 5355–5368.

Broecker, W. S. 1971. A kinetic model of the chemical composition of sea water. *Quaternary Research* **1**: 188–207.

Clarke, W. B., Beg, M. A. & Craig, H. 1969. Excess ^{3}He in the sea: evidence for terrestrial primordial helium. *Earth and Planetary Science Letters* **6**: 213–220.

Codispoti, L. A. & Richards, F. A. 1976. An analysis of the horizontal regime of denitrification in the eastern tropical North Pacific. *Limnology and Oceanography* **21**: 379–388.

Corliss, J. B., Dymond, J., Gordon, L. I., Edmond, J. M., von Herzen, R. P., Ballard, R. D., Green, K., Williams, D., Bainbridge, A., Crane, K. & van Andel, T. H. 1979. Submarine thermal springs on the Galápagos Rift. *Science*, New York **203**: 1073–1083.

Emiliani, C. 1972. Quaternary paleotemperatures and the duration of the high-temperature intervals. *Science*, New York **178**: 398–401.

Fanale, F. P. 1971. A case for catastrophic early degassing of the earth. *Chemical Geology* **8**: 79–105.

Garlick, G. D. 1974. The stable isotopes of oxygen, carbon, and hydrogen in the marine environment, pp. 393–425 *in* Goldberg, E. D. (Ed.) *The sea, volume 5, marine chemistry*. New York: Wiley-Interscience.

Garrels, R. M. 1965. Silica: role in the buffering of natural waters. *Science*, New York **148**: 69.

Garrels, R. M. & Perry, E. A. Jr. 1974. Cycling of carbon, sulfur, and oxygen through geologic time, pp. 303–336 *in* Goldberg, E. D. (Ed.) *The sea, volume 5, marine chemistry*. New York: Wiley-Interscience.

Goldberg, E. D. 1971. Atmospheric dust, the sedimentary cycle and man. *Comments on Earth Sciences: Geophysics* **1**: 117–132.

Goldschmidt, V. M. 1933. Grundlagen der quantitativen Geochemie. *Fortschritte der Mineralogie, Kristallographie und Petrographie* **17**: 112–156.

Goldschmidt, V. M. 1954. *Geochemistry*. Oxford: Oxford University Press.

Holland, H. D. 1962. Model for the evolution of the earth's atmosphere, pp. 447–477 *in* Engel, A. E. J., James, H. L. & Leonard, B. F. (Eds.) *Petrologic studies: a volume to honor A. F. Buddington*. New York: Geological Society of America.

Holland, H. D. 1964. On the chemical evolution of the terrestrial and Cytherean atmospheres, pp. 86–101 *in* Brancazio, P. J. & Cameron, A. G. W. (Eds.) *The origin and evolution of atmospheres and oceans*. New York: John Wiley.

Holland, H. D. 1965. The history of ocean water and its effect on the chemistry of the atmosphere. *Proceedings of the National Academy of Sciences of the United States of America* **53**: 1173–1183.

Holland, H. D. 1972. The geologic history of sea water – an attempt to solve the problem. *Geochimica et Cosmochimica Acta* **36**: 637–651.

Horn, M. K. & Adams, J. A. S. 1966. Computer-derived geochemical balances and element abundances. *Geochimica et Cosmochimica Acta* **30**: 279–297.

Humphris, S. E. & Thompson, G. 1978. Hydrothermal alteration of oceanic basalts by sea water. *Geochimica et Cosmochimica Acta* **42**: 107–125.

Lafon, G. M. & Mackenzie, F. T. 1974. Early evolution of the oceans – a weathering model, pp. 205–218 *in* Hay, W. W. (Ed.) *Studies in paleo-oceanography*. Tulsa: Society of Economic Paleontologists and Mineralogists.

Mackenzie, F. T. 1975. Sedimentary cycling and the evolution of sea water, pp. 309–364 *in* Riley, J. P. & Skirrow, G. (Eds.) *Chemical oceanography*, Vol. 1. London: Academic Press.

Mackenzie, F. T. & Garrels, R. M. 1966. Chemical mass balance between rivers and oceans. *American Journal of Science* **264**: 507–525.

Millero, F. J. 1974. Seawater as a multicomponent electrolyte solution, pp. 3–80 *in* Goldberg, E. D. (Ed.) *The sea, volume 5, marine chemistry*. New York: Wiley-Interscience.

Millero, F. J. 1975. The physical chemistry of estuaries, pp. 25–55 *in* Church, T. M. (Ed.) *Marine chemistry in the coastal environment*. Washington D.C.: American Chemical Society.

Olausson, E. 1961. Sediment cores from the Mediterranean Sea and the Red Sea. *Report of the Swedish Deep Sea Expedition 1947–48* **8**: 337–391.

Rasool, S. I. & McGovern, W. E. 1966. Primitive atmosphere of the earth. *Nature*, London **212**: 1225–1226.

Rubey, W. W. 1951. Geologic history of sea water. *Bulletin of the Geological Society of America* **62**: 1111–1147.

Schilling, J. G., Unni, C. K. & Bender, M. L. 1978. Origin of chlorine and bromine in the oceans. *Nature*, London **273**: 631–636.

Sillén, L. G. 1961. The physical chemistry of sea water, pp. 549–81 *in* Sears, M. (Ed.) *Oceanography*. Washington D.C.: American Association for the Advancement of Science.

Sillén, L. G. 1963. How the sea got its present composition. *Svensk Kemisk Tidskrift* **75**: 161–177.

Sillén, L. G. 1967. The ocean as a chemical system. *Science*, New York **156**: 1189–1197.

Smith, A. G., Briden, J. C. & Drewry, G. E. 1973. Phanerozoic world maps. *Special Papers in Palaeontology* **12**: 1–42.

Towe, K. M. 1978. Early Precambrian oxygen: a case against photosynthesis. *Nature*, London **274**: 657–661.

Van Valen, L. 1971. The history and stability of atmospheric oxygen. *Science*, New York **171**: 439–443.

J. D. Burton
D. Wright
Department of Oceanography
The University
Southampton SO9 5NH

CHAPTER 7

Deep ocean sediments

D. R. C. Kempe

A glance at the simplest map of the world makes it immediately clear that there are at least three large areas of sea comprising some 70 per cent of the Earth's surface, which we call oceans; a closer look reveals one or perhaps two more. But once we go back in geological time a more sophisticated method of enquiry is required. The reconstruction of palaeogeography has occupied earth scientists for over a century and their discussions are described elsewhere in this volume. However, most of the best data are applicable only to the Mesozoic and Cenozoic eras; when we turn to the Precambrian and the Palaeozoic the discussion becomes more and more theoretical. In the Archaean, for example, which parts of the Earth were covered with continent and which with ocean? What was the nature of the Archaean oceanic crust and what do we know of the overlying sediments? The answer to both questions is uncertain, although attempts have been made recently to reconstruct the early configuration of the surface of our planet (Goodwin, 1974). This subject is also discussed in Chapters 1 and 2, but before embarking on an account of the ocean sediments that we *can* sample and examine, it is useful to remind ourselves that we are dealing only with the sediments laid down during the last 200 million years of the Earth's history, out of a total of some 4500 million years.

One model for the early development of the Earth is that of Hargraves (1976), who proposed an Earth whose concentric shells of sial and overlying hydrosphere evolved slowly through the Archaean until, in the Proterozoic, large scale mantle convection resulted in rifting and sea-floor spreading, leading to the progressive segregation of these layers into continents and oceans. Other workers prefer a single Precambrian supercontinent, resembling the configuration of the late Palaeozoic Pangaea proposed by Wegener in 1912 (see Chapter 2, fig. 2.11).

A number of reconstructions have been made of the stages of continental drift within the Palaeozoic period and, to a lesser extent, within the Proterozoic, when magnetic evidence indicates that intracontinental dislocations were present. The continents had coalesced by the end of the Palaeozoic (Permian) into the Pangaean supercontinent; this continued up to the Early Jurassic and from then on the patterns of continental drift are clearer and have been the subject of a massive literature, some of which is discussed in Chapters 13 and 14. Prominent amongst the proponents of a uniformitarian approach are Smith *et al.* (1973), who hold that as early as 570 million years ago the continents were by no means united. They suggest that several discrete land masses existed then and that plate movement continued through to the present.

The oldest period for which samples of sediment have been found in the present oceans are of Jurassic age, but where are the fossil sediments formed prior to this? The answer must be in the obducted and subducted masses that have been overthrust onto continental blocks are underthrust below oceanic trenches. Ophiolite sequences occur in the Troodos massif, Cyprus; Greece; Turkey; Oman; the Himalayas; north-east Canada; California and elsewhere. Ophiolites consist of radiolarian cherts and other pelagic sediments; ferruginous muds (ochres), manganiferous muds (umbers), and massive sulphide deposits; basaltic pillow and flow lavas; and sheeted dykes, gabbros, and ultramafic igneous rocks (containing mainly iron and magnesium silicates). They are regarded as the only surviving masses of material representing former oceanic crust, most of which has been destroyed in ancient subduction zones. These zones are to be found today below the trench systems or perhaps partly buckled up onto island arcs, such as the Marianas, Philippines, and other circum-Pacific systems, and the East and West Indies; much of the older sediment, however, has disappeared as a result of major crustal movements.

In this chapter we are chiefly concerned with the sediments forming the ocean floors of the present day. Each land mass is surrounded by a gently sloping area known as the continental shelf; this margin forms part of the continental block and its sediments are dealt with in

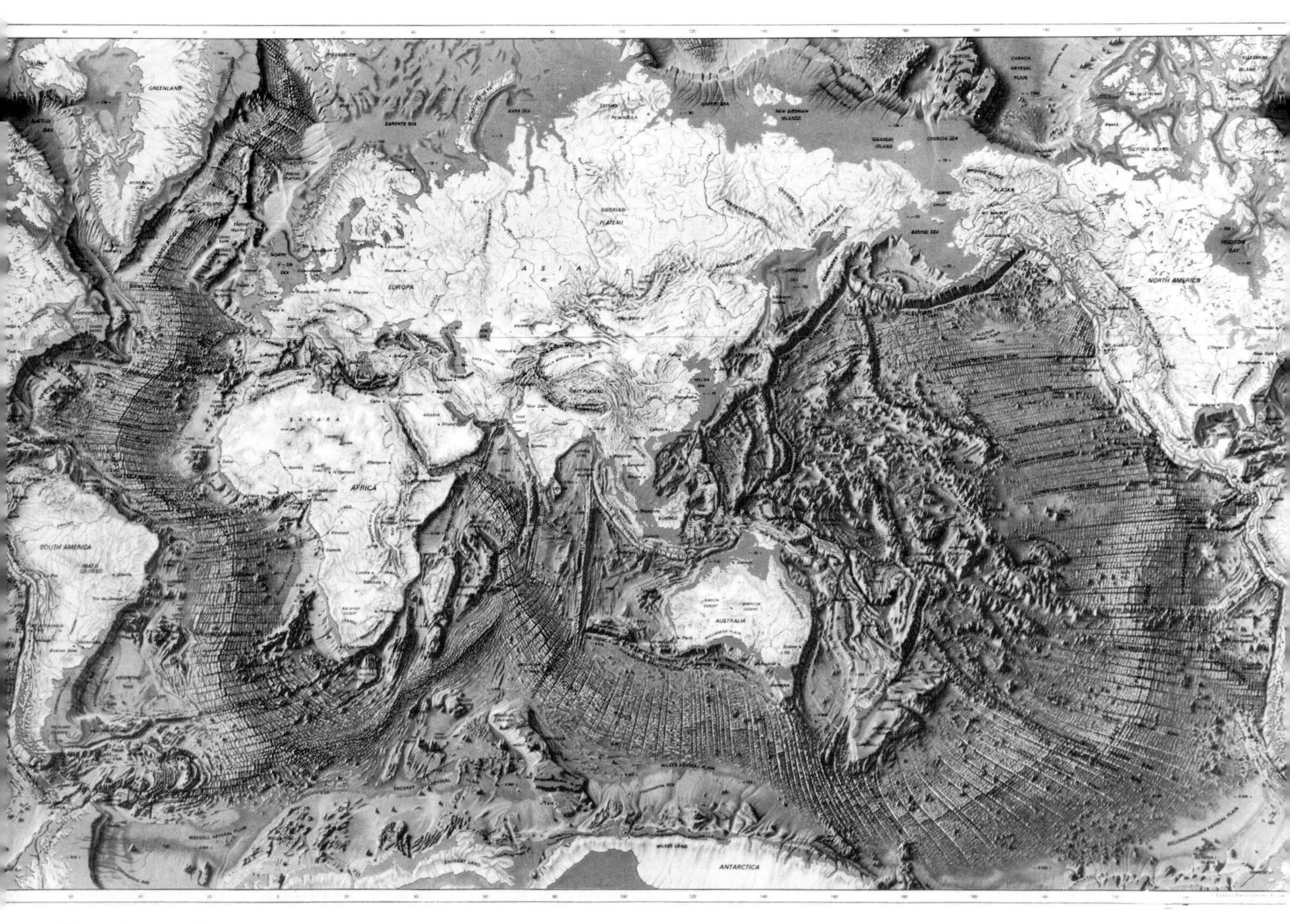

Figure 7.1 The floor of the world's oceans. (Reproduced by courtesy of the US Navy Office of Naval Research, and of B. C. Heezen and Marie Tharp).

Chapter 4. The edge of the shelf is anything from a few kilometres to more than 300 km from the coast and averages about 80 km; beyond the shelf edge the sea bed angles more steeply to the deep ocean floor. This area – the continental slope and rise – forms part of the ocean crust and is considered here, together with the abyssal plains and basins forming the deepest parts of the ocean. Figure 7.1 shows the map of the world's oceans and their floors, whilst Figure 7.2 gives a small portion of the North Atlantic on a larger scale and Figure 7.3 illustrates diagrammatically the different physiographic provinces of the Atlantic Ocean.

What makes up the oceanic sediments? There are only two major source areas, the land or continental regions, and the oceanic crust, in the form of volcanoes or hydrothermal (hot water solutions) vents. Material from both sources can be introduced either as solids or in solution. A third, extra-terrestrial, source can be considered negligible. The processes whereby materials are introduced by vertical deposition, oceanic current action, and organic and occasionally inorganic precipitation, are our concern here. The dynamic nature of the total oceanic body, with the same physical parameters operating throughout, then ensures that the sediments formed are heterogeneous only in so far as their environments are radically different. There is the same range of environments within each ocean with little variation from one ocean to another.

Modern oceanography began with the voyage of H.M.S. *Challenger* which lasted from 1872 to 1876, although the collections of sea-floor samples made by the ship's scientists were not the earliest of all – the US Coast Survey Brig *Washington*, for example, dredged samples from off the

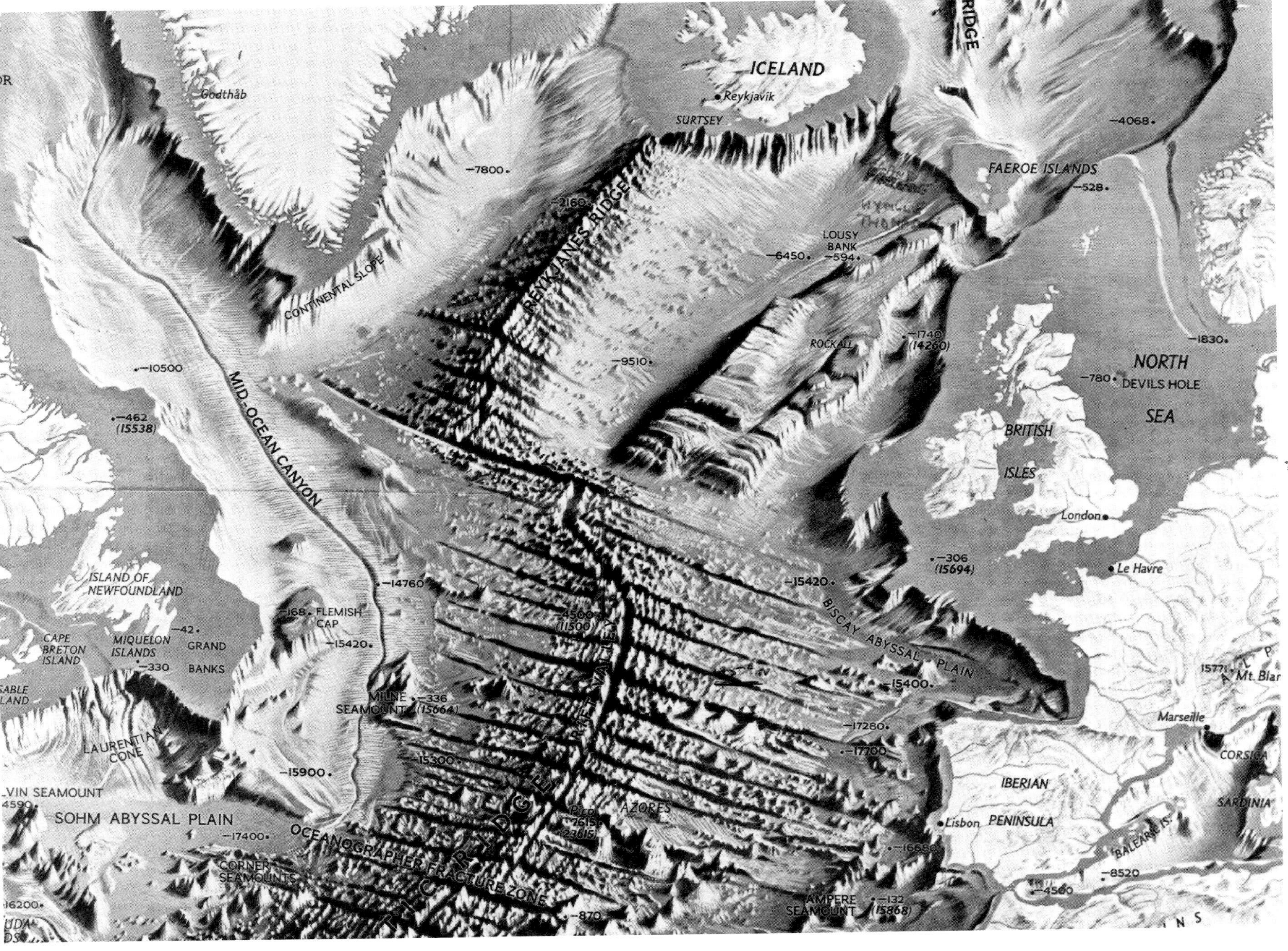

Figure 7.2 Detail from the floor of the North Atlantic Ocean. (Reproduced by courtesy of the National Geographic Magazine).

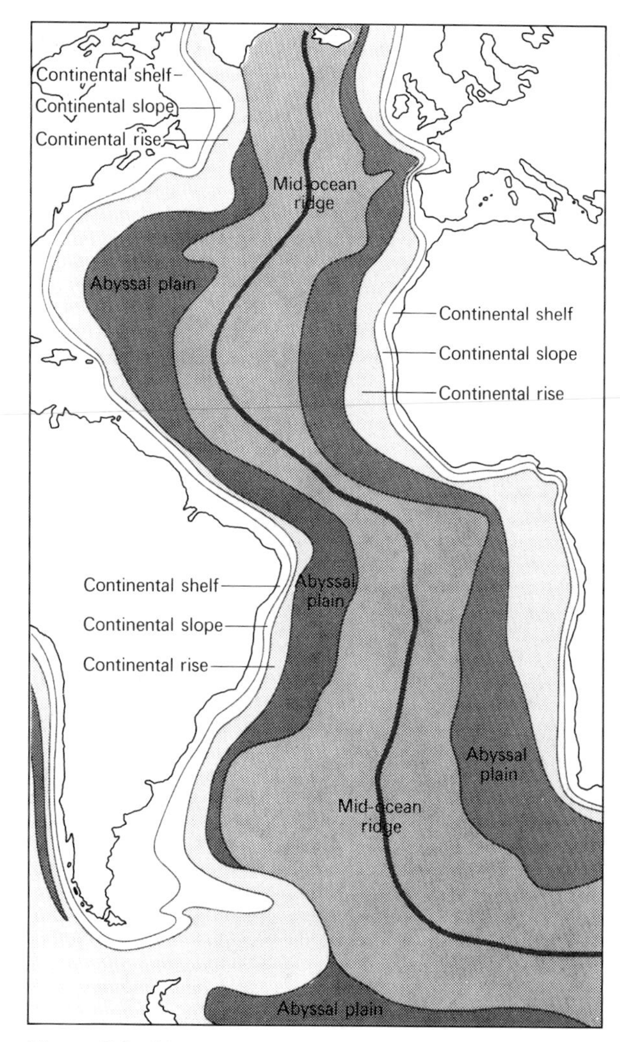

Figure 7.3 Physiographic provinces of the Atlantic Ocean (after Gass *et al.* 1977).

west coast of America in 1844. The *Challenger*'s samples of deep-sea deposits, described by Murray & Renard (1891), were obtained by dredging and sounding – the weight at the end of the sounding line incorporated a short tube which filled with sediment as the line drew the weight across the sea bed. Murray & Renard classified the samples into pelagic oozes and red clay, and terrigenous muds, and made a sediment distribution map (Fig. 7.4): both their general classification and the broad outlines of their map are still valid today. The *Challenger* samples were mostly taken from the surface of the sea-floor, and thus included only Recent deposits or older sediments exposed by erosion. With the development of gravity coring devices, longer cores could be taken, from the 120 cm cores of Schott in the *Meteor* from 1925–27, and the longer cores of Piggot in 1933, to the piston corer of Kullenberg, which in 1947 took core samples 3 metres long. Thus the Pleistocene sedimentary record became accessible. This, together with the development of the theory of turbidites (see below), have required some modification of the Murray & Renard (1891) classification.

Deep gravity coring was expected to reveal the nature of oceanic sediments dating far beyond the Pleistocene. However this proved difficult and, although much progress was made, it remained for the Deep Sea Drilling Project (DSDP), beginning in 1968, to drill cores of sufficient depth to reveal finally the complete sedimentary record in the ocean floor. This was sometimes found to be as old as Jurassic, and the cores penetrated through all the sediments down to the igneous basement. Detailed logs of these cores, including lithological, palaeontological, and other data, have been published. The various chief types of ocean-floor sediments will now be discussed in turn.

Non-pelagic sediments

The term 'pelagic' sediments is used for deep ocean material 'containing no significant terrigenous components', although in practice most sediments are mixtures; pelagic sediments mainly comprise the carbonate-rich and siliceous oozes and brown or red clay of the deep ocean. Thus 'non-pelagic' sediments contain a high proportion of land-derived material produced by physical and chemical weathering, and are found in the shallower, peripheral parts of the ocean. Hemipelagic sediments form an intermediate group.

The continental slope provides a 'non-pelagic' environment, extending a variable distance into the ocean, of which 150 km is a typical figure. The slope begins at the edge of the shelf and is inclined at angles of up to 20° or more where continental rifting has recently occurred, but averaging perhaps 4°. The final descent to the abyssal plain is usually more gentle, with an average angle of slope of around 1°, and is sometimes called the continental rise.

The chief non-pelagic sediments are given in Table 1. The commonest are the terrigenous muds, which contain more than 30 per cent of silt and sand of certain terrigenous origin. These muds vary in colour from black to red, green, and blue and occasionally white – their minerals are mainly quartz and clay minerals. Black muds, known as sapropels, have a high organic content and are rich in hydrogen sulphide; they accumulate in poorly ventilated basins (such as the Mediterranean or Black Sea) in anaerobic, and therefore strongly reducing, conditions produced by bacterially-induced decomposition of organic remains. Red muds are highly oxidised, leading to the conversion of their iron from the ferrous to the ferric state; such oxidation may have occurred before the sediments reached the sea, perhaps in rivers. In contrast, green muds contain iron in the ferrous state which are formed in mildly reducing conditions, and include some organic matter. The less common blue muds have a similar origin, whilst white mud is derived from coral reef debris. Salt deposits, together with black shales and cherts, are the oldest sediments in the Atlantic Ocean, occurring locally on or near the continental margins.

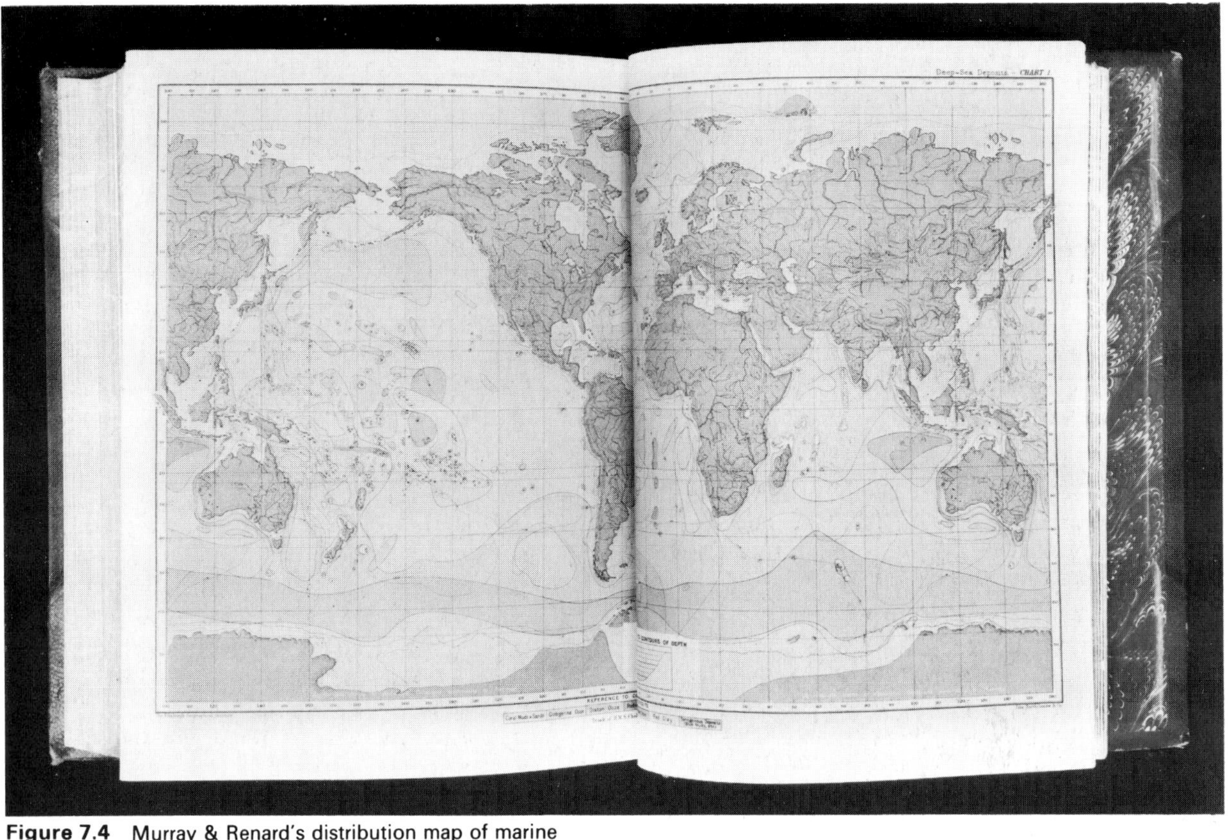

Figure 7.4 Murray & Renard's distribution map of marine deposits (1891).

Turbidity currents, ice-rafted, volcanic, and air-borne sediments

Turbidity currents and other gravity flows result when masses of sediment, resting unstably on the continental shelf or slope, begin to slide down it. Whether or not they alone are responsible for initiating submarine canyons, they certainly enlarge existing canyons by the scouring action of their sediment load. Turbidity currents can reach a speed of 25 ms^{-1} and the heterogeneous sediments they carry, varying from coarse to fine grain sizes, can travel as far as 1500 km from the coast (see Chapter 4), finally coming to rest chiefly on the continental rise. Island reef and forereef debris from islands can also become caught up in turbidites or may even form them. The uppermost layer of a turbidite is fine-grained and results from slow deposition from the turbid (nepheloid) layer of bottom water, up to a few hundreds of metres thick. The turbid layer itself is caused partly by turbidity currents. Also important on the continental rise are sediments deposited by slow-moving bottom currents. An example is Feni Ridge, a 600 km feature near the Rockall trough, off Ireland.

One of the most interesting and variable sedimentary sequences occurs along an active continental margin, where the slope extends down to a trench. Examples occur along the Aleutian and Peru–Chile trenches (von Huene, 1974); such a trench-fill sequence is shown in Figure 7.5. The sediments consist of perhaps 500 to 1000 m of muds and silts, coarsening upwards into sand and silt turbidites which thicken to the landward side. Under the fill sequence

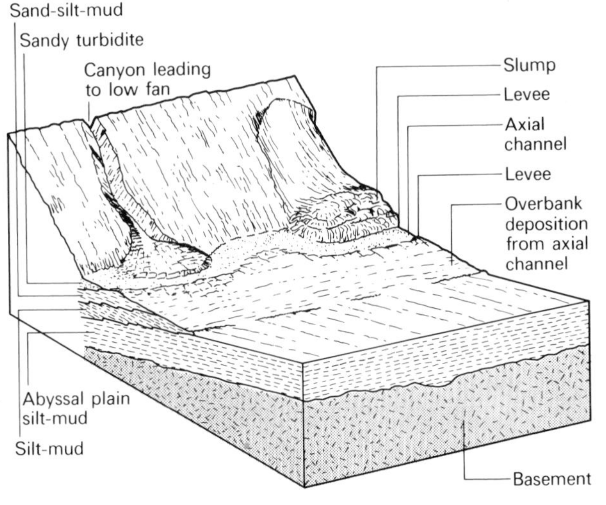

Figure 7.5 Facies model of Aleutian trench sedimentation (after von Huene, 1974).

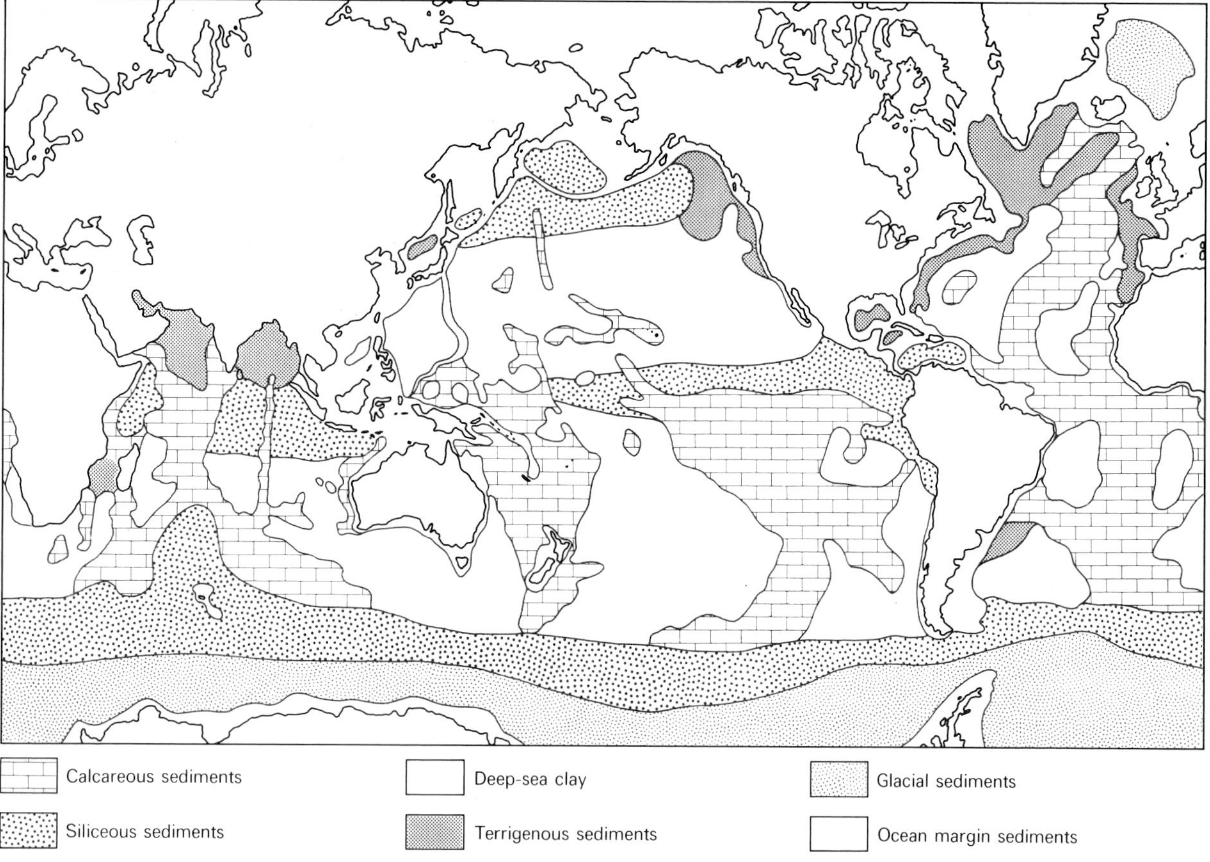

Figure 7.6 Distribution of the principal types of sediment on the floors of the oceans (after Davies & Gorsline, 1976, fig. 24.7).

lies an abyssal plain sequence of silts and muds, overlying claystone and chalk. This abyssal plain sequence, perhaps 500 m thick, itself overlies a basement of igneous rocks; it contains evidence of thrust faulting and perhaps consists of oceanic sediments scraped from the sea-floor. Along or near to the base of the landward wall, between levees, runs the axial channel, some 2·5 to 6 km wide, penetrated by canyon fan and slump sediments. These relatively coarse and heterogeneous deposits appear to be re-distributed along the trench, evening up the fill, and were probably deposited by density currents flowing in an axial direction. Not surprisingly, the rate of sedimentation is very much greater in trenches than in oceanic basins, averaging some 2000 mm per thousand years (ty) (see Table 4). This rate may have reached 3500 mm ty^{-1} during glacial and dropped to as low as 200 to 300 mm ty^{-1} during interglacial periods. The fossil trench sequences, marking the subduction zones of earlier periods, now lie mainly beneath thick eugeosynclinal sequences. Good examples of these are the piles of sediment and volcanic materials which have filled up marginal seas.

In high latitudes the Pleistocene sediments locally contain abundant coarse material, irregularly stratified, which has been ice-rafted, and was originally of glacial origin. Another localised sediment type is derived from volcanoes, both subaerial and submarine, at varying distances from the source. Layers of altered volcanic glass, ash, and tuff, with local concentrations of glass, can be found near oceanic islands, island arcs and other areas of volcanic activity. An exceptional example of submarine volcanic deposit is afforded by the holohyaline (glassy) and lapilli tuffs that form a 400-metre sequence on the aseismic Ninety East Ridge, eastern Indian Ocean (Fleet & McKelvey, 1978). Deep-sea drilling is revealing many more such instances of thick sequences of pyroclastics or volcanic sands, but the ubiquitous occurrence of small fragments of pumice, spread by drifting made possible by its porous nature and consequent ability to float, was noted as long ago as 1891 by Murray & Renard.

In addition to clastic sediments produced by the erosion of continental rocks, pyroclastic components, and glacially rafted material, there is a small additional contribution from aeolian dusts and cosmic spherules. Aeolian (wind-borne) dust normally makes up only a few per cent of sediments but can exceed 50 per cent (on a carbonate-free basis) off the Atlantic coast of equatorial Africa and the

Table 1 Deep ocean sediments

Non-pelagic sediments
Terrigenous muds (having more than 30 per cent of silt and sand of definite terrigenous origin).
- Black muds
- Red muds
- Green muds
- Blue muds
- White muds

Turbidites (derived by turbidity currents from the land areas or from submarine highs).
Slide deposits (carried to deep water by slumping).
Glacial marine (having a considerable percentage of allochthonous ('foreign') particles derived from iceberg transportation).
Volcanic debris (pyroclastic ash, tuff, and volcanic glass).
Aeolian dust and cosmic spherules.

Hemi-pelagic sediments
Pelagic sediments with some non-pelagic sediments.

Pelagic sediments
Brown or red clay (having less than 30 per cent biogenous material).
Diagenetic deposits (consisting dominantly of minerals crystallised in sea-water, such as zeolites and manganese nodules).
Biogenous deposits (having more than 30 per cent material derived from organisms).
Foraminiferal and/or Coccolith ooze (having more than 30 per cent calcareous biogenous material, largely foraminifera and/or coccoliths, particularly common in cores penetrating the Tertiary.
Pteropod ooze (having more than 30 per cent calcareous (aragonite) biogenous material, largely pteropods).
Diatom ooze (having more than 30 per cent siliceous biogenous material largely diatoms).
Radiolarian ooze (having more than 30 per cent siliceous biogenous material, largely radiolarians).
Dinoflagellates; pollen; spores; silicoflagellates; and other planktonic debris. Sponge spicules; fish teeth; and other benthonic debris.
Coral reef debris (derived from slumping around reefs).
- Coral sands
- Coral muds (white)

Largely from Shepard (1963, p. 402).

Gulf of Guinea, where the seasonal dust cloud known as the 'Harmattan haze' occurs. Cosmic spherules – thought to be the glassy and metallic droplets formed as meteorites melt and burn up on entering the Earth's atmosphere – also make up a small proportion. Spherules and aeolian dust are also present in deep ocean pelagic sediments.

Pelagic sediments

Any sediment distribution map of the world's oceans, from the time of Murray & Renard (1891; see Fig. 7.4) to the present, will show that the greatest areas are dominated by pelagic brown or red clay and calcareous ooze, mainly comprising the calcareous tests of foraminifera, coccoliths and discoasters, as well as ostracods and other nannoplankton (Fig. 7.6 and Table 1); a subsidiary group consists of the polymorphous aragonite-shelled molluscs known as pteropods. Less important than calcareous plankton, but ubiquitous in the deep oceans and also of great significance, are the opaline silica-shelled groups – radiolaria, diatoms, and lesser groups such as silicoflagellates and sponge spicules (Fig. 7.7). Dinoflagellates, pollen, spores, fish fragments, and other organic material are also found in oceanic oozes.

The distribution of sea-floor deposits dominated by red or brown pelagic clay, a calcareous ooze, or a siliceous ooze, or combinations of these types, results from the delicate balance between regions of high fertility, such as the equatorial band, the distance from sources of terrigenous material, and the depth of water and thus of the carbonate compensation depth (CCD). The latter is the depth at which solution of carbonate exactly balances input at that level, so that no net increase of calcareous sediment takes place. The Deep Sea Drilling Project has confirmed geophysical evidence in showing that the pelagic pile typically reaches a maximum thickness of about 1000 m, although thicknesses of half or double this amount are not uncommon.

Calcareous ooze

Calcium carbonate, derived from the weathering of limestones and from dead shells, is delivered in solution to the oceans at such a rate that the total amount in the oceans is equivalent to the input for only one million years. As saturation is approached, the carbonate is precipitated near the surface where, in high fertility areas, it is absorbed by shell-building organisms and crystallises out in their growing tests. The shells of the dead organisms sink to the bottom and form the familiar calcareous ooze. However, if the depth of water is in the region of 4000 m or more, the solubility of $CaCO_3$ in sea water is increased by the greater pressure and lower temperature, and dissolution of the calcareous shells begins. Murray (in Murray & Renard, 1891) first noticed the general decrease in carbonate percentage with depth, and over the years recognition has developed of the lysocline and the carbonate compensation depth. The concept of the lysocline is more complex than that of the CCD. Originally intended to indicate a major facies change between well- and poorly-preserved calcareous assemblages on the ocean floor, it can be regarded as the zone between the carbonate dissolution depth and the carbonate compensation depth, at which there is the maximum change in the composition of calcareous fossil assemblages due to differential dissolution. The lysocline and CCD (Fig. 7.8) vary with time, from ocean to ocean, and also within oceans; a good discussion of these phenomena is given by Berger (1974).

The effect of fertility or productivity can be paradoxical and is linked with the supply of organic matter (Berger, 1974). In highly fertile continental slope areas, highly increased benthic activity, and CO_2-rich interstitial waters, result from the rich supply of organic matter. This leads to a shallowing of the lysocline, and shells will start

Figure 7.7 Planktonic organisms in oceanic oozes. **Calcareous**: (A) Foraminiferan. ×120; (B) Coccolith, ×9000; (C) Discoaster, ×6200. **Siliceous**: (D) Radiolarian, ×1000; (E) Diatom, ×1000; (F) Silicoflagellate, ×2200. SEM photographs by H. A. Buckley, S. Fairman and P. A. Sims.

to dissolve at only a few hundred metres depth. At the equator, on the other hand, high fertility results in depression of the lysocline because the small supply of organic matter leads to a high shell/organic carbon ratio and thus a low carbon dioxide content. The sequence of dissolution of the calcareous microfossils and the level at which dissolution takes place – in the water or on the sea-bed – are both complex questions. In general, however, coccoliths will resist dissolution longer than robust foraminifera, which in turn outlast the more delicate species.

Siliceous ooze

Siliceous planktonic organisms are notably abundant in a band running round the Antarctic region at around 60 °S. A second diatom ooze-rich strip follows the north Pacific coastline from Alaska to the Sea of Okhotsk. Radiolarian ooze predominates in an equatorial band across the Pacific, following up and down the west coast of America and merging with diatom muds, to link up with the high latitude diatom oozes (Table 2). It also occurs in the Gulf of Guinea, off West Africa, and in the central Indian Ocean (Fig. 7.6). The siliceous oozes dominate the areas of upwelling, where the nutrient content of the cold upwelled water is high, and consequently promotes high biological productivity. Phytoplankton predominate in high latitudes, zooplankton in the equatorial belt. Coastal regions are favoured, due mainly to the circulation of water offshore, which is replaced by water from below, enabling nutrients to reach the sunlit shallow water. Silica-rich belts thus tend to develop around the coastlines of oceans and also across the equatorial Pacific, with its high productivity, but not the Atlantic. Siliceous planktonic organisms are present throughout all the oceans, however, and it is largely relative abundance, controlled mainly by the depth of the CCD, which decides the dominant sea-floor deposits. It is quite common to find radiolarian clay in which these two relatively slowly-accumulating constituents are the most abundant in the absence of calcareous material. The siliceous fossil content in the peripheral oceanic areas,

Table 2 Percentage of deep ocean floor covered by pelagic sediments

	Ocean			
	Atlantic	*Pacific*	*Indian*	*World*
Calcareous ooze*	65·1	36·2	54·3	47·1
Pteropod ooze	2·4	0·1	—	0·6
Diatom ooze	6·7	10·1	19·9	11·6
Radiolarian ooze	—	4·6	0·5	2·6
Pelagic clays	25·8	49·0	25·3	38·1
	100·0	100·0	100·0	100·0
Relative size of ocean (% of total)	23·0	53·4	23·6	100·0

* Mainly foraminifera and coccoliths.
From Gass *et al.* (1978a, p. 32).

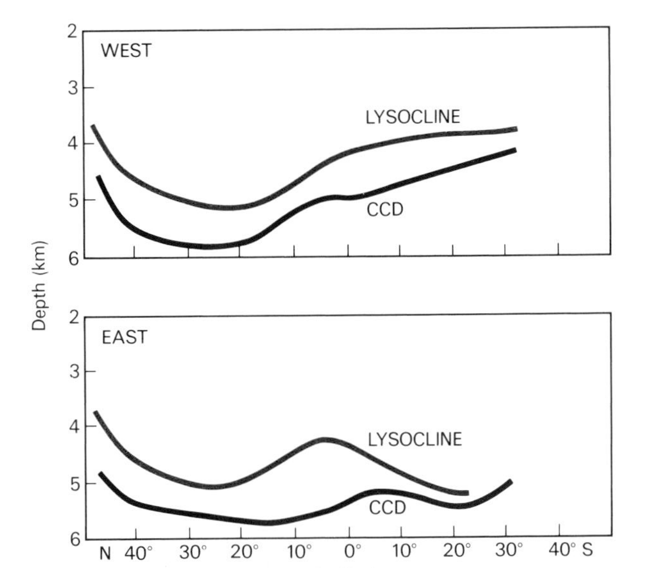

Figure 7.8 Depth distribution of lysocline and carbonate compensation depth (CCD) in the Atlantic Ocean, west and east of the Mid-Atlantic Ridge (after Berger, 1974).

however, where the production of terrigenous clastic material is also high, is sufficient to resist dilution by the inorganic component.

Except in the equatorial band, high production of silica shells does not usually follow high production of carbonate shells. The pH values ideal for preservation are markedly different – silica dissolving in an alkaline and carbonate in an acid environment. The CCD is important – where it is deep, as in the equatorial Pacific, mixed calcareous and siliceous oozes result. Along the edge of the Antarctic, where the CCD is shallow, chiefly siliceous oozes are found. Thus the relative enrichment of siliceous fossils at sub-lysocline levels, at the expense of calcareous remains, determines the nature of the sediment. The special environment afforded by peripheral oceanic regions of high fertility encourages siliceous, but inhibits calcareous, growth. Further, the tendency for silica corrosion to occur in the silica-undersaturated upper waters, and for dissolution of calcareous organisms to take place at depth, below the lysocline, all support a strong negative correlation between ideal growth conditions for the two types of organism and thus accentuate their relatively clear-cut facies boundaries.

The source of silica in the oceans is mainly from river supply, but appreciable amounts are also derived locally from the breakdown of glass in volcanic ash and glassy basalts and by the alteration and disintegration of volcanic glass, including hydration (palagonitisation), where its dispersal is aided by hydrothermal circulation. Just as foraminifera and coccoliths secrete calcite, siliceous organisms precipitate opaline silica which is integrated into their tests. This process is independent of the dissolution rate which is also analogous with that of carbonate, but with one difference – silica dissolution is not depth-controlled but takes place only when undersaturation in the upper waters creates a demand; carbonate, on the other hand, invariably dissolves at depth in any case. Undissolved opaline silica, in the shells of planktonic organisms, and concentrations of silica generated by the alteration of volcanic deposits remain on the sea-floor to be buried and subsequently form chert.

Bedded chert (and porcellanite) bands of Cretaceous and Eocene age in deep-sea oozes greatly surprised geologists when they were first revealed in the Atlantic, Pacific, and Indian Oceans by deep-sea drilling. They are thought to have developed by the mobilisation and recrystallisation of silica, leading ultimately to a diagenetic layered rock of microcrystalline quartz with the residual fossils also filled with quartz. Intermediate stages involved the development of tridymite and disordered cristobalite quartz minerals, sometimes found associated with the formation of the clay minerals sepiolite and palgorskite, and the zeolite clinoptilolite (Calvert, 1974). Together with montmorillonite clay this assemblage, when present, argues for a volcanic provenance for much of the silica, although the greatest source of supply is thought to be biogenic. The mobilised silica appears to have migrated horizontally along bedding planes, or vertically: vertical gradients may develop where organic remains generate a silica-poor reduced layer into which silica migrates from the relatively silica-rich oxidised horizons (Heath & Dymond, 1973). The division of cherts, into pelagic (both 'clean', containing planktonic remains and authigenic minerals only, and 'dirty', with volcanic detritus, clay minerals and iron oxides) and hemipelagic (containing much terrigenous silt and iron sulphide), correlates with their mode of formation: pelagic cherts occur in nodular form in carbonates but bedded in clay-rich sediments, whilst hemipelagic cherts are always bedded.

Pelagic clay

The brown pelagic clay, or red clay of Murray & Renard (1891), which forms the sea-floor in all the abyssal parts of the ocean, has been the subject of considerable discussion since the days of the *Challenger* expedition. It exists as a major sediment because of the dissolution of carbonate at and below the lysocline, for brown clay forms very slowly and above the CCD would constitute only a minor component in the rapidly forming calcareous ooze. The radiolarian clay of the Indian Ocean illustrates this point: there a mixture of brown clay and siliceous ooze is found below the CCD, with accretion of the radiolaria and some diatoms taking place slowly enough to allow the clay fraction to account for about a half of the total sediment.

Normally, however, brown clays are low in both silica and carbonate; in texture they are soft, plastic, and greasy to touch, with a mean grain size close to one micron. They were thought by Murray & Renard (1891) to result mainly from the decomposition of volcanic pyroclastic material, but this idea was later rejected because of the failure to find progressive changes in clay cores such as would be expected from the weathering of volcanic layers. It is now accepted that pelagic clays, which have a very constant chemical composition (Table 3), result from the very slow accumulation of volcanic dust and the products of localised volcanic activity, cosmic spherules, and aeolian dust. They are also thought to contain rather more fine-grained products of continental erosion than the original definition of pelagic sediments would suggest. These detrital components are augmented by authigenic minerals, including ferromanganese oxides and hydroxides which often form spots and micronodules; fish teeth; and siliceous fossil debris: radiolarian and some arenaceous foraminiferal tests, and spicules. Brown clays are frequently burrowed and show colour mottling. Carbonate fossils are virtually absent and support the accepted view that the clays form very slowly, at great distance from land, and at depths where dissolution of silica is nearly total and of carbonate complete. The clay minerals provide most of the evidence concerning provenance: they form some two-thirds of the clay-size fraction of brown clay, which forms some 90 per cent of the whole. Of these minerals, montmorillonite (largely derived from basic volcanic material) and illite (of continental origin) form two-thirds to three-quarters of the whole; the remaining common clay minerals – kaolinite and chlorite – are also continental in origin. Kaolinite is produced by warm weather weathering, whilst chlorite is characteristic of high latitude and colder climates.

Table 3 Average chemical composition of pelagic clay

SiO_2	55·34
TiO_2	0·84
Al_2O_3	17·84
Fe_2O_3	7·04
FeO	1·13
MnO	0·48
CaO	0·93
MgO	3·42
Na_2O	1·53
K_2O	3·26
P_2O_5	0·14
H_2O	6·54
$CaCo_3$	0·79
$MgCO_3$	0·83
Org. C	0·24
Org. N	0·016
Total	100·366

From Chester and Aston (1976, p. 290).

Most of the detrital and biogenic material that forms pelagic clay occurs in particles of diameter $< 2\ \mu m$. Such particles, if obeying Stokes' law, would take 70 years to reach bottom at 5000 m. Whilst some, especially aeolian, particles do take even centuries to settle, the majority are aggregated at the surface by flocculation induced by wave turbulence, or by filter-feeding organisms also at the surface which, having ingested the particles, aggregate them as faecal pellets. Thus the clay minerals and planktonic tests are assembled into aggregates in excess of 100 μm diameter, enabling them to sink more rapidly. Whilst this size fraction may form more than 75 per cent of the vertical flux, it remains a tiny proportion (4 to 5 per cent) of the total concentration of suspended particles.

Rates of sedimentation

Sedimentation rates are measured by dividing the thickness of the sediment at a particular area of crust by its age, determined from palaeontological or palaeomagnetic evidence. However, they are affected also by ocean-floor topography and a number of other factors already mentioned such as areas of high fertility due to upwelling and other factors; physico-chemical conditions such as the CCD and rate of silica solution; and detrital contributions (from the continents, volcanic sources, aeolian dusts, and extra-terrestrial sources). Table 4 gives the sedimentation rates from several different environments, in which the values reflect the factors mentioned above. Active continental erosion environments, usually involving turbidity currents, range up to 2000 mm ty^{-1}; highly productive areas of calcareous ooze formation can reach 60 mm ty^{-1}. At the bottom end of the scale, below the CCD and silica solution level, brown or red clay can accumulate as slowly as 0·1 mm ty^{-1}.

Table 4 Rates of accumulation of Recent and sub-Recent sediments

Facies	*Area*	*mm* ty^{-1}
Terrigenous mud	California borderland	50–2000
	Ceara abyssal plain	200
Calcareous ooze	North Atlantic (40–50° N)	35–60
	North Atlantic (5–20° N)	40–14
	Equatorial Atlantic	20–40
	Caribbean	~ 28
	Equatorial Pacific	5–18
	Eastern equatorial Pacific	~ 30
	East Pacific Rise (0–20° S)	20–40
	East Pacific Rise (~ 30° S)	3–10
	East Pacific Rise (40–50° S)	10–60
Siliceous ooze	Equatorial Pacific	2–5
	Antarctic (Indian Ocean)	2–10
	North and equatorial Atlantic	2–7
	South Atlantic	2–3
Brown clay	Northern North Pacific (muddy)	10–15
	Central North Pacific	1–2
	Tropical North Pacific	0–1

From Berger (1974, p. 216).

Secondary deposits

In this section the products of some further processes are described. In the oceanic environment where pressure and, especially, temperatures are so much lower, they are only rarely associated with metamorphism and are known by several other terms.

Diagenesis and authigenesis

Compaction and dehydration of oceanic oozes and clays, by hydrostatic (burial) pressure alone, leads to the formation of thin limestones or chalks (sometimes when fine-grained called micrite or micarb); cherts; and stiff clays, shales, and claystones. The recrystallisation of the carbonate commonly involves some loss of strontium to the sea- or pore-waters with which it is in contact, and a considerable gain in magnesium, resulting in partial dolomitisation. A slow change in the environment, from oxidising to anaerobic, may occur, resulting in the development of pyrite (FeS_2) from reduced iron and sulphate.

Commonly known as diagenesis (see also Chapter 4), these processes are often accompanied even at very low temperatures by authigenesis – the growth of new minerals, formed within the sediment (or even at the interface with the water) by mineral–water reactions. The minerals formed, such as zeolites, alkali feldspar, and some clay minerals (Fig. 7.9), depend on such factors as depth of burial, the mineralogical and chemical composition of the detrital content of the sediment, and the composition of the interstitial water. Of the clay minerals, montmorillonite and other smectites are chiefly authigenic, whilst kaolinite, illite, and chlorite are generally detrital. The zeolite phillipsite, first recognised in deep-sea deposits on the *Challenger* expedition, is one of the most common authigenic zeolites. Clinoptilolite (-heulandite) is the other widespread mineral in this group and has already been mentioned as a product of the alteration of volcanic pyroclastic material which, together with montmorillonite and silica minerals, can lead to the development of chert. Analcite and other, rarer, zeolites, may also develop from altered volcanic material. The onset of the development of such zeolites as natrolite and laumontite defines the change from diagenesis to low grade metamorphism.

The decomposition of minerals and recombination of their chemical constituents leads also to the formation of montmorillonite and other smectites, sepiolite and palygorskite, as well as some illite and chlorite (among the clays); alkali feldspar; micaceous minerals such as glauconite and celadonite; carbonates such as dolomite and siderite; sulphates such as baryte; phosphates; and the sulphides of iron and other base metals. Phillipsite, clinoptilolite, and montmorillonite, the principal authigenic minerals formed from the alteration of volcanic material, are most common in pelagic sediments from the Pacific Ocean. This can be explained by the high incidence of volcanic activity, relatively low input of river sediment, and low productivity coupled with a shallow CCD and a high rate of re-solution of silica (Gass *et al.*, 1978). The

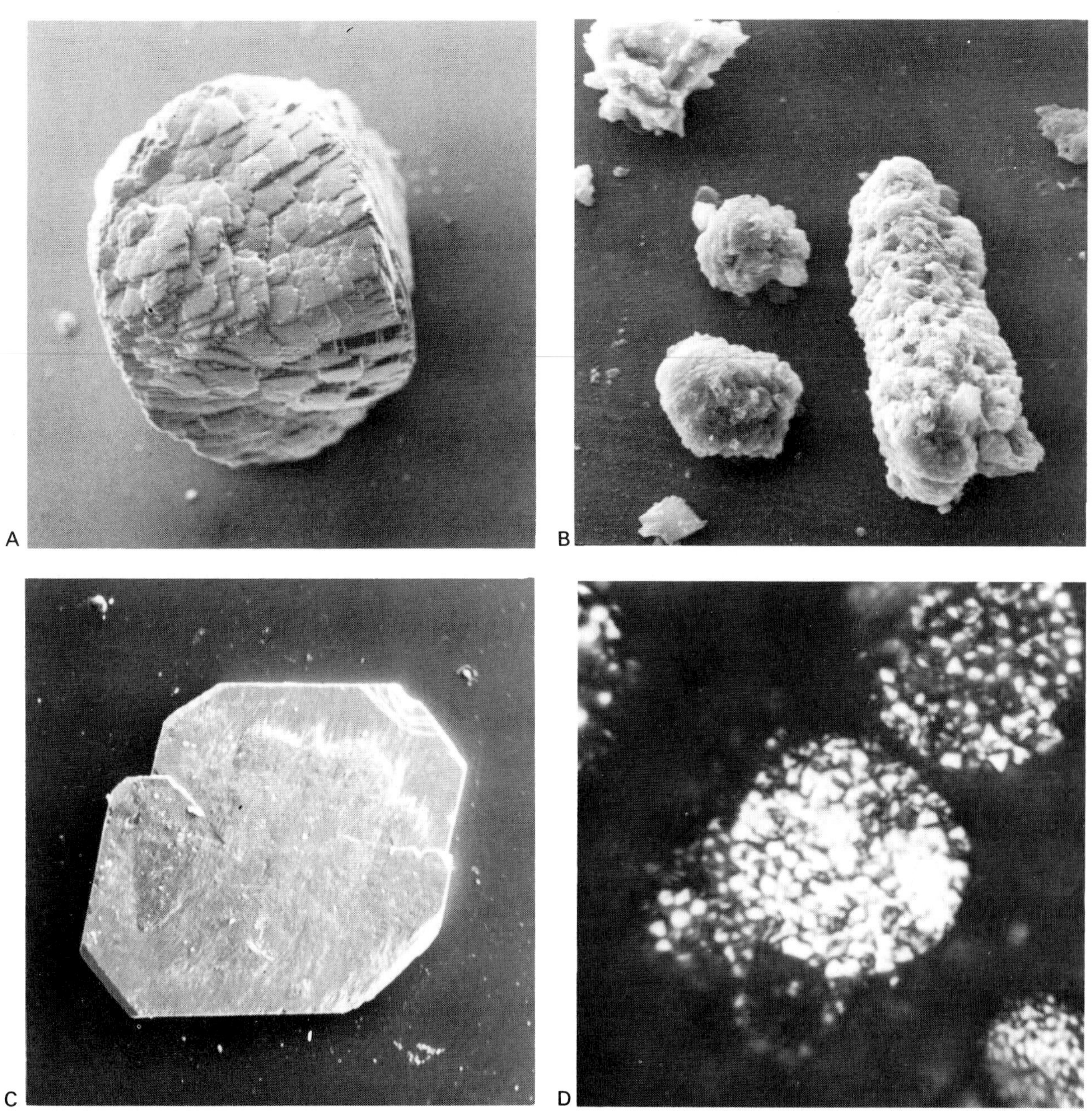

Figure 7.9 Authigenic, diagenetic and other secondary minerals in oceanic sediments. (A) Rhodochrosite (Mn carbonate), ×100; (B) Siderite (Fe carbonate), ×1000; (C) Baryte (Ba sulphate), ×50; (D) Octahedral and framboidal pyrite (Fe sulphide), ×10000; (E) Clinoptilolite (zeolite) on lepispheres of opal – CT, ×1100; (F) Palygorskite needles in porcellanite, ×4000; (G) Lepispheres of opal – CT (disordered cristobalite-tridymite mixture), ×5500; (H) Metasomatic garnet, ×8000. SEM photographs by H. A. Buckley (A, B, C, H) and U. von Rad (E, F, G); D is a reflected light optical photomicrograph by A. J. Criddle. All are reproduced by kind permission of the Deep Sea Drilling Project.

formation of clinoptilolite or phillipsite, both derived from volcanic glass, depends on the ratio of silica to aluminium. The high Si/Al ratio zeolite clinoptilolite results from the slow alteration of acidic glass, and therefore derives from older volcanic material, whereas the low Si/Al ratio phillipsite results from the rapid alteration of young basic material. Analcite usually results from the reaction between hot pyroclastic material and sea water, but may also be derived from the burial metamorphism of pre-existing phillipsite or clinoptilolite (Petzing & Chester, 1979).

A result of such authigenic growth is the fluxing of elements between decomposing mineral, pore-water and sea-water; much of the freed material, such as silica and

calcium, will be taken up by the new minerals but some will be released into the ocean. Compensating absorption of elements such as magnesium and potassium in the newly-formed minerals, sometimes termed 'sinks', then helps to regulate the chemical composition of the oceans.

Phosphorite is a highly localised but important authigenic mineral where it occurs, mainly on the the continental shelf and slope at depths greater than about 150 m. It is confined to areas of upwelling and high fertility, where the sea-floor is subjected to a 'rain' of phosphatic source material from the organisms living in the sea, and it usually forms as nodules or encrustations in association with or by replacement of carbonate (Baturin & Bezrukov, 1979). Jones & Goddard (1979) have shown how such replacement off the north-west coast of Africa is restricted to limited depth zones, which were controlled by subsidence of the sea-floor. Thus phosphatisation had taken place at shallow levels, resulting in replacement rocks with up to 40 per cent P_2O_5; below about 1000 m the carbonate was unaffected and dredge hauls include both kinds of material. Phosphorites also occur on seamounts; they are often associated with glauconite, siliceous sediments, and volcanic material, as well as carbonates, and are found mainly off the west coast of North and South America, and on Chatham Rise, east of New Zealand; however, they seem to be forming today only off the coasts of Peru, Chile, and Namibia.

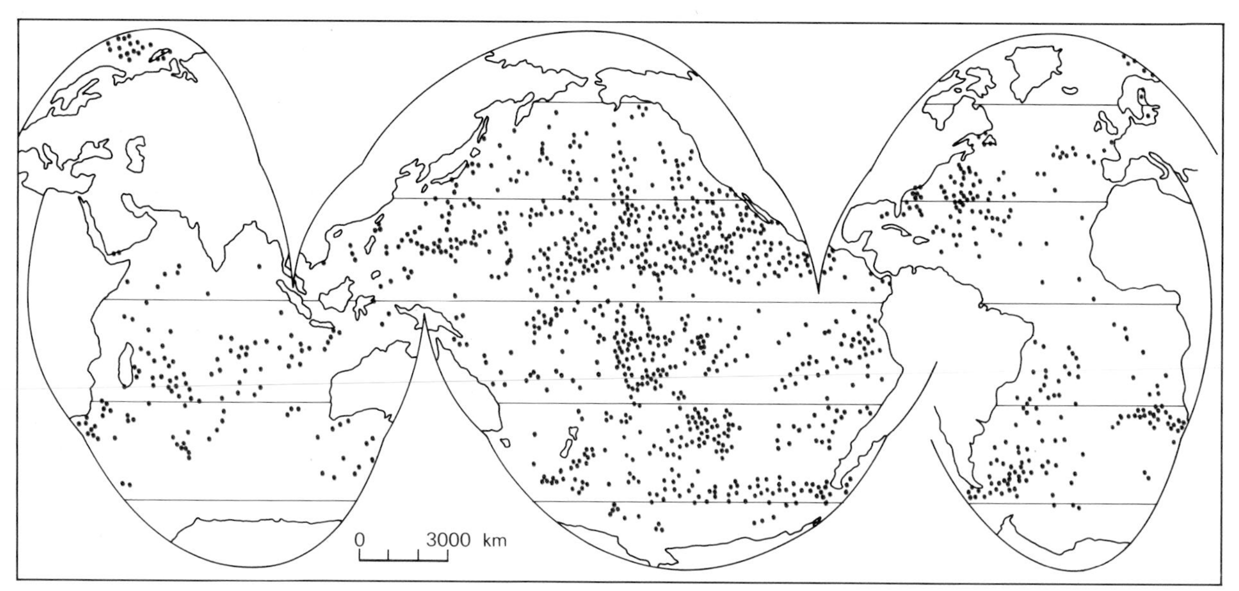

Figure 7.10 Areas from which ferromanganese nodules and concretions have been reported (after Berger, 1974).

Ferromanganese deposits

The production of some oxides and hydroxides of iron and manganese can also be regarded as authigenic, although many of them are also hydrothermal products. Authigenic ferromanganese deposits include the well-known, problematical, and potentially valuable manganese nodules. The first nodules were probably dredged from the Kara Sea by the *Sofia* Expedition in 1868. However, they were first recovered from the deep ocean by the *Challenger* near the Azores, where they were first thought to be coagulated lumps of tar or pitch. Chemical analysis soon revealed their true composition and the ship went on to dredge many more from the Pacific. The origin of manganese nodules has occupied many pages in the literature, which has been reviewed by Cronan (1976). Murray (in Murray & Renard, 1891) regarded decomposed volcanic materials as the ultimate source of the manganese, whilst Renard considered that it was derived from solution in the sea water.

Manganese nodules occur in all the oceans (Fig. 7.10) and vary in size from micronodules measuring up to 1 mm, through the 0·5 to 20 cm range, to slabs or 'pavements' and crusts of wide extent. Many sea-bottom photographs show potato-sized nodules evenly dispersed across the sea-floor (Fig. 7.11). Most, but not all, contain solid nuclei of rock or mineral fragment, or perhaps a shark's tooth or other organic detritus. Around these nuclei the manganese minerals are deposited in concentric layers of oxide (δ MnO_2), todorokite, and lesser birnessite. The slab type of deposit encrusts hard rocks such as basalt, and there is no indication of elemental migration between growing crust and substrate. A feature of economic importance is the minor element content of nodules, mainly cobalt, nickel, and copper. High copper and nickel accompany a high Mn/Fe ratio, whilst cobalt is richer in nodules of higher iron content; the latter type, with less manganese, are more commonly found at shallower depths where they grow more rapidly and are more highly oxidised.

Of the many hypotheses put forward to explain the origin of manganese nodules, the derivation of manganese from solution in sea water (hydrogenous) is the most widely accepted. Other theories include diagenetic precipitation within the wet sediment and subsequent migration to the surface in the same way as iron; however, iron is less mobile and tends to form sulphides preferentially. A third theory is the derivation of manganese (and iron) from hydrothermal activity (see below) and by the weathering of basalt. The nodules then accrete at the surface at a rate estimated as several millimetres per million years, which is similar to the estimated rate of accumulation of the metals in sediments.

Nodules are found less frequently at more than two metres or so below the ocean floor, but the reason for this is not understood: in the past it was thought that they only occurred at the surface, those below becoming redissolved after burial. Biological processes have also been invoked in partial explanation of nodule growth and also in demonstrating a correlation between the composition of nodules and the biological productivity of the regions where they occur. Thus manganese nodules possess interest and pose problems, and, because of their trace element content, are of potential economic value out of all proportion to their abundance on the ocean floor.

Metasomatism

The second major group of ferromanganese or metalliferous sediments are regarded as hydrothermal, or metasomatic,

Figure 7.11 (A) Manganese nodule 'field'; (B) single potato-sized nodule; and (C) cut and polished section of nodule, showing concentric growth around nucleus. Photograph (A) supplied by the Institute of Oceanographic Sciences, others by D. S. Cronan.

in origin. It is now widely accepted that hydrothermal convection cells operate through the Earth's crust and where such cells impinge on recently erupted basalt, as on the flanks of a mid-ocean ridge, much alteration and leaching of the basalt takes place down to considerable depths. Some elements, for example manganese and iron, oxidise and precipitate more readily than others at the surface of the ocean floor, at the sediment–water interface, and are deposited relatively rapidly as metalliferous sediments. These oxides and hydroxides of iron and manganese, as well as zinc, lead, copper, nickel, cobalt, silver, gold and other metals, were first recognised on the East Pacific Rise by Böstrom & Peterson (1966). Such activity takes place in areas of high heat-flow, usually on the flanks of spreading ridges but also off-ridge. Perhaps the best examples occur at the Galapagos hydrothermal mounds; in the FAMOUS area of the Mid-Atlantic Ridge; and in the Red Sea, where they are overlain in three deeps by hot brines (salt-enriched sea water). A recent discovery of massive deep-sea sulphides on the East Pacific rise, off Baja California (Francheteau *et al.*, 1979), is of special interest. The sulphides of zinc (containing 29 per cent metal), copper (6 per cent), and iron occur as the minerals sphalerite and pyrite, with lesser marcasite and chalcopyrite, and also in amorphous form. They are directly analogous with the sulphides found on land in the ophiolite sequences of, for example, the Troodos massif of Cyprus. However, sulphides, which also occur in the Red Sea, are minor compared with the oxide and hydroxide deposits.

An unusual form of hydrothermal activity is known from the Mid-Indian Ocean Ridge (Kempe & Easton, 1974). There, a hydrothermal cell operated on the ridge flank in an area where nearly 20 m of nannoplankton ooze had been deposited. Instead of bulk precipitation of Fe–Mn sediments, iron, aluminium and silicon are thought to have been introduced into the wet lime mud ($CaCO_3$) of the ooze, where they reacted to form microscopic CaFe garnets (Fig. 7.9H). Oxygen isotope measurements on these sediments suggest a temperature of formation for the garnets of about 170 °C.

Metamorphism

Baking, or low-grade contact metamorphism, takes place when basalt flows are extruded over, or (as sills) intruded within, pre-existing ooze. A micarb or micrite chalk is then formed, in which the foraminifera and nannofossils are destroyed by recrystallisation and overgrowth. Growth of 'authigenic' minerals, such as zeolites, occurs in such an environment if there is a substantial clay component in the ooze. Oceanic metamorphism can also lead to the formation of fluid hydrocarbons.

Palaeoenvironments

What can deep ocean sediments tell us about tectonic environments or movements in the ocean basin? Different sedimentary environments will develop at different types

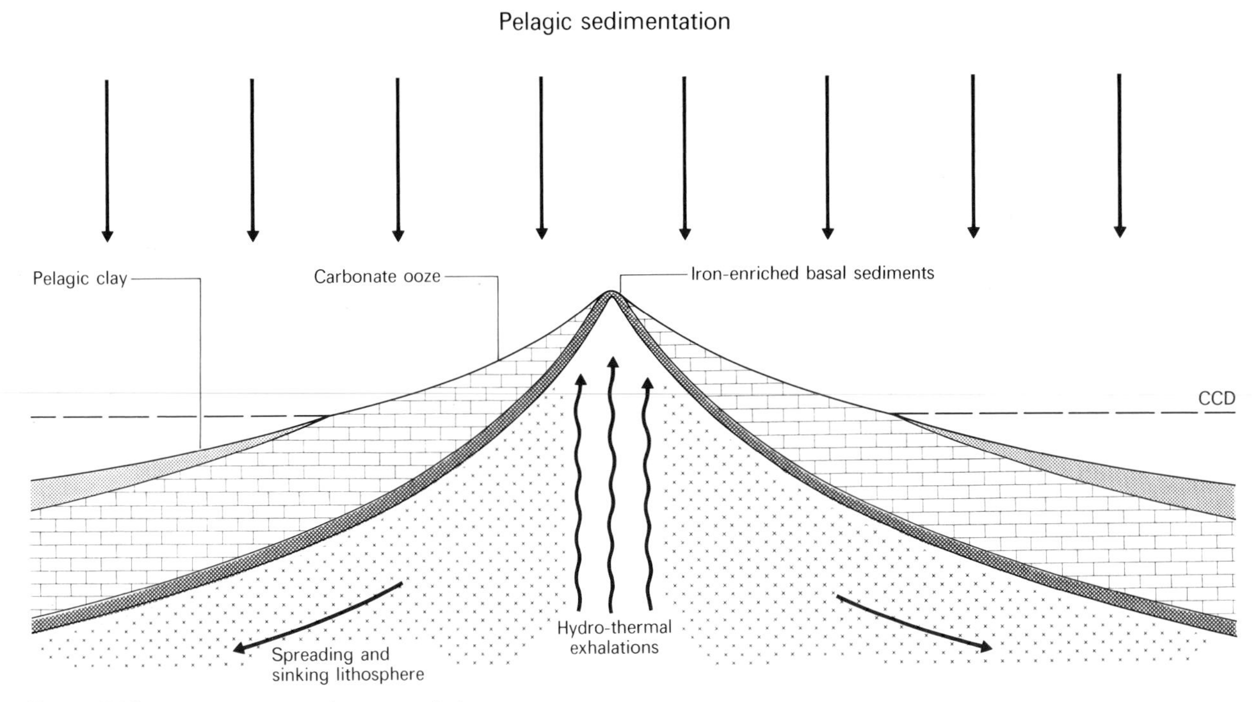

Figure 7.12 The distribution of sediment facies accumulating on lithospheric plates spreading from a mid-ocean ridge (after Davies and Gorsline, 1976, fig. 24.30).

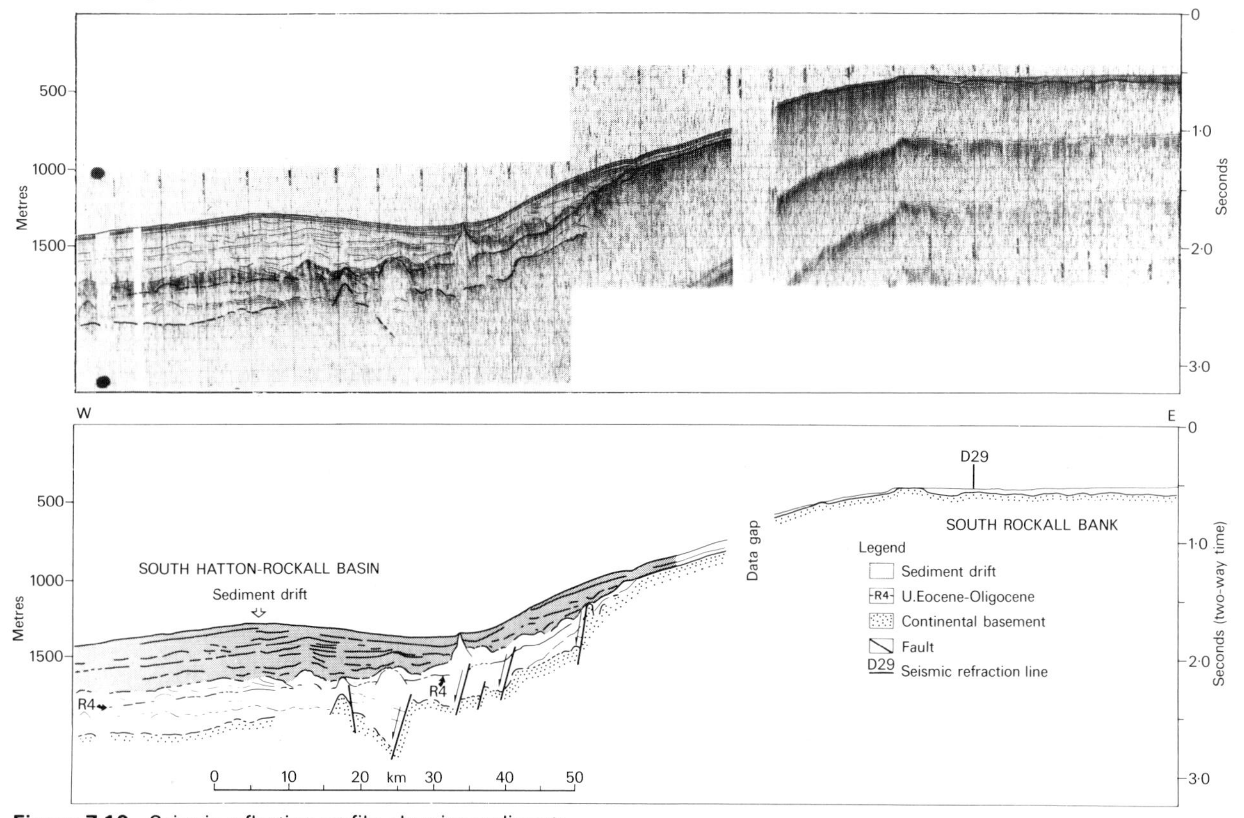

Figure 7.13 Seismic reflection profile, showing sediments and structure of the ocean floor north-west of Scotland. Reproduced by permission of D. G. Roberts and the Institute of Oceanographic Sciences.

of continental margin (see also Chapter 11). Continent–ocean (aseismic or passive) and ocean–continent (seismic or active, with subduction or obduction) plate boundaries lead to highly complicated environments that are poorly understood and sampled. The cycle of rifting–igneous activity–separation at a passive margin is perhaps simpler than the sequence of events at a subduction zone: trench–Benioff zone–island arc. An island arc, such as Indonesia, may be oceanic in that it is separated from the continent by a marginal basin and possibly a secondary back arc but its sediments are different from those of the open oceans. An ocean–ocean plate boundary at a spreading ridge gives rise to the relatively simple sedimentary environment commonly encountered in the oceans (Fig. 7.12). Away from the ridge on both sides are the basins, whose further sides are bounded by the next ridge or margin. A large basin of uniformly flat topography is known as an abyssal plain, whilst small depressions containing sediments in the flanks of a ridge system are named ponds. The contraction due to cooling that takes place after the oceanic lithosphere leaves a spreading centre, involving first uplift and subsequent subsidence, plays a major role in determining the topography of the sea-floor; this is a response to the need for isostatic readjustment as the new crust cools, and has reached a rate of as much as 60 to 100 m per million years. Thus there is a correlation between the age of the crust and the depth of the basin or plain as subsidence has proceeded. Such movement will result in diachronism and non-conformities within the sediments deposited on the oceanic plate (Fig. 7.13).

The ages of the oldest sediments vary considerably in each of the oceans, and are a record of the opening stages of the development of that particular ocean. The ages vary from Jurassic at both margins of the simple and symmetrical Atlantic to quite recent examples of new rifting. The Pacific is asymmetrical, and there the active East Pacific Rise and Juan de Fuca Ridge are displaced to the eastern side of the ocean, by over-riding of the Pacific plate by the American plate, with no sediments older than Eocene age present at their eastern margin. The sediments on the western margin, on the other hand, are as old as Upper Jurassic just seaward of the bordering trenches. The most complex of the three major oceans is the Indian, which also has Upper Jurassic sediments in the north east, against the Indonesian trench. On the western margin, near the triple point junction where the active and asymmetric Mid-Indian Ocean Ridge diverges into two branches, the oldest sediments are Upper Cretaceous. In each ocean the fossil ages generally agree with the ages obtained from magnetic stripes and isotopic dating, discussed elsewhere (see Chapters 3 and 11).

The ocean basins are not static. To demonstrate an ever-changing ocean, we can again consider the Pacific. As well as in its asymmetry, this ocean is marked by a northward drift of the crust in addition to its east–west spreading. Cores show that as the plate moved over the Equator, the high fertility in that region resulted in diachronous deposition of a belt of organic oozes in water depths normally sufficiently close to the CCD to result in the formation of abyssal clay. This northward drift is also demonstrated by the progressive increase in age northwards of the Hawaiian Islands and, to their north, of the submerged Emperor seamounts (Heezen & McGregor, 1973). The Mediterranean Sea is another example of a changing ocean. Once part of the Tethys Ocean, it became closed at the eastern end when the Arabian and Eurasian plates collided. Left with only narrow straits connecting it, much as at present, with the open ocean to the west, and with a drier climate, it rapidly dried up, with the deposition of thick evaporite beds. At the end of the Miocene (see Chapter 14), evaporite deposition ceased with a change in the climate, but the Mediterranean remained as an inland and practically closed sea. A similar history applies to the Red Sea, once linked with the Mediterranean, before the opening into the Indian Ocean developed, also at the end of the Miocene.

Acknowledgements

Thanks to D. S. Cronan, A. J. Fleet, and E. J. W. Jones for making many valuable suggestions.

References

Baturin, G. N. & Bezrukov, P. L. 1979. Phosphorites on the sea floor and their origin. *Marine Geology* **31**: 317–332.

Berger, W. H. 1974. Deep-sea sedimentation, pp. 213–241 *in* Burk, C. A. & Drake, C. L. (Eds.) *The geology of continental margins*. Berlin: Springer-Verlag.

Böstrom, K. G. & Peterson, M. N. A. 1966. Precipitates from hydrothermal exhalations on the East Pacific Rise. *Economic Geology* **61**: 1258–1265.

Calvert, S. E. 1974. Deposition and diagenesis of silica in marine sediments. *Special Publication of the International Association of Sedimentologists* **1**: 273–299.

Chester, R. & Aston, S. R. 1976. The geochemistry of deep-sea sediments, pp. 281–390 *in* Riley, J. P. & Chester, R. (Eds.) *Chemical oceanography, Vol. 6*. London: Academic Press.

Cronan, D. S. 1976. Manganese nodules and other ferro-manganese oxide deposits, pp. 217–263 *in* Riley, J. P. & Chester, R. (Eds.) *Chemical oceanography, Vol. 5*. London: Academic Press.

Davies, T. A. & Gorsline, D. S. 1976. Oceanic sediments and sedimentary processes, pp. 1–80 *in* Riley, J. P. & Chester, R. (Eds.) *Chemical oceanography, Vol. 5*. London: Academic Press.

Fleet, A. J. & McKelvey, B. C. 1978. Eocene explosive submarine volcanism, Ninetyeast Ridge, Indian Ocean. *Marine Geology* **26**: 73–97.

Francheteau, J., Needham, H. D., Choukroune, P., Juteau, T., Séguret, M., Ballard, R. D., Fox, P. J., Normark, W., Carranza, A., Cordoba, D., Guerrero, J., Rangin, C., Bougault, H., Cambon, P. & Hekinian, R. 1979. Massive deep-sea sulphide ore deposits discovered on the East Pacific Rise. *Nature*, London **277**: 523–528.

Gass, I. G. & course team. 1977. *Oceanography: introduction to the oceans, unit 1*. 79 pp. Milton Keynes: The Open University Press.

Gass, I. G. & course team. 1978. *Oceanography: sediments, unit 12.* 50 pp. Milton Keynes: The Open University Press.
Hargraves, R. B. 1976. Precambrian geologic history. *Science*, New York **193**: 363–371.
Heath, G. R. & Dymond, J. 1973. Interstitial silica in deep-sea sediments from the north Pacific. *Geology* **1**: 181–184.
Heezen, B. C. & MacGregor, I. D. 1973 (November). The evolution of the Pacific. *Scientific American* **229**: 102–112.
Jones, E. J. W. & Goddard, D. A. 1979. Deep-sea phosphorite of Tertiary age from Annan Seamount, eastern equatorial Atlantic. *Deep-Sea Research* **26**: 1363–1379.
Kempe, D. R. C. & Easton, A. J. 1974. Metasomatic garnets in calcite (micarb) chalk at Site 251, southwest Indian Ocean. *Initial Reports of the Deep Sea Drilling Project* **26**: 593–601.
Murray, J. & Renard, A. F. 1891. Deep-sea deposits. *Report on the scientific results of the voyage of H.M.S. Challenger.* 525 pp. London: Stationery Office.
Petzing, J. & Chester, R. 1979. Authigenic marine zeolites and their relationship to global volcanism. *Marine Geology* **29**: 253–271.
Piper, D. J. W., von Huene, R. & Duncan, J. R. 1973. Late Quaternary sedimentation in the active eastern Aleutian trench. *Geology* **1**: 19–22.
Shepard, F. P. 1963. *Submarine geology*. 557 pp. New York: Harper & Row.
Smith, A. G., Briden, J. C. & Drewry, G. E. 1973. Phanerozoic world maps. *Special Papers in Palaeontology* **12**: 1–39.
von Huene, R. 1974. Modern trench sediments, pp. 207–211 *in* Burk, C. A. & Drake, C. L. (Eds.) *The geology of continental margins.* Berlin: Springer-Verlag.

D. R. C. Kempe
Department of Mineralogy
British Museum (Natural History)

CHAPTER 8

The development of the atmosphere

G. Turner

The atmosphere of the Earth is, in terms of amount, a rather minor feature, accounting for a little short of one millionth of the total mass of the Earth. Nevertheless this apparently trivial gaseous envelope has had a profound, and in many ways dominant, influence on the evolution of the Earth's surface. As a geological weathering agent, as an agent for the erosion and transport of weathered rock debris, and, above all, as a major component in the cycle of life, the atmosphere is of supreme importance. The importance of the atmosphere in all these respects has been emphasised particularly in the last decade by the ability to compare the Earth's surface rocks with those of our companion in space, the Moon (Taylor, 1975). Devoid of an atmosphere the lunar surface is lifeless and its basaltic rocks, although several thousand million years old, are more pristine and unweathered than modern terrestrial basalts. Moreover, the only feeble instrument of erosion and transport on the Moon is the slow influx of meteorites and micrometeorites, which results in all but a small fraction of the very old thin surface 'soil' layer being locally derived.

The purpose of the present chapter is to look at those aspects of the atmosphere that are especially relevant to the geological and biological evolution of the Earth. To achieve this we shall begin with a brief description of the main features of the present atmosphere and follow this with a discussion of the ways in which the atmosphere interacts chemically with the Earth's surface and biosphere. Having thus established a frame of reference for the present day we shall look back in time at ideas of how the atmosphere originated and finally the way in which its composition may have evolved through time, mainly as a result of the growth and development of living organisms.

Today's atmosphere

In contrast to the solid Earth, the atmosphere is remarkably simple in composition, consisting almost entirely of three elements – nitrogen, oxygen and argon. Together these account for 99·97 per cent of the atmosphere, the remainder being made up of carbon dioxide, the other inert gases He, Ne, Kr, Xe, and compounds of the major constituents (see Table 1). Apart from photochemically induced variations – principally the existence of the ozone layer – the relative proportions of atmospheric gases remain constant up to a height of around 100 km. Above this height the mean free path of the molecules becomes so large and mixing processes so inefficient that gravitational separation on the basis of molecular weight occurs, with the proportion of lighter gases increasing with height.

Atmospheric pressure and density decrease approximately exponentially, with a scale height between 8 and 6 km in the lower atmosphere. The scale height (the change in height corresponding to a factor e ($= 2{\cdot}718$) decrease in pressure or density) is equal to (kT/mg), where

Table 1 The average composition of the atmosphere

Gas	*Composition by volume* (ppm)*	*Total mass (kg)*
N_2	780 900	$3{\cdot}865 \times 10^{18}$
O_2	209 500	$1{\cdot}184 \times 10^{18}$
Ar	9300	$6{\cdot}55 \times 10^{16}$
CO_2†	330	$2{\cdot}6 \times 10^{15}$
Ne	18	$6{\cdot}4 \times 10^{13}$
He	5·2	$3{\cdot}7 \times 10^{12}$
CH_4	1·5	$4{\cdot}3 \times 10^{12}$
Kr	1	$1{\cdot}5 \times 10^{13}$
N_2O	0·5	$4{\cdot}0 \times 10^{12}$
H_2	0·5	$1{\cdot}8 \times 10^{11}$
O_3‡	0·4	$3{\cdot}1 \times 10^{12}$
Xe	0·08	$1{\cdot}8 \times 10^{12}$

* The volume concentrations give directly the relative numbers of different molecules and atoms by virtue of Avogadro's law.

† The amount of CO_2 in the atmosphere is currently increasing at a rate of 0·7 ppm per year.

‡ The ozone is concentrated as a layer between 15 and 30 km.

k is Boltzmann's constant, T the absolute temperature, m the mean molecular weight of the atmosphere and g the gravitational acceleration.

Structurally the atmosphere is divided into a number of zones that are related to the temperature profile (Fig. 8.1) and the effects of the absorption of energy from the Sun. The lower part of the atmosphere, the troposphere, is dominated by convection as a result of heating of the Earth's surface by solar radiation in the visible and near infra-red wave-bands. Because of convection the temperature decreases adiabatically with height.

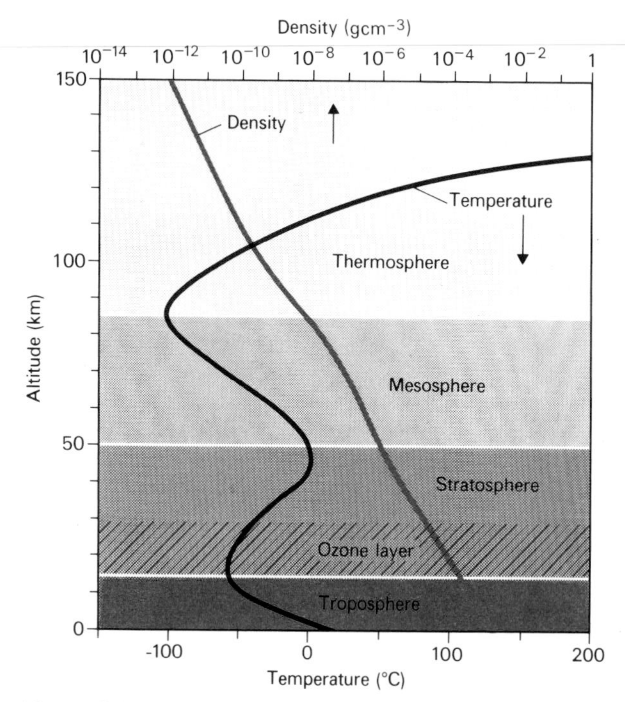

Figure 8.1 Variation of temperature and density with altitude in the Earth's atmosphere. The temperature rises with height in the stratosphere due to the absorption of low energy solar ultra-violet and in the thermosphere due to the absorption of higher energy solar ultra-violet. This leads to stable stratified regions with little vertical mixing. In contrast major energy absorption below the troposphere and mesosphere causes large-scale vertical circulation (convection) in these regions, and results in an adiabatic temperature profile.

At a height of around 10 to 15 km the troposphere gives way to the stratosphere, so called because of the absence of any strong vertical circulation. The stratification arises because of atmospheric heating due to the absorption of low energy solar ultra-violet rays. This causes the temperature to fall at a rate below the adiabetic rate and eventually to increase with height, thereby inhibiting convective motion and vertical mixing. The absorption of solar ultra-violet also produces small chemical changes, the most significant being the production of a layer of ozone (the ozonosphere) from around 15 to 30 km. The ozone layer and the associated absorption of harmful ultra-violet rays are vital to the existence of most forms of life on the Earth's surface.

The rate of absorption of energy reaches a maximum at 50 km and thereafter, in a region known as the mesosphere, decreases up to a height of 85 km. With reduced energy influx the temperature also decreases with height. When the rate of decrease becomes equal to the adiabatic lapse rate the atmosphere once again becomes unstable and convection occurs, leading to efficient vertical mixing within the mesosphere as within the troposphere.

Higher in the atmosphere more energetic ultra-violet rays and X-rays are absorbed and produce ionisation and rising temperatures in a complex region known to atmospheric physicists as the thermosphere and to plasma physicists as the ionosphere. The ionosphere is itself subdivided into a number of layers or regions that were characterised originally by their ability to absorb and reflect radio waves (see Boyd, 1974).

In the D-region from 60 to 90 km ionisation is mainly due to the absorption of H Lyman α radiation (121·6 nm) by minor molecular species such as nitric oxide. Because of the relatively high neutral gas pressure the major negative species is NO_2^- with electrons being a minor component, at least in the lower part of the region. The major positive ions are NO^+ and O_2^+.

NO^+ is also the major positive species in the E-region, from 90 to 160 km, where ionisation is mainly by soft X-rays and ultra-violet capable of ionising atomic and molecular oxygen. Electrons are the major negative species and remain so in the regions above. The temperature of these electrons begins to rise well above the ion and neutral gas temperature in the upper part of the E-region owing to the reduced pressure and the inefficiency of kinetic energy exchange between particles of widely disparate mass.

The F-region is the region of maximum ionisation, of the order of 10^{12} electrons per cubic metre at 300 to 400 km. Because of the gravitational separation of O from the more massive N_2, NO^+ is replaced by O^+ as the dominant positive ion species. During the day the F layer is divided with a small peak in electron concentration (F_1) below 200 km and a major peak (F_2) at 300 to 400 km. The division results from the dependence of the ionisation production rate on the scale height of O while the loss process depends on the (smaller) scale height of N_2. The low pressure and large mean free path in the F-region lead to a growing importance of diffusion of O^+ in determining ionisation levels in addition to the local production and loss processes, while above the F_2 maximum, electromagnetic forces on the plasma have a dominating influence.

This then very briefly is the present-day structure and composition of the terrestrial atmosphere. The dynamics of this atmosphere, that is to say the weather and climate, will in part be the subject of the next chapter (see also Houghton, 1977). For the remainder of this chapter I shall look at our attempts to understand how the atmosphere came into being and what processes have caused it to evolve to its present state. A number of leading questions have directed the thoughts of geochemists interested in the problem and it is probably worthwhile listing a few of these to set the scene for the discussion which follows: How and when did the Earth's atmosphere originate? What was its original composition? Has the total amount of atmosphere changed with time and, if so, how? What are the processes by which atmospheric gases may be added or removed? How has the chemical composition changed over the four and a half thousand million years of Earth history? A major aspect of this last question is one with an important bearing on the evolution of life, namely, how and when did the high concentration of oxygen appear in the atmosphere? Intimately connected with all questions concerning the atmosphere is the problem of the origin and evolution of the oceans and the other 'volatile' components of crustal rocks, the most important of the latter being the carbon dioxide locked up in limestone. Moreover, the problem cannot be discussed seriously without also considering the whole question of the chemical differentiation of the Earth and the formation of the crust. To understand our atmosphere it is clear that we need to look into many apparently unrelated areas. Let us begin by considering the chemical processes that operate on the major constituents of today's atmosphere to maintain (or possibly change) its composition.

Chemistry of the atmosphere's major constituents

Today the major constituents of the atmosphere are oxygen and nitrogen. To these we must also add carbon dioxide, for although it accounts for only 0·03 per cent by volume of the atmosphere, it is a significant component of biological processes. Moreover it is *the* major constituent of the atmospheres of our neighbouring planets, Venus and Mars, and, as we shall argue later, was probably the major constituent of the Earth's primeval atmosphere. Argon being inert is excluded from discussions of chemistry. We shall return to it later when considering the origin of the atmosphere.

The abundances of the major atmospheric gases are determined largely by the way in which they interact with the surface of the Earth. The range of processes operating is extremely complex and in order to reduce the problem to manageable proportions it is necessary to make many generalisations and simplifications. One of these simplifications is the idea of classifying the (temporary) storage locations of atmospheric gases on or near the Earth's surface in terms of a small number of reservoirs. The problem of accounting for the abundances of atmospheric gases then becomes one of determining the rates of processes, again suitably generalised, which transfer elements between the reservoirs in a cyclic fashion. Critical parameters are the sizes of the reservoirs, the average length of time an element stays in a given reservoir, and the rates of transfer of elements between communicating reservoirs (the perennial and much-maligned schoolboy problem of the leaking bath!).

Examples of 'reservoirs' are: the atmosphere itself, the living organic matter on the surface of the Earth, and the reservoir of sedimentary rocks. Examples of processes that transfer elements from one reservoir to another are: photosynthesis, which adds molecular oxygen to the atmospheric reservoir at the expense of carbon dioxide, and weathering which, among other things, removes oxygen from the atmospheric reservoir. The major cycles involving oxygen, nitrogen and carbon dioxide are represented pictorially in Figure 8.2. Some of the parameters appropriate to the major reservoirs are also summarised in the figure and are discussed in more detail below. The reader should realise that many of these estimates are rather uncertain and that the picture presented in Figure 8.2 is an over-simplified one. For a more detailed treatment the reader is referred to Walker (1977).

Oxygen

In comparison to the other planets, the Earth is unique in having an atmosphere with a large proportion of molecular oxygen. As mentioned above, carbon dioxide is the predominant atmospheric gas on Mars and Venus, with somewhat smaller amounts of nitrogen. The presence of free oxygen on the Earth is of course associated with the presence of living matter and more specifically, the process of photosynthesis, whereby green plants use the energy of sunlight to convert carbon dioxide and water to oxygen and reduced carbon in the form of plant tissue. A modern estimate of the rate of production of atmospheric oxygen and surface organic matter by photosynthesis is 10^{16} moles per year. It has been estimated that land plants contribute approximately four times as much to this figure as oceanic organisms.

A second source of atmospheric oxygen is the dissociation of water vapour high in the atmosphere by solar ultra-violet radiation, followed by the escape of some of the hydrogen into space before recombination can occur. The estimated rate of this process is very slow, of the order of 10^{10} moles per year. However, in contrast to photosynthesis, it is a non-cyclic process, and, since the hydrogen escapes from the Earth, constitutes a net source of atmospheric or surficial oxygen over geological time.

The annual production by photosynthesis represents 0·026 per cent of the atmospheric oxygen reservoir and, were it not to be balanced by an equally large removal mechanism, this would correspond to a doubling of the amount of oxygen every 3800 years. This time is geologically very short and represents a characteristic residence time or turn-over time for oxygen in the atmosphere. It provides a measure of how rapidly (or slowly depending on one's point of view) the composition of the atmosphere would respond to a major change in the production or removal rate of oxygen.

The mechanisms that remove oxygen from the atmos-

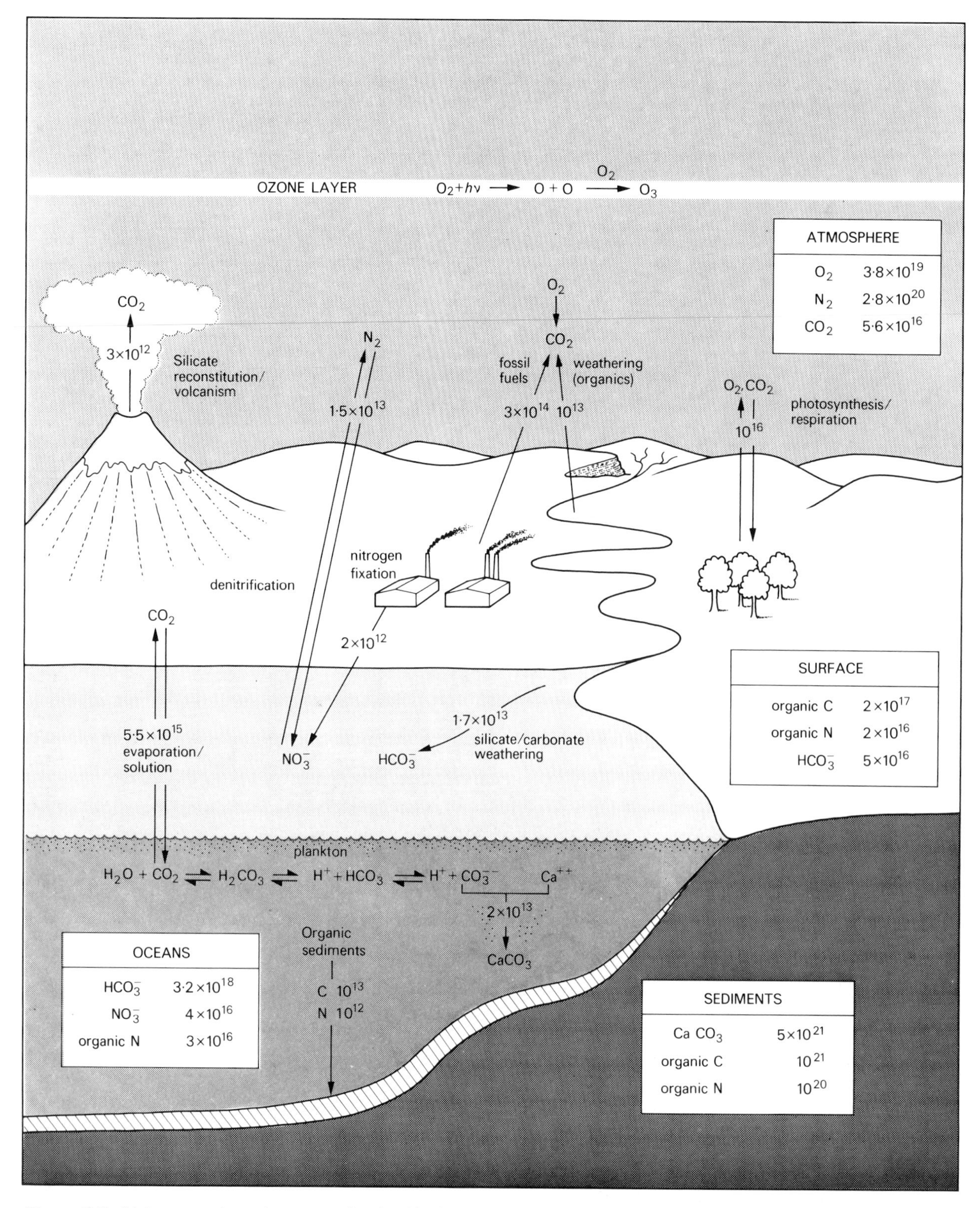

Figure 8.2 Major reservoirs and processes involved in the oxygen, nitrogen and carbon dioxide cycles. The contents of the reservoirs, shown in the rectangular boxes, are expressed in moles for O_2 and CO_2 and in g atoms for N and are taken from Walker (1977). The rates of transfer between the reservoirs by the different processes are expressed in moles (or g atoms) per year. The surface waters of the oceans and the land surface are both included in the surface reservoir. The true situation has been greatly over-simplified in this figure.

phere and maintain the balance are the processes of respiration and decay and also, at a much slower rate, the weathering of rocks. Respiration and decay may be thought of as the reverse of photosynthesis in that the net result is the removal of atmospheric oxygen and organic carbon, and the production of an equivalent amount of water and carbon dioxide. To maintain a balance, on a time-scale of a few hundred years, the rate of these processes must equal the rate of photosynthesis, that is about 10^{16} moles per year. However, Figure 8.2 indicates that the size of the surface organic carbon reservoir (2×10^{17} moles) is only one two-hundredth of the atmospheric oxygen reservoir, and consequently the average time spent by a carbon atom in this reservoir is a mere 20 years ($= 2 \times 10^{17}/10^{16}$).

From the small size of the surface organic carbon reservoir it is clear that this cannot be regarded as the sole source of atmospheric oxygen. If for example photosynthesis were suddenly to cease, the ensuing decay of living matter and the total removal of organic carbon from the Earth's surface would only consume a half of one per cent of the atmospheric oxygen. This apparent dilemma is resolved when it is realised that a much greater, photosynthetically produced, reservoir of organic carbon exists locked up within sedimentary rocks. A recent estimate indicates a value of around 10^{21} moles for this reservoir, that is 26 times as much as the atmospheric oxygen. Thus it is only necessary to assume that a few per cent of the oxygen produced in the generation of the total organic carbon budget is retained in the atmosphere. The sedimentary carbon is part of a much slower cycle than the surface carbon. Sediments are laid down in the oceans and may accumulate for periods of millions of years before geological processes cause them eventually to be incorporated as part of a continent, whereupon they may then undergo the reverse process of destruction by weathering and erosion.

Sediments are found throughout the geological column, the oldest known being 3800 million years old. Their average age is much less than this, and, in view of the observed fact that the oceanic floor is renewed by the mechanism of plate tectonics on a time-scale of the order of 100 million years, we would expect the average recycling time for sedimentary carbon to be of this order. In order to maintain a reservoir of 10^{21} moles of carbon with a cycle time of 10^8 years it is therefore necessary to supply carbon from the surface reservoir at a rate of around 10^{13} moles per year; that is 0·1 per cent of the total surface production of carbon must find its way into sediments for 'long-term storage'. Isotopic evidence suggests that most of this carbon is supplied by marine organisms.

To maintain a long-term balance in the oxygen composition of the atmosphere, the burial of carbon (equivalent to the release of free oxygen) must be offset by the recombination of reduced carbon and oxygen during the weathering process; this must therefore also proceed at an average rate of around 10^{13} moles per year. The time-scale over which this balance must be maintained is actually rather short in geological terms, of the order of a few million years. That this is so can be seen by noting that the time required to weather an amount of sedimentary carbon equivalent to the total oxygen content of the atmosphere is $3{\cdot}8 \times 10^{19}/10^{13}$ i.e. 3·8 million years. There is evidence that the level of oxygen has been maintained at or near its present level for much longer periods than this. The combination of this observation with the short time-span over which major fluctuations could be expected to arise from variations in sedimentation rates and weathering rates, suggests very strongly that a stabilising feedback mechanism operates to maintain the rates of carbon burial and weathering approximately equal.

In looking for a feedback mechanism one would anticipate that either the weathering rate or the rate of incorporation of sedimentary carbon, or both, depend on the oxygen content of the atmosphere in such a way as to force the two rates to become equal. One might, for instance, expect that weathering would be enhanced by increased oxygen content so that if, for some reason, sedimentation rates increased and the oxygen thereby increased, weathering would also increase so counterbalancing the change in oxygen level. However, present evidence appears to indicate that this is *not* the mechanism. Provided the oxygen level is not too small, weathering rates are independent of oxygen concentration, and are probably determined by tectonic processes. In effect the present oxygen level is such that *all* organic material raised above sea level by tectonic forces and exposed to the forces of erosion becomes *totally* weathered.

In fact control of the oxygen level appears to be exercised through the other half of the cycle, namely through the rate of incorporation of sedimentary carbon, in the following way. A rise in oxygen level in the atmosphere would lead to a higher concentration of dissolved oxygen in the oceans. The consequence of this would be a greater overall tendency for organic matter to be subject to decay, i.e. to be oxidised. This in turn would lead to a reduction in the rate of incorporation of sedimentary carbon and a corresponding decrease in atmospheric oxygen.

At equilibrim the rate of photosynthesis almost exactly balances the rate of oxidation, with only a very small proportion of unoxidised carbon surviving to be incorporated in sediments. Now the rate of photosynthesis is limited by the nutrient content of surface waters of the ocean and in particular by the phosphate content. On the other hand the rate of oxidation is dependent on the dissolved oxygen which in turn is proportional to the oxygen content of the atmosphere and the solubility of oxygen in sea water. The requirement that these two rates be equal implies that the equilibrium oxygen content of the atmosphere is determined principally by the amount of phosphate dissolved in the oceans and by the solubility of oxygen.

One aspect, ignored above but which is currently of some interest, is the extent to which the burning of fossil fuels may, by accelerated oxidation of the sedimentary carbon reservoir, change the level of oxygen in the atmosphere. Present consumption of fossil fuel is occurring at a rate of around 3×10^{14} moles per year, some 30 times greater than the natural weathering rate of sedimentary carbon, but only $\frac{1}{30}$ of the respiration and decay rate of the surface carbon. The fossil fuel reservoir is, however, quite small, possibly as much as 10^{18} moles, and is insufficient, even if totally consumed, to affect the oxygen content of the atmosphere significantly, even in the absence of the regulating mechanism discussed above. The response of this regulator to the enhanced consumption of oxygen would be anticipated as an increase in the rate of deposition of sedimentary carbon for the geologically short period that fossil fuels survive.

Nitrogen

The interaction of atmospheric nitrogen with the surface of the Earth is, as for oxygen, largely the result of biological processes. However, these interactions, and more specifically the rates at which they occur, are less well established than for oxygen, where photosynthesis is the dominant process.

The chemistry of nitrogen is rather complex, not the least because of the number of valence states. To simplify discussion, a distinction is usually made between the relatively inert forms of nitrogen (molecular nitrogen (N_2) and nitrous oxide (N_2O)), which reside mainly in the atmosphere, and the more reactive compounds which are susceptible to incorporation in the various reservoirs at the Earth's surface. These compounds, for example nitric acid (HNO_3), are referred to collectively as fixed nitrogen, and processes that convert inert molecular nitrogen in the atmosphere to the more reactive forms are referred to as the fixation of nitrogen. The reverse process is described by the term denitrification. The basic problem in understanding the geochemistry of terrestrial nitrogen is to understand how, where and at what rate these two processes occur, and whether there are feedback mechanisms that regulate them.

The size of the major nitrogen reservoirs are indicated in Figure 8.2. 'Fossilised' nitrogen in sedimentary rocks (10^{20} gm atoms) is roughly one third of the amount of nitrogen in the atmosphere and is estimated as such on the basis of the size of the sedimentary carbon reservoir and an assumed value of 10 for the average (C/N) ratio in organic matter. The rate of transfer of nitrogen between the surface reservoir and the sedimentary reservoir is similarly estimated from the weathering rate of sedimentary carbon, and is presumably governed by the same burial and weathering processes discussed in connection with the oxygen budget. As a consequence of the (C/N) ratio of 10 in the sediments and the (N_2/O_2) ratio of 4 in the atmosphere the atmospheric nitrogen abundance is only one-fortieth as sensitive as oxygen to fluctuations in the rates of burial and weathering.

The surface reservoir of fixed nitrogen (9×10^{16} gm atoms) is less than 0·05 per cent of the atmospheric reservoir. It is made up in almost equal parts of nitrate ions (NO_3^-) dissolved in the oceans and of organic material (mainly as the amino radical, NH_2) much of which is also in solution in the oceans. Living organisms account for an estimated 6 per cent of surface nitrogen.

Denitrification is thought to occur mainly in oxygen-deficient environments in the oceans, where organisms make use of nitrate ions as a source of oxygen. The global rate of denitrification has been estimated to be of the order of $1{\cdot}5 \times 10^{13}$ gm atom N per year. Rates of nitrogen fixation, which occurs mainly through biological processes, are difficult to estimate, but based on the assumption of equilibrium should be equal to the denitrification rate. From these figures and from the reservoir sizes, the average residence time of a nitrogen atom in the atmosphere before undergoing fixation is of the order of 20 million years. Its average residence time in the surface reservoir is a much shorter 6000 years. The short lifetime of the major part of the fixed (surface) nitrogen implies that the assumption of equilibrium is a reasonable one.

In recent times the rate of nitrogen fixation has been artificially increased as a result of industrial activity and the production of inorganic fertilisers. This increase corresponds to around 10 per cent of the natural fixation rate. The way in which the natural system will respond to the increase is not clear, although it seems likely that a new equilibrium will be reached either by a decrease in natural fixation or an increase in denitrification. The atmospheric reservoir is so large that no detectable change in it is likely for the foreseeable future. Failure of the natural system to respond to the changed rate of fixation would double the surface nitrogen reservoir in 60000 years. The problem of nitrate pollution therefore seems not to be urgent on a global scale, although it is certainly a subject of which improved understanding is necessary.

There is, for instance, a need to establish the nature of feedback mechanisms that may be operating to maintain a balance between fixation and denitrification. A possible mechanism is suggested by the observation that the nitrogen to phosphorus ratio in sea water is approximately the same as that in the dominant organisms (plankton) that live there. To account for this the suggestion has been made that nitrogen-fixing organisms fix nitrogen only when they encounter water that does not contain enough nitrogen. In this context 'enough' nitrogen means enough nitrogen in relation to the other nutrients present. Phosphorus, rather than carbon, is usually the limiting nutrient, and therefore 'enough' nitrogen also means enough nitrogen in relation to phosphorus. Thus the nitrogen-fixing organisms will continue to fix nitrogen in response to a shortage, until the population of the dominant organism has grown to such an extent that further growth is prevented by the availability of phosphorus. The pressure to fix more nitrogen is then removed.

So far we have only considered the major chemical interactions of the atmosphere with the Earth's surface.

Within the atmosphere itself a large number of reactions occur leading to the formation of minor chemical species. Many of these reactions are photochemical, involving the absorption of energetic solar photons. Some of the most important, and the only ones we shall have space to discuss here, are those leading to the formation of the ozone layer. Since they involve both oxygen and nitrogen this is a convenient place to refer to them.

The major, though not the only, source of ozone in the atmosphere arises within the stratosphere as a result of the photodissociation of molecular oxygen to form atomic oxygen

$$O_2 + h\nu \rightarrow O + O$$

This reaction is followed by the recombination of atomic oxygen with molecular oxygen –

$$O + O_2 + M \rightarrow O_3 + M$$

The molecule M may be any other atmospheric molecule and is required to take up the energy released in the reaction. There are a number of possible reactions that lead to the destruction of ozone, but the most efficient, and the only ones that appear able to account for the equilibrium concentration of ozone, make use of the oxides of nitrogen; for example:

$$NO + O_3 \rightarrow NO_2 + O_2$$

followed by –

$$NO_2 + O \rightarrow NO + O_2$$

The source of nitric oxide within the stratosphere is thought to be a reaction between metastable oxygen atoms and nitrous oxide (N_2O), which in turn diffuses up from the troposphere having been formed initially at the Earth's surface by denitrification! This important example should, if nothing else, make clear to the reader the tortuous reaction chains which govern much of the chemistry, ultimately so important to our very existence. It also illustrates the reason for recent interest and concern over the possible but inadequately understood effects of adding trace pollutants to the stratosphere, in that these could conceivably enhance the destruction rate of ozone thereby increasing the harmful ultra-violet influx to the Earth's surface.

Carbon dioxide

The abundance of carbon dioxide in the atmosphere is, as already noted, rather small: 0·03 per cent or $5{\cdot}6 \times 10^{16}$ moles in total. However, this small abundance belies its true importance since its interactions involve a number of much larger and closely linked reservoirs at the Earth's surface. These are (Fig. 8.2): the surface organic reservoir already discussed in connection with oxygen (2×10^{17} moles); the ocean waters which in effect contain carbon dioxide in the form of bicarbonate ions (HCO_3^-) in solution ($3{\cdot}2 \times 10^{18}$ moles); the organic carbon reservoir in sedimentary rocks (10^{21} moles); and, the largest reservoir of all, carbonate rocks (5×10^{21} moles).

The interaction of atmospheric carbon dioxide with these reservoirs takes place at different rates. In the short term the major interactions are those involving the surface organics and the oceans. In addition to these we must also consider an artificial perturbation of the system caused by the burning of fossil fuels at a rate greatly in excess of the natural weathering rate. Fortunately the short-term behaviour of carbon dioxide can be quite accurately measured by the use of naturally occuring ^{14}C. This is produced in the atmosphere at a more or less steady rate by the action of cosmic rays on ^{14}N. It decays with a half-life of 5700 years and so, by measuring the $^{14}C/^{12}C$ ratio in the surface reservoirs, it is possible to determine the rate at which they exchange carbon with the atmosphere, provided the time-scales involved are less than a few tens of thousands of years.

The involvement of carbon dioxide in photosynthesis at a rate of 10^{16} moles per year implies a residence time of only six years in the atmosphere before it is incorporated into surface organic material. Respiration and decay of these organics return the carbon dioxide to the atmosphere in 20 years, and *to the extent that the organic reservoir remains fixed in size* the cycle maintains a natural balance. Local variations of a few tens of parts per million in the atmospheric abundance of carbon dioxide occur both in space and time due to the fact that photosynthesis only proceeds during daylight hours and also exhibits a seasonal variation at high latitudes. Respiration and decay, on the other hand, proceed at a more or less steady rate independently of the presence of sunlight. The resulting variations in carbon dioxide abundance provide information on the effectiveness of mixing within the troposphere.

Carbon dioxide also exchanges at a fairly rapid rate with the oceans, but it is necessary to make a distinction between surface waters and the deep ocean. The surface waters, to an average depth of around 70 m, consist of a relatively warm well-mixed layer and form a reservoir containing approximately 5×10^{16} moles of dissolved inorganic carbon, mostly as bicarbonate ion (HCO_3^-). Below this layer the temperature falls quite rapidly in a region known as the thermocline, and thereafter more slowly to the ocean bottom which may be only a few degrees above zero Celsius. This temperature profile is stable and inhibits deep mixing. The only source of mixing within the deep ocean is therefore the large-scale circulation brought about by the temperature difference between poles and Equator. The deep ocean reservoir contains the major part of the dissolved inorganic carbon, some $3{\cdot}2 \times 10^{18}$ moles.

Based on ^{14}C measurements the average age of the carbon in deep water is 800 years in the Atlantic and 2100 years in the Pacific, which can be used to infer an average exchange rate for carbon between the deep water reservoir and the surface reservoir of around 2×10^{15} moles per year. The $^{14}C/^{12}C$ ratio of surface water is much closer to that in the atmosphere and implies an exchange rate between surface water and atmosphere of $5{\cdot}5 \times 10^{15}$ moles per year. Comparing this with the size of the atmospheric reservoirs we see that it corresponds to an average residence time for carbon dioxide in the atmosphere of ten years before it dissolves in the oceans.

The effect of burning fossil fuels over the last century has been to perturb the atmospheric carbon dioxide budget, which is currently increasing at an average rate of 0·7 ppm per year. The perturbation is also evident in the atmospheric $^{14}C/^{12}C$ ratio, which from measurements on tree rings of known age is known to have decreased by 2 per cent over the last 100 years due to the addition of fossil carbon, which is devoid of ^{14}C. This decrease is a factor of ten lower than would have been the case if the extra carbon dioxide had remained in the atmosphere, and reflects the rapid exchange discussed in the previous paragraph between the atmosphere and the much larger oceanic reservoir. Since the time constant for equilibration with the surface reservoir is quite short, and the rate of transfer to the large deep ocean reservoir of a similar order, it might be anticipated that the atmospheric carbon dioxide would rapidly attain a new equilibrium concentration and cease to increase. The reason that the concentration continues to increase is that the *rate* at which fossil fuels are being consumed is itself increasing.

Over periods of the order of a thousand years, the circulation time for the deep ocean, ocean and atmosphere are in equilibrium, and on this time-scale the content of carbon dioxide in the atmosphere is determined by the carbonate chemistry of the oceans (see for example Turekian, 1976). The reactions involved in establishing this equilibrium are: the solution of carbon dioxide to form carbonic acid; the dissociation of carbonic acid to form bicarbonate ions; the further dissociation of bicarbonate ions to form carbonate ions; and the precipitation of solid calcium carbonate to form limestone deposits (either directly or through the intermediary of the shells of living organisms). These reactions are all reversible, and the equilibrium concentrations of dissolved ions are easily calculable. When this is done it can be shown that the concentration of carbon dioxide in the atmosphere depends only on the acidity of the oceans. What is not clear is whether the acidity is controlled by other factors (which therefore ultimately control the amount of carbon dioxide in the atmosphere) or whether it is the carbon dioxide in the atmosphere which mainly controls the acidity and is itself controlled by other factors: this we shall now discuss.

There is yet another cycle of carbon dioxide that may be the controlling factor. This is a cycle that operates on a longer, geological time-scale and involves, on the one hand, the weathering by carbon dioxide of carbonate and silicate rocks, and, on the other, the addition of carbon dioxide to the atmosphere by volcanic activity. Weathering involves a large number of reactions specific to the minerals being weathered. They can be summarised for simplicity by the reactions:

$$CaCO_3 + H_2O + CO_2 \rightarrow Ca^{2+} + 2HCO_3^-$$

for carbonate rocks, and by:

$$CaSiO_3 + 2CO_2 + H_2O \rightarrow Ca^{2+} + 2HCO_3^- + SiO_2$$

for silicate rocks.

The effect of the first reaction is simply to transfer carbonate deposits from the continents to the ocean floor without changing the overall equilibrium. The effect of the second reaction is the conversion of silicate minerals to carbonate minerals. Were it not for a process operating in the opposite direction it is possible that silicate weathering would, through a slow increase in the alkalinity of the oceans, bring about a corresponding change in the abundance of carbon dioxide in the atmosphere. The reverse process that completes the cycle is the return of carbon dioxide to the atmosphere as a result of volcanic activity or metamorphism. The process is known as silicate reconstitution and can be represented by the reaction:

$$CaCO_3 + SiO_2 \rightarrow CaSiO_3 + CO_2$$

The rate at which carbon dioxide is entering the atmosphere from volcanic activity is of the order of 5×10^{12} moles per year. (Some of this carbon dioxide may be entering the atmosphere for the first time from the Earth's interior, as opposed to being recycled material. However, the proportion of this 'primordial' carbon dioxide is probably quite small.) Silicate weathering is occurring at a similar rate, although the extent to which the processes are in balance is by no means clear. Dividing the size of the carbonate reservoir by the rate of silicate reconstitution we see that the cycle time is of the order of a thousand million years. This is less than the age of the Earth, but sufficiently long (and uncertain) to allow long-term changes in the carbon dioxide budget of the atmosphere.

Later in this chapter we shall discuss the ways in which the evolution of life has affected the atmosphere. From what we have learned above we may summarise the ways in which the composition of the present atmosphere is dependent on life. We have seen that the presence of large amounts of oxygen is entirely dependent on living organisms through the mechanisms of photosynthesis, although the actual abundance is possibly determined by the concentration of phosphate nutrients in the oceans. We have seen, in contrast, that the abundance of nitrogen is only marginally influenced by living matter. In the absence of life, approximately a quarter of the Earth's nitrogen budget currently locked up as organic material in sediments would reside in the atmosphere, more or less balancing the oxygen lost. The abundance of carbon dioxide in the atmosphere is also relatively insensitive to the presence or absence of life. Although it plays an intimate part in the biological cycle, its abundance in the atmosphere is determined by inorganic processes, in the short term, principally carbonate equilibrium reactions in the oceans, and in the long term the competing effects of silicate weathering and volcanic activity.

The origin of the atmosphere

It should be apparent by now that the origin and evolution of the atmosphere cannot be understood without also considering other volatile chemical species found in the crust and the oceans. The relative abundances of the major ones are summarised in Table 2. Also included in the table are the relative abundances in the Sun and two important classes of stone meteorite. The abundances have been expressed relative to the element silicon, which is an involatile element common throughout the solar system. This is a widely used practice which is particularly convenient when discussing the origin and distribution of volatiles.

Table 2 Relative abundance of volatile elements in the Earth, meteorites and the Sun

	*Relative abundance (Si = 100)**			
	H	*C*	*N*	^{36}*Ar*
Earth†	0·68	0·042	0·0011	$2{\cdot}0\times10^{-8}$
Carbonaceous chondrites‡	4·0	60	4·0	$1{\cdot}0\times10^{-6}$
Ordinary chondrites‡	⩽0·08	1·5	0·05	$1{\cdot}5\times10^{-8}$
Sun	$3{\cdot}0\times10^{6}$	1000	275	19

* Abundances are expressed relative to silicon because this is an abundant involatile element.
† H is predominantly in the oceans, C in carbonate rocks, N and ^{36}Ar in the atmosphere.
‡ These are two major stony meteorite classes. The abundance of volatiles in meteorites shows an appreciable variation; the figures quoted are representative.

Hydrogen resides predominantly (97 per cent) as water in the oceans, with the remainder in ground water rivers, lakes and polar ice. As already discussed, carbon occurs principally in carbonate deposits and nitrogen in the atmosphere; smaller amounts of each are present as organic material in sediments. An additional component of nitrogen, present in igneous rocks, is not included and would somewhat increase the figure quoted.

The major isotope of argon in the atmosphere is ^{40}Ar which has been produced over geological time by the radioactive decay of ^{40}K. The importance of ^{40}Ar in telling us about the continuing release of gases from the interior of the Earth will be referred to later. The isotope ^{36}Ar listed in Table 2 makes up only 0·337 per cent of atmospheric argon but is more relevant to our present discussion because, since it is not a product of radioactive decay, it is representative of the Earth's original inventory of volatiles. Being inert, ^{36}Ar and the other noble gases have a history that is not complicated by chemical interaction with the Earth. For this reason the noble gases, as we shall see later, provide information about atmospheric evolution that is quite out of proportion to their apparently negligible abundance.

Ideas about the origin of the Earth's volatile species are closely linked to ideas about the origin of the Earth itself. These are discussed in Chapter 1, in which the prevailing view that the Sun and planets formed from a contracting nebula of gas and dust was argued. Starting from such a picture the planets could have acquired their atmospheres and other volatiles in two ways. They could have been retained as original gaseous envelopes by the gravitational field of the planet if it formed sufficiently rapidly in the presence of a gaseous nebula. Alternatively they could result from later release, by heating, of gases dissolved in the solid objects (planetesimals) which are hypothesised to have grown by gravitational attraction and mutual collisions into the planets (see Dermott, 1978).

The first mechanism is unquestionably responsible for the formation of the outer planets. Based on their low density, the composition of the 'gas giants' – Jupiter and Saturn – must be quite similar to that of the Sun, that is, predominantly hydrogen. Neptune and Uranus formed in a similar way, but the fact that they have higher densities than Jupiter while being considerably smaller than Saturn indicates that they failed to acquire a full 'solar' complement of hydrogen and have a higher proportion of heavier species – carbon, nitrogen and oxygen. The lower abundance of hydrogen in Uranus and Neptune is understood to be the result of their slower growth which, in spite of the lower temperatures at this distance from the Sun, permitted more hydrogen to escape. Growth by planetesimal collisions becomes slower with increasing distance from the Sun as a natural consequence of Kepler's Third law, which indicates that oribital velocity, and therefore collision probability, decreases as orbital distance increases.

What then of the inner planets, and particularly the Earth, for which volatiles are depleted relative to their solar abundance by factors of between 10^4 and 10^9 when comparison is made on the basis of an involatile species such as silicon (Table 2)? The current view is that the second mechanism is responsible for the atmospheres of Earth, Mars and Venus. Reference once more to Table 2 will illustrate the kind of arguments that can be used to support this view. Meteorites, and in particular the class known as chondrites, have been clearly established as primitive objects that were formed at the time the solar system itself was formed. It is also established that they have always existed as parts of small bodies, of asteroidal rather than planetary dimensions, which would be unable to retain gravitationally bound atmospheres. Nevertheless, meteorites do contain volatiles and, in the case of the (aptly named) carbonaceous chondrites, volatiles are present in appreciable amounts. These are held in a variety of ways, for example in carbonaceous chondrites, carbon, nitrogen and hydrogen are present as organic polymers. Hydrogen is also present as structural OH in 'low temperature' clay minerals or as water of crystallisation in other mineral species. In ordinary chondrites, carbon and nitrogen are present in solid solution in 'high-temperature' iron-nickel

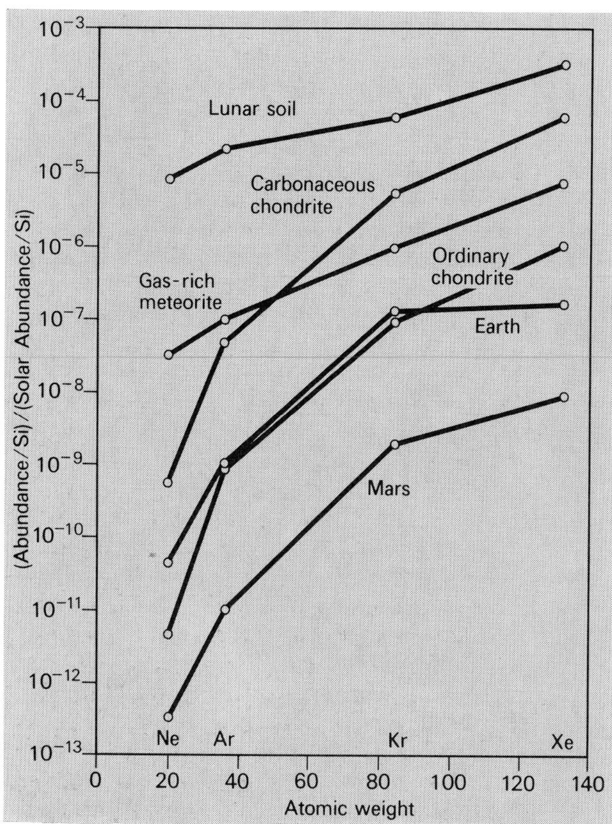

Figure 8.3 Relative abundances of the noble gases in the atmospheres of Earth and Mars, in typical lunar soil and in three important meteorite types. The abundances are calculated relative to the common involatile element silicon, and have been divided by the corresponding abundance in the Sun. Calculated in this way the abundances represent the factors by which the gases are depleted relative to involatile species in the Sun. Gas of solar composition would plot as a horizontal line. The gas in lunar soil and gas-rich meteorites originates from solar wind implantation. The similarity of the patterns for chondrites and the atmospheres of Earth and Mars supports the view that the atmospheres resulted from secondary outgassing of the planetary interiors.

metal. Hydrogen may be present but in very small amounts and probably combined with carbon as an organic polymer.

If, as seems likely, the Earth grew from the aggregation of smaller bodies with composition similar to that of the meteorites it is clearly possible, on the basis of the figures in Table 2, to account for its present inventory of volatiles. It is argued that they were released partly by heating and melting during the collisions associated with the accretion process and partly by subsequent melting and associated volcanism during the geological history of the Earth. There appears to be no need to assume an original gravitationally captured gaseous component, although at the moment a contribution from this source for the heavier gases cannot be entirely excluded.

The noble gas abundances in the atmosphere are particularly important in leading us to the above conclusions. They are illustrated in Figure 8.3 along with representative abundances in meteorites and lunar soil and with recent measurements of the Martian atmosphere. The abundances are plotted in the figure as 'depletion factors' relative to the solar abundance. The depletion factor is the abundance relative to the involatile species silicon divided by the corresponding ratio for the Sun. The composition of the Sun would plot as a horizontal line with an abundance ratio of one. Gas of solar composition, but depleted relative to silicon, would plot as a horizontal line with an abundance ratio less than one. Noble gases are found in high abundance in lunar soil and are known to be implanted there from the solar wind which is, in effect, the outer atmosphere of the Sun streaming out into space as an ionised plasma. For this reason the lunar soil line is almost, but not quite, horizontal. The high abundance arises from the fact that the lunar soil is very fine grained with a large ratio of surface area to volume. It is continually stirred by micrometeorite impacts and therefore is able to capture the solar wind very effectively. A group of meteorites of various types, but known collectively as the gas-rich meteorites, shows a similar abundance pattern. Various features indicate that they too were once part of the surface region of a planetary or asteroidal body and acquired this 'solar' pattern in a similar way to the lunar soil. In fact the lunar observations confirmed an earlier suggestion that the gas-rich meteorites contained entrapped solar wind.

However, the bulk of primitive meteorites show a pattern that is greatly depleted in the light noble gases. The gas appears to be dissolved in a few very minor phases and the mechanism by which it got there from the primitive collapsing nebula is not yet known. What is clear is that the pattern reflects quite closely the pattern seen in the atmosphere of the Earth, and also apparently of Mars, and supports the view that the atmosphere was formed by the 'outgassing' of the primitive planetesimals. Attempts to match the Earth's inventory of volatiles with those of the meteorites have led to the view that the Earth formed in the main from volatile-poor planetesimals, comparable in composition to the ordinary chondrites, together with the addition of a few per cent of volatile-rich material, comparable to the carbonaceous chondrites. The low abundance of rare gases in the atmosphere of Mars suggests that either it acquired a much lower contribution of volatile-rich material or it has been outgassed to a lower degree, or both.

This is a greatly simplified outline, and there are many problems to be overcome in trying to understand details. For example, the abundance of xenon in the atmosphere is too low by a large amount compared to krypton (see Figure 8.3). There are also significant isotopic differences in the heavy rare gases. Krypton in the atmosphere is slightly enriched in the light isotopes by comparison with meteoritic krypton. In contrast xenon is strongly enriched (several per cent) in the heavy isotopes. The abundance differences may be partly accounted for by supposing that the Earth has not fully outgassed its original inventory and that outgassing has been most effective for the lighter gases. Furthermore there is some observational evidence that a major proportion of terrestrial xenon is absorbed in sedimentary rocks. The isotopic differences are harder to understand, and may ultimately require a revival of the gravitational capture hypothesis to account for the 'heavy' xenon (with the lighter noble gases coming from later incomplete outgassing).

Let us accept the idea that the atmosphere and other surface volatiles were produced by outgassing of the Earth and attempt to answer some important questions which this raises: When did the outgassing occur? Is it continuing at the present day? How completely is the Earth outgassed? The last question is the easiest to answer, based on the abundance of ^{40}Ar in the atmosphere. As mentioned above, this isotope is produced almost entirely by the radioactive decay of ^{40}K. In addition to generating ^{40}Ar, the decay of ^{40}K, along with ^{235}U, ^{238}U and ^{232}Th, provides the heat source that drives the Earth's geological engine. Because of this the total abundance of ^{40}K within the Earth can be estimated moderately well by measuring the average heat flow from the crust, and corresponds to an average abundance for elemental potassium of roughly 250 ppm. Over the 4500 million-year life of the Earth this amount of potassium has produced roughly three times the amount of ^{40}Ar currently in the atmosphere and we therefore conclude that, for this gas, the Earth is around 33 per cent outgassed. Furthermore we may conclude that the period over which this outgassing occurred must have been extensive, since it would have taken at least 700 million years from the time of formation of the Earth just to generate the total amount of ^{40}Ar now in the atmosphere.

What of the volatile elements *already* present within the Earth at the time of formation? It seems likely that these must have been outgassed *at least* as thoroughly as ^{40}Ar. Indeed if, as we shall argue below, the rate of outgassing was highest soon after the Earth formed they will be *more* thoroughly outgassed. On the basis of arguments like this it is generally believed that the Earth has outgassed somewhere between 50 and 100 per cent of its volatiles.

What of the present day? It is virtually impossible to draw any definitive conclusions from the present day release of volatiles from volcanic activity, since the bulk of these undoubtedly represent recycled material from within the upper mantle or from hydrothermal activity within the crust. Some recent evidence of excess ^{3}He in deep ocean water and within the chilled glassy margins of ocean ridge basalts indicates that the mantle of the Earth is still releasing 'new' volatiles, but at a slower rate than in the past.

Many arguments have been put forward in support of the view that the release of volatiles was much higher in the past than at present and that, for example, the oceans of the Earth are ancient. There is certainly evidence of an extensive early water cover in that some of the oldest rocks, dating from as long ago as 3800 million years, have a sedimentary origin.

Consideration of the mechanisms by which volatiles may escape from the interior of the Earth also lead to this conclusion in a fairly convincing way and are worth a brief discussion. It is popularly supposed that if a planet such as the Earth went through a molten phase its inventory of volatiles would be rapidly released into the atmosphere. That this is not necessarily the case can be seen by considering the two processes of convection and diffusion that must operate in parallel to transport the gas molecules to the surface. Diffusion is an extremely inefficient transport mechanism. For example, a molecule free to diffuse at random through a molten but non-convecting Earth would in a time of one million years travel a distance of roughly 10 metres, and in the course of the 4500 million years of Earth history less than a kilometre! (The distance diffused increases as the square root of time). For volatiles to escape rapidly then it is necessary for them to be carried first of all by convection within the mantle to a region closer to the surface. Clearly they cannot be brought to within a few metres of the surface by convection and so another process of 'mass transport' must be invoked at this stage, namely volcanism. From the point of view of the volatiles the effect of volcanism is to generate, by partial melting, a separate mobile phase into which they can readily diffuse, *provided the melting is initially on a sufficiently fine scale.* In considering this chain of processes, by which the Earth is able to release its volatiles, the reader might like to draw an analogy with the way in which evolution has provided vertebrates with complexly subdivided respiratory organs of great surface area in order to match up the efficient mass transport systems of breathing and blood flow with the inefficient intermediary of diffusion across cell membranes. If such complex organs are required by an animal as small as a few centimetres, how much more difficulty for a body as large as the Earth to breathe.

From the above arguments we see that the release of volatiles is intimately connected with the process of volcanism and the generation of the Earth's crust. To a first order therefore it is reasonable to expect that the rate of release would be proportional to the rate of volcanism and also to the amount of volatile material still remaining within the Earth. The radioactive heat sources responsible for driving the volcanic process have decreased in intensity by a factor of eight since the Earth formed, largely due to the decay of ^{40}K and ^{235}U, which have half-lives of the order of 10^9 years. The volatile content of the Earth has also decreased by the very act of outgassing, and therefore it is reasonable to conclude that the rate of outgassing has decreased substantially since the Earth formed. It seems quite feasible that the bulk of the volatiles now at the surface of the Earth were released within the first thousand million years of its history.

Evolution of the atmosphere

In this final section we will look at the way in which the atmosphere is thought to have developed since its first beginnings. Three major questions will direct our thoughts: What was the composition of the primitive atmosphere?

How, if at all, did that composition evolve prior to the development of life? How did it evolve as a result of the development of life?

The composition of the early atmosphere has been covered in the preceding section in all but one important respect. We have seen that a good idea of the elemental composition of the escaping gases can be inferred from the present inventory of volatiles in the crust. What we have not so far considered is their oxidation state. At its crudest the question can be expressed thus: were the atmospheric gases and other volatiles highly reduced (that is consisting predominantly of gases such as methane, ammonia and hydrogen) or were they more oxidised (consisting largely of carbon dioxide, nitrogen and water)? Many arguments have been produced to support one or other viewpoint but we shall restrict this discussion to just two.

Table 3 The composition of typical Hawaiian volcanic gases

Constituent	*Volume* (%)
H_2O	77·0
CO_2	11·7
SO_2	6·5
N_2	3·0
H_2	0·5
CO	0·5
S_2	0·3
Cl_2	0·05
Ar	0·05

The first argument relates to the composition of present-day volcanic gases associated with regions of active sea-floor spreading (see Table 3), and the oxidation state of the upper mantle. We have already argued that most of the volatiles now being released are essentially recirculated surface volatiles; however, it can be shown that the oxidation state of these gases (specifically the ratios H_2O/H_2 and CO_2/CO) is that which would be expected for gas in equilibrium with basalt melts of the kind that are produced in the upper mantle. Thus gas escaping from great depth at the present day would be expected to have a similar composition. In effect this composition reflects the oxidation state of the region of the mantle with which the volatiles have equilibrated. Now the Earth *as a whole* is more reduced than the upper mantle, as witnessed by the existence of a large iron core. But, provided the core was in existence very early in Earth history and the free iron therefore unable to affect the oxidation state of the escaping gases, it is likely that the composition of the surface volatiles has been largely determined by the oxidation state of the mantle. On this argument the original atmosphere was probably dominated by carbon dioxide, nitrogen and water.

The second important argument relates to the origin of life. The most widely accepted view at present is that in the early history of the Earth the basic building blocks of living matter, namely complex organic molecules such as proteins, were synthesised from the primitive atmosphere. The energy sources necessary to drive the chemical reactions are presumed to be either energetic solar ultraviolet rays or electrical discharges (lightning). These processes have been extensively studied in the laboratory, since the pioneering experiments of Miller in 1953. Proteins have been produced in the laboratory from a wide variety of possible energy sources and atmospheric compositions (including carbon dioxide and nitrogen), but their production is most favoured by the more reducing mixtures (methane and ammonia) and totally inhibited by quite minor concentrations of oxygen.

Combining the two quite distinct lines of argument, it seems most likely that the primitive atmosphere was dominated by carbon dioxide and nitrogen but was probably more reducing than present-day volcanic gases and in particular contained a higher proportion of hydrogen, methane and ammonia. The subsequent evolution of this atmosphere will now be discussed. The major features are summarised in Figure 8.4.

Let us consider first how these volatiles might interact with each other and the surface of the Earth in the absence of life. Consider first the carbon dioxide and let us begin, purely for the sake of argument, by considering a hypothetical situation in which the present-day mass of surface carbon is converted to carbon dioxide and confined in the atmosphere. It would exert a pressure of about 27 atmospheres, somewhat less than, but of the same order as, the present-day atmospheric pressure on Venus. In the presence of an ocean some of it would dissolve, the actual amount depending on the acidity and other dissolved chemicals. Again for the sake of argument let us consider an extreme case in which the ocean has its present volume but contains no dissolved solids, i.e. a distilled water ocean! It is a simple matter to calculate that roughly one quarter of the carbon dioxide would dissolve, reducing the surface pressure to 20 atmospheres, and producing a very acidic ocean with a pH of around 3·2.

The value of this illustration is that it allows us to place a limit, albeit a rather extreme one, on our discussion of the actual situation and also it helps us to understand qualitatively what might actually have happened. It suggests that the effect of outgassing carbon dioxide, together with other 'acidic' gases such as chlorine and hydrogen chloride, would be to produce an initial ocean considerably more acidic than the present one (pH $\cong$ 8) which would in consequence react very rapidly with (i.e. weather) the surface rocks on the ocean floor and, if any existed, on continental areas. The net result of these interactions would be to reduce the proportion of carbon dioxide in the atmosphere, to reduce the acidity of the ocean and lead to extensive precipitation of carbonates from the dissolved Ca^{2+} and Mg^{2+} ions. With volcanism continuing to release more carbon dioxide into the atmosphere, and also recycle the existing carbonates, we would anticipate that an equilibrium would be established in which the partial pressure of carbon dioxide in the atmosphere was considerably greater than at present and in which the oceans were more acidic than at present. Unfortunately it is not yet possible to go beyond these rather qualitative statements and the problem of doing so is aggravated by the absence of rocks older than 3800 million years.

Another feature of the chemistry of the primitive atmosphere that is of some importance is the possibility that the dissociation of water vapour by solar ultra-violet, followed by the escape of the hydrogen into space, would lead to the production of molecular oxygen. Several attempts have been made to estimate the oxygen concentrations which would ensue, with somewhat different conclusions (Berkner & Marshall, 1964; Walker, 1977). Two processes operate to remove the oxygen and it is the effectiveness of these that determines the equilibrium level. On the one hand is the oxidation of surface rocks, and on the other the reaction with hydrogen gas released by volcanism. We have already argued that the early volcanic emissions were probably more reducing than those of the present day and it seems quite likely that a relatively large flux of hydrogen would therefore be able to restrict the oxygen abundance to very low levels. The arguments used in connection with the synthesis of complex organic molecules seem to require this conclusion.

Following the somewhat obscure period of Earth history in which the atmosphere first formed, the major cause of chemical change has been the origin and development of living organisms and it is the effect of these which we shall now attempt to chart. This is a field in which current research is very active and about which a substantial amount of information is present in the geological record, both in the form of fossils and also as geochemical and isotopic evidence. It will be possible to outline here only the major steps now understood in the development of life. Fortunately it is a field about which much has been written, both at a popular and a scientific level, and the reader is therefore encouraged to supplement this account by reading elsewhere the fascinating details of the story of the biochemical and biological changes that have led from the earliest primitive cells to the life of today (McAlester, 1977; Schopf, 1976, 1978).

Until quite recently the history of life could be traced

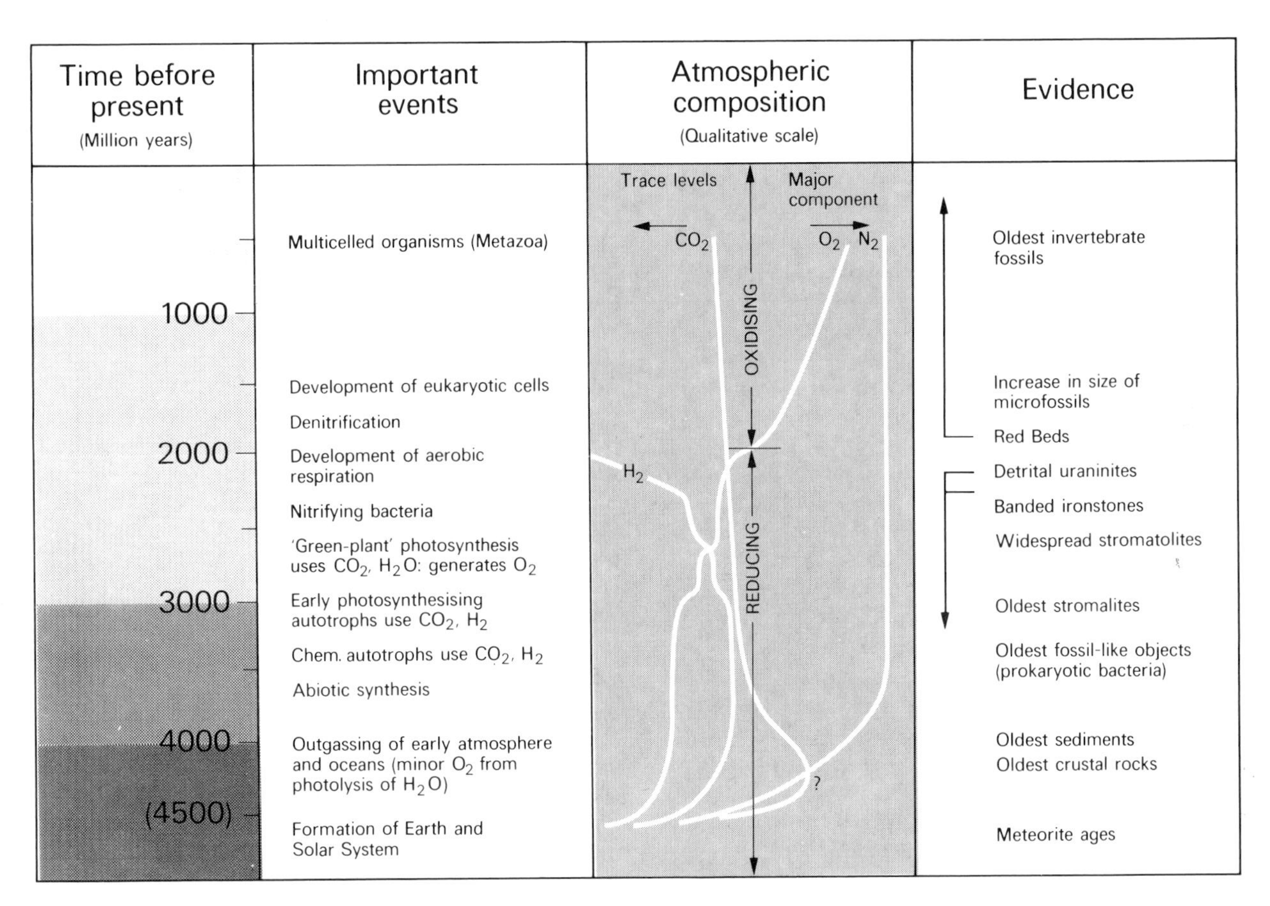

Figure 8.4 Major events affecting the evolution of the Earth's atmosphere represented as an approximate chronological sequence. The sequence is based on interpretation of the fossil and geochemical record (right-hand column) and from studies of the morphology, metabolism and biochemistry of modern organisms. The transition from an oxygen-poor to an oxygen-rich atmosphere took place roughly 2000 million years ago.

back only as far as the Cambrian period, a mere 500 million years of the Earth's existence. Why the abundant fossil record should begin in the Cambrian, with organisms as complex as for example the trilobites, was recognised to be a major problem by Darwin long before the absolute time scale involved was known. It was largely resolved in the 1950s by the discovery, in much older Precambrian rocks, of microscopic fossils resembling modern bacteria and blue-green algae. The latter occur as layered mat-like structures called stromatolites (Fig. 8.5) and were known to geologists long before their true significance was recognised. Over the past two decades these fossils have been identified in some 50 or so localities, the oldest of which date back more than 3000 million years, and morphological studies provide a basis for tracing the evolution of this primitive life.

Additional evidence of a different kind can be obtained by studying the biochemistry of *modern* organisms. One approach is to look for possible analogues of the early cells in the most primitive of present-day bacteria and algae and, by studying these, to attempt to understand how the early bacteria might have functioned and evolved. A second approach involves the study of the biochemistry of more advanced organisms, coupled with attempts to understand how complex chemical pathways could have evolved from earlier and more primitive predecessors, since the biochemistry of organisms usually evolves by the addition of new reaction pathways to existing ones rather than by starting again. An example, which may make this idea clearer, arises in the metabolic chain by which modern cells extract energy from glucose and oxygen – the process of respiration. The earliest reactions in this chain do not in fact make use of oxygen at all but release a small amount of energy by a process akin to fermentation. The reactions involving oxygen, which release much larger amounts of energy, occur later in the reaction sequence and therefore appear to have been added to the earlier ones as part of the evolutionary process.

The earliest organisms, formed more than 3000 million years ago, were almost certainly primitive bacteria. How they originated is a major question in biology we shall not attempt to answer here. The early evolution of these cells, and the way in which they may have influenced the atmosphere, is also somewhat speculative but, based on biochemical arguments, is thought to be as follows. The first organisms were probably of very limited metabolic capability and utilised organic molecules (synthesised abiotically in the atmosphere) as building material and, by fermentation reactions, as a feeble energy source. Because of the lethal ultra-violet rays falling on the Earth's surface (in the absence of the screening provided by the present ozone layer) the habitat of these organisms must have been a suitably shielded location. They may, for example, have existed in the oceans at depths greater than a few metres. The major effect on the atmosphere of their existence was presumably to provide an additional sink or loss mechanism for atmospheric carbon and hydrogen in the form of organic debris added to ocean sediments.

Figure 8.5 Precambrian stromatolite built up from a succession of thin layers of blue-green algae. This specimen is from Canada.

The next major step that is postulated to have occurred is the evolution of bacteria, referred to as autotrophs, that were capable of using the abundant inorganic chemicals, present in the atmosphere and dissolved in the oceans, to synthesise complex organic molecules directly, rather than relying on the vagaries of an external supply. Modern methane bacteria provide an example of such an organism and they derive their energy from a reaction between carbon dioxide and hydrogen:

$$CO_2 + 4H_2 \rightarrow CH_4 + 2H_2O$$

With abundant carbon dioxide and hydrogen in the primitive atmosphere it is likely that once such organisms evolved, their population would rapidly expand until limited by the availability of hydrogen. The effect on the atmosphere of this development may therefore have been a reduction in the amount of molecular hydrogen and a corresponding increase in the trace levels of oxygen produced by photolysis of water. By using chemical energy to synthesise organic molecules these autotrophs would provide an additional food source for the fermentation bacteria.

The development of photosynthesis, by which certain bacteria acquired the capability of using sunlight directly as an energy source, is thought to have been the next step forward. The earliest photosynthesisers probably made use of hydrogen, carbon dioxide and organic molecules as their raw materials as opposed to water and carbon dioxide, which are the raw materials of the familiar but more complex process of 'green-plant' photosynthesis. In this way they probably contributed to a further reduction in the equilibrium level of hydrogen in the atmosphere.

The evolution of photosynthesisers capable of using

water as a raw material, rather than hydrogen, is thought to be represented in the fossil record by the stromatolites which, as mentioned earlier, are interpreted as being fossil mats of blue-green algae. This development was of profound significance. Not only did it free life from dependence on the rather limited supply of volcanic hydrogen in exchange for an essentially unlimited commodity, water, but also it marks the first introduction of a biological source of atmospheric oxygen. The oldest stromatolites date back almost 3000 million years suggesting that this was when the development of 'green-plant' photosynthesis took place. It must certainly have been in full swing by 2300 million years ago since several widespread occurrences of stromatolites date from that time.

To the earliest photosynthesisers the oxygen was a potentially lethal waste product capable, if not controlled, of attacking and rapidly destroying the cell. Modern organisms control the level of oxygen and oxygen radicals within the cell by the use of enzymes. The early organisms must have developed similar mechanisms in parallel with the development of photosynthesis. However, the generation of free oxygen presented the evolving organisms with opportunities as well as problems and these opportunities were eventually realised with yet another major development, that of respiration. The significance of respiration as an energy source is profound. By adding new biochemical pathways they were able to increase the energy yield from the photosynthesised organic matter by a factor of 18. The efficiency with which this allowed them to use their organic raw materials presumably led to a major expansion of living organisms.

Figure 8.6 Banded ironstone from the Precambrian rocks of Australia (Specimen width about 10 cm). The dark bands are composed of haematite (Fe_2O_3) and the light coloured bands are of chert (SiO_2). Both compounds were probably precipitated biogenically. The oxidation of Fe^{2+} to Fe^{3+} and the consequent precipitation of Fe_2O_3 is thought to have been brought about by photosynthetically produced oxygen.

The photosynthetic source of oxygen also led to the development of nitrifying bacteria which, for the first time, introduced signficant amounts of nitrate to the oceans. Prior to this the major source of fixed nitrogen was ammonia. The availability of nitrate led on to the evolution of denitrifying bacteria, which make use of the nitrate ion as a source of oxygen and, as we saw earlier, are a feature of the present-day nitrogen cycle.

By this stage the major biochemical pathways of life were established, and we may now pause to look at the geological record to time the sequence of events and to provide evidence of their effect on the atmosphere. Three lines of geochemical evidence are usually considered as providing the major clues concerning the transition of the atmosphere from a weakly reducing one to a strongly oxidising one. Evidence of the early reducing nature of the atmosphere is provided by the existence of pyrite (FeS_2) and uraninite (U_3O_8) grains in old conglomerates and sandstones that were probably laid down as river deposits. They are referred to as detrital grains, which means that they are thought to have been eroded (in the presence of the atmosphere) from older rocks and then redeposited. In the presence of an oxygen-rich atmosphere (1 per cent oxygen or greater) they would be expected to be unstable, and their presence in old deposits, but virtual absence in rocks younger than 2000 million years old, seems to imply a rise in atmospheric oxygen at that time. Some doubt has been cast on the strength of these conclusions by the observation that some detrital pyrite may be present even in modern sediments. More convincing evidence is provided by the existence in the sedimentary record of banded ironstone formations.

Banded ironstone formations (Fig. 8.6) are sedimentary rocks consisting of alternating layers of iron oxide and silica with very little detrital material. This absence of detrital material indicates that the iron and silica were deposited from solution probably on the ocean floor. They are common in the geological record between 3000 and 2000 million years ago, but are absent in younger rocks. In the early reducing environment, iron would be present in the oceans in solution as the ferrous ion (Fe^{++}). The banded ironstone formations are thought to have arisen as a result of the gradual oxidation of this ferrous iron to the insoluble ferric oxides (Fe^{+++}) by oxygen produced by the early photosynthesisers. A major proportion of the banded ironstone formations was produced in a period of a few hundred million years, around 2000 million years ago. In this relatively short time the oceans rusted. The coincidence in time of the earliest abundant stromatolites and the banded ironstones reinforces this interpretation.

The disappearance of the banded ironstones is marked by the appearance of sedimentary rocks, such as sandstones, containing abundant oxidised iron and known collectively as 'red beds' because of their colour. They are in many cases river deposits and have clearly been laid down under oxygen-rich aerated conditions. Prior to the appearance of red beds the oxygen produced in the oceans was being consumed in oxidising the abundant reduced compounds in solution. Only when this process was complete was it possible for oxygen to enter the atmosphere in significant amounts. The level of free oxygen may also have been influenced around this time by the development of respiration, which would have brought about a major and rapid increase in the size of the population of living organisms.

There is evidence arising from studies of the sizes and morphology of microfossils that has some bearing on the development of respiration. This evidence is based on the existence of a major division among living organisms, between those whose cells have nuclei and those which do not. The former, known as *eukaryotes* (from the Greek *eu* – meaning well, and *karyon* – meaning kernel), have a complex internal structure and contain within the nucleus DNA organised into chromosomes. They are capable of sexual reproduction, whereby genetic variations can be passed from parent to offspring in new combinations, and their metabolism is exclusively aerobic, that is they require oxygen for the purpose of respiration. The others without nuclei are known as *prokaryotes* (from the Greek *pro* – meaning before). They exist mainly as single cells and include only two major groups – the bacteria and blue-green algae. The earliest organisms were exclusively of this type. Typical eukaryotes are much larger than the less evolved prokaryotes, and based on this characteristic are thought to be first recognisable in the fossil record of about 1500 million years ago. Eukaryotes may have developed from a symbiotic interrelation between, and later fusion of, more primitive prokaryotic cells. The fact that they are exclusively aerobic indicates that the process of respiration was operating at that time.

The development of eukaryotic cells with the capacity for sexual reproduction led to a dramatic increase in the rate of evolution and ultimately to the development of highly complex multicellular animals and plants that first appear in the geological record about 700 million years ago. However, it seems that all the major biochemical processes responsible for the present oxygen-rich atmosphere were operating well over 1000 million years ago, and that the growth of atmospheric oxygen had begun 2000 million years ago. How rapidly the level grew and when it reached that of today has been a matter of speculation but one about which little is certain.

Attempts have been made to place limits on past oxygen levels by studying the ability of present-day organisms to cope with reduced oxygen pressures. The basic conclusion is that the oxygen abundance since the Cambrian has probably been somewhere between a few per cent and its present value of 20 per cent. A much lower concentration than this would be sufficient to produce an ozone layer capable of screening the Earth's surface from ultra-violet, radiation and it is therefore possible that the ozone layer is quite ancient.

Based on our earlier simplified discussion of the factors affecting the present oxygen abundance, and in particular the limiting effect of the nutrient phosphorus, it might be supposed that we could make theoretical estimates of the past abundance. In practice the situation is far too complex and poorly understood for this to be attempted at present. The phosphate in sea water has a residence time of the order of only 100000 years and its abundance in the surface layer of the oceans may therefore have varied considerably in response to geological changes that have affected such things as sea level, ocean circulation patterns, shores lines, and rates of erosion etc. It seems quite likely that the oxygen abundance has fluctuated significantly since becoming established as a major component of the atmosphere, but what the magnitude of those fluctuations is we cannot yet say. Past concentrations of carbon dioxide in the atmosphere are similarly uncertain; as we have seen, the ocean has acted as a buffer to smooth out short-term fluctuations, while a gradual reduction in the rate of volcanism over geological time has probably led to a similar reduction in the equilibrium abundance.

References

Berkner, L. V. & Marchall, L. C. 1964. The history of growth of oxygen in the Earth's atmosphere. *In* Brancazio, P. J. & Cameron, A. G. W. (Eds) *The origin and evolution of oceans and atmospheres.* 234 pp. New York: J. Wiley & Sons.

Boyd, R. L. F. 1974. *Space physics, the study of plasmas in space.* 100 pp. Oxford: Oxford Physics Series, Clarendon Press.

Dermott, S. F. (Ed.) 1978. *Origin of the solar system.* 668 pp. New York: J. Wiley & Sons.

Houghton, J. T. 1977. *The physics of atmospheres.* 203 pp. Cambridge: Cambridge University Press.

McAlester, A. L. 1977. *The history of life.* 152 pp. New Jersey: Prentice-Hall.

Schopf, J. W., 1976. Evidence of Archaean life: a brief appraisal. pp. 589–593 *in* Windley, B. F. (Ed.) *The early history of the Earth.* New York: J. Wiley & Sons.

Schopf, J. W. 1978. The evolution of the earliest cells. *Scientific American* **239**, no. 3: 84–102 (issue devoted to Evolution).

Taylor, S. R. 1975. *Lunar science: A post* Apollo *view.* 372 pp. New York: Pergamon Press Inc.

Turekian, K. K. 1976. *Oceans.* 150 pp. New Jersey: Prentice-Hall.

Walker, J. C. G. 1977. *Evolution of the atmosphere.* 278 pp. New York: Macmillan.

G. Turner F.R.S.
Department of Physics
University of Sheffield
Sheffield S3 7RH

CHAPTER 9

The evolution of climates

M. J. Hambrey and W. B. Harland

Weather is a phenomenon that is experienced, and so far as possible explained and predicted in the science of meteorology. Climate is a generalisation of weather for studying regional differences and long-term change; hence the subjects climatology and palaeoclimatology. The Sun's energy is the dominant influence or control. There is no net gain of solar energy at the Earth's surface as whatever is received is subsequently lost somewhere into space. This loss is roughly proportional to the area on the Earth's surface that radiates, while the energy received per unit area varies with the angle of incidence of the Sun's rays, but both loss and gain are subject to many other factors, for example the circulation of the atmosphere and hydrosphere. The angle between the Earth's axis of spin and its orbital plane around the Sun determines both the broad pattern of climatic zones and the seasonal variations within these zones. Such a pattern is modified by the distribution of land and water.

The interest of climates for this volume lies in their variation with time, and they form one factor in the understanding of the evolution of the Earth and its biosphere. Early ideas of the different conditions that had prevailed on the Earth culminated in the debate concerning Diluvium, and the importance of the Biblical Flood, followed around 1840 by an understanding of the Ice Age from which the concept of climatic change evolved. Subsequent stratigraphical studies and the interpretation of palaeoclimatic indicators yielded more evidence of past climatic anomalies, for example the occurrence of glaciated rocks in present-day temperate and tropical latitudes, supposed tropical forests or hot desert conditions in present-day temperate latitudes, and not least deposits formed in temperate and even tropical climates but now located in polar latitudes. By the late nineteenth century such work led to a clear notion of climatic fluctuations and the necessity for the pole to have changed its position relative to continents (polar wandering), as shown in Kreichgauer's (1902) sequence of maps with present-day geography and by Köppen and Wegener's (1924) world climate maps, in which continental drift through later Phanerozoic time was assumed.

Since then many attempts have been made to assess the climatic history of the Earth on a large time scale (Precambrian through to Recent), with units in millions of years, on an intermediate scale in tens of thousands of years (Quaternary fluctuations), or on a human scale in tens and hundreds of years as recorded in historical records (Frakes, 1979; Schwarzbach, 1963; Manley, 1961; Lamb, 1972).

Variable factors influencing climate

Galactic motions

Various speculative mechanisms that could have influenced past climates have been proposed. As the solar system circulates in the galaxy it might pass through the spiral arms of the galaxy and thus variations in the concentration of interstellar matter are important factors. Greater concentrations of matter might result in an increase in the energy output of the Sun, which would probably more than compensate for the filter effect on the amount of radiation reaching the Earth. The duration of such a passage could be in the order of 50000 years. The solar system travels in a galactic orbit (of around 30 kiloparsecs) with an eccentricity of about 10 and a period estimated as decreasing from 400 million years to 274 million years. However, the uncertainties have led astronomers to depend on the Earth's climatic sequence for constraints in their calculations, especially in connection with the alleged periodicity of ice ages.

Planetary motions

The total incident solar radiation and its latitudinal distribution are influenced by three types of perturbation of the Earth's motion. They were first suggested by Croll (1875), much developed by Milanković (1930, 1938) and recently re-evaluated by Evans (1971) and by Pearson (1978). They are as follows:

1 The obliquity of the ecliptic is the angle that the plane of the equator makes with the Earth's orbital plane. This wobbles between 22° and 24·5° every 40000 years.
2 The eccentricity of the Earth's elliptical orbit results in the Earth being 4·8 million kilometres further from the Sun in July than in January. The eccentricity is variable, with a cycle of about 90000 years.
3 The precession of the equinoxes is due to a wobble of the elliptical orbit, so that the major diameter rotates through 360° with a periodicity of 21000 years.

The combined effect of these three factors, which result in different amounts of solar radiation reaching the northern and southern hemispheres, produces a relatively complex pattern, and the resemblance of this pattern to that of ice ages has been emphasised. The effect of these factors is now well known in Pleistocene studies and, although few attempts to consider them prior to Pleistocene time have been made, there is no reason to suppose that they had no influence throughout geological time; indeed ancient glacial deposits suggest that they did.

Two long-term mechanisms have been suggested. It is virtually certain that the length of the day has increased as a result of the tidal effect generated mainly by the Moon from early in Precambrian time, and possibly also as a result of expansion of the Earth. About 400 million years ago the day may have been 21·9 hours long. A more radical but entirely speculative proposal by Williams (1975), to account for the low latitude Varangian ice age of around 670 million years ago, was to assume an obliquity of the ecliptic of 54° or greater, which would result in an absence of glaciations in high latitudes.

Geomagnetic fluctuations

Fluctuations in the intensity of the geomagnetic field affect the incidence of ionised material of solar origin. There is a complex inter-relationship between the ionisation of gases and their transparency to different radiation wavelengths. Correlations suggest that in the short term higher magnetic intensity would be associated with colder climates. For longer-term fluctuations, the reduction in intensity of the Earth's magnetic field at times of magnetic pole reversal might theoretically have some effect on climates. Pearson (1978, pp. 62–97) has discussed these and other effects.

Geothermal influences

Different rates of heat loss from the Earth's interior have had an important influence on weather and climate. Regions of vulcanicity and hot springs have only a local effect, but in early Archaean time the heat generated in the Earth and being issued through volcanic activity may well have warmed both oceans and atmosphere. However, whether or not occasional peaks of magmatic activity throughout the Earth's history have had any major effect is not yet firmly established.

Composition of the atmosphere

A more important cumulative effect of vulcanism has been the generation of the Earth's early atmosphere and hydrosphere, especially regarding the output of carbon dioxide. Atmospheric CO_2 is more transparent to solar radiation of shorter wavelengths than to the radiant heat which is reflected from the Earth's surface (and so retained). This is known as the 'greenhouse effect' which could be of great significance in climatic change. Throughout most of Archaean time (until 2500 million years ago) the CO_2 concentration may have been higher and so the atmosphere warmer than subsequently. Deposition of carbonaceous and carbonate deposits would have removed CO_2, and it has been suggested that ice ages correlate with extensive limestone and dolostone formation (Roberts, 1971). However, it is not certain that increase of CO_2 would always lead to higher temperatures because of other compensating mechanisms.

Water vapour production and the resultant cloud formation increased the Earth's albedo (the reflectivity of the Earth to the sun's radiation), but this is perhaps a self-regulating effect because evaporation and precipitation rates are so dependent on temperature. Widespread volcanic dust is often considered to have had a cooling effect and volcanic activity might relate to tectonic history. However, little is known of the magnitude of volcanic episodes in the past, and in any case basic volcanic activity is not especially explosive and therefore unlikely to produce significant amounts of atmospheric dust. The evolution of the atmosphere is considered further in Chapter 8.

Biological evolution

The balance between photosynthesis and respiration has affected the CO_2/O_2 balance of the atmosphere, so that free oxygen was probably available for oxidising sediments throughout most of Proterozoic time from 2500 million years ago onwards.

The colonisation of land by vegetation must have had a profound climatic effect through the growth of extensive forests and later of grasslands. This is evident today in some areas since, when the vegetation is removed, it can be rapidly replaced by deserts. Thus before Devonian time climates were probably systematically different.

Distribution of land and sea

Several distinct variables relate to the configuration of the solid Earth's surface in relation to the hydrosphere, and they all influence climate significantly. A distinction has been made between (1) general properties of the Earth's relief at any particular time (hypsometric factors) and (2) particular properties at any time of the local and regional distribution of land and sea (configurational factors) (Harland & Herod, 1975).

Hypsometric factors. Continentality is the overall ratio of the area of land to sea. Geological changes of continentality are identified as transgressions and regressions of the sea in particular areas, while net global sea level changes are termed eustatic. Continentality is a difficult quantity to measure as sea level is relative and global sea level can only be considered in relation to an average land level or to a solid earth parameter.

Related to, but not the same as, continentality is the degree of relief. The hypsometric curve is obtained by plotting depth and height relative to sea level against the area for successive elevation intervals. While continentality can be measured from the hypsometric curve as a direct percentage of land area, the relief is taken to be the average height of the land above sea level. The average depth of the sea in relation to its area should give a relatively slowly changing ocean volume. Since only the present-day hypsometric curve is known, past measures of relief can only be estimated roughly from independent evidence recorded in the Earth's sedimentary record.

Configurational factors. There are at least three climatic effects of variations in the particular distributions of land and sea, namely:

1 *Position of spin axis.* The Earth's poles, as a result of the differential shift of the lithospheric plates with respect to latitude (polar wandering), may at any time fall in the middle of large areas of land or sea. The present position of the South Pole in the middle of Antarctica and of the North Pole in the Arctic Ocean show different effects. In particular, ice generated on a continent will continue to build up and not dissipate and melt as rapidly as will polar pack ice formed in the sea. Thus a large ice sheet will tend to influence climate rather more than will a cover of sea ice.

2 *Pattern of seaways.* The shapes and sizes of oceans and continents also influence climates by their effect on ocean currents. In this respect latitudinal or meridional routes are important. There must be a critical size (width and depth) above which a seaway provides an effective circulation. For example, the build up of sea ice in the present Arctic Basin is enhanced by the lack of circulation. Also the distribution of rain forests and deserts at the same latitudes on opposing coasts of a north–south orientated ocean may depend on ocean current circulation. A north–south route provided by the opening Atlantic Ocean was suggested by Ewing & Donn (1956) as an oscillating mechanism responsible for glacial and interglacial periods in late Cenozoic time.

3. *Concentration of land and sea.* The relative size of land masses and oceans also influences climate, for example the monsoon effect at the margins and the reduction of precipitation towards the interior of a continent. It has been argued that a supercontinent is incapable of developing a large ice-sheet, but the Gondwana glacial deposits within Pangaea would seem to refute this view.

Discussion

Of the above factors that might be relevant to the climatic history of the Earth, the galactic motions, if effective at all, are long term, while the planetary motions are certain and calculable in their effect and are relatively short term. Both, however, are cyclic. Geomagmatic and geothermal effects are uncertain and probably slight. The effects of biological evolution which would both affect and be affected by the composition of the atmosphere, would tend to be long term and irreversible. For example sufficient atmospheric CO_2 might inhibit ice ages altogether, as may have been the case in Archaean time. It has been proposed that the Late Precambrian ice ages relate to diminished CO_2 which was removed during the deposition of extensive carbonate sediments well known in the North Atlantic region. However, these regions were then tropical and equivalent carbonates in other latitudes have yet to be established. The most effective variable of medium-term effect is the land–sea distribution which is chiefly controlled by tectonic activity.

Indicators of past climates

Great difficulties often exist in establishing the environments of deposition in which sedimentary rocks have formed, so perhaps it is not surprising that climatic conditions at different stages of the Earth's history are difficult to determine. Although there are few simple and reliable indicators of old climates of any sort, several factors taken together can be used to build up a picture. These indicators are biological, lithological, geochemical, morphological and palaeomagnetic (Schwarzbach, 1963; Nairn, 1961 and 1964 gave comprehensive summaries). Certain fossils have particular ecological or physiological characteristics that reflect the climate, for example the size and colouring of some organisms, the shape of leaves (drip-points), annual rings or a reef structure. Fossils may also have a close relationship with modern climate-contained species, such as reindeer, musk oxen or pollen in the Quaternary Period, and palms in Tertiary time. However, biological evidence is often indefinite; the significance of many fossils has been only approximately evaluated, and sometimes even misinterpreted, and organisms of similar structure may adapt to different environments, for example elephant and mammoth. It is often difficult to distinguish climatic and non-climatic factors; for example a meagre coral reef fauna may be the result of low salinity rather than low temperature.

Lithological evidence includes the conditions of sedimentation reflected in bedding or the lack of it, ripple marks, rain splash marks, or the distribution of loess; the mineralogy and petrology of sediments, such as the presence of limestone, salt, gypsum or glacial deposits; and post-depositional effects, notably weathering, including laterite development or silicification within the sediment. However, completely different geological processes can result in the same rock type. As an example unbedded, unsorted boulder beds, which have been behind much controversy concerning past ice ages, can be produced not only by glacial deposition but also by subaqueous mass gravity flows.

The application of geochemical techniques to palaeoclimatological study is a relatively recent development. These techniques are principally concerned with determining the $^{18}O/^{16}O$ isotope ratio in calcium carbonate taken from marine fossil shells in ocean cores. This ratio is temperature-dependent and reflects the temperature of the water in which the carbonate shell was originally deposited during the life-time of the animal. The ratio also depends on the original composition of the sea water. This may vary, for example because of the abstraction of the lighter isotope in continental ice. Indeed in the Pleistocene Period the original composition of the sea water was probably more important than temperature. $^{18}O/^{16}O$ ratios of ice in Greenland and Antarctica have also been successfully used in determining temperature variations for a much shorter period (about 100000 years) although in great detail. Where melting is unimportant the snow and resulting ice in cores taken from these ice sheets reflect the temperature at which the snow originally fell, and in the last few hundred years even seasonal variations can be detected. These isotopic methods depend on very precise measurements with the mass spectrometer, and this, and the difficulty in selecting reliable experimental material, mean that care is necessary in assessing the results. Nevertheless the method has been widely used with considerable success.

Morphological evidence of climate is represented by familiar landforms such as U-shaped valleys, striated pavements, cirques, moraines, eskers, tors, inselbergs, river terraces, raised beaches, amongst others, all reflecting past climates in some way. Certain of these features are occasionally preserved in ancient rocks.

The application of palaeomagnetism to palaeoclimatology is different. The study of the Earth's geomagnetic field in the geological past is based on the determination of permanent magnetisation in the rocks, which records the orientation of the Earth's magnetic field at the time of rock formation. Such data enable the mean position of the magnetic pole to be determined for a particular time and, since these when averaged appear to coincide with the geographical poles, the ancient latitudes of the sampling sites can also be established. This information is not directly climatic but information about latitude assists in the interpretation of climatic zonation.

Evidence of cold climates

Glacier ice today covers about 10 per cent of the Earth's land surface and at the height of the Pleistocene ice ages probably covered as much as 29 per cent (Flint, 1971). Extensive ice ages are well documented in the geological record as far back as middle Precambrian times, some 2500 million years ago. Thus the varied effects of glacial erosion and deposition are widespread, and constitute the most important means of identifying cold climates of the past.

Erosional landforms occur on all scales. Valleys, approximately U-shaped in cross-section and often overdeepened and containing lakes or fjords, are characteristic of glaciated mountain areas. Cirques or corries are often located above the floors of these valleys. With prolonged erosion the walls of adjacent cirques may intersect to produce arêtes and horns. Medium-scale features include the whale-back shaped roches moutonnées. Small-scale features include striations, crescentic gouges, crescentic fractures and lunate fractures which, along with roches moutonnées, are useful for determining the direction of ice flow. Erosional landforms are well developed in areas formerly occupied by glaciers during the Pleistocene Period, but some of these features are also preserved in the geological record. For example U-shaped valleys formed by Permo-Carboniferous glaciation have been reported in Namibia (South West Africa) and Australia, while striations have often been used to establish the extent and direction of movement of former ice sheets (Fig. 9.1).

A wide variety of sediments is deposited in glacial environments, including material that has been passively transported at the glacier surface, subjected to grinding at the glacier base, and reworked by subglacial and proglacial streams. Material deposited directly by ice (till) is variable in composition, stone shape and sedimentary structure. On land till tends to mask the underlying topography and can reach a considerable thickness. There is generally little argument concerning the recognition of Pleistocene tills but their lithified equivalents, tillites, are often confused with other types of conglomerates, such as deposits formed by sub-marine gravity-induced flow. Observed pre-Pleistocene tillites, more often than not, tend to have been deposited in marine environments (Fig. 9.2), and thus have a different appearance from terrestrially deposited tills, especially when the direct glacially derived component is small. The bulk of the sediment may have been deposited offshore from rivers carrying much glacial material and subjected to sorting and redeposition by turbidity currents. Coarser materials, often including boulders, may be released from icebergs, giving rise to distinctive dropstone structures in the perhaps finely stratified sediments on the sea floor.

Many types of landforms of deposition exist. Morainic ridges, the most widespread, are occasionally preserved in the geological record; perhaps the most famous is a moraine described late last century by Reusch in the Late Precambrian glacigenic rocks of northern Norway. Other depositional landforms – eskers, kames, kettleholes, drumlins and erratics – although useful in determining Pleistocene climatic events, are rarely recognisable in earlier deposits.

Periglacial phenomena, like glacial landforms, are good indicators of present and earlier climates. These include block fields or 'Felsenmeer' which are the products of intense physical weathering; various types of patterned ground, notably non-sorted stone stripes, circles and polygons and all gradations between; ice wedges which eventually become replaced by sand or other fine material; solifluction lobes and pingos. Fossil patterned ground is widespread in areas close to the margins of former

Figure 9.1 (above) The Nooitgedacht striated pavement near Kimberley, South Africa. A Precambrian volcanic surface eroded by glaciers of the Permo-Carboniferous Gondwana glaciation (Photography courtesy of V. von Brunn.)

Figure 9.2 Tidewater glacier in Smeerenburgfjorden, north-western Spitsbergen. Calving of icebergs from such glaciers can result in the transport of glacial debris by ice rafting for considerable distances. The distinctive deposits thus formed are the most important glacial sediments preserved in the geological record.

Pleistocene ice sheets, particularly in central Europe, Britain and North America. It has also been reported in much older rocks such as those associated with Late Precambrian tillites in Scotland and Spitsbergen.

Some light may be thrown on past climates by the lithological characteristics of the sediments. Chemical weathering is slow in cold climates and kaolinite is unable to form, while unstable ferromagnesium minerals may remain. On the other hand physical weathering, particularly frost action, is rapid, and the constituent particles of the new deposits will closely resemble the rocks from which they are derived. The temperate climates of interglacial phases are indicated by the leaching of till which will give rise to colour variations from top to bottom, as in soil profiles. This may have occurred to a depth varying from 10 cm to several metres, depending on the length of the interglacial phase.

Plants are of great help in recognising cold climates in Tertiary and, more especially, in Quaternary time. However, few pre-Tertiary plants are useful in establishing cold climates. Possible exceptions are the fern-like flora *Glossopteris* and *Gangamopteris* which may have flourished in a cold climate since they are associated with glacial deposits in Gondwanaland. In the glacial phases of Quaternary time, Central Europe was colonised by Arcto-Alpine species such as *Dryas octopetala*, arctic willow (*Salix polaris*) and dwarf birch (*Betula nana*). The southern limit of these species today approximately coincides with the 10 °C isotherm for the warmest month, suggesting that temperatures during glacial phases were 6–10 °C lower than those of today in areas where these tundra species occurred. Conifers also tend to indicate cool climates. In times of favourable climate species such as juniper, spruce, larch and pine spread into the polar regions, but by contrast extended much further south than at present during cool phases. Many species can be recognised when Quaternary deposits are subjected to pollen analysis, and the resulting composite diagrams enable one to detect quite small climatic fluctuations.

Animal remains are sometimes useful in establishing climatic conditions in Quaternary time. Those typical of a tundra climate are white grouse (*Lagopus*), lemmings (*Myodes*), arctic fox (*Alopex lagopus*), arctic hare (*Lepus variabilis*), reindeer (*Rangifer tarandus*), musk ox (*Ovibos moschatus*), mammoth (*Mammuthus primigenius*) and woolly rhinoceros (*Coelodonta antiquitatis*). The distribution of Pleistocene mammals is considered further in Chapter 15. Marine invertebrates have also been used for determining past climatic conditions in Quaternary time. Distinctive assemblages dependent on climate have been recognised in the raised beach deposits of lands around the Arctic Ocean. Cold water foraminifera include *Globigerina bulloides* and *Globigerina inflata*. *Globigerina pachyderma* develops different coiling directions in cold and warm water species; for the former it is sinistral.

Brief mention should also be made of the indirect indicators of cold climates, namely raised beaches and river terraces, which reflect a balance between changes in sea level and the response of land to its burden of ice. Lower sea levels imply an increase in the amount of ice but where this accumulates it will depress the land. However, the latter is much the slower process and thus the two effects need not cancel each other out. The height of the sea level also controls the gradient of a river and its ability to erode, and this is reflected in its irregular long profile and associated river terraces. These characteristics have been particularly useful in reconstructing Quaternary climatic change, but are difficult to observe in the earlier geological record.

Evidence of hot climates

The degree of weathering in warm areas depends to a considerable extent on the rainfall. Reddish horizons in soils, resulting from iron staining, are regarded as characteristic of hot climates, although they tend to occur in areas where the rainfall is seasonal, rather than in deserts. Red coloration is said to occur where the mean annual temperature exceeds 16 °C, and the rainfall 1000 mm in areas of silicate rocks or over 625 mm in areas where carbonates are found. These conditions prevail in the tropical savannas where hard red horizons of laterite develop. Lateritic soils are almst completely leached of SiO_2, while there is enrichment of Al_2O_3 and Fe_2O_3. Redeposition of silica, often by silicification, takes place at depth. Weathering may extend to considerable depths (50 m or more) in hot humid regions and the rate of chemical alteration is rapid. Red loams are typical of subtropical savannas, while brown loams occur in the rain forests.

Red beds are widely distributed in the geological record. Bauxite, one type of red bed and the principal ore of aluminium, can be regarded as a fossil laterite, while kaolin is often developed in the deeper parts of the laterite profile. Laterites are only common after the spread of terrestrial vegetation in the Devonian Period, suggesting that vegetation may be important for laterite development. However, other types of red bed are common in older rocks, and although traditionally regarded as evidence of hot arid climates, the actual climatic significance is uncertain. The red pigment in sandstones has sometimes been shown to form by the redistribution of iron away from the decomposing grains by oxygenated pore waters during diagenesis; thus the climate during deposition of the sandstones may be irrelevant to the formation of haematite-cemented sandstones.

Other indicators of hot climates are a reflection on different aspects of organic activity. Particularly important in the geological record is the relationship between limestone deposition and marine organisms. The calcium carbonate shells dissolve in water containing carbon dioxide. Unlike most substances, the amount of $CaCO_3$ that can be dissolved increases as the temperature decreases. Thus $CaCO_3$ precipitates out as calcareous mud or oolites when the waters become warmer, such as in the shallow parts of tropical areas; most fine grained, bedded limestone formations are believed to form in this way. Such chemically precipitated limestones are unknown in present-day high latitude seas. Coral reefs are regarded as good indicators of tropical climates although small accumulations occur in more temperate latitudes. Other creatures such as archaeocyathids, sponges, bryozoa, thick-shelled bivalves and calcareous algae have built up reefs as well as corals, but these are less important. Modern coral reefs generally need a minimum temperature of at least 21 °C, as well as clear, aerated, shallow water of normal salinity, and are thus confined to littoral zones between latitudes 30° N and 30° S. Fossil reefs often form squat hills of unbedded limestone. The growth of individual corals is also controlled by climate, and is revealed by a structure resembling annual rings but, because non-climatic factors may be involved, such information is of little use on its own. Although one cannot assume that limestone deposition occurs principally in warm water conditions, these do seem to be the most likely, especially when considered along with other palaeoclimatic evidence. However, carbonates are also closely associated with tillites in late Precambrian rocks. It is interesting to note that recent work on the sediments of the icy waters of the Barents Sea, between Norway and Svalbard, has revealed significant modern deposition of carbonates and this has implications for the interpretation of palaeoclimates elsewhere (Bjørlykke *et al.* 1978), although it should be noted that deposition there is bioclastic rather than by precipitation.

Since warm-blooded land mammals are adaptable, their fossils are of little use as palaeoclimatic indicators unless they are identical to their modern equivalents. Cold-blooded creatures like reptiles, on the other hand, are less tolerant, and their young could hardly have survived arctic conditions. The optimum conditions for reptile life are hot, humid climates, a preference which is reflected in the increase in size and number of species of reptile towards the equator today. Some reptiles of the past, however, may have been warm-blooded, so the rapid extinction of the large Mesozoic reptiles (around 65 million years ago) may not have been due entirely to climatic deterioration.

As with both terrestrial and marine faunas, the number of plant species increases towards the equator. For example in the Central European woodlands there are only 10–15 varieties of trees, compared with over 500 in the tropical rain forests of the Cameroons. Certain trees, particularly those of the Carboniferous coal forests such as tree ferns, were once regarded as tropical, but the evidence for this is flimsy. The most that can be said is that the plants belonged to a warm, but more importantly to a wet, climate. Comparative studies between fossil plants and similar present-day species can only be used from Tertiary time onwards and they are most useful when plant assemblages rather than single species are examined. Palms are perhaps the most indicative; in Tertiary time they were far more extensive than today and are often well preserved in the sedimentary record.

Evidence of arid climates

The best evidence of hot, arid climates is the presence of evaporite deposits, such as salt, gypsum and potash. Precipitation of salt today occurs principally in salt lakes, for example the Caspian, Aral and Dead Seas, which are located in hot arid belts both north and south of the equator. Salts can also precipitate out in or below the surface of soils in arid regions. Thick beds of evaporites occur throughout the geological record, although they are rare in Precambrian strata.

Silicification also occurs under dry climates; this might affect organic matter such as trees, and some fine examples have been preserved in the Triassic rocks of Arizona. Crusts of calcite with a proportion of silica, known as caliche or calcrete, also result from chemical precipitation and occur in areas where the rainfall is below 625 mm and the temperature for most of the year above 25 °C, i.e. where evaporation exceeds precipitation. These crusts form at depths in the soil profile in semi-arid areas and can reach more than 50 m in thickness. Ancient concretionary carbonates (cornstones) are sometimes preserved in the geological record.

Other deposits forming under arid conditions bear evidence of the effects of wind. Sand dunes are particularly important in desert environments; so are ripples, and it is often possible to differentiate wind from water ripples by, for example, the long low profile of the latter. Sometimes when dealing with ancient rocks morphological characteristics may be insufficient indicators and terrestrial and subaqueous depositional environments can only be distinguished on other criteria. Prior to Devonian time terrestrial sand dunes are unlikely to be good climatic indicators, as they are likely to have been present also in wet climates because of the lack of vegetation and absence of cohesive soils. Wind-transported sand grains are often well-rounded and have a matt surface. Sand blasting of pebbles produces characteristic faceted ventifacts (typically 3-cornered, when they are known as dreikanter). All the features just discussed can be found in both cold and arid environments, but finer-grained silty sediments known as loess seem to be restricted to cool or cold lands. Loess is apparently of glacial origin and is widely distributed immediately to the south of the area occupied by the major Quaternary ice sheets, particularly in central Eurasia. Loess has also been recorded occasionally in association with pre-Pleistocene glacial deposits as far back as Precambrian time. Rain pits are also, surprisingly, usually an indicator of an arid climate. They are the imprint of isolated showers falling onto sand or mud, and are sometimes preserved in the geological record.

Certain erosional landforms are typical of desert regions, of which 'Inselbergs' are the most impressive. These are hills rising abruptly above flat plains, but are not a reflection of the structure or hardness of the rocks. Inselbergs have occasionally been recognised in the geological record, as in Triassic and Miocene (Tertiary) sandstones in Spain, and Permian rocks in Czechoslovakia.

Some plants and animals can be used as indicators of arid climates. Salt-loving (halophytic) plants are particularly characteristic but they sometimes also grow in non-arid areas. Animals are generally sparse in arid regions, but a few, such as scorpions, occur and may survive as rare fossils in rocks such as desert sandstones. The use of animals as palaeoclimatic indicators is limited. The best fossil evidence is confined to Quaternary deposits, particularly those representing dry, continental climates, as in the Russian steppes of today where the marmot (*Spermaphilus*), saiga antelope, and Irish elk (*Megaceros hibernicus*) have been found.

Evidence of humid climates

One of the best indicators of a humid climate is kaolin and its lithified equivalent kaolinite (Al_2 (Si_2O_5) $(OH)_4$) which form in soils by extreme chemical weathering, particularly of feldspar-rich granites and other igneous rocks, when the temperature is not too low. If the rainfall is seasonal kaolin occurs within the laterites of warm climates. Exceptionally, kaolin-rich soils are associated with swamps where humic acids result in efficient decay of other mineral constituents. Occasionally, evidence of former humid climates, generally of Pleistocene age, is seen in the form of old lake terraces in areas where present-day lakes have no outlets, e.g. Lake Bonneville in the Great Basin of Utah.

Plants are sometimes good indicators of a wet climate as their leaves may develop 'drip points' to enable the rain water to drain off rapidly. These are common in tropical and subtropical rain forests, but become rarer towards the drier areas, and have been used to show how rainfall conditions varied during Tertiary time in Central Germany. Coal seams and peat bogs also reflect a wet climate, but if they are formed in upland regions may also reflect a cool climate and they develop mostly as a result of the growth of *Sphagnum* moss.

The climatic record

We have outlined above the principal lines of evidence that can be used to infer the nature of the Earth's climate in the geological past. Few criteria alone are sufficient proof of a particular climate but a reasonable assessment can often be made if several are considered together. It should, however, be borne in mind that time spans of five million years represent only a tiny fraction of Earth history and yet that is often the limit of accuracy in correlation. In view of what we know about the many considerable climatic changes that have occurred in the last two million years, care is necessary in generalising about climates either areally or stratigraphically. Even the relatively well-known Quaternary record is in need of critical reassessment. The further one goes back in geological time the less complete becomes the evidence, so that by Precambrian time palaeoclimatic data are extremely sparse, yet Precambrian time represents 88 per cent of the Earth's history. Convenient accounts of climates through time are by Schwarzbach (1963) and Frakes (1979). Attention is also drawn to compilations by Hallam (1977) and Habicht (1979).

Precambrian climate

Archaean. The Precambrian record may be divided at 2500 million years, which is now agreed to separate Archaean from Proterozoic time. Relatively few Archaean supracrustal sedimentary rocks are known that preserve any recognisable climatic record. Life is known to have existed, so temperatures well below the boiling point of rocks may be assumed, and indeed some stromatolitic rocks suggest that environmental conditions were similar to those that existed in some later geological periods. It might be argued that, with a generally reducing environment and little oxygen in the atmosphere, CO_2 might have had the effect of preventing low temperatures; certainly no unequivocal tillites have been identified for most of Archaean time. Towards the end, however, the Witwatersrand tillites may represent the earliest known glacial record.

Proterozoic. An early Proterozoic event appears to have been the genesis of red beds around 2200 millions years ago. These indicate not so much a dry arid climate as an abundance of oxygen. At about the same time (2400–2100 million years), some long-established tillites were formed. Evidence for the three early Proterozoic Huronian glaciations, named after Lake Huron in Canada, is among the best preserved of any of the 'older' glaciations, and includes fine examples of striated boulders and dropstones. Tillites of early Proterozoic age have also been recorded in the USA, southern Africa, India and Australia, which suggests that this was a time of one or more of the Earth's major ice ages. Fine-grained or oolitic dolomites and limestones, supposedly indicators of fairly warm conditions, are widely developed in later Proterozoic rocks. For example, towards the end of Proterozoic time such rocks were deposited in areas now as widely scattered as Greenland, Scotland, Siberia, North Africa and Australia. Late Precambrian red beds are also common and have traditionally been held to be indicative of arid conditions, examples being the Torridonian Sandstone of north-western Scotland, the Keweenawan of Lake Superior and Canada, the Sala and Jotnian Sandstones of Sweden, and the Vindhyan Supergroup of India. However, detailed sedimentological observations in recent years have shown that arid conditions did not necessarily prevail when these rocks were deposited. For example, much of the Torridonian Sandstone is of fluvial origin. Unequivocal evaporite deposits are unusual in the Precambrian record, and warm, wet conditions cannot be ruled out, especially as the land was yet to be colonised by plants, and floral climatic indicators are therefore absent.

There is also abundant evidence of at least one extremely widespread glaciation during latest Precambrian time (about 670 million years ago) which was first established in the last century in northern Norway, named the Varangian glaciation, and subsequently recognised in other parts of the world. To account for this it has been argued that the extent of ice was localised and that its widespread distribution is a function of rapid polar wandering and non-synchroneity of glaciation. However, increasing evidence from all continents except Antarctica suggests that a world-wide glaciation reached very low latitudes. Furthermore, with deposits in areas like Svalbard, East Greenland, northern Norway and Scotland, which are thought from palaeomagnetic evidence to have been near the equator, it is difficult to accept anything other than a major ice age. Probably most of the extensive ice had disappeared in latest Precambrian time when the Ediacaran metazoan marine fauna developed.

Palaeozoic climates

The most remarkable change at the beginning of the Cambrian Period (around 600 to 570 million years) was the rapid increase in complex organisms in the oceans, notably animals with skeletons capable of preservation. Although parts of the Earth may have been ice-covered, it is known that Earth's climate rapidly ameliorated, and this may have contributed to rapid evolution. At the same time marine transgressions were widespread so that the land area was greatly reduced and vast areas of shallow, warm, sunlit seas had developed. Carbonates and evaporites were widely deposited, often following on quite sharply from the Late Precambrian tillites. For example, several hundreds of metres of limestone and dolomite were deposited in tropical latitudes which are today found in the North American Rockies, Alaska, Greenland, north-western Europe and even Antarctica. These limestones often contain reef-forming archaeocyathids and other warm marine indicators. Associated continental areas are generally assumed to have had a hot, dry climate, and latitudinal temperature gradients were apparently not extreme.

In the succeeding Ordovician Period climatic zones appear to have become strengthened, a hypothesis based on the wide variety of faunas and sediments present. In addition to the continued deposition of carbonates and evaporites in the Ordovician equatorial zone that now lies in present-day Arctic and sub-Arctic regions, one finds elsewhere late Ordovician glacial deposits and evidence of glacial erosion, notably in the Sahara. Palaeomagnetic evidence suggests that the Ordovician south pole was located in northern Africa, then part of the vast supercontinent of Gondwanaland. The evidence for Ordovician glaciation in North Africa is impressive, and the fresh-looking striated pavements are often more reminiscent of Quaternary glaciations than older ones. The enhancement of climatic zones is also indicated by the distribution of different species of brachiopod, conodont, trilobite and graptolite.

Melting of the late Ordovician land-ice led to a rise in early Silurian sea level. The south pole had migrated to Argentina as Gondwanaland had turned around on the globe. Reef limestones and dolomites were deposited in central parts of northern Europe and North America, as well as a strip of Asia between Afghanistan and Malaysia, and in parts of Australia.

In Devonian time there was a surge in the growth of reefs (stromatoporoids) in the tropical seas, as in north and central Europe. Shorelines fluctuated widely at this time and the sedimentary record shows evidence of extensive deltas (some with coal forests), coastal plains and beaches. Red beds are characteristic of the continental deposits of this period, formed as a result of extensive erosion of the massive Caledonian mountain range that included the Appalachians, the north-western highlands of Britain, Scandinavia and east Greenland. Evidence for glaciation is poor; but tillites have been recorded near the south pole of that time in the area north of Buenos Aires, in adjacent areas of South America and in South Africa.

The Devonian Period ended with a rise in sea level, resulting in early Carboniferous time in the development of wide shallow seas in which marine life flourished and developed. Extensive coral reefs formed and limestone deposition was widespread in sub-tropical areas, including much of North America and Europe: the low-lying, swampy land was colonised by luxuriant forests from which coal eventually developed. However, Gondwanaland, surrounding the south pole hemisphere and consisting of Antarctica, South America, Africa, Australia and India, experienced a major ice age, and continental and marine tillites were deposited over the whole continent. Climatic evidence is abundant in all these areas, and the reconstruction by Köppen & Wegener in 1924 for that time has not been improved very much since (Fig. 9.3).

The Permian Period was characterised by considerable plate-tectonic activity, whereby the major continents at the time combined to form a single supercontinent (Pangaea). The Tethyan waters between Laurasia (Eurasia and North America) and Gondwanaland closed, resulting in a radical alteration of the global climatic pattern. Extensive glaciation (on the scale of that of the Pleistocene Period) continued into the Permian Period, centred on the south pole which then, as now, was located in Antarctica.

Hot, dry conditions prevailed over much of Europe, the Asian USSR and North America, and northern Gondwanaland. Deserts were widespread and evaporites were deposited in many areas. The humid warm-temperate zones were rather less extensive. The meeting of the northern and southern continents set off a chain of events ending with a melting of ice and rise of sea level. There were considerable changes in fauna, reptiles gaining at the expense of amphibians when the Palaeozoic Era came to a close.

Mesozoic climates

The Mesozoic Era, comprising the Triassic, Jurassic and Cretaceous Periods, was characterised by relatively stable climates. The Triassic Period was generally tectonically quiet except in some Pacific margins (eg. South-east Asia), but in middle and late Mesozoic time Pangaea split apart as the new Atlantic and Indian oceans opened up, while in late Mesozoic and on into Cenozoic time, continental collision of the northern and southern margins of the Tethys Ocean resulted in the Alpine, Himalayan, Burman and Indonesian mountain chains.

The Mesozoic Era (especially mid-Triassic to mid-Cretaceous) was typified by relatively warm climates, and climatic belts migrated markedly polewards. In Europe warm, relatively arid climates appear to have dominated the Era. In the Triassic Period thick coral reefs, shelf limestones and dolomites were deposited. These provided much of the rock sequence in the Alps and elsewhere. Desert sands and evaporites, indicative of hot, arid climates, make up the bulk of the continental deposits of Britain and north-western Europe, where salt and gypsum are of great economic importance. In North America Triassic rocks indicative of an arid climate are also widely developed, in particular sandstones and gypsum. Widespread red bed deposits of this time are thought to have formed under similar conditions. A similar hot arid belt lay in the southern hemisphere, in part taking over from areas where coals had developed in Late Palaeozoic time (Fig. 9.4).

Away from the equator these belts gave way to a warm, wet climate with lush vegetation. Triassic coals therefore developed in, for example, north-central Siberia, Germany, Virginia and North Carolina, and the forests were colonised by warmth-loving reptiles. Indeed, Triassic reptiles have been recorded from as far north as Svalbard, now at 78° N but then located about 40° N.

Similar conditions prevailed in the Jurassic Period, a time when sea levels were relatively high. Limestones continued to be deposited extensively in Europe and to a lesser extent in North America. It has been suggested on the basis of $^{18}O/^{16}O$ determinations on Jurassic sediments that ocean temperatures were in many places 10°C higher than those of today. Coral reefs lay over England, and a strip of reefs extended almost the whole length of Japan. Evaporites and red beds were again prominent in Europe, central Asia and the western USA. They are also widely distributed in the southern hemisphere. The warm humid climatic belt of Jurassic time is represented by coal in southern Sweden, Svalbard, Greenland, Scotland, the Caucasus Mountains and in a belt through Asia from the east of the Urals to Japan. In the southern hemisphere carbonaceous strata occur in Queensland and Victoria. Extensive kaolinisation also took place in some areas, notably in Sweden.

Early Cretaceous climates continued warm but later underwent marked cooling, although still warm compared with the present. The presence of large reptiles in high

A CARBONIFEROUS

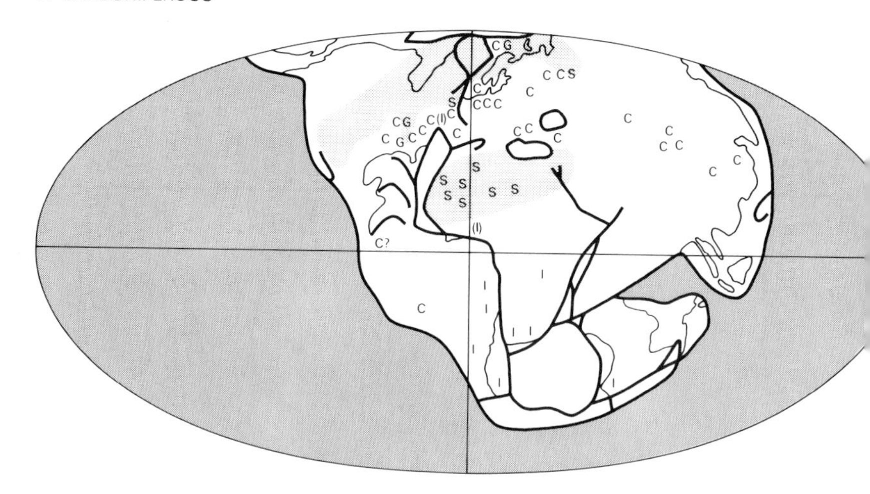

B PERMIAN

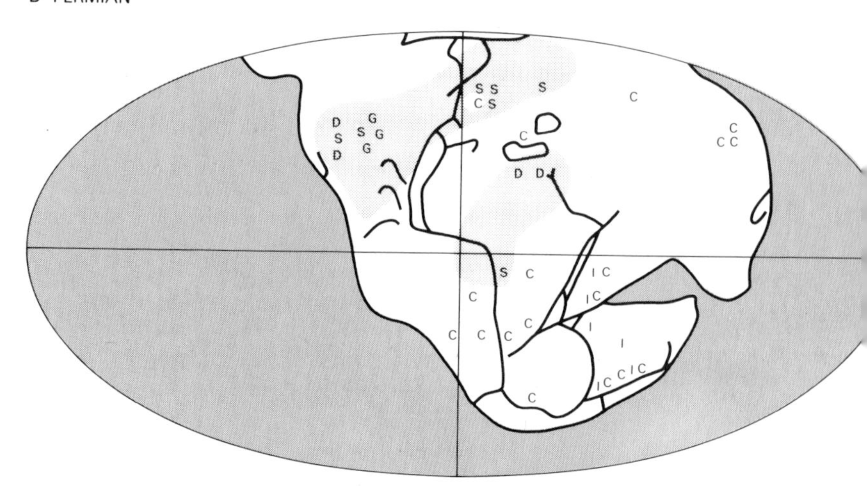

Figure 9.3 The palaeocontinental reconstructions of Köppen & Wegener (1924, Figs. 3 & 4) illustrating distribution of various palaeoclimatic indicators: C = coal, D = desert sandstone, G = Gypsum, I = traces of ice action, S = salt. Arid areas are shaded.

palaeolatitudes, and other fossil evidence suggest, as earlier in Mesozoic time, that the climatic gradients were not pronounced.

However, the cooling gave rise in late Cretaceous time to boreal climates. Widespread regression of the sea occurred in late Jurassic–early Cretaceous time, resulting in erosion and non-marine sedimentation in many parts of the world. A probable surge in the rate of sea-floor spreading led to a reversal of this process, and before long the Earth experienced possibly the greatest marine transgression since Precambrian time, with a vertical amplitude of perhaps 600–700 m. However, the effect such changes had on the climate is not certain. Climatic changes during the latter part of the Mesozoic Era appear as a gradual shift of climatic belts southwards corresponding to a polar wandering effect as the main land masses moved

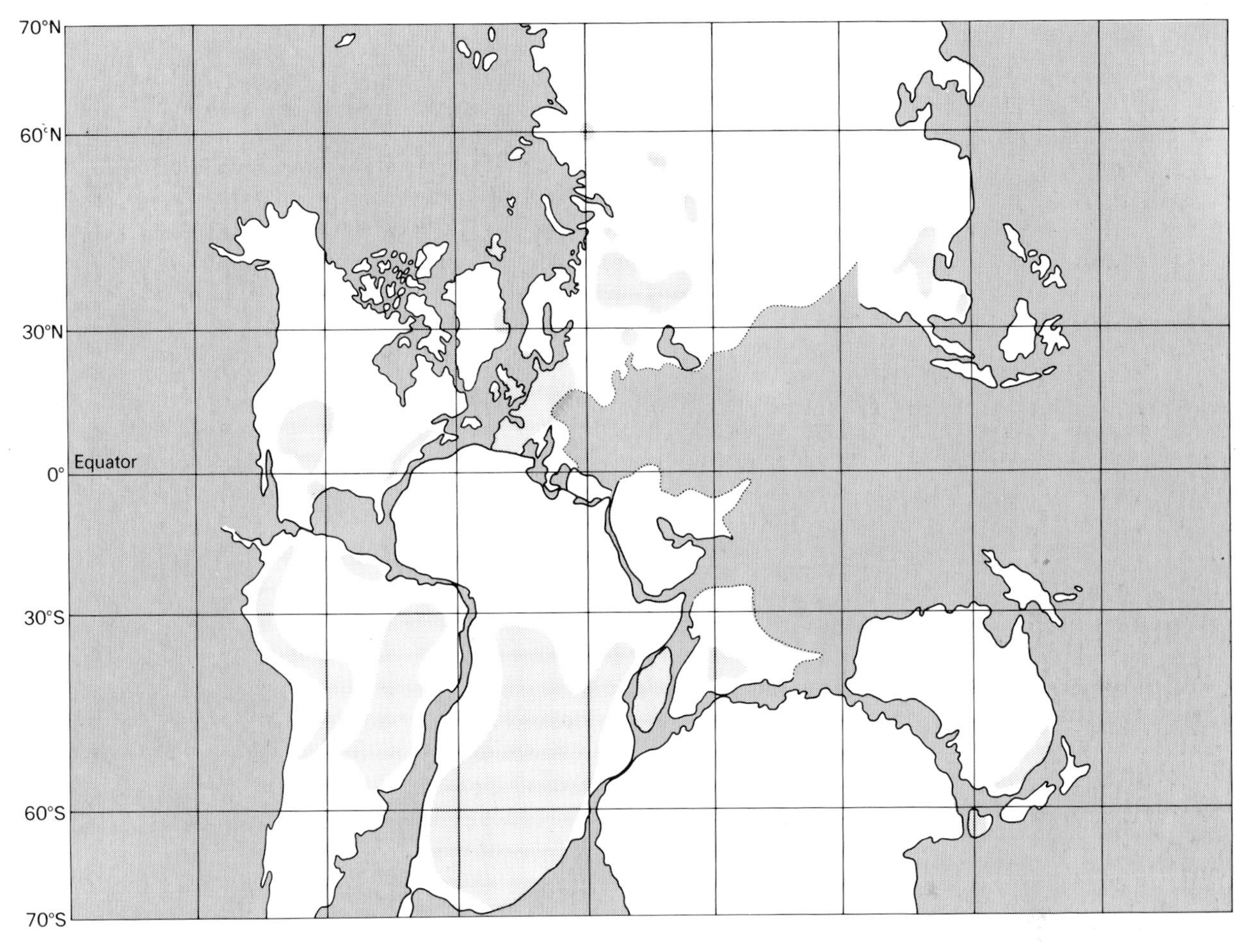

Figure 9.4 Approximate distribution of Permo-Triassic red beds, from van Houton (1961, Fig. 7), plotted on a palaeocontinental reconstruction (Mercator projection) for early Triassic time (220 million years) by Smith & Briden (1977). Dotted lines are postulated continental margins prior to collision.

northwards. The reef belt migrated from about 55° N in the Triassic and Jurassic Periods to about 45° N in the Cretaceous Period. The evaporite belt followed a similar trend. The Era as a whole seems to have been warm. Whilst Mesozoic glacial deposits have occasionally been reported, none has yet stood up to critical examination.

Cenozoic climates

Tertiary (Palaeogene and Neogene Periods). From the beginning of the Cenozoic Era (about 65 million years ago) to the present, it becomes increasingly possible to determine climatic variations in detail, partly because the flora, especially angiosperms, and fauna often resemble that of today and thus can often be used as direct climatic indicators; in addition almost complete records of oceanic sedimentation are available.

In Tertiary time modern mammals gained their rapid ascendency. Concurrently, continental opening of the north Atlantic took place (with Greenland becoming separated from Europe) and Australia split off from Antarctica. Major periods of widespread orogenic activity are exemplified in the Western Cordillera of North America, the Andes and elsewhere in late Mesozoic and Palaeocene time, and again by the Alpine–Himalayan orogenies both in Oligocene and Plio-Pleistocene time. These were periods of relatively extensive land in contrast to the major marine transgressions of mid-Eocene and early to mid-Miocene time. It has been thought that these continental episodes corresponded with more extreme arid and possibly colder conditions in contrast to times when the areas of the sea were larger.

Changes in world climate during Cenozoic time led to fluctuations in the width of the tropical belt. Cenozoic $^{18}O/^{16}O$ determinations on shells in Victoria, Australia, reveal first a rise, then a gradual but significant fall in temperature (Fig. 9.5). However, until about 2 million

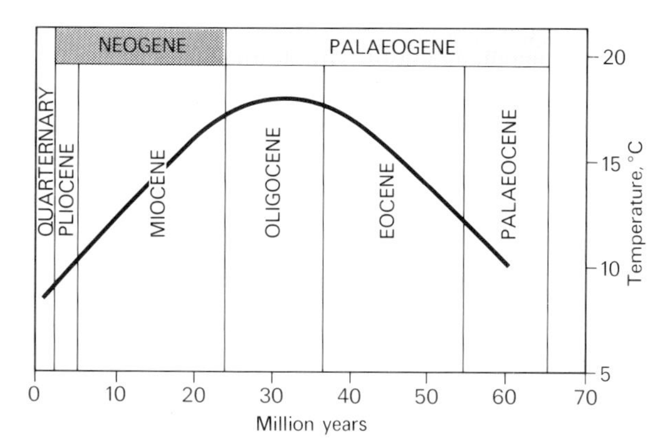

Figure 9.5 Mean isotopic temperature curve for the Cenozoic Era based on $^{18}O/^{16}O$ ratios in marine shells (scallops, *Glycymeris* and oysters) from Victoria, Australia (after Gill, 1961, Fig. 2).

years ago Tertiary climates were warmer than those of today and Palaeogene sub-tropical vegetation grew in areas that are now temperate. For instance palms grew as far north as Alaska (62° N), while the rich temperate vegetation growing in Svalbard (78° N) led to the formation of extensive coal deposits. In north-western Europe the flora, and sometimes the fauna, illustrate a progressive cooling, a somewhat different pattern in the first half of the Tertiary Period compared with that recorded by marine shells of Australia. Genera now found only near the equator predominated in central Europe in early Tertiary time. Estimates of mean annual temperature in the London Basin indicate a fall from 21 °C in Eocene (early Tertiary), through 18 °C in Upper Oligocene and 16 °C in Miocene, to 14 °C in Pliocene (late Tertiary) time, compared with 10 °C, at present. The flora of the southern hemisphere also indicates warm, wet climates in much of Tertiary time. This applies to areas that are now temperate, for example Southern Chile and New Zealand, although the limited extent of present land means that results are not as conclusive as those from the northern hemisphere. Lands with warm climates in early Tertiary time are also indicated by red-weathering products, such as bauxites in Arkansas and Yugoslavia, and red beds in South Germany, Switzerland and the Yenesey district of Siberia.

Coral reefs continued to develop in Europe to a limited extent, especially in the Mediterranean area and in southern USSR, but in eastern North America coral reefs extended only a little further north than at present. In the southern hemisphere coral reefs continued to form in New Zealand and South Australia, but the displacement of reefs away from the poles was less marked than in the northern hemisphere.

Further evidence for gradual cooling during Tertiary time is provided by $^{18}O/^{16}O$ ratios of deep oceanic sediments. Bottom temperatures in the eastern equatorial Pacific ocean dropped from 10 °C in mid-Oligocene, through 7 °C in the Miocene to 2 °C in the late Pliocene epoch, the last figure being the same as that of today.

It is difficult to generalise concerning rainfall in Tertiary time as this was extremely variable. Nevertheless, some indications of world-wide trends are provided by a few areas. In southern and central Europe widely-distributed evaporite deposits of early Tertiary age indicate aridity, whereas subsequent extensive lignite deposits of Oligocene and Miocene age suggest considerable rainfall. Late Miocene time was possibly another period of marked aridity. In the mid-USSR the climate of the Tertiary Period was often wetter than it is today and the present-day treeless steppes were covered by woodland, at first containing evergreen tropical species but giving way to deciduous forests and later to mixed coniferous-deciduous forests. Steppe floras only appeared near the end of Miocene time. The effects of the development of a mountain range on climate are exhibited clearly in western North America. Lush tropical forests in Palaeogene time in what are now intermontane desert areas gave way to an arid flora in mid-Miocene time, then deciduous hardwoods and conifers indicating a late Miocene moist, cool climate. As slight mountain uplift took place in the west (in mid-Pliocene time) the coastal slopes carried a humid fauna while a semi-arid climate with shrubs and grassland developed to the east. When the main uplift of the coastal mountains subsequently occurred, conditions began to resemble those of today, and deserts developed widely to the east. The uplift of the Andes modified the climate of South America in a similar way.

Evidence for Tertiary glaciations, recognised only relatively recently, is building up rapidly and the time for the onset of glaciation is gradually being pushed back. Extensive Pliocene glacial deposits, both terrestial and marine, have been found in the USSR, Alaska, Iceland, South America and Antarctica, but the extent of the ice cover is disputed. The earliest evidence of ice-rafting of till in Antarctica is in Late Oligocene time (around 25 million years ago) although it has been suggested that the ice sheet of Greater Antarctica did not begin to build up until 14 million years ago; only local glaciers are said to have existed before this date.

Quaternary. Although Quaternary history, lasting less than two million years and still in progress, spans a relatively short geological time, it is unparalleled in the apparent magnitude and abruptness of climatic change in Earth's history as we know it. It is the development of a succession of ice ages that is the hallmark of the Quaternary, and although glaciers had returned to the Earth as much as 20 million years earlier, it was the dramatic cooling at the end of Pliocene time and the subsequent spread of ice sheets that led to the use of a separate geological division.

The Quaternary Period may be divided into Pleistocene and Holocene epochs, the latter covering only the last 10 000 years or so and often referred to as post-glacial time, although this is something of a misnomer as there is no evidence to suggest that we are not simply in an interglacial period. Much effort has been devoted to the study of

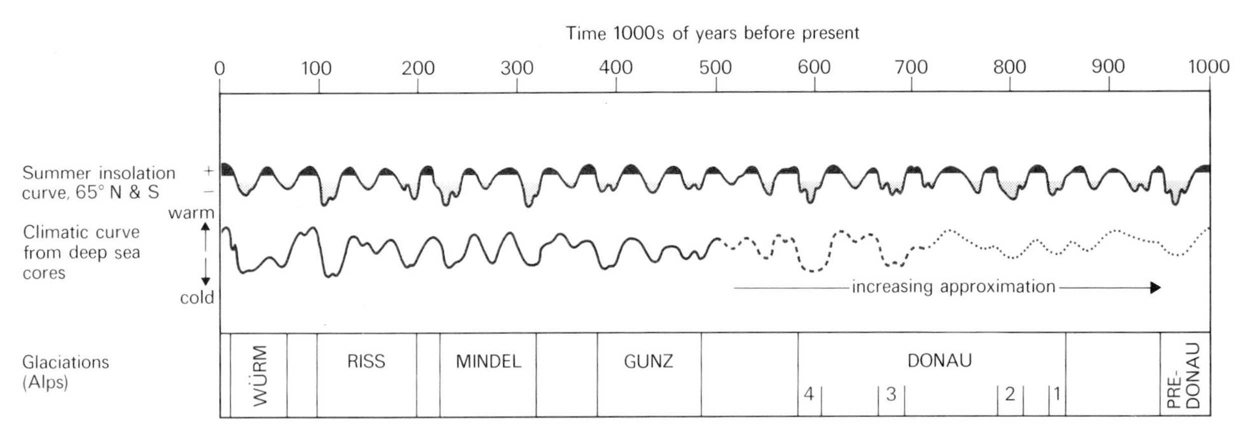

Figure 9.6 Climatic change and tentative chronology for the last million years (after Evans, 1971, Plate 1).

Quaternary climates, and a wide range of techniques has been applied to establishing a Quaternary chronology, but unfortunately many of these are unsuitable for use in older deposits (see John, 1977 and Imbrie & Imbrie, 1979 for useful summaries). The classic approach was to consider the stratigraphical sequence of glacial and interglacial deposits, but this has been fraught with difficulties. One of the most commonly employed techniques used today is that of pollen analysis, which involves the determination of the proportions of different types of pollen in non-glacial sediments, enabling one to recognise particular assemblages that are characteristic of warm or cold climates. Faunal remains, ranging from large mammals to beetles, have also been frequently employed in establishing patterns of climatic change. In the last decade foraminifera have been widely used to determine associated ocean temperatures using $^{18}O/^{16}O$ techniques and these are useful for time scales extending back into Tertiary time and earlier. Other techniques have been developed for shorter time scales. For example $^{18}O/^{16}O$ ratios of glacier ice from cores taken from the continental ice sheets of Greenland and Antarctica have enabled dating back to more than 100000 years. Radiocarbon dating of organic material is generally effective back to 30–50000 years, depending on the quality of the sample, but other radiometric techniques have so far been of very limited use. For time scales of a few thousand years or less, studies of lichens, tree rings and varves have been variously employed.

Table 1 Various land-recognised glacial periods.

Alps	*Northern Europe*	*British Isles*	*North America*	*USSR*
Würm	Weichselian	Devensian	Wisconsin	Valdaian
Riss	Saale	Wolstonian	Illinoian	{ Moscovian Dnepr
Mindel	Elster	Anglian	Kansan	Oka?
Günz	–	–	Nebraskan	
Donau	–	–	–	
Biber?	–	–	–	

For approximate ages see Figure 9.6.

There have been many attempts since the beginning of the century to sub-divide the Quaternary Period into glacial and interglacial stages, yet after years of controversy the chronology that was widely adopted has in recent years been found to be misleading if not grossly inaccurate and incomplete. The principal problem seems to be the incompleteness of the stratigraphic record on land, leading to wide-ranging correlations based on insufficient data. The classic names applied to glacial periods in the northern hemisphere are given in Table 1, youngest at the top. However, these glacial episodes have never been satisfactorily placed in an adequate time scale, and it is now known that reliable correlation can only be made between the last glaciation in each area. The frequent lack of synchroneity between glaciations in different areas, the incompleteness of the record and the absence of absolute dating techniques for the older deposits means that correlation is increasingly uncertain the older the deposit.

The recent critical reassessment of Quaternary chronology (e.g. Kukla, 1977) based on the analysis of oxygen isotope ratios in deep-sea sediments, where the stratigraphical record is relatively complete, has shown that the traditional chronology should be treated with a great deal of circumspection. Indeed Kukla goes as far as to say that the classical terminology in all inter-regional correlations should be abandoned and that stratigraphical subdivision of the Pleistocene Epoch be based on the oxygen isotope record from foraminifera of deep-sea sediments. The study of deep-sea sediments has shown that there have been at least 17 'glacials' and 17 'interglacials' in Europe in the last 1700000 years.

The climatic curve for ocean sediments shows wide variations, which for about the last million years correspond closely to the curves of summer insolation (solar energy received per unit area of Earth, Fig. 9.6). The latter, calculated according to the Köppen–Milankovitch hypothesis, is a function of astronomical causes, such as the tilt

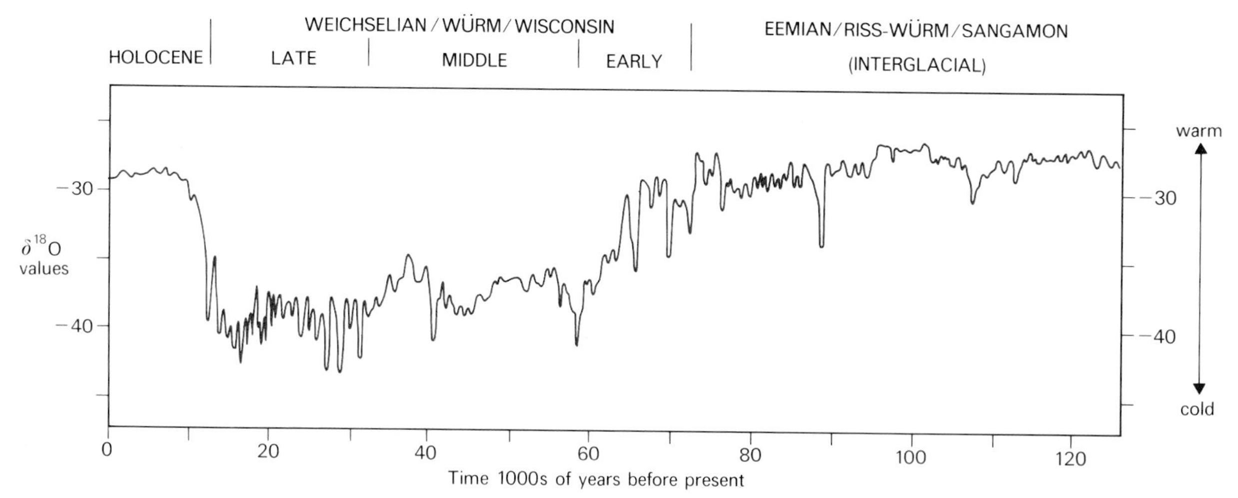

Figure 9.7 Variations in oxygen isotope ratio ($^{18}O/^{16}O$) through the Greenland Ice Sheet at Camp Century for the last 120000 years. Higher values represent higher temperatures (after Dansgaard *et al.*, 1973).

of the Earth's axis or the Earth's ellipticity. There seems to be no evidence to suggest that the number of Quaternary glaciations was restricted to four or five, but we still have a long way to go before we can establish precisely the number, extent, duration and intensity of glaciations *per se*, as opposed to cold periods alone. Nevertheless after years of debate a remarkable agreement between the Milankovitch solar radiation curves and the periodicity of cold periods is now generally accepted, especially from the evidence of the deep-sea sedimentary record (eg. Evans, 1971; Imbrie & Imbrie, 1979).

Oxygen isotope determinations on glacier ice from the cold continental ice sheets, taken during the last decade, have provided us with a record of relative temperature variations in detail for the last 120000 years (e.g. Fig. 9.7 from Dansgaard, *et al.*, 1973). The curve represents the interglacial period preceeding the last (Würm/Weichselian) glaciation, the glacial period itself and the interglacial period in which we now live. Of particular note are the wide climatic fluctuations in the last glaciation followed by a dramatic and sudden warming.

Although little can be said concerning the climatic conditions prevailing at any particular time interval in the Quaternary Period, some comments can be made about glacial and interglacial periods in general. At its maximum extent, ice covered the whole of Scandinavia and the northern USSR, extending south over the northern European lowlands and west to meet an ice sheet over the British Isles. In North America the land became almost entirely covered as far south as Kentucky and probably also joined the Greenland ice-sheet. Large ice-fields associated with high mountain ranges developed in the Alps, Andes, Himalayas, in New Zealand and even equatorial Africa. By contrast the Antarctic ice sheet, confined by the surrounding oceans, was probably not much greater than it is today (Fig. 9.8).

The existence of large ice-sheets has a profound effect on atmospheric and oceanic circulation, which in the past has pushed the climatic belts up to 2000 km towards the equator, resulting in a marked narrowing of the equatorial zone during the cold periods. Thus major changes in environmental conditions and the distribution of animals occurred (see also Chapter 15). During glaciations the present-day temperate zones were occupied by mammals of a tree-less tundra environment, such as mammoth, woolly rhinoceros, musk ox and reindeer. The extreme climatic changes resulted in intense selection and rapid evolution. The most successful species were those able to adapt to the changing conditions. The Quaternary Period has therefore seen the rise of man, a species uniquely able to successfully adapt to and even alter his environment.

Postscript – the future

It is still outside the realm of man's ingenuity to make reliable predictions about the climate of the future. With the need to take into account so many variables in determining the causes of climatic change, it is not surprising that there is wide divergence of opinion concerning the Earth's future prospects. Our future climate is bound up in the supply of solar energy from the sun, the study of which relies on the precise techniques of astronomers and meteorologists. Geologists rely on a varied, comprehensive but incomplete record for determining past climates, although when good data are available they are able to make predictions about the future. Combining astronomical with geological evidence has led to the prediction that we will have 1000 more years of relative warmth, to be followed by 22000 years of cooling with the return of another glacial episode (Imbrie & Imbrie, 1979). However this excludes consideration of the changes that man is inflicting on his environment. It is to

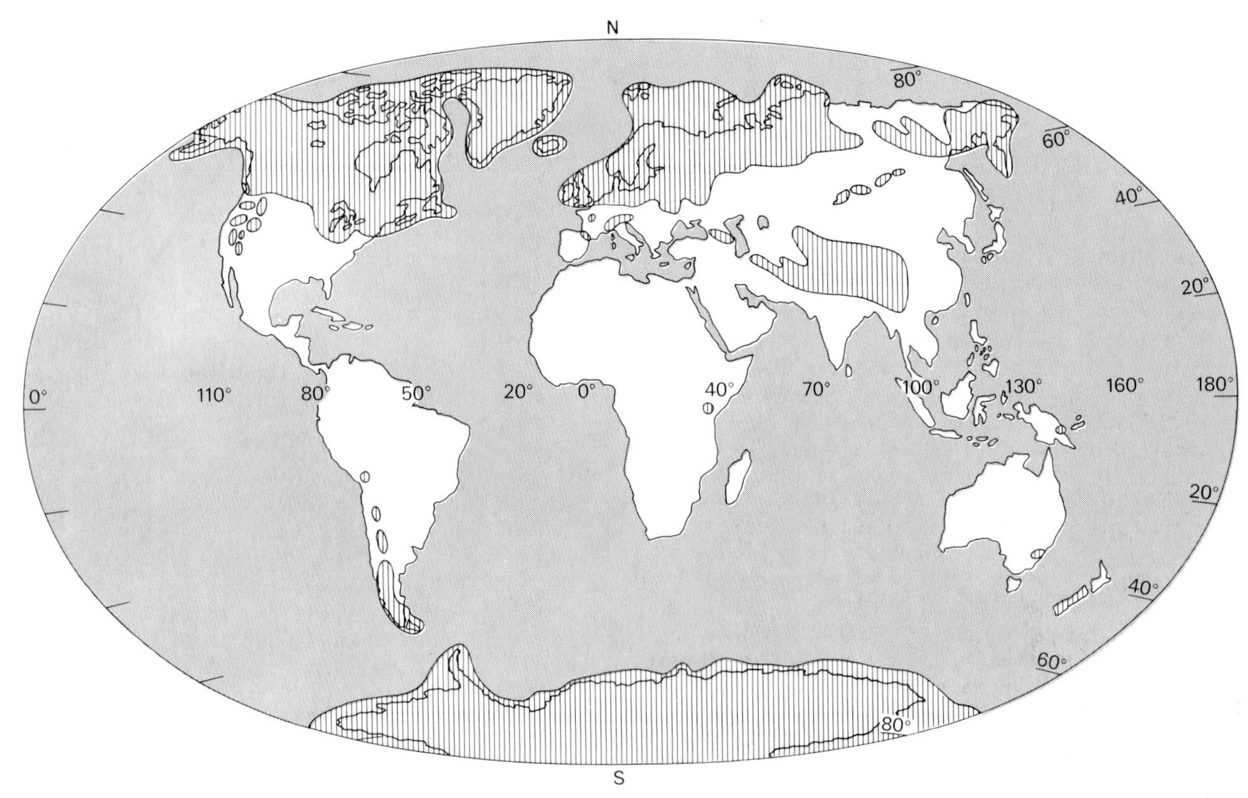

Figure 9.8 Areas formerly covered by ice sheets in Quaternary time. Marine limit of Antarctica ice sheet largely defined by the edge of the continental shelf. Much of northern and eastern Eurasia is largely hypothetical. Based principally on Flint (1971) and Hallam (1976, p. 256).

the understanding of the effects of our own activities that perhaps our efforts should be most urgently directed. Uppermost in mind is the question: could the increasing carbon dioxide content of the atmosphere reverse any trend there might be towards a new glacial period? Present calculations suggest that if we continue to burn fossil fuels on the present scale, the CO_2 content of the atmosphere could double in about 50 years. The destruction of forests, notably those of the Amazon Basin, is removing a major sink for atmospheric CO_2. Scientists like Mercer (1978) argue that the increased greenhouse effect will lead to a temperature increase at 80° S sufficient to cause catastrophic distintegration of the West Antarctic Ice Sheet and a resulting 5 m rise in sea level, with drastic effects on many of the major cities and food producing regions of the world. Other workers question a simple greenhouse effect of increased atmospheric CO_2. And what of other pollution, such as the output of sulphur dioxide: does this behave like volcanic dust and have the effect of absorbing carbon dioxide? What effects do agriculture and irrigation have on local climatic conditions, or how might such projects as diverting ocean currents affect global climatic circulation? We need answers to these questions if we are not to be taken by surprise by sharp climatic changes in the next few hundred years.

Acknowledgements

The authors are grateful to Ian Fairchild and Clive Pickton for commenting on a draft of the manuscript.

References

Those marked * are referred to in the text. The remainder are a selection of other significant works concerned with the development of climates.

***Bjørlykke, K., Bue, B. & Elverhøi, A.** 1978. The Quaternary sediments in the north-western part of the Barents Sea and their relation to the underlying Mesozoic bedrock. *Sedimentology* **25** (2): 227–246.

Berkner, C. V. & Marshall, L. C. 1964. The history of the growth of oxygen in the Earth's atmosphere, pp. 102–126 *in* Brancnzi, C. J. & Cameron, A. G. W. (Eds) *The origin and evolution of atmosphere and oceans.* New York: Wiley.

Blatt, H., Middleton, G. & Murray, R. 1972. *Origin of sedimentary rocks.* 634 pp. New Jersey: Prentice Hall.

Clough, H. W. 1933. The eleven year sunspot cycle, secular periods of sunspot activity and synchronous variations in terrestrial phenomena. *Monthly Weather Review*. April.

*Croll, J. 1875. *Climate and time in their geological relations, a theory of secular change of the Earth's climate*. London: Stanford.

Crowell, J. C. & Frakes, L. A. 1970. Phanerozoic glaciation and the causes of ice ages. *American Journal of Science* **268**: 193–224.

***Dansgaard, W., Johnson, S. J., Clausen, H. B. & Gundestrup, N.** 1973. Stable isotope glaciology. *Meddelelser om Grønland* **197** (2) 53 pp.

Egyed, L. 1961. Temperature and magnetic field. *Annals of the New York Academy of Sciences* **95**: 72–77.

Emiliani, C. 1966. Paleotemperature analysis of the Caribbean cores 6304–8 and 9, and a generalised temperature curve for the last 435000 years. *Journal of Geology* **74**: 109–126.

***Evans, P.** 1971. Towards a Pleistocene time-scale, pp. 121–356 *in The Phanerozoic timescale. A Supplement Part 2. Special publications of the Geological Society of London* No. 5.

***Ewing, M. & Donn, W. L.** 1956. A theory of ice ages. I. *Science* **127**: 1159–1162.

***Ewing, M. & Donn, W. L.** 1958. A theory of ice-ages. II. *Science* **123**: 1061–1066.

***Fairbridge, R. W.** 1964. The importance of limestone and its Ca/Mg content to palaeoclimatology, pp. 431–478 *in* Nairn, A. E. M. (Ed.) *Problems in paleoclimatology*. London: Interscience.

Fairbridge, R. W. & Bourgeois, J. 1978. *The encyclopedia of sedimentology*. 901 pp. Stroudsberg: Dowden, Hutchinson & Ross.

***Flint, R. F.** 1971. *Glacial and Quaternary geology*. 892 pp. New York: Wiley.

***Frakes, L. A.** 1979. *Climates throughout geologic time*. 310 pp. Amsterdam: Elsevier.

***Gill, E. D.** 1961. The climates of Gondwanaland in Kainozoic Times, pp. 332–353 *in* Nairn, A. E. M. (Ed.) *Descriptive paleoclimatology*. New York: Interscience.

***Habicht, J. K. A.** 1979. *Paleoclimate, paleomagnetism, and continental drift*. 32 pp. *Association of American Petroleum Geologists*. Studies in Geology No. 9, Tulsa.

***Hallam, A.** 1977. *Planet Earth. An encyclopedia of geology*. 319 pp. Oxford: Elsevier Phaidon.

***Harland, W. B. & Herod, K. N.** 1975. Glaciations through time, pp. 189–216 *in* Wright, A. E. & Moseley, F. (Eds) *Ice ages ancient and modern*. Liverpool: Steel House Press.

Herschel, J. 1830. On the astronomical causes which may influence geological phenonema. *Transactions of the Geological Society of London*. 2nd Edn. **3**: 293.

***Van Houten, F. B.** 1961. Climatic significance of red beds. pp. 89–139 *in* Nairn, A. E. M. (Ed.) *Descriptive paleoclimatology*. New York: Interscience.

***Imbrie, J. & Imbrie, K. P.** 1979. *Ice ages: solving the mystery*. 224 pp. London: Macmillan.

***John, B. S.** 1977. *The Ice Age*. 254 pp. London: Collins.

King, J. W. 1974. Weather and the Earth's magnetic field. *Nature*. London **247**: 131–134.

***Köppen, W. & Wegener, A.** 1924. *Die Klimate der geologischen Vorzeit*. 256 pp. Berlin: Borntraeger.

***Kreichgauer, D.** 1902. *Die Aquatorgrage in der Geologie*. 394 pp. Steyl. Zwanshoek.

***Kukla, G. J.** 1977. Pleistocene land–sea correlations. I. Europe. *Earth Science Reviews* **13**: 307–74.

***Lamb,** 1972. *Climate: present, past and future*. 2 volumes. London: Methuen.

McCrea, W. H. 1976. Glaciations and dense interstellar clouds. *Nature*, London **263**: 260.

***Manley, G.** 1961. Late and postglacial climatic fluctuations and their relationship to those shown by the instrumental record of the last 300 years. *Annals of the New York Academy of Sciences* **95**: 162–172.

***Mercer, J. H.** 1978. West Antarctic ice sheet and CO_2 greenhouse effect: a threat of disaster. *Nature*, London **271**: 321–325.

***Milankovič, M.** 1930. Mathematische Klimalehre und astronomische Theorie der Klimaschwankungen, *in* Köppen, W. & Geiger, R. (Eds) *Handbuch der Klimatologie*. Berlin: Borntraeger.

***Milankovič, M.** 1938. Astronomische Mittel zur Erforschung der Erdesgeschichlichen Klimate. *Handbuch der Geophysik* **9**: 593–698.

***Nairn, A. E. M. (Ed.)** 1961. *Descriptive paleoclimatology*. 380 pp. New York: Interscience.

***Nairn, A. E. M. (Ed.)** 1964. *Problems in paleoclimatology*. 705 pp. London: Interscience.

***Pearson, R.** 1978. *Climate and evolution*. 274 pp. London: Academic Press.

Plass, G. N. 1961. The influence of absorptive molecules on the climate. *Annals of the New York Academy of Sciences* **95**: 61–71.

***Roberts, J. D.** 1971. Late Precambrian glaciation: an anti-greenhouse effect? *Nature*, London **234**: 216–217.

Sawyer, J. S. 1972. Man-made carbon dioxide and the greenhouse effect. *Nature*, London **239**: 23–26.

Schwarzbach, M. 1963. *Climates of the past*. (Translated from German by P. O. Muir). 328 pp. London: Van Nostrand.

Simpson, G. C. 1929. Past Climates. *Memoirs of the Manchester Literary and Philosophical Society* **74**: 1–34.

***Smith, A. G. & Briden, J. C.** 1977. *Mesozoic and Cenozoic palaeocontinental maps*. 63 pp. Cambridge: Cambridge University Press.

Steiner, J. & Grillmair, E. 1973. Possible galactic courses for periodic and episodic glaciations. *Bulletin of the Geological Society of America* **84**: 1003–1018.

Turekian, K. L. (Ed.) 1971. *Late Cenozoic glacial ages*. New Haven, Connecticut: Yale University Press.

Urey, H. C., Lowenstam, H. A., Epstein, S. & McKinney, C. R. 1951. Measurement of the paleotemperature of the Upper Cretaceous of England, Denmark, and the south eastern United States. *Bulletin of the Geological Society of America* **62**: 399–416.

***Williams, G. E.** 1975. Late Precambrian glacial climate and the Earth's obliquity. *Geological Magazine* **112**: 441–465.

M. J. Hambrey
W. B. Harland
Department of Earth Sciences
Cambridge University
Downing Street
Cambridge CB2 3EQ

PART IV

Continental drift and plate tectonics

Introduction

In no other area of the Earth Sciences has there been such a revolution of ideas in the last 20 years as in the history and dating of the oceanic crust. It has led to the reality of continental drift being fully accepted by most scientists. In the late 1950s a group of American scientists proposed to drill a borehole to the Mohorovicic discontinuity that separates the Earth's crust from the underlying mantle. The oceanic crust is only about one-quarter as thick as the continental crust, and a suitable oceanic site was to be selected where a hole through about 6000 metres of rock below about 4500 metres of sea water would reach the mantle. Although a sample of the rocks of the mantle would have been the ultimate objective of the Mohole Project, as it came to be known, a continuous core through the sediments on the ocean floor was an intermediate step that was expected to yield equally interesting and valuable information.

Writing in 1959, the secretary of the Mohole Committee said that these sediments 'could contain an uninterrupted record of the earth's development for two thousand million years...if palaeontologists can obtain cores of fossil-bearing strata considerably older than Cambrian sediments, one of their fondest dreams will be realized' (Bascom, 1959, pp. 46, 48). Such views seem remarkable only 21 years later. The Mohole Project to drill a single main hole to the mantle became too expensive, and died for political and financial reasons, but out of the ideas behind the Mohole grew the Deep Sea Drilling Project. This operated from 1965 to 1975, then the International Phase of Ocean Drilling started in 1975 and is still continuing. Over 500 boreholes have now been drilled in all the more accessible parts of the Earth's oceanic crust. The major result, entirely unexpected 20 years ago, has been to show that the oldest sediments on the ocean floor are only Jurassic in age. In fact the sediments and the underlying crust of the world's oceans vary from Recent to Jurassic in age, and are nowhere older than 200 million years. By careful mapping of the floors of the Atlantic and

Indian Oceans, it has been shown that these two oceans did not exist 200 million years ago.

This is the most direct proof of continental drift, the history and development of which is outlined in Chapter 10. The original idea, put forward as a scientific theory by Wegener, was rejected by most scientists, partly for lack of an acceptable mechanism, and partly because some of the supporting evidence could be faulted. It was not until about 1960 that an entirely new line of evidence, derived from an analysis of the direction of magnestism induced in suitable rocks in continental areas that had cooled from a molten state, persuaded some scientists that the continents had moved. When the technique was applied to the ocean floor, series of magnetic stripes of different ages were found, ages that were confirmed when the sea-bed sediments and crust below were sampled as outlined above. Movement was not of continental crust over the top of oceanic crust, as Wegener thought, but of plates consisting of both types of crust moving away from ocean-floor spreading centres such as the mid-Atlantic ridge. One of the main difficulties in accepting the original concept of continental drift, that no known mechanism was powerful enough to drag continental crust over oceanic crust, was thus removed. The theory of plate tectonics is outlined in Chapter 11, and it forms a background to Section E, in which the changing geography of the past 200 million years is considered stage by stage.

So it is knowledge of the relative ages of the different parts of the Earth's crust that has advanced so much in recent years. Rather than believing that the continental and oceanic crusts are similar in age, and basically both very old, we now believe that the continental crust formed about 4000 million years ago and has remained relatively stable except for sedimentary recycling, local melting, magma emplacement and mountain building, whereas the oceanic crust has been entirely formed or renewed in the last 200 million years. It is such wholesale renewal of the 70 per cent of the Earth's surface that is occupied by oceanic crust in less than one-twentieth of the geological age of the Earth, that is the most fundamental change in our view of the evolution of the Earth.

Perhaps it is necessary to be clear exactly what is meant by the geological age of rocks. The Earth was formed as a single mass approximately 4600 million years ago, and the surface was solid and cool at least by 600 million years later. In one sense this is the age of all the material of which the Earth is composed. Geological age is, however, different and refers to the age of solid rocks in the surface crust. In the case of an igneous rock that has cooled from a molten or plastic state, the age is the date at which it solidified. The rocks forming the oceanic crust have all been formed in this way and have cooled to surface temperatures within the last 200 million years. It is thought by most geologists that the rock material is constantly recycled through the mantle in immense convection current cells.

Continental crust is very different: igneous rock that cooled from a molten state only slightly less than 4000 million years ago can still be found. However, much of the rock of the continental crust is igneous or metamorphic rock that has cooled more recently, or it is sedimentary rock. The geological age of the latter is the date at which it was aggregated as a sediment in the form in which it is now found. The oldest sedimentary rocks found so far have remained unchanged since their deposition about 3700 million years ago. Sedimentary rocks formed at most other dates up to the present are known, and where they entomb fossils, relative ages can often be given more easily than absolute dates in millions of years. Absolute dates are obtained from analysis of suitable intrusive or interbedded igneous rocks (see Chapter 3). Viewed like this, the much older and relatively more stable continental crust is markedly different from the young, constantly-changing oceanic crust.

In Chapter 12 the possibility of the Earth's expansion during geological time is discussed. From careful cartography, Owen has deducted that the Earth may have been 80 per cent of its present diameter 200 million years ago. One of the factors that has featured most frequently in debates about continental drift and earth expansion is the accuracy of the fit of the continental edges of Africa and South America. In this Section two reconstructions are given: Figure 10.6 is the very close fit that emerged in 1965 when Bullard, Everett and Smith programmed a computer to produce an objective geometrical solution to the problem. In work newly done for this book, Owen claims (Figure 12.1*B*) that such a fit is not possible, and the only way to obtain an accurate fit of the continents is on a globe 80 per cent of the present diameter (Fig. 12.1*C*). Readers might complain that the shape of the two continental edges is surely known with sufficient accuracy for an objective assessment of their fit to be made, but as so often in scientific controversy, they are left to decide for themselves which reconstruction to believe. The possibility of global expansion during geological time is discounted by most geologists, firstly because of the lack of an acceptable mechanism, and secondly because of problems in extrapolating backwards in time. At a constant rate of expansion, the Earth would have been 55 per cent of its present diameter 700 million years ago. This gives a volume of only one-sixth of that of the Earth today, and seems improbable. Most physicists believe that the mechanisms proposed so far for expansion are impossible in a body as small as the Earth. However, the topic is one of interest and is generating discussion amongst astronomers as well as Earth scientists. The lack of a known mechanism for expansion should be treated with caution, for continental drift was once discounted for this reason, until it was shown that the 'impossible mechanism' was not required.

Reference

Bascom, W. 1959. The Mohole. *Scientific American*, **200** (4): 41–49.

M. K. Howarth
British Museum (Natural History)

CHAPTER 10

The development of the concept of continental drift

A. C. Bishop

From earliest times man has been aware of the power of natural forces. Even modern man with all his knowledge and technology feels helpless before the energy of volcanic eruptions, earthquakes, or even violent atmospheric storms and, though such phenomena may produce local changes in topography, their effect is relatively small taken on a global scale and over man's short life time. The overall impression is of a durable, permanent Earth that changes but little from generation to generation – a concept that is virtually as old as man himself. Ancient writers saw the Earth as an unchanging place that had stood firm since its original creation. Impressive though the power of natural phenomena may be, it would certainly have been considered unreasonable to entertain the idea that entire continents might have moved relative to one another and to the Earth's rotational and magnetic axis. As long as man's knowledge of the Earth remained relatively limited there was no reason why the idea should ever be entertained but as travel became progressively easier, so his understanding of the Earth became correspondingly more comprehensive.

For centuries travel for the most part was limited to the distance a man could walk or perhaps ride in a day, until the coming of the railways brought a dramatic and permanent change. Following the discovery of new continents by Columbus and the other great explorers, the accurate charting of the continents and oceans developed gradually through the sixteenth to nineteenth centuries and with it the knowledge of the land areas of the Earth. Until the shapes of the major land masses were known, ideas of continental movement were either inconceivable or belonged to the realm of fantasy. A century or so after the discovery of America by Columbus, the configuration of the opposing coasts of the Atlantic was sufficiently well known for Bacon (1620) to comment on the similarity of the coastlines of Africa and South America. Snider's book *La Création et ses mystères dévoilés* (1859) is generally regarded as being the first expression of the idea that major continental masses were once united and had subsequently moved apart. He illustrated his ideas with two maps that have become famous beyond the merits of the book itself which was far from a serious scientific work even at the time of its publication. The diagrams show the main continental masses in contact (Fig. 10.1) as Snider envisaged them to have been before the great Noachian deluge, with South America (designated Atlantide) in contact with Africa, and the very diagrammatic North America in contact with Europe. Snider had few geological reasons for postulating his separation but sought proof in the similarities between human cultures in the old and new worlds and in historic volcanic and earthquake activity. He is on somewhat safer ground when he draws attention to the fit between the coasts of Africa and South America though, as Bullard (1975) has pointed out, Snider's maps are greatly distorted and the coastlines do not in fact fit – a point to which we shall return later. Clearly Snider envisaged the separation as a cataclysmic event coincident with Noah's flood and so his book is all the more remarkable for having hit upon the right idea for quite the wrong reasons.

The flood and its supposed catastrophic effects had been invoked by others to explain continental movement, notably T. C. Lilienthal in the eighteenth century, who concluded that the Earth had been disrupted by the flood and cited the fit of opposing coastlines as evidence of this, and similar ideas prompted Alexander von Humboldt to suggest early in the nineteenth century that the Atlantic had been carved by the erosive action of water on a gigantic scale.

Such catastrophist theories passed with the nineteenth century as it came to be more widely realised that in attempting to decipher the past it was pertinent to observe those processes that are presently taking place. The nineteenth century was a time of rapid and remarkable development in geological sciences. It saw the establishment of the principles of stratigraphy by William Smith and their application in geological mapping to the definition of the major periods of Earth history and the economic

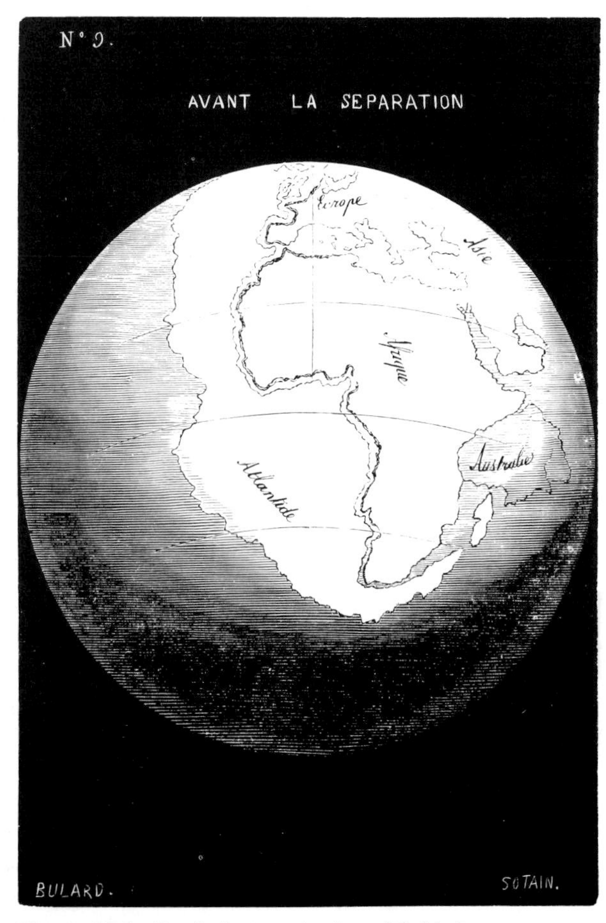

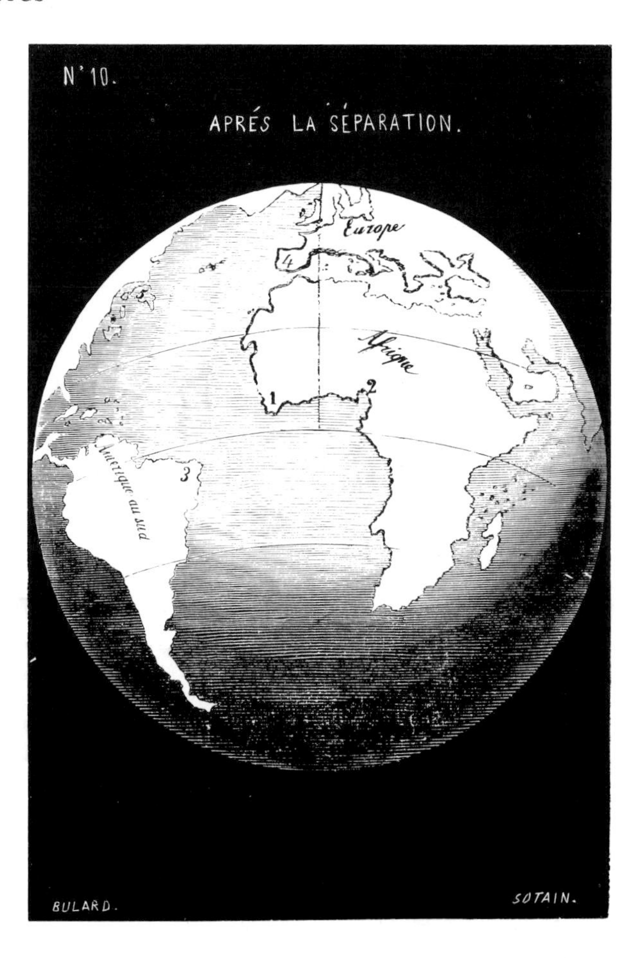

Figure 10.1 Facsimile reproduction of Snider's maps showing his concept of the separation of the Americas from Africa and Europe.

needs of the Industrial Revolution. Much work was concentrated on the elucidation of local geological details so essential to the winning of coal and other raw materials but the broad view of Earth history was not altogether neglected. Richard Owen observed in 1857 that the Earth's major land masses are opposite major oceans, in much the same way that one corner of a tetrahedron is opposite each face, and his tetrahedral hypothesis embraced both continental displacement and the possibility of an expanding Earth. Although the general view was that oceans were ancient features that had remained from a primeval contracting Earth, this view was complicated by the presence of mid-oceanic ridges which became known by the end of the nineteenth century. Although by modern standards information was relatively meagre, it was sufficient for there to be some speculation. As we shall see, Wegener was able to draw on earlier ideas of continental displacement, many of them coming from his own countrymen.

There were a few geologists who attempted large-scale syntheses which involved the lateral movement of land masses. One of the foremost of these was F. B. Taylor (1910) who published a theory to explain the distribution and origin of mountain ranges. He proposed that the Moon had been captured by the Earth during the Cretaceous period and that the resulting tidal force caused continents to move from the poles towards the Equator, throwing up belts at their leading edges as they moved. Continental movement was thus incidental to mountain building. Holmes (1944) presented objections to Taylor's hypothesis, pointing out that if the capture of the Moon was responsible for Cretaceous and Tertiary mountain building then there is no explanation for earlier mountain building movements and that if the tidal force resulting from the Moon's capture was strong enough to move continents, it would have slowed the Earth to a standstill within a very short period of time.

The contribution of Alfred Wegener

Although several geologists had advocated large-scale displacements of the continents, Wegener was the first to assemble evidence in support of the idea that the presently dispersed land masses were at one time a single large continent, which he called Pangaea. Wegener was born in Berlin in 1880 and trained as an astronomer, obtaining his doctorate in 1905. He then became interested in meteorology and, being an active balloonist, he undertook experiments himself at the Royal Prussian Aeronautical Observatory at Lindenberg using both balloons and kites. His major interest now turned to polar exploration and he joined the Danish expedition to Greenland in 1906 and spent two winters there making meteorological observations. On his return to Germany in 1908 he was appointed lecturer in meteorology at the University of Marburg.

Wegener first became interested in the idea of continental drift in 1910, having been impressed, like so many before him, by the similarity of the shapes of the coast of the continents on either side of the Atlantic Ocean, but he dismissed the idea of displacement as being improbable. A collection of references which came to his notice by accident in 1911 led him to consider the palaeontological evidence for a former land connection between Africa and Brazil and started his research into the geological evidence for drift. On 6 January 1912 he lectured to the Geologische Vereinigung in Frankfurt-am-Main on 'Die Herausbildung der Grossformen der Erdrinde (Kontinente und Ozeane) auf geophysikalischer Grundlage' and on 10 January he gave a second lecture, this time to the Gesellschaft zur Beförderung den gesamten Naturwissenschaften zu Marburg, on 'Die Entstehung der Kontinente' (Wegener 1912a, b).

He was unable to follow up his ideas immediately because he went again to Greenland in 1912–13 to participate in the crossing of the ice cap and then he was called to active service in the German army in the First World War. He was twice wounded and his active service ended in 1915. He used his convalescence to expand the articles based on his two lectures into what was to become his classic book *Die Entstehung der Kontinente und Ozeane* (1915) which was the first comprehensive account of continental drift. His published lectures and his book illustrate the evolution of Wegener's ideas on continental drift, for revised editions appeared in 1920, 1922 and 1929. The third edition (1922) was translated into several languages including English (the English edition was published in 1924) and had a profound effect on geological opinion because it challenged the established views of the Earth's structure and evolution. The accepted view at the time of the First World War was that the Earth had solidified from an initially molten mass and, in so doing, had both differentiated so that the lighter materials, broadly granitic in composition, collected at the surface, leaving the denser basaltic rocks and then a metallic core beneath, and contracted so as to form mountain ranges in much the same way as wrinkles form on the skin of a drying apple.

Wegener argued that the accepted view was untenable because the dominant presence of shallow water sediments on the land masses indicated that the continents and oceans were not interchangeable areas as had been maintained. Moreover, there was evidence that the oceans were underlain by denser rocks (called mnemonically Sima on account of the presence of silica and magnesia) than the continents (or Sal – later Sial – on account of the dominance of silica and aluminium) and so, because the lighter material in effect floated on the denser, it was most unlikely that continental masses could have foundered beneath the sea. The ocean basins, it seemed, were permanent features of the Earth. Wegener was bold enough to suggest that his super-continent Pangaea broke up during the Mesozoic Era and that the separate fragments moved apart. This movement he described as 'Horizontale Verschiebung der Kontinente' and this was translated into English as horizontal displacement of continents. Only later did it come to be known as continental drift.

In support of this theory, Wegener marshalled geophysical, geological, palaeontological and biological, and palaeoclimatic arguments which are not described in detail here having been reviewed elsewhere (e.g. Hallam, 1973; Marvin, 1973). He argued that if sialic continents 'float' on a denser simatic substratum in an approximation to gravitational equilibrium, or isostasy, and so give rise to vertical isostatic movement, then it is plausible that horizontal movement is also possible, given a sufficiently powerful force to cause the movement. He further argued that repeated geodetic determinations of longitude using radiotime transmissions would be likely to show that Greenland was moving westwards away from Europe.

It was the palaeontological evidence for a former land connection between Africa and Brazil that first inspired Wegener seriously to consider continental displacement. He found that the distribution of both past animals and plants was usually explained by palaeontologists in terms of land bridges which spanned the oceans and which had subsequently sunk beneath them. He drew particular attention to the small Permian reptile *Mesosaurus* which is confined to South Africa and Brazil, and to the *Glossopteris* flora of the southern continents. He further cited the present distribution of marsupials and suggested that continental drift was a more plausible hypothesis in accounting for their distribution than the foundering of land bridges spanning widely separated continents.

Although many have commented on the apparent fit of the opposing coasts of the Atlantic, Wegener took this line of reasoning further by showing that the geological structures on opposite sides of the Atlantic matched if the continents were juxtaposed. For example, gneiss complexes in Africa resembled those of Brazil, the Karroo sediments of South Africa were remarkably like those of the Santa Catharina System of Brazil and he noted that the Caledonian and Variscan orogenic belts of Europe had their counterparts in North America.

Wegener used his meteorological training in assembling the geological evidence for past climates in support of his theory. He was able to show that on his reconstructed Pangaea in the Permo-Carboniferous period, coal and evaporites lay on or close to the old Equator, whereas tillites in India, Australia, South America and South Africa lay at or near the old pole (Fig. 10.2). From this it was also apparent that the position of the land masses had moved not only relative to one another but also relative to the poles. In general the movement had been away from the poles and this was called by Wegener '*Polflucht*'.

Developments in the inter-war years

The publication of Wegener's book in 1915 and the appearance of subsequent editions (the last appeared in the year before his death) opened the earth sciences to debate and controversy that was to remain inconclusive and to last until a Second World War turned the attentions of most of mankind to other matters. It is easy, blessed with hindsight, to wonder why the geological community in those inter-war years did not pay more attention to Wegener's ideas, for his case was well argued and researched. Though the book initially appears to have made little impact, its reception was not wholly unsympathetic. However, opposition gradually grew as the leading geologists and geophysicists spoke out against Wegener's ideas, and by the outbreak of the Second World War geological opinion as a whole had virtually rejected them. There were several reasons why this came about. Firstly, geological knowledge was limited. Travel in the 20s and 30s was relatively expensive and slow; simply getting to South America and South Africa took over a week and travel within these continents was far from easy. Much of the criticism came from geologists who had not seen the evidence and, because Wegener himself was in the same position, it was easy to argue that sometimes his evidence was thin and possibly overstated. For example, his main and original idea came from the fit of the continents on either side of the Atlantic, but the maps he published, though illustrating well enough the large-scale features such as the way the bulge of West Africa fits into the Caribbean area, were little more than sketch maps and his critics were able to challenge him fairly on the grounds of inaccuracy. It is somewhat ironic in the light of what happened subsequently that much was made of the fact that it was the shorelines that had been matched for, it was argued, these were so subject to modification by uplift and subsidence that a close match was hardly to be expected. We shall return to this topic later.

Other authoritative critics questioned the similarities stated by Wegener to exist between the rocks and structures on either side of the Atlantic but even his strongest lines of evidence – the distribution of the evidences of Permo-Carboniferous glaciation and the similarities between the faunas and the floras of the rocks in South Africa and South America – were challenged or explained away by geologists with little first hand knowledge of the rocks. Further, Wegener's contention that Greenland had separated from Europe only during the last 100000 years implied a rate of drift (about 30 m per year) that geologists

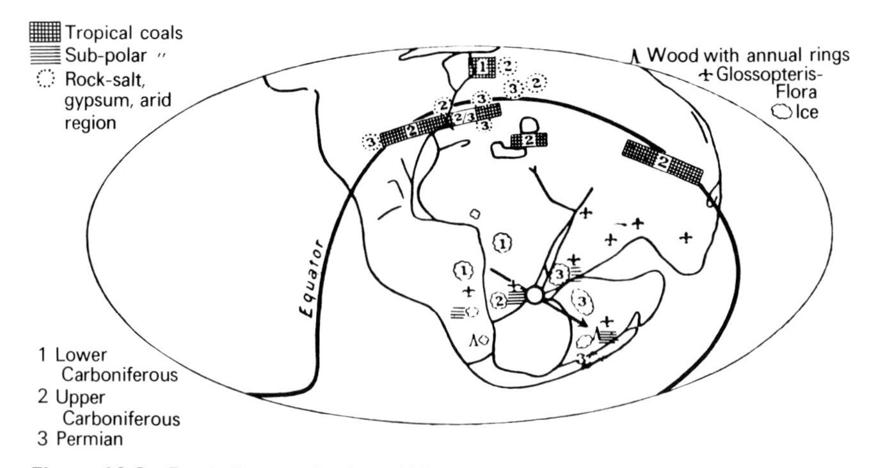

Figure 10.2 Facsimile reproduction of Wegener's map showing the distribution of different types of climate in the Permo-Carboniferous period.

simply could not accept, and his attempts to demonstrate the westward drift of Greenland were at best not very convincing.

Bullard's (1975) assessment of the criticism faced by Wegener is that it had two main contributory factors: the first was that Wegener enlisted weak or fallacious arguments among his strong ones, so that when the former were either refuted or shown to be of doubtful validity it was easy to believe that the whole position had been lost; and secondly that there is always an inherent tendency for a body of professionals to endorse orthodoxy and to reject a contrary view, the more so when orthodoxy is supported by the weight of authority – and there were many authorities who opposed the idea of drift. Of the many who contributed to the debate few carried more weight than the geophysicist Sir Harold Jeffreys. In 1924 he published the first of five editions of his book *The Earth* which was a statement of what was then known of the application of mathematics and physics to the study of the Earth. In the 1920s the understanding of the structure of the Earth was dominated by two concepts: firstly the recognition of the fundamental structural difference between continental and oceanic crust and the principle of isostasy, by which the lighter continents are in equilibrium on the denser basaltic substratum, and secondly the continents were considered to be essentially static and were constrained from lateral movement by the finite strength of the substrate.

Wegener had sought to suggest a mechanism for his Polflucht and continental drift. He argued that the gravitational forces resulting from the Earth's equatorial bulge (the Earth is an oblate spheroid) caused the Polflucht, and that westward drift of continents resulted from drag on the continental masses produced by tidal friction induced by the gravitational attraction of the Sun and the Moon. Wegener put forward these ideas only tentatively: he says, 'the question as to what forces have caused these displacements, folds and rifts cannot yet be answered conclusively' (English edition, p. 194). Tentative though these suggestions were they proved to be the Achilles' heel of his theory of continental drift. Jeffreys and others were able to show that the forces Wegener had proposed were quite inadequate to produce the movements he envisaged. The Earth's strength was such that it was most unlikely to have been capable of deformation by tidal forces; nor was it feasible to imagine that very small forces would result in significant deformation even if they operated for long periods of time. In short, Jeffreys dealt systematically and sometimes none too kindly with the evidence Wegener put forward in support of his theory; and what ultimately proved to be the most enduring of his arguments, namely those based on geological evidence, were simply dismissed.

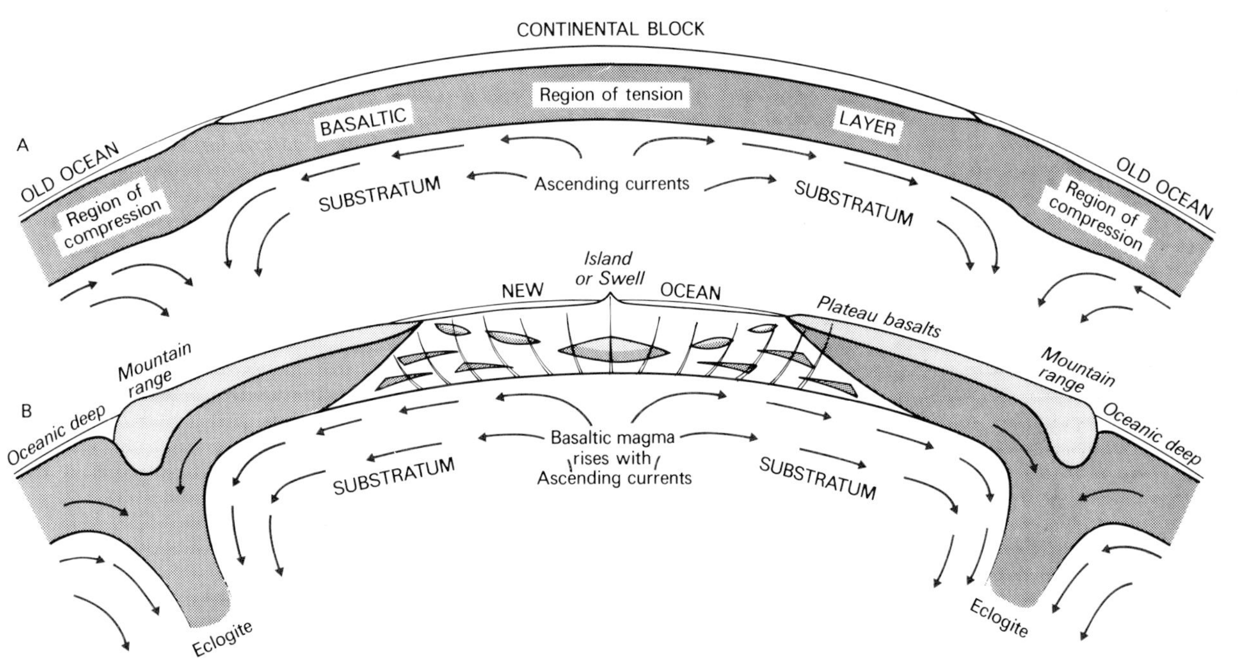

Figure 10.3 The disruption of a continental block and the lateral displacement of fragments by convection currents as envisaged by A. Holmes. Note also the situation of oceanic deeps and mountain ranges over descending currents. (After Holmes, 1944)

The real difficulty was that Wegener's broad proposal could not be discussed within the framework of the geological data available at that time. The weak arguments such as the search for a driving mechanism which Wegener himself realised to be a great difficulty, were swiftly despatched by the geophysicists who argued that because no feasible mechanism could be envisaged, drift could not therefore have taken place. It is interesting that the reawakening of interest in continental drift after the Second World War, described below, came largely as a result of geophysical research. In any case it is faulty to reason that geological phenomena are implausible simply because an acceptable mechanism to account for them is wanting. Ice ages are no less real for the want of an acceptable mechanism to explain them. Feelings sometimes ran so deep as to cause some to call into question Wegener's standing as a scientist. He was even dismissed as one who was so blinded by the advocacy of his theory that it impaired his impartial search for the truth. Certainly, Wegener at the time of his death on the Greenland ice cap in 1930 could have derived little comfort from the reception accorded to his views. He was not without his supporters however, among whom the most active were Arthur Holmes in Britain and A. L. du Toit in South Africa.

Arthur Holmes devoted the last chapter of his classic book *Principles of Physical Geology* (1944) to continental drift, and, having reviewed objectively the geological evidence by Wegener, du Toit and others, he concluded that the evidence is compelling and adds (p. 504) 'The only serious argument advanced against the above solution [that drift had occurred] is...the difficulty of explaining how continental drift on so stupendous a scale could have been brought about.' He went on to propose a mechanism for the transport of sialic slabs (the continents) on convection currents within the Earth's mantle generated as a result of heat transfer from the core. He had first proposed these ideas some fifteen years earlier (Holmes, 1929). A rising current, spreading out beneath the crust, would disrupt a continent and transport the fragments in much the same way as scum is transported on the surface of a saucepan of porridge (Fig. 10.3). Ocean deeps and fold mountains

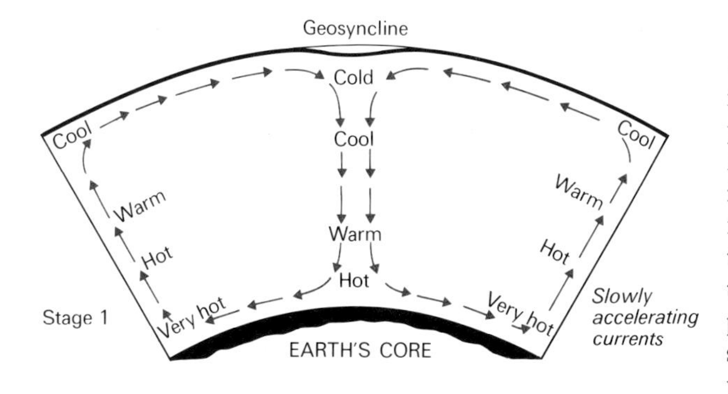

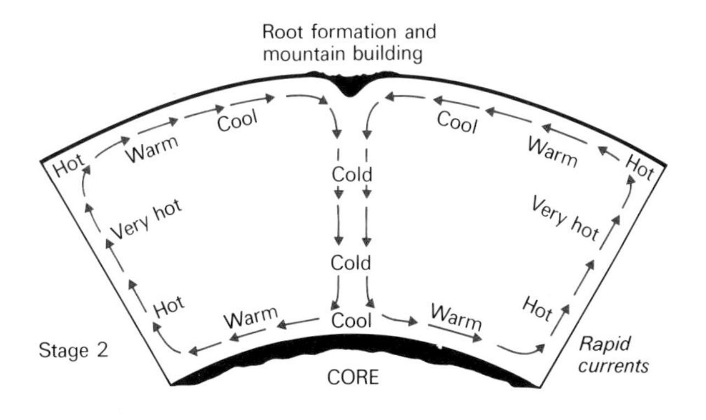

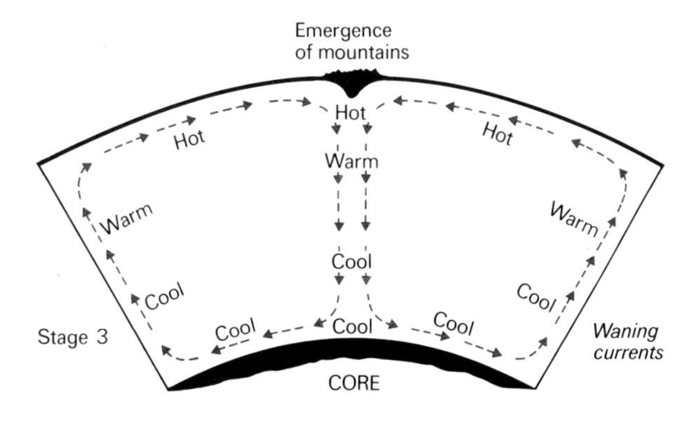

Figure 10. 4 Holmes' concept of the development of orogenies as a result of convection currents within the mantle. (After Holmes, 1944)

occur where currents meet and turn down (Fig. 10.4) and some evidence for this had recently come from investigations of the oceans. Because of the greater radioactivity of continental rocks when compared with oceanic rocks, the temperature would be likely to be greater beneath the continent and give rise to ascending convection currents. These proposals did far more than provide a mechanism for continental drift that was independent of external forces of gravitation, they went a long way towards explaining the distribution of other features such as recent orogenic belts and rift valleys.

Alexander du Toit brought to the continental drift controversy an extensive knowledge of the geology of his native South Africa and the other southern continents, particularly South America. Whereas European and North American geologists had been content to dismiss as inferences what Wegener and others had claimed to be facts, du Toit brought his first hand knowledge of the rocks to bear on the problem and strongly supported Wegener's theories in his book *Our Wandering Continents* (1937). He provided much additional factual geological evidence in support of continental drift and showed the inadequacy of what was then geological orthodoxy to explain such major features of the Earth as fold mountains, rift valleys and igneous intrusions. Du Toit was among the first to argue that the fit of continents should be made not at the present coastline but at the edge of the continental shelves and he argued further that drift was not necessarily a relatively recent occurrence but that older orogenic episodes such as the Variscan and Caledonian had probably also involved some kind of continental displacement.

One of his great contributions was the reassembly of Gondwanaland, based on his knowledge of the geology of the southern continents and particularly on the evidence of a Palaeozoic geosyncline (called by him the Samfrau geosyncline – the name being an acronym of South America, Africa and Australia) in Argentina, South Africa and Eastern Australia. Du Toit was utterly convinced of the reality of continental drift but expressed his theories with an enthusiasm that was not always beneficial.

Wegener died knowing the hostility of the reception accorded to his ideas. Far from being acclaimed, he found it difficult to obtain an academic post in Germany – he finally got one at the University of Graz in Austria – largely because his interests were so wide. He did not conform to the accepted idea of a respectable man of science, but his great contribution derived from the breadth of his thought in that, at a time when science was rather compartmentalised, he was ahead of his time in marshalling evidence from disciplines as disparate as geophysics, palaeoclimatology and palaeontology. The debate that his ideas provoked continued for a decade or so after his death but was inconclusive. In the United States the opposition was such that the idea of continental movement was virtually abandoned. Pockets of enthusiasm existed elsewhere but for the most part the concept was ignored by geophysicists and by most geologists. Stalemate had been reached: the verdict, to borrow from Scottish law, was at best 'not proven'.

Revival of interest

It is significant that the revival of interest in continental drift came largely as a result of geophysical research. During the late 1930s there began an investigation of the ocean basins which, at that time, were largely unknown. Britain had taken a lead in oceanographical exploration with the voyage of H.M.S. *Challenger* between 1873 and

1876, but there was no systematic continuation of this early work. It was not until the years immediately before the Second World War that serious exploration began, initially of the continental shelves using seismic methods, backed up by the echo sounder, corer, and dredge. At this time the Dutch geophysicist, F. A. Vening Meinesz extended the successful measurements of gravity he had made on land with a pendulum gravity meter, by measuring gravity at sea. To do this he mounted his apparatus in a Dutch submarine so that he could make the measurements when it was submerged and free from the pitch and roll of surface vessels. It gradually became clear that the oceans differed greatly and significantly from the continents. The sediments on the sea bed were found to be thin (usually less than 1 km thick) and relatively young, being of Cretaceous or younger age. All the igneous rocks recovered were basaltic and the depth to the Mohorovicic discontinuity, taken as being the boundary of crust and mantle, was found to be much nearer the surface beneath the ocean floors (about 10 km) than beneath the continents (30 km). The gravity measurements indicated that the ocean basins could not be sunken continental masses, and the idea that 'land bridges' once spanned the oceans became less and less tenable.

Bathymetric surveys continued after the war and there emerged the now familiar extensive world-wide pattern of mid-ocean ridges and rifts. The nature of these ridges was soon to assume a great importance for Holmes (1929) had postulated that they occurred where ascending convection currents parted to leave behind continental fragments poised above them (see Fig. 10.3). But the ridges proved not to be fragments of continent but, like the rest of the oceans floors, were made essentially of basic rocks – and basic rocks which had been serpentinised. Hess (1954) discussed the nature of the mid-Atlantic Ridge and sought an explanation in a rising mass of basalt carried up by mantle convection, becoming serpentinised, and lifting the areas to form a ridge. We now know that these mid-ocean ridges are the places where new crustal material is added from the underlying mantle and that they are the active areas of sea-floor spreading (see Chapter 11).

The second contribution made by geophysicists was the investigation of the magnetism of rocks. It had long been known that when igneous rocks which contain iron-bearing and magnetically susceptible minerals such as magnetite cool from temperatures near their melting point, they acquire, at a temperature known as the Curie Point, a permanent but weak magnetisation in the direction of the Earth's magnetic field for that particular locality. It was known that modern lavas are magnetised in the direction of the present-day field, but it was also known that some older igneous rocks sometimes had a reversed magnetisation, as though the north and south magnetic poles had changed places, and others were magnetised in directions other than that of the present field. P. M. S. Blackett, and later many other workers, investigated the magnetism of rocks, confirmed the reality of reversed magnetisation, and also showed that when rocks with a remnant magnetism from widely scattered localities were arranged in chronological order, they grouped to reveal periods of either normal or reversed magnetisation indicating that there had been periodic reversals in the Earth's magnetic field. The period of time during which a particular polarity regime prevailed was found to vary greatly, from several tens of thousands of years to a million years or more. These polarity 'epochs' each contain short events of opposite polarity, and this history of polarity reversals has been extended back gradually over geological time. This calibration of magnetic polarity with time has only been possible since methods have been developed for the determination of the age of the magnetised rocks by measuring in them the amount of certain isotopes produced by radioactive decay. Gradually it was found that many different types of rock contained this magnetic record – rocks as diverse as lavas, red sandstones, and even oceanic sediments.

These palaeomagnetic studies, in addition to demonstrating that the changes in polarity reflected changes in the Earth's magnetic field, showed also that the greater the geological age of the specimen, the more its 'frozen in' magnetic field departed from the present pole position. By assembling the pole position data for specimens arranged chronologically there emerged a series of points that defined a polar wandering path; in other words it appeared as if the Earth's magnetic pole had moved systematically and significantly with respect to the pole position of the area from which the specimens were collected. It was known that the Earth's field behaves as though there were a magnet in the Earth's core aligned along the axis of rotation and that, although the magnetic pole is not coincident with the geographic pole, the two are statistically the same over a period. So it could be demonstrated that the pole had moved with respect to a particular continent, but for some time it was not clear whether the same polar wandering path obtained for all the continents or whether each continent had its own unique polar wandering path. Because certain rocks act as fossil compasses, if it were to be established that each continent in fact had different polar wandering paths then there would be strong proof of relative movement between the continents: in other words a demonstration of continental drift was in sight.

The first results which indicated that Europe and North America had different polar wandering tracks were those of Runcorn (1956) and the later compilation of Blackett *et al.* (1960) indicated still more clearly that different continents had different polar wandering paths and that continental drift had in fact occurred (Fig. 10. 5). Disparate polar wandering paths were found to obtain for the southern continents, and the divergences of these paths with time could again be resolved if the continents were reassembled into Gondwanaland though, in the early stages, the data were open to more than one interpretation. For the first time, therefore, geologists had access to new information that was independent of the evidence advanced by Wegener

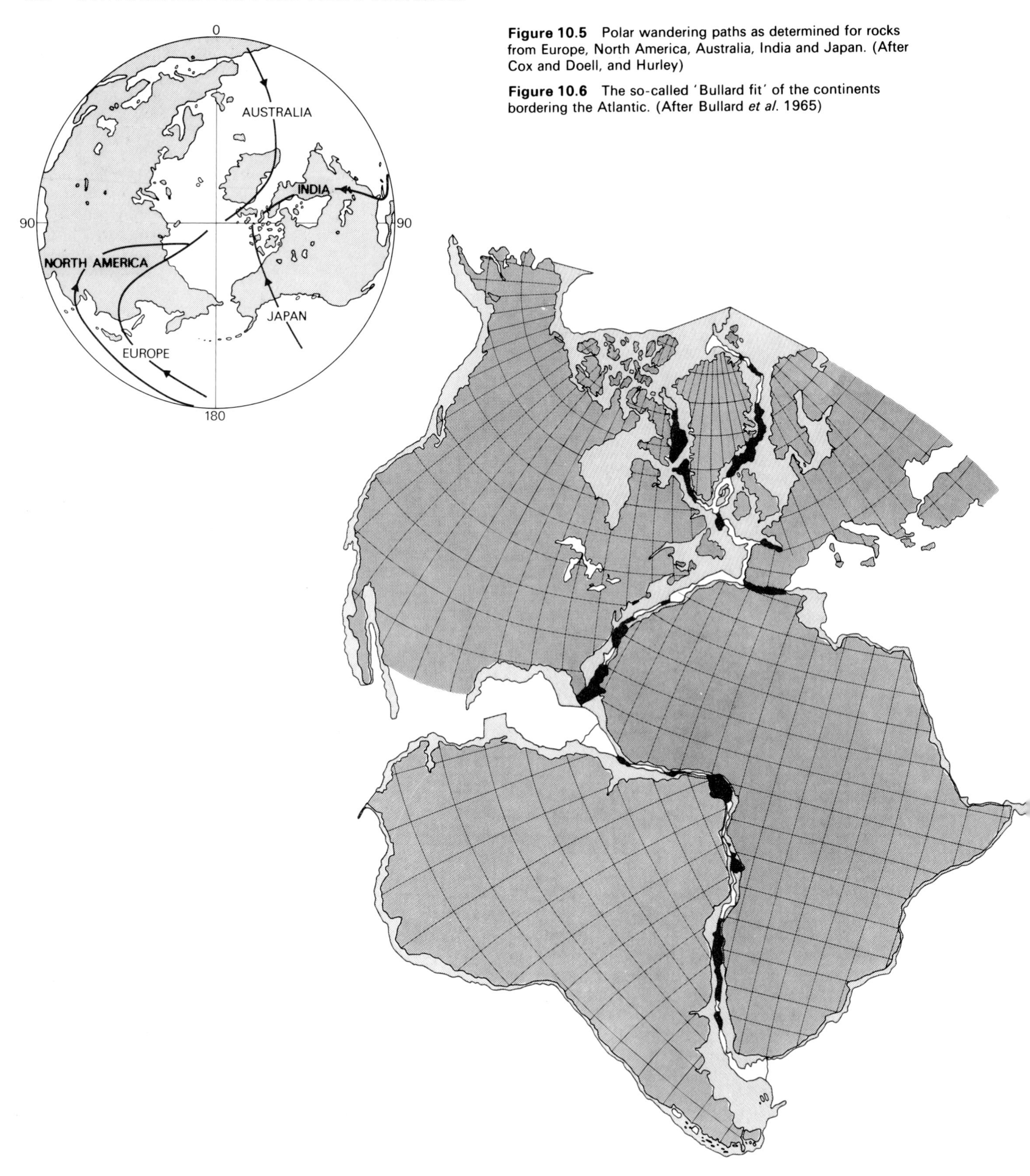

Figure 10.5 Polar wandering paths as determined for rocks from Europe, North America, Australia, India and Japan. (After Cox and Doell, and Hurley)

Figure 10.6 The so-called 'Bullard fit' of the continents bordering the Atlantic. (After Bullard *et al.* 1965)

and that had been debated so inconclusively for so long, and it resulted in a general change of opinion in favour of continental drift.

Concomitant with research into palaeomagnetism was the investigation of the ocean basins, and this work confirmed in a remarkable and quite unforeseen way the conclusions of palaeomagnetism. The result was that the earth sciences were, by 1960, poised for a revolution in geological thought. The proof of continental drift was to lead to a new concept of global dynamics and plate tectonics – a broad synthetic approach that linked sea-floor spreading, volcanism, earthquake activity, mineralisation and mantle evolution into one all-embracing theory. The 1960s closed with man's landing on the Moon and this also enabled us to see the Earth as never before – in perspective and from a distance – sharpening the realisation that this is an active planet far different from its satellite.

But what of those objections to continental drift that carried such weight in the years of the debate? One of the most consistently used objections, apart from the inadequacy of the proposed mechanism, was that Wegener had overstated his case for the match between continental margins. Jeffreys argued that, far from fitting snugly together, the continental margins in places showed a mismatch of over 1000 miles (1600 km). First Warren Carey (1955) and later Bullard tackled the problem more rigorously than before, both realising, as du Toit had said, that the match has to be made not at the present coastline but at the edges of the continental shelves. Bullard, Everett and Smith (1965) by the application of a geometrical technique and a computer, produced a fit for the Atlantic that was quite remarkable: it reduced the mismatch (overlaps and gaps) to about 50 km (Fig. 10.6) and made an immediate impact. This 'Bullard Fit', as it soon became known, was virtually to silence arguments about the alleged inaccuracy of proposed reconstructions by removing the subjective human element and substituting an objective geometrical solution to the problem. The work was later extended to other oceans but much less was heard of arguments about poor fits thereafter. By the late 1960s the ideas of continental drift and sea-floor spreading had been incorporated into the concept of plate tectonics. Had he lived a long, but not an exceptionally long, life, Wegener would have seen his ideas, on which so much scorn was poured at the time of his death, come to be vindicated and reach the point of almost universal acceptance. The qualifier is important, however, for today there are those who have reservations about large-scale horizontal displacement.

Sir Harold Jeffreys, so long an opponent of horizontal movements of the crust, remains unconvinced of their reality mainly on the grounds of the finite strength of the material of which the Earth is composed. Many of the objections to drift stated in earlier editions of *The Earth* have now been resolved, particularly the misfit of the continental coastlines. The greatest problem seems to be the record of magnetic reversals and sea-floor spreading, Jeffreys maintaining that the magnetic record is explicable in some other, but unspecified way that does not imply continental displacement. Another who is less than convinced as to the reality of continental displacement is the Russian V. V. Beloussov, who is influenced by the importance of vertical movements in continental areas and by the way in which new information that has been obtained in recent years from the study of the ocean basins has tended to take precedence over the very detailed data obtained over long periods from the continental areas. He contends that the concept of great horizontal movements has overwhelmed geological thought and that vertical movements implicit in sedimentation and mountain building, studied for so long and in such detail on land, have been reduced to the level of the broad and incomplete data from the oceans. He discounts horizontal motion and has exerted a powerful influence on Soviet geological opinion whose attitude has been in general to wait and see whether the plate tectonic theory will endure, rather than, like most western scientists, to jump aboard the band wagon.

Conclusion

There is still debate as to the mechanism by which plates become separated. The concept of a shrinking Earth has long since been abandoned, though it was the subject of a Presidential Address to the Geological Society of London as recently as 1952 (Lees, 1953). The diametrically opposite view, that the Earth is expanding, has been advocated by Warren Carey and more recently by Owen (1976). Owen, by careful cartographic reassembly of the land areas surrounding the Pacific Ocean, has shown that they fit closely and the present continental masses would cover about half the surface of a globe of 80 per cent of the present Earth's diameter. He envisages that the Earth had such a size in the late Triassic to early Jurassic and that the drift apart of the various continental components has resulted as much from expansion as from conventional sea-floor spreading on an Earth of constant size. Owen's ideas are based on painstaking work and should not be dismissed without careful consideration. However, this explanation of drift is not widely accepted because it is hard to see why the Earth's basic crust and mantle should have increased in volume whereas the continental, sialic parts, have not. But the history of ideas about continental drift has shown that it is unwise to discount theories just because they seem improbable. It is interesting, and not a little sobering, to reflect how completely the climate of geological opinion has changed since 1950, so that the heterodoxy of the acceptance of continental drift has become the present orthodoxy of plate tectonics. The change, and particularly the rapid change in the late 1960s, has been called, with good reason, a revolution in geology – a revolution that has affected the thoughts and concepts of all who study this planet and its inhabitants past and present, human and otherwise. It is sad that Wegener did not live to see it.

References

Bacon, F. 1620. *Instauratio Magma* (*Novum Organum*). 360 pp. London: Billium.

Beloussov, V. V. 1968. Some problems of development of the Earth's crust and upper mantle of oceans, pp. 449–59 *in* Knopoff, L. (Ed.) *The crust and upper mantle of the Pacific area.* American Geophysical Union Monograph **12**.

Blackett, P. M. S., Clegg, J. A. & Stubbs, P. H. S. 1960. An analysis of rock magnetic data. *Proceedings of the Royal Society of London* Series A **256**: 291–322.

Bullard, E. C. 1975. The emergence of plate tectonics: a personal view. *Annual Review of Earth and Planetary Science* **3**: 1–30.

Bullard, E. C., Everett, J. E. & Smith, A. G. 1965. The fit of the continents round the Atlantic. *Philosophical Transactions of the Royal Society of London* Series A **258**: 41–51.

Carey, S. W. 1955. Wegener's South America–Africa assembly, fit or misfit? *Geological Magazine* **92**: 196–200.

du Toit, A. L. 1937. *Our wandering continents.* 366 pp. Edinburgh: Oliver and Boyd.

Hallam, A. 1973. *A revolution in the Earth sciences.* 127 pp. Oxford: Clarendon Press.

Hess, H. H. 1954. Geological hypothesis and the Earth's crust under the oceans. *Proceedings of the Royal Society of London* Series A **222**: 341–48.

Holmes, A. 1929. Radioactivity and earth movements. *Transactions of the Geological Society of Glasgow* **18**: 559–606.

Holmes, A. 1944. *Principles of physical geology.* 532 pp. London: Nelson.

Jeffreys, H. 1924. *The Earth.* 278 pp. Cambridge: Cambridge University Press.

Lees, G. M. 1953. The evolution of a shrinking earth. *Quarterly Journal of the Geological Society of London* **109**: 217–257.

Marvin, U. B. 1973. *Continental drift – the evolution of a concept.* 239 pp. Washington, D.C.: Smithsonian Institution Press.

Owen, H. G. 1976. Continental displacement and expansion of the Earth during the Mesozoic and Cenozoic. *Philosophical Transactions of the Royal Society of London* Series A **281**: 223–291.

Runcorn, S. K. 1956. Palaeomagnetic comparison between Europe and North America. *Proceedings of the Geologists' Association of Canada* **8**: 77–85.

Snider, A. 1859. *La création et ses mystères dévoilés.* 487 pp. Paris: A. Franck, E. Denton.

Taylor, F. B. 1910. Bearing of the Tertiary mountain belt on the origin of the Earth's plan. *Bulletin of the Geological Society of America* **21**: 179–226.

Von Humboldt, A. 1801. Esquisse d'un tableau géologique de l'Amerique Méridionale. *Journal Physique de Chemie d'Histoire Naturel et des Arts*, Paris **53**: 130–60.

Wegener, A. 1912a. Die Entstehung der Kontinente. *Petermann's Geographische Mitteilungen* **58**: 185–195, 253–256, 305–309.

Wegener, A. 1912b. Die Entstehung der Kontinente. *Geologische Rundschau* **3**: 276–292.

Wegener, A. 1915. *Die Entstehung der Kontinente und Ozeane.* Vieweg, Braunschweig, 135 pp. (Later editions 1920 to 1962.) An English translation of the 1922 edition by J. G. A. Skerl was published by Methuen in 1924 as *The origin of continents and oceans*; and a translation of the 1929 edition by J. Biram was published by Dover in 1966.

A. C. Bishop
Department of Mineralogy
British Museum (Natural History)

CHAPTER 11

Plate tectonics

A. L. Graham

Much of the revolution in thought that occurred in the earth sciences in the 1960s can be ascribed to the resolution of the continental drift controversy. The background to the sometimes virulent arguments between the drifters and the non-drifters, as the protagonists were called, has been traced in the preceding chapter. Here the story is taken up at the point at which more relevant information became available, enabling geologists to construct testable hypotheses on the reality, or otherwise, of continental drift. This information resulted from the development and application of two geophysical methods of investigation. The first of these, seismology, is the study of earthquakes and the shock, or seismic, waves generated by them. The second is palaeomagnetism, the study of the history of the Earth's magnetic field as indicated by the direction of magnetisation of certain crustal rocks. The analysis of earthquake shocks provides us with data on the structure and subdivision of the crust and upper mantle of the Earth at present, whereas palaeomagnetism allows us to go back in time and reconstruct the relative positions of elements of this crust to each other in the past, and relative to the rotational axis of the Earth. Contemporary with these developments was the intensified topographic surveying of the ocean floors which showed them not to be flat plains but to have rugged areas and extensive mountain ranges.

The early belief that the ocean floors were regions of unchanging calm, being slowly covered by fine-grained sediments, was replaced by a concept that changed the ocean basins into areas of very large-scale activity and connected the oceanic regions to the continents by a single, unifying hypothesis. This concept progressed through a series of names – initially it was known as continental displacement or drift, then as sea-floor spreading, global tectonics and finally as plate tectonics. Though this chapter is concerned with plate tectonics, the other terms are frequently met in the literature and involve essentially the same idea and refer to the same processes.

The beginnings of the re-awakening of the geological community to new ideas on the formation of the crust of the Earth occurred in 1960, when Professor H. H. Hess of Princeton University expounded the hypothesis that he called sea-floor spreading (Hess, 1962). In it he attempted to synthesise the results that had been obtained from seismic refraction surveys at sea. These showed that the oceanic crust was about 6 km thick, in contrast with the typical thickness of continental crust of around 30–40 km. From this he argued that it was physically impossible for the continental crustal blocks to have moved through the oceanic crust, as some older continental drift hypotheses would have it. Much more reasonably, he suggested that the continents and the oceans must be connected and moving together. Where they are moving to, and how quickly, were questions also raised by Hess. Both of these can now be answered in part but before we attend to them we should look at the surface of the Earth and examine the configuration of the moving pieces of crust and how they interact.

Surveys of the topography of the ocean floor show that there are long linear ridges within the ocean basins. In the Atlantic the ridge is located approximately centrally between the Americas and Europe and Africa and runs in a generally north–south direction. Similar ridges have been found in all the oceans of the world; their crests are all raised 2–3 km above the level of the deep ocean floor and are connected together to form a global network. Hess suggested that these ridges are sited on the upwelling limbs of convection currents in the Earth's mantle and that it was here that new oceanic crust was actively forming. It was known that these ridges have very little sedimentary material covering them and palaeontological evidence showed these sediments to be very young. Another remarkable discovery was that the sediments on the deep ocean floors away from the ridges are very thin when compared with the vast thicknesses of sedimentary strata exposed on the continents. Oceanic sediments are rarely more than 1·6 km thick and 200–300 metres is more usual; the rate of sediment accumulation on the deep ocean floors is around 1–20 mm per 10^3 years and if these floors were a

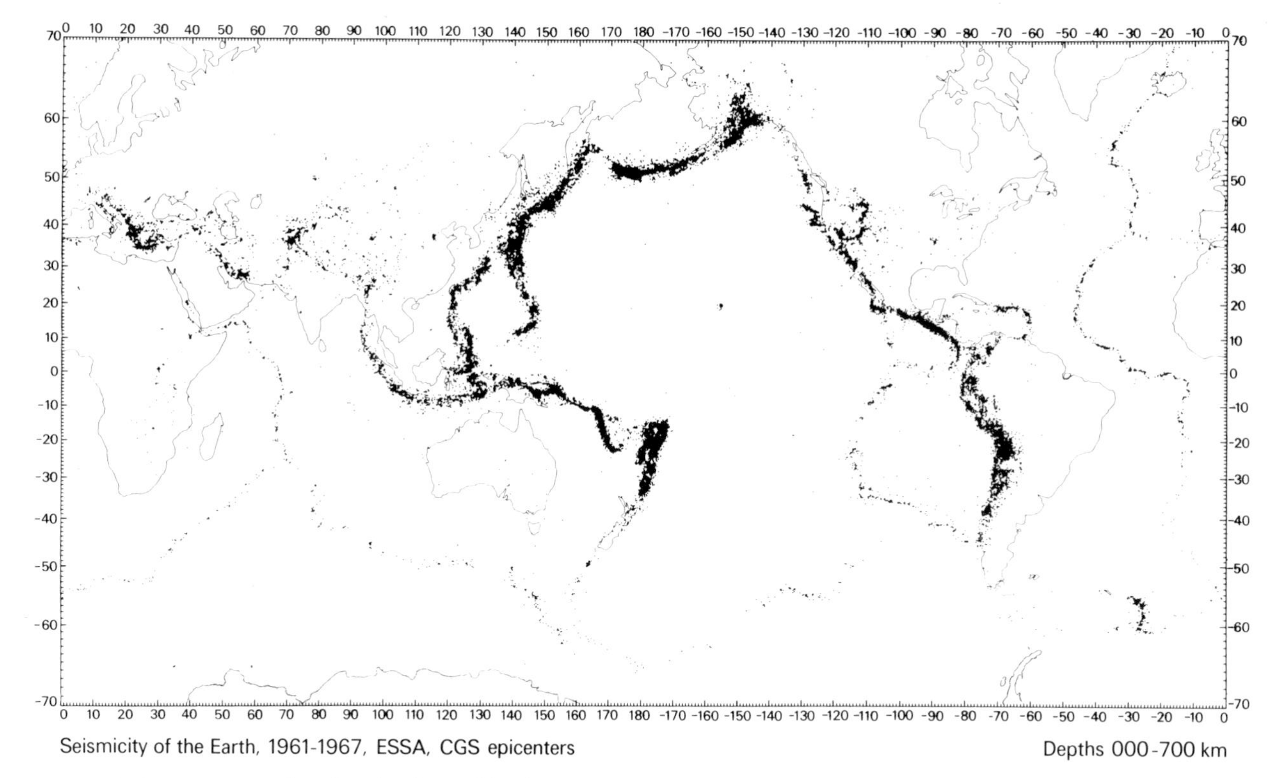

Figure 11.1 Epicentres of earthquakes located by the world-wide standardised seismic array (ESSA) for 1961–1967 (from Barazangi & Dorman, 1969). Note the continuity of the seismic belts that outline large stable blocks.

permanent feature of the Earth's crust a much greater thickness of sedimentary deposits would be expected. These findings implied not only that the oceanic ridges are geologically very young but that so also are all the present ocean floors. If, as it was suggested, new crust to form the floors of these ocean basins was being generated at the ridges, what was happening to the older crust? The hypothesis created crustal rocks in oceanic regions, and required either that the Earth expand to accommodate this new crust or that a mechanism of crustal 'destruction' be found if the Earth were to remain of constant size. A discussion of the possibility of an increase in the volume of the Earth with time is given in Chapter 12 and is not considered further; it is assumed here that the Earth has remained approximately constant in size, at least for the last 200 million years. We are thus left with the problem of how crustal material is 'destroyed' or recycled, thereby making room for new oceanic crust, and how the necessary balance is maintained between the two processes, the one consuming what the other produces.

The emphasis on the primary subdivision of the surface of the Earth has moved away from that of continental and oceanic units to blocks of crustal material bounded by zones of tectonic activity. At present most of these blocks are composed of both oceanic and continental material, though a few are entirely oceanic. This change came as a result of the publication of the data obtained from a world-wide seismic network; it was from this that the location, depth of focus, and sense of movement of earthquake shocks could be calculated. The network was principally designed to detect underground nuclear tests and thereby monitor the observance of nuclear test-ban agreements. When the data collected between 1961 and 1967, from about 30000 earthquakes, were plotted on a globe it was obvious that by far the majority of these shocks occurred in narrow, linear zones (Fig. 11.1). It was also shown that the depths of the earthquake foci from the surface were generally less than 70 km. Quakes with foci depths of up to 700 km were recorded but these occurred only in areas associated with deep ocean trenches. This distribution of earthquake epicentres within narrow zones, while the rest of the crust was largely stable, led to the suggestion that the surface of the Earth should be considered as composed of a number of rigid blocks with abutting edges. It is at these edges that the relative movement between the rigid blocks takes place and earthquakes occur. These blocks are now referred to as crustal plates (Fig. 11.2) and the whole concept of their relative movement is known as plate tectonics. The global distribution of the edges of these crustal plates forms a network of narrow active zones. Where three of these zones meet, for example in the Afar region south-west of the Red

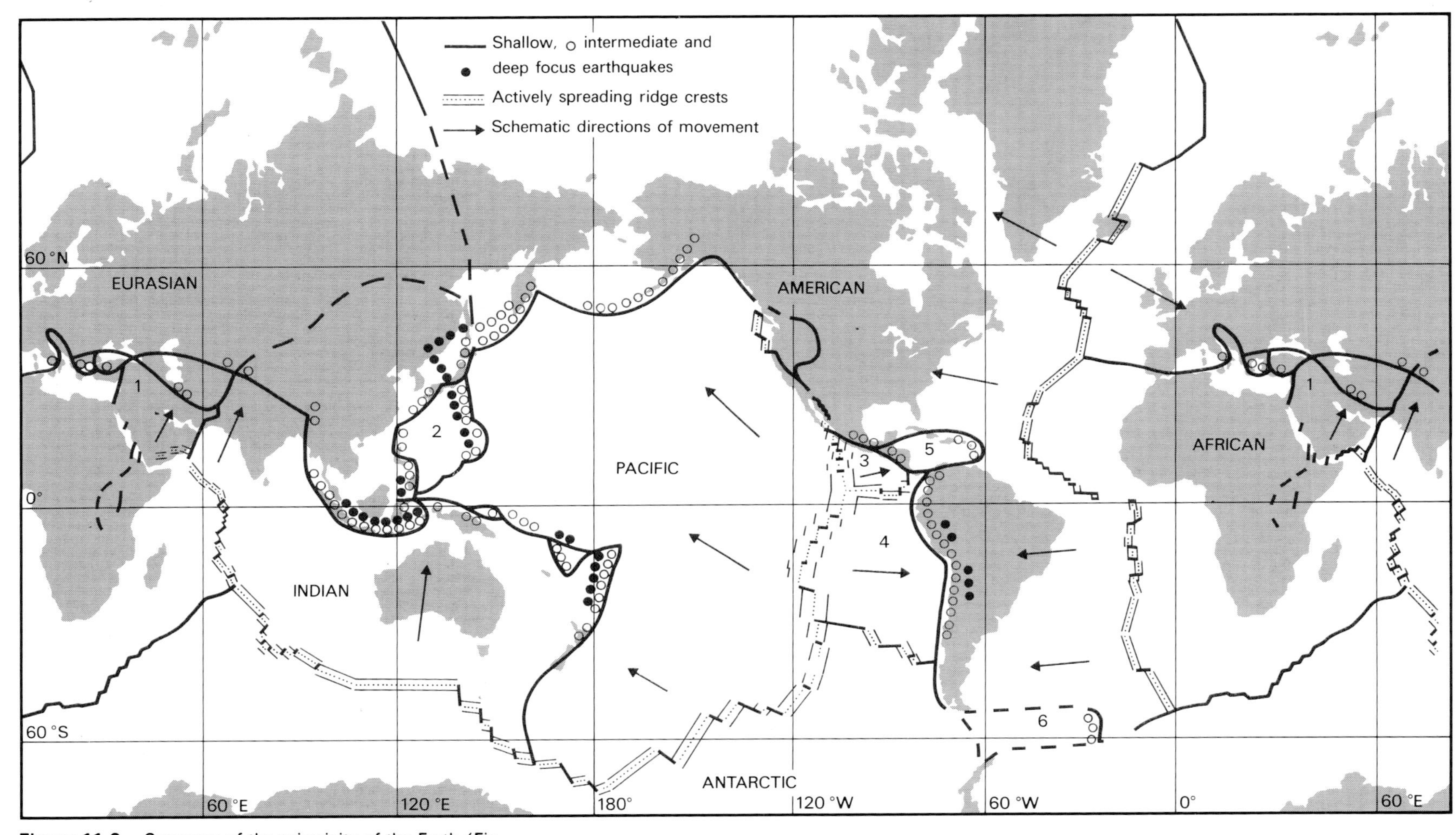

Figure 11.2. Summary of the seismicity of the Earth (Fig. 11.1), and the lithospheric plates. The six major plates are named; minor plates are numbered: (1) Arabian; (2) Philippine; (3) Cocos; (4) Nazca; (5) Caribbean; (6) Scotia (after Gass *et al.* 1972).

Sea, south-west of Southern Africa or in the Indian Ocean east of Madagascar (Fig. 11.2), the structure is called a triple junction (McKenzie and Morgan, 1969). It is at such junctions that the activity of the three zones, or plate margins, combine at a point which is not necessarily fixed in relation to any one plate. Triple junctions are themselves mobile and not fixed at any specific position on the globe, but they may move as the plates move.

While we now have some idea about the surface expression of plate tectonics, what happens at depth within the crust and upper mantle? The principal source of information here is the variation in the speed of seismic signals with depth (Fig. 11.3). The reasons for this are not clear; two likely possibilities which have been proposed are differences in the composition of the crustal materials, and their response to the increasing pressure at depth. These observations have led to the subdivision of the outer layers of the Earth into the lithosphere, the asthenosphere and the mesosphere. The rigid crustal plates with high seismic velocities make up the lithosphere, which generally extends to a depth of about 70–100 km below the Earth's surface; it is within this shell that the majority of earthquakes occur. Below this there is a region between 50 and 100 km thick which is characterised by lower seismic velocities and is known as the asthenosphere, or low velocity zone. It is within this zone that the differential movement of the lithosphere and mesosphere is thought to be accommodated. Underneath this is the mesosphere, in which the rigidity of the mantle is greater than in the asthenosphere.

The subdivision of the crust and upper mantle is reflected in the earthquake foci depths. Shallow focus earthquakes are those occurring at depths of less than about 70 km, that is within the lithosphere. Intermediate (70–200 km depth) and deep (200–700 km) focus earthquakes generally occur only where rigid and relatively cold lithospheric material is being forced down to these depths by crustal movements. An example of the seismic results obtained from a study of earthquakes along a line normal to a plate boundary is given for Amchitka Island in the Aleutians, (Fig. 11.4). This shows that the thickness of the zone of seismic activity varies with depth. For near-surface, shallow focus, earthquakes the zone is about 150 km wide in this example but for the intermediate and deep focus earthquakes the zone is only about 50 km thick.

From analysis of the first motions of the earthquakes occurring at plate margins, stress vectors can be obtained which indicate the direction of relative movement of the participating lithospheric plates. This enables us to map out not only the extent of the plates themselves but also their present relative movement. It is not an easy task to conceive the surface of the Earth as being composed of crustal plates, all moving about and constrained only by each other. Not only are the plates themselves in relative movement but so also are the plate boundaries. Generally the movements of the plates are slow, varying between 1 and 18 cm per year, and to assess the effect of this movement we need to consider a longer time interval than

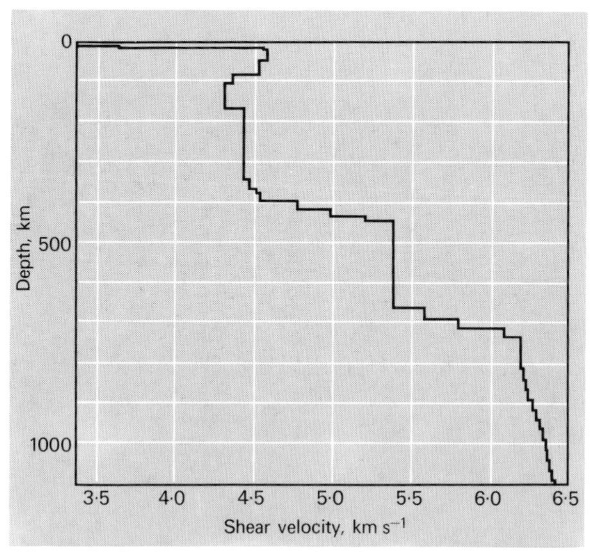

Figure 11.3 Shear wave velocity distribution with depth for oceanic areas, derived from surface wave studies. The step-like changes in velocity result from the computation technique; in reality the changes are smooth (after Gass *et al.* 1972).

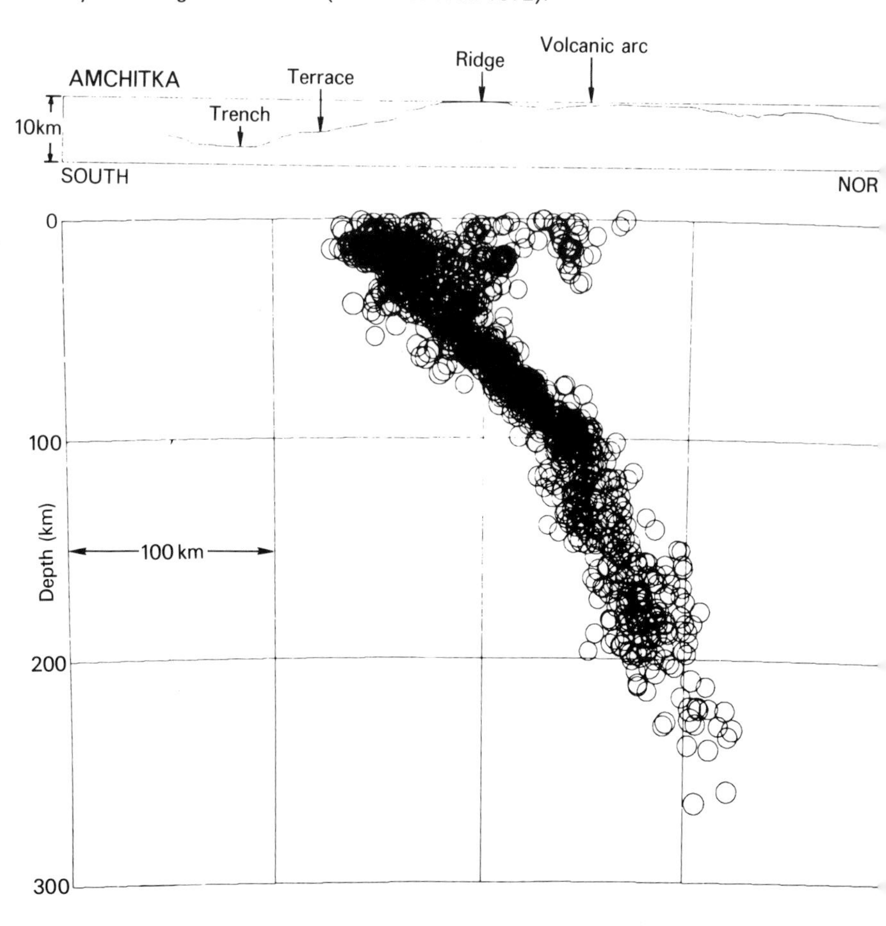

Figure 11.4 Projection of earthquake foci on a vertical plane through Amchitka Island, Aleutians. It is this distribution of earthquake foci which defines the Benioff zone, see text (after Enghdahl, 1977).

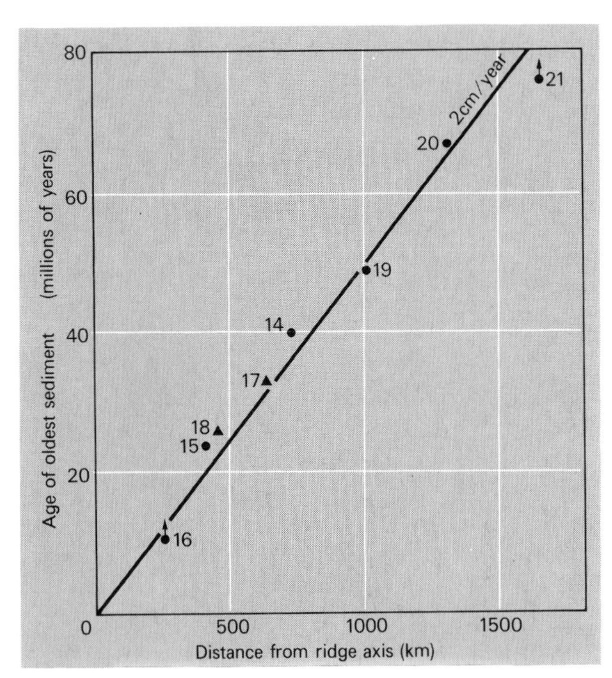

Figure 11.7 The age of the oldest sediment at different numbered drilling sites, as discovered on leg III of the *Glomar Challenger*, plotted against distance from the ridge axis. There is close agreement with the line drawn which is based on a spreading rate of 2 cm/year (after Maxwell *et al.* 1970).

of the crust is unable to keep pace and provide space for the basalt at the sea-floor. A volcanic pile is then built up above the ridge, which in some cases reaches sea-level; Iceland has formed as a result of excessive lava production within a limited area of the Mid-Atlantic Ridge. The rate of opening, or spreading, at the crest varies from ridge to ridge but is of the order of 1–9 cm per year on each side. Often the spreading rate is similar for the two sides but this is not always the case, nor need it be. Ridges may move relative to each other and some irregularities in the crustal 'tape recorder' have been explained in terms of sudden movements of a section of the active ridge centre by 100 km or so in a direction normal to the ridge axis.

As the new oceanic crust moves away from the ridge crest it cools and in so doing subsides. Since the rate of subsidence is principally dependent upon cooling rate, which is fixed mainly by the thermal conductivity of the crustal rock (basalt) there is a regular relationship between the depth of water above the oceanic crust and the crustal age of that point on the deep ocean floor (Fig. 11.10). Variation in the spreading rate does not change this relationship but results in different areas of sea-floor under a particular depth of water. It has been suggested that changes in the rate of spreading would cause changes in sea-level, faster rates leading to higher sea-levels than slower rates. Calculations show that this mechanism for producing changes in sea-level, and consequent marine

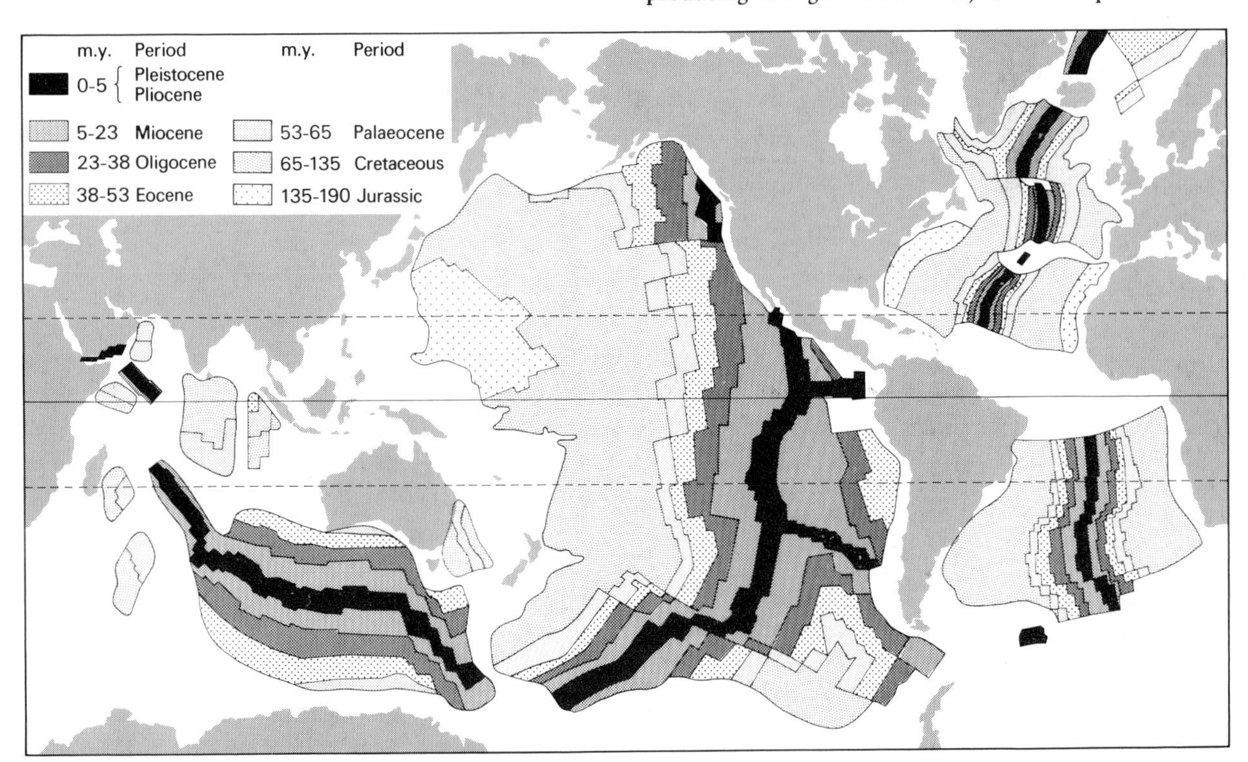

Figure 11.8 The distribution of oceanic crust according to age. The map was compiled in 1974 from the information then available about the magnetic anomaly pattern on the ocean floor. Unmarked areas: incomplete data (after Turekian, 1976).

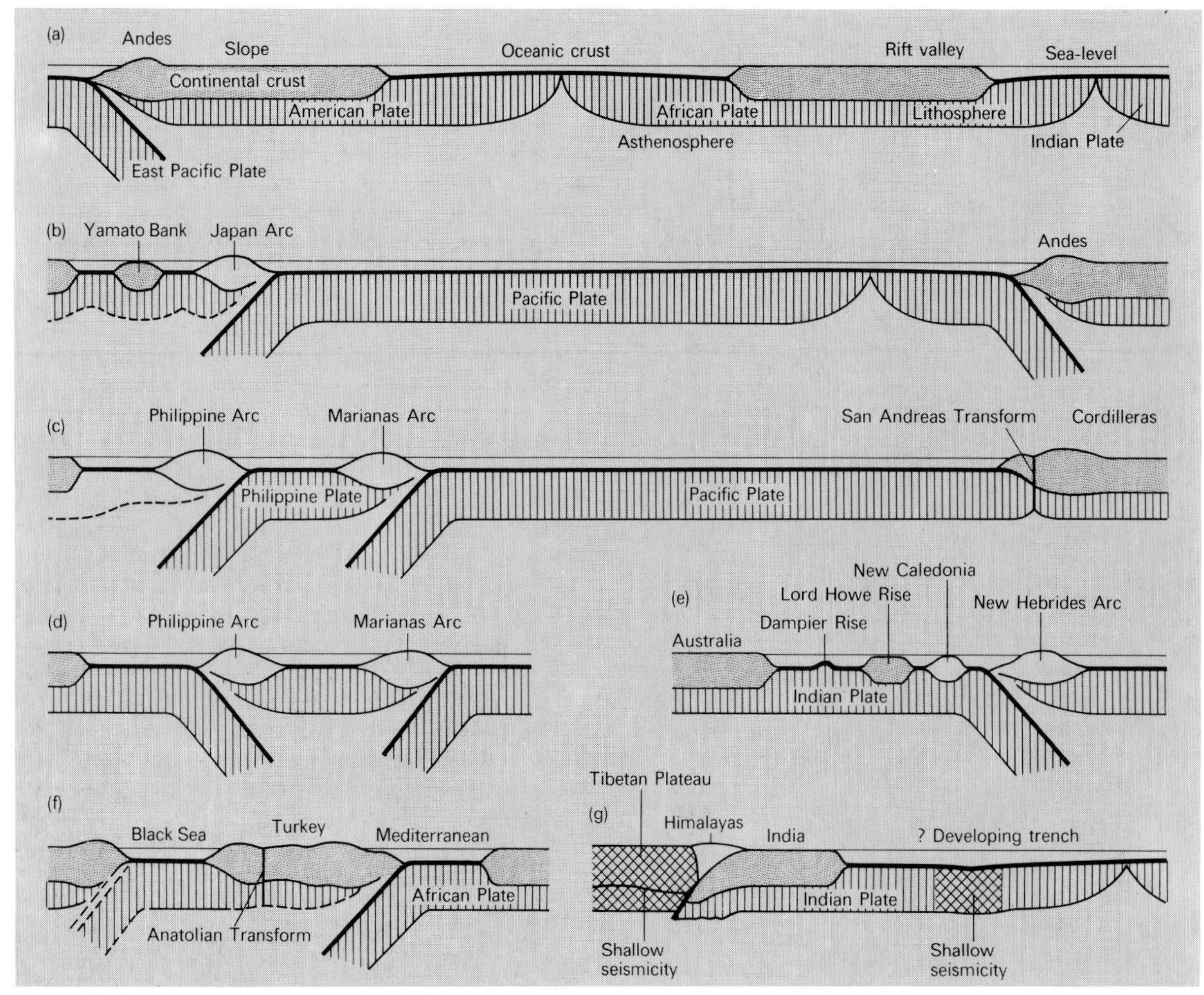

Figure 11.9 Schematic sections showing plate, ocean, continent, and island-arc relationship. Black = oceanic crust; stipple = continental crust; vertical bars = lithospheric plates (after Dewey & Bird, 1970).

transgressions and regressions over land surfaces, is much less effective than continental glaciation but more effective than any other suggested mechanism. The subsidence of the oceanic crust with age is well illustrated by the guyots and seamounts of the central Pacific. In many cases slow subsidence has resulted in their being covered with very great thicknesses of coral and the formation of atolls, despite the inability of the coral polyp to live actively at a depth greater than about 50 m (Fig. 11.11).

Crustal formation at the oceanic ridges, the dominant constructive process, requires a concomitant destruction or remobilisation (reconstruction) of crustal material to maintain an approximately constant surface area of the Earth. Subduction is the term applied to the process by which crustal material is returned to the upper mantle. Both subduction and spreading rates have been measured and are similar in magnitude (Fig. 11.12). Some confusion may be caused by the method of reporting a spreading rate. In some cases the half spreading rate is given, that is the rate at which one limb of the ridge moves away from the ridge crest. In other cases the full spreading rate is quoted. The latter is the rate at which the two sides of the spreading centre, the ridge crest, separate or in the case of a subduction zone, converge, and it is this which is given in Figure 11.12. The activity at destructive plate margins is dictated mainly by the density of the colliding crust. Oceanic crust has a higher density than continental material; consequently when the two meet at a destructive margin the continental crust rides over the oceanic crust, which is forced down into the mantle. One example of this occurs at the eastern Pacific margin, where the western coast of South America, part of the American plate, is riding over the Pacific oceanic crust formed at the East Pacific Rise. The descending slab of oceanic crustal material is cooler and more rigid than the mantle material it is sinking into and this causes stress and creates a zone

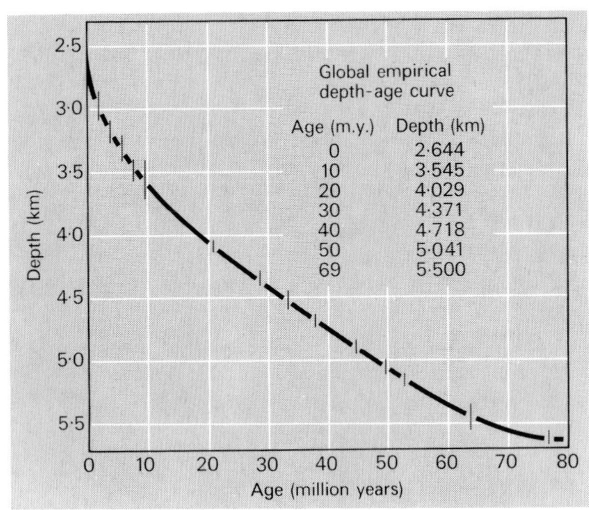

Figure 11.10 The age versus depth for the oceanic crust. Vertical bars indicate standard deviation for averaged data. The table gives age versus depth data at 10-million year intervals which have been interpolated from the graph. Data from Sclater *et al.*, 1971 (after Pitman, 1978).

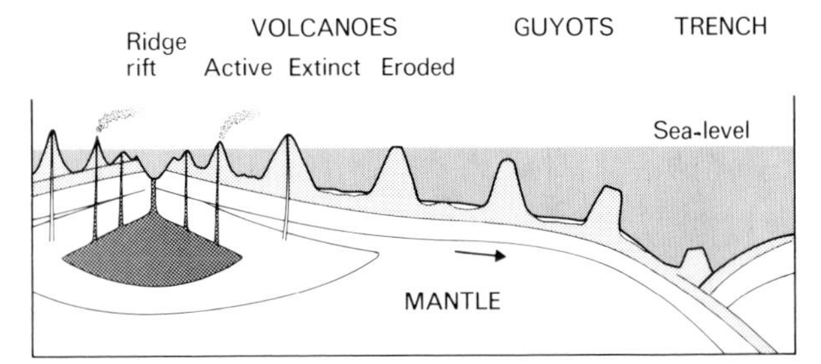

Figure 11.11 A diagram illustrating the growth, submersion, erosion, and destruction of volcanoes on a spreading sea floor. A guyot with a tilted platform sits on the lip of the Tonga trench; the Kodiak guyot sits upright in the bottom of the Aleutian trench (after Marvin, 1973).

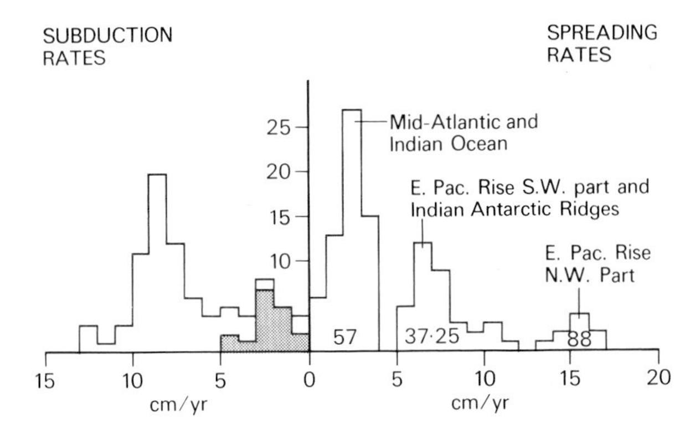

Figure 11.12 A histogram of world spreading and subduction rates. A mean rate has been calculated for each 5° length of ridge or arc; from the data of Chase (1972). Spreading rates are full rates of opening. The stipple indicates boundaries where continent/continent interaction is occurring (after Oxburgh, 1974).

of seismic activity known as the Benioff zone, named after the seismologist Hugo Benioff (Fig. 11.4). Such zones, more usually termed subduction zones, are often associated with deep oceanic trenches such as the Java Trench, and the Peru–Chile Trench.

Other destructive plate margins are marked by deep oceanic trenches associated with arcs of volcanic islands; for example Tonga, the Aleutians, and the Marianas (Fig. 11.13). Here the crustal structure is termed an island arc, the rocks forming the islands being dominated by calc-alkaline volcanic rocks. The oceanic crust is subducted below the islands and the heating and dehydration of the descending slab causes volcanicity by producing lavas which ascend through the overlying lithosphere to the surface to form the islands (Fig. 11.14A). The chemistry of these lavas, on extrusion, indicates that they were not derived from the descending slab but mainly from the lithosphere above it. Similar volcanic activity has occurred in Chile and Peru. Here, in an ocean–continent collision, the descending oceanic lithosphere has de-hydrated at depth and stimulated lava production in the overlying lithosphere, producing volcanic activity on the continental crust of South America (Fig. 11.14B). Among the commonest of these calc-alkaline volcanics are andesites, so called since they form most of the Andean mountain chain, the backbone of South America. Associated with these volcanics and coarser igneous rocks such as granodiorites are very large tonnages of copper ores. The discovery of the association of subduction zones with particular types of ore body has led to a new rationale for mineral exploration and remarkable re-assessments of the mineral resources of a number of countries.

Plate collisions involving oceanic crust usually result in the subduction of this material. However it is believed that true oceanic crust is exposed subaerially in the so-called ophiolite complexes, for example at Troodos in Cyprus, Vourinos in Greece, Liguria in Italy, New Caledonia and elsewhere. These complexes consist of lavas extruded in the form of pillows lying above sheeted dykes, which in turn overlie gabbros and peridotites. Above the pillow lavas are iron- and manganese-rich sediments and cherts. This sequence is very similar to that occurring in the present oceanic crust, inferred from dredging, drilling and seismic results. Another important type of destructive margin involves the collision between two plate margins with continents at their edges. Continental crustal material cannot be subducted since it is lighter than the underlying mantle, and piles up on itself forming high mountains with deep roots, all of continental crust. An excellent example of this is the collision between the Indian plate moving northwards into the Eurasian plate, which has resulted in the formation of the Himalayas and the associated high mountain chains.

So far we have considered only volcanic and seismic activity occurring at plate margins. There are a few areas of volcanic activity which are located far away from these margins and perhaps the best known of these is the

Hawaiian–Emperor chain of volcanic islands and sea mounts, whose present-day volcanic activity is situated close to the middle of the Pacific plate (Fig. 11.15). Intraplate activity is not uniquely explained by the plate tectonic hypothesis, nor does it predict the position of such activity. One possible explanation is referred to as the 'hot spot' or 'mantle plume' hypothesis (Morgan, 1972). Such a spot or plume is the source of excess heat in the mantle and may also even provide a portion of the liquid that is erupted above the hot spot or plume.

We have little indication of the distribution of the convecting cells in the upper mantle other than at plate margins where the movement of the plate is that of the upper limb of the cell. In some areas, particularly the eastern Mediterranean, the relationships between the cells are very complex. While we might prefer simple, large cells under the Pacific lithospheric plate, the hot spot hypothesis argues that such cells cannot explain the presence of the Hawaiian islands. Under certain and as yet not well defined conditions, the hypothesis holds that very localised

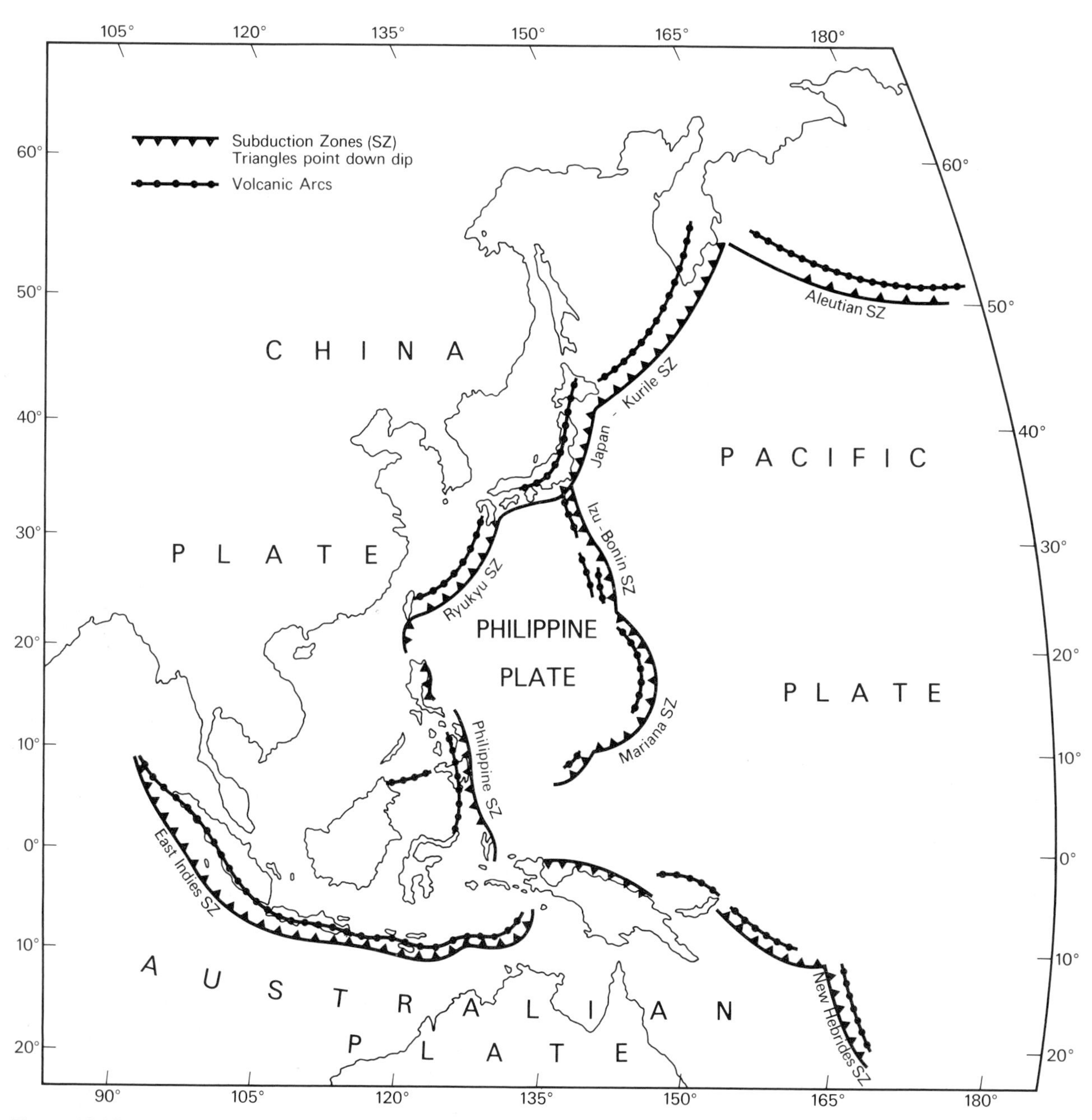

Figure 11.13 The crustal structure of the western Pacific showing the extensive development of island arcs and their associated subduction zones, SZ (after Condie, 1976).

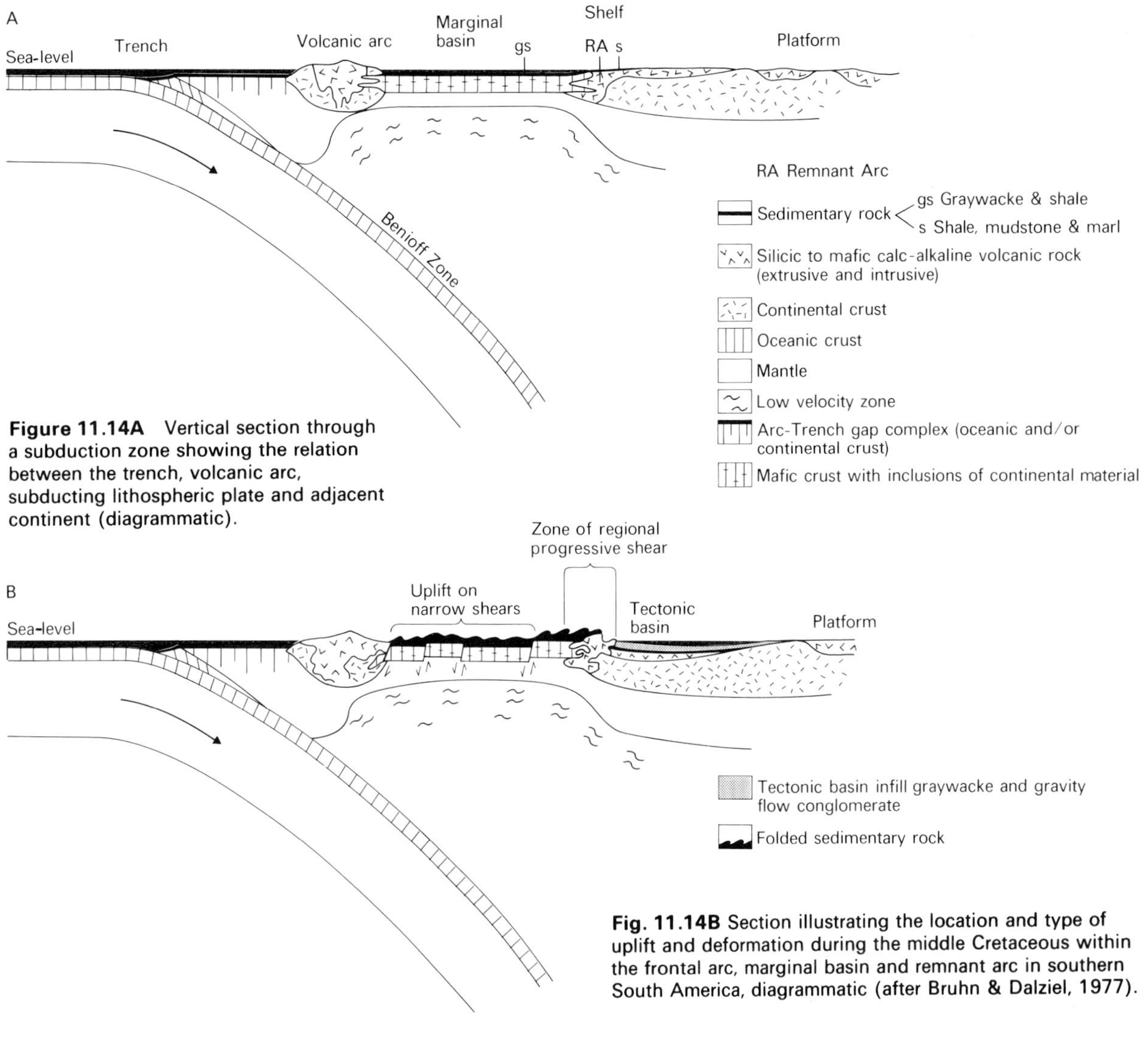

Figure 11.14A Vertical section through a subduction zone showing the relation between the trench, volcanic arc, subducting lithospheric plate and adjacent continent (diagrammatic).

Fig. 11.14B Section illustrating the location and type of uplift and deformation during the middle Cretaceous within the frontal arc, marginal basin and remnant arc in southern South America, diagrammatic (after Bruhn & Dalziel, 1977).

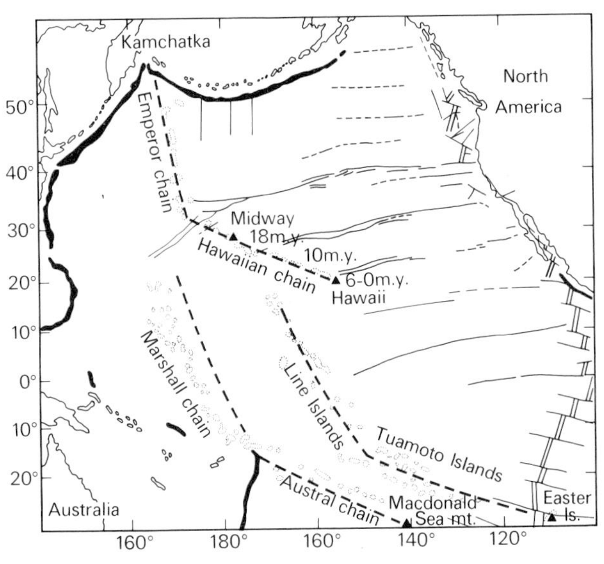

Figure 11.15 The linear island chains of the Pacific Ocean. Dark tone indicates oceanic trenches; double line, ridge crests; single fine line, transform faults and fracture zones. The heavy dashed lines indicate the expected trends of the two other island chains if they were generated by motion of the Pacific plate parallel to the Hawaiian–Emperor trend, over Hawaii, Easter Island and MacDonald seamount as fixed mantle 'hot spots'. Some ages are given along the Hawaiian–Emperor chain (after Grommé & Vine, 1972).

areas of high heat flow, the 'hot spots', become active and cause within-plate volcanism. It has been argued that hot spots result from plumes or blebs of material rising from the core–mantle boundary. If the plume rises underneath the middle of a large lithospheric plate, such as the Pacific, it may cause rifting within the plate and produce a new oceanic ridge or, if less active, merely result in isolated volcanism forming oceanic islands. The 'hot spot' is believed by some to be fixed in the upper mantle while the lithospheric plate moves over it. At intervals this spot stimulates activity that forms chains of volcanic islands such as the Hawaiian group. The intermittent nature of the activity is reflected in the distribution of islands along the chain which extends for over 3000 km. The age of the activity increases progressively and regularly in a north-westerly direction from Hawaii to Midway, beyond which the chain changes course and continues in a more northerly direction with the Emperor seamounts. These seamounts continue to increase in age northwards with distance along the chain in the direction indicated by the Hawaiian islands.

Though this example would seem to be good confirmation of the hot spot hypothesis, other island chains in the Pacific do not agree with it. The Line Islands, for example, do not vary in age systematically along the chain as would be expected. Taken in a global context it seems that some hot spots must have migrated with respect to each other if the volcanic features of the deep ocean floor are indeed generated in this way (Molnar & Atwater, 1973). Unfortunately the initial simplicity of the hot spot concept is not maintained; while the hypothesis is an attractive one, there appear at present to be so many variations required to explain intraplate volcanism that it is by no means widely accepted without modification. While the formation of volcanic islands by mobile 'hot spots' is generally accepted as a working model, the extension of this hypothesis to the chemistry of the volcanic rocks forming these islands being dictated by the chemistry of the mantle 'hot spot' is a contentious point.

So far we have considered only those oceanic ridges which are defined by the seismic activity and are present-day plate boundaries. There is a second type that shows little or no seismic activity and is referred to as an aseismic ridge; examples are the Walvis Ridge in the south Atlantic and the Ninety East Ridge in the Indian Ocean. A further distinction between the two ridge types is on the age of the basal sediment at the crest. Along the crest of an active ridge, or one that was active in the past, the age of these sediments does not change. This is not the case with an aseismic ridge, for example the age of the crest sediments of the Ninety East Ridge increases northwards. Also this ridge is at a much shallower depth than would be expected from its age; its depth/age relationship does not plot on the normal curve, (Fig. 11.10). These ridges remain, however, 1·5 to 3 km below sea-level but it is possible for volcanic islands to form on them.

The Ninety East Ridge cannot have been a normal active ridge which has since become quiet. One suggestion for its formation invokes a hot spot under the ridge which, instead of being sporadic in activity, as happened in the Pacific for the Hawaiian chain, was continuously active. As the Indian plate moved over this spot, its continuous submarine volcanic activity formed the Ninety East Ridge.

Occasionally the rate of lava production has been large enough to build up the sea floor and form the isolated islands in the southern Indian Ocean, for example Amsterdam, Kerguelen and St Paul. Another explanation for the formation of the Ninety East Ridge is as follows. In the past the Indian and Australasian plates moved northwards at different speeds and the differential movement was accommodated along the line of the Ninety East Ridge which was, at that time, a transform fault and a plate boundary. However, at present the Australian and the Indian plates are moving northwards at practically the same speed and so the Ninety East Ridge is no longer a plate boundary and this would explain why very little seismic activity is recorded from it.

Although the broad concept of plate tectonics has been convincingly established, there are many finer points that are not yet understood and inconsistencies that are not fully explained. In particular, much remains to be learned when applying the concept to the Palaeozoic. We do not know when plate tectonics first became the dominant process shaping the surface of the Earth. It may be that in the Archean and Palaeozoic different processes were active in promoting crustal movements and that plate tectonics as we now know it evolved from them during the late Palaeozoic (see also Chapter 2). However, plate tectonics does provide an explanation for the remarkable climatic changes required by the palaeo-environmental evidence from the stratigraphic column, particularly since the Jurassic. If the continents and their associated intra-continental basins have migrated over the surface of the Earth with the passage of geological time, it should come as no surprise to find deposits formed in tropical regions in what is now Antarctica, or tillites and glaciated pavements in continental areas now in the tropics. The flora and fauna associated with the mobile land masses must similarly have changed in response to the variation in latitude of the continental plates over geological time.

References

Barazangi, M. & Dorman, J. 1969. World seismicity map of ESSA Coast and Geodetic Survey epicenter data for 1961–1967. *Bulletin of the Seismological Society of America* **59**: 369–380.

Bruhn, R. L. & Dalziel, I. W. D. 1977. Destruction of the early Cretaceous marginal basin in the Andes of Tierra del Fuego, pp. 395–405 *in* Talwani, M. & Pitman, W. C. III (Eds) *Island arcs, deep sea trenches and back-arc basins*. Washington, D.C.: American Geophysical Union.

Chase, C. G. 1972. The N plate problem of plate tectonics. *Geophysical Journal of the Royal Astronomical Society* **29**: 117–122.

Condie, K. C. 1976. *Plate tectonics and crustal evolution*. 288 pp. New York: Pergamon Press Inc.

Cox, A. 1969. Geomagnetic reversals. *Science* **163**: 237–245.

Dewey, J. F. & Bird, J. M. 1970. Mountain belts and the new global tectonics. *Journal of Geophysical Research* **75**: 2625–2647.

Engdahl, E. R. 1977. Seismicity and plate subduction in the central Aleutians, pp. 259–271 *in* Talwani, M. & Pitman, W. C. III (Eds) *Island arcs deep sea trenches and back-arc basins*. Washington, D.C.: American Geophysical Union.

Gass, I. G., Smith, P. J. & Wilson, R. C. L. (Eds) 1972. *Understanding the Earth*. 383 pp. Open University Press.

Grommé, S. & Vine, F. J. 1972. Palaeomagnetism of Midway atoll lavas and northward movement of the Pacific plate. *Earth and Planetary Science Letters* **17**: 159–168.

Hess, H. H. 1962. History of ocean basins, pp. 599–620 *in* Engel, A. E. J., James, H. L. & Leonard, B. F. (Eds) *Petrologic studies: a volume to honor A. F. Buddington*. New York: Geological Society of America.

McKenzie, D. P. & Morgan, W. J. 1969. Evolution of triple junctions. *Nature*, London **224**: 125–133.

Marvin, U. B. 1973. *Continental drift, the evolution of a concept*. 239 pp. Washington D.C.: Smithsonian Institution Press.

Maxwell, A. E., von Herzen, R. P., Hsü, K. J., Andrews, J. E., Saito, T., Percival, S. F. Jr., Milow, E. D. & Boyce, R. E. 1970. Deep sea drilling in the south Atlantic. *Science*, New York **168**: 1047–1059.

Menard, H. W. 1969. Growth of drifting volcanoes. *Journal of Geophysical Research* **74**: 4827–4837.

Molnar, P. & Atwater, T. 1973. On the relative motion of 'hot spots'. *Nature*, London **246**: 288–291.

Morgan, W. J. 1972. Deep mantle convection plumes and plate motions. *Bulletin of the American Association of Petroleum Geologists* **56**: 203–213.

Oxburgh, E. R. 1974. The plain man's guide to plate tectonics. *Proceedings of the Geologists' Association* **85**: 299–358.

Pitman, W. C. III 1978. Relationship between eustasy and stratigraphic sequences of passive margins. *Bulletin of the Geological Society of America* **89**: 1389–1403.

Turekian, K. 1976. *Oceans*. 2nd edition. 150 pp. Prentice Hall.

Sclater, J. G., Anderson, R. N. & Bell, M. L. 1971. The elevation of ridges and the evolution of the central eastern Pacific. *Journal of Geophysical Research* **76**: 7888–7915.

Vine, F. J. & Matthews, D. H. 1963. Magnetic anomalies over oceanic ridges. *Nature*, London **199**: 947–949.

Wilson, J. T. 1965. A new class of faults and their bearing on continental drift. *Nature*, London **207**: 343–347.

A. L. Graham
Department of Mineralogy
British Museum (Natural History)

CHAPTER 12

Constant dimensions or an expanding Earth?

H. G. Owen

In science, as in everyday life, an hypothesis is only as good as the observations upon which it is based. It is reasonable to assume that if you turn to the next page of this book, you will find it printed with the continuation of this chapter. Before you turn the page, your assumption is a reasonable working hypothesis, because you can test it by turning the page. Suppose, however, when you turn to the next page you find it blank or printed with a continuation of another chapter and not this one, you have no alternative but to abandon your earlier hypothesis and replace it with another which fits more readily the facts. You will then return the book to the shop to be replaced.

The same philosophy applies to the study of continental displacement. As new data have become available, so former hypotheses have had to be modified or abandoned and replaced with others which accord with the field data. Unfortunately, one cannot take the Earth back to the Creator to be replaced. The history of the controversy between scientists who were essentially uniformitarian in outlook and those who recognised that there was evidence for major displacement of the continents relative to each other (or continental 'drift') has been outlined in Chapters 10 and 11 and in detail by Harland (1969) and Marvin (1974). Continental displacement supported by ocean-floor spreading data and palaeomagnetic observations is now an established concept even without a strict physical explanation of how it has occurred. However, uniformitarian principles are still very much in evidence in a controversy which has run concurrently with that of continental displacement and continues unabated today; this concerns the possibility of variations in the Earth's radius during its history.

In the middle and latter years of the nineteenth century, it was thought by some (such as Dana, 1846) that the Earth had contracted as a result of cooling of the original hot mass and/or the separation from it of the mass of the Moon, and that the present continents and oceans were features fixed from the earliest times. In support of this, they explained the Earth's orogenic mountain belts as having been caused by the shrinkage in area of the crust over its contracting core and mantle. The American, Richard Owen in 1857 (see Marvin, 1974 pp. 40–42) introduced the concept of convulsive changes in the Earth, culminating in a final expansion of it from a tetrahedral shape to a sphere which involved major displacement of continental regions and the ejection of the mass of the Moon.

As more geological data were accumulated and the nature of the orogenic belts, isostatic movements and the tensional nature of much of the epeirogenic movements were discovered and established, both the contraction hypothesis and the convulsive expansion idea had to be abandoned. They were largely replaced by the concept of an essentially molten Earth, with or without a 'solid' core, the outer layers of which was subject to convection cells with currents rising beneath mid-ocean ridges and descending under the continents (Fisher, 1889, 1891).

The contraction hypothesis is not entirely without supporters, even in the latter half of the twentieth century. Bucher (1933) has speculated on the possibility of alternate phases of global expansion and contraction due to periodic changes in heat flow within the mantle. Steiner (1967) has also suggested periodic expansion and contraction of the Earth as a response to cyclical changes in the position of the solar system relative to the centre of the Milky Way galaxy and associated changes in the value of galactic gravity. This is calculated against a background of a decreasing value of Universal gravity. Certain palaeomagnetic studies have led some to suggest that there has been a slight contraction of the Earth during the last 400 million years (McElhinny *et al.* 1978, p. 317). However, all of these hypotheses of a contracting Earth are based essentially on speculation or upon assumptions of questionable physical theory; they are elegant models, but the direct evidence by which the concept may be tested is lacking.

Much the same could be, and indeed was, said of the concept of continental displacement proposed by Taylor (1910) and Wegener (1912). The geological and palaeontological data upon which Wegener based his hypothesis of a Carboniferous supercontinent Pangaea, which subsequently broke up with the development of the intervening ocean basins, although providing strong *prima facie* evidence that the continents had once been joined together, could not prove it in the absence of age and structural data from the crust of the oceans. The oceanic crust occupies some 65 per cent of the Earth's present surface area and if to this is added the continental margins, the geology of 70 per cent of the Earth's total surface area was virtually unknown in 1912. The hypothesis of continental displacement was largely rejected on the grounds that it contravened elements of geophysical theory and that there was no direct geological evidence to disprove the concept that the ocean basins were permanent features of great antiquity.

During the intervening years up to the late 1960s, palaeomagnetic vector determinations from the continents, together with further stratigraphic and tectonic studies of matching continental margins, provided strong evidence that Wegener's supercontinent Pangaea had once existed and commenced to break up from Triassic time onward. By 1965, continental displacement was largely accepted as fact. Concurrent with these studies were others that speculated upon the possible expansion of an essentially fluid body, albeit very viscous, such as the Earth in response to a functional reduction in the value of Universal gravity (G). Other workers examined the possibility that the Earth's continental crust had once formed a complete outer shell of light sialic rocks, the material of which, by internal differentiation and convection, had migrated to the surface of an Earth of much smaller diameter than that of today. Yet others examined tectonic and palaeogeographical evidence which they considered accorded more readily with an expanding Earth. This earlier work on global expansion is well reviewed by Carey (1975, 1976).

Carey as early as 1958 demonstrated that if the continents were reassembled into the Pangaea configuration on a globe representing the Earth's modern dimension, the fit was reasonably precise at the centre of the reassembly, the common margins of north-west Africa and the United States east coast embayment, but became progressively imperfect away from this centre. He realised that the fit of the reassembly could be made much more precise if the diameter of the Earth was assumed to have been smaller at the time of Pangaea. Unfortunately, Carey was fighting for the recognition of continental displacement itself at a time when the concept was still considered by many to be nothing short of ludicrous, let alone attempting to establish the concept of global expansion. The same problem militated against the acceptance by geologists of global expansion as had acceptance of the continental displacement hypothesis, namely that little geological information was available to indicate the age and structure of the Earth's oceanic crustal regions.

Magnetometer traverses of oceanic crust had shown a curious alternation of magnetically north and south oriented 'stripes' or anomaly lineations (see Chapter 11). Vine & Matthews (1963) recognised that the sequence of magnetic stripes was symmetrical each side of mid-ocean ridges and realised that each corresponding pair of these stripes represented a series of dykes generated simultaneously from the mid-ocean ridge and which had in them imprinted the orientation of the Earth's magnetic field at the time of their intrusion. The sequence of magnetic stripes, therefore, becomes older away from the generating ridge. Here for the first time was direct field evidence which would prove or disprove the continental displacement hypothesis and the concept of ocean-floor spreading proposed originally by Taylor (1910) and rediscovered by Hess (1962). With the identification and dating of the lineations world-wide, the hitherto major problem of determining the age and nature of the Earth's oceanic crust became capable of solution.

Since 1968 when the Journal of Geophysical Research published a suite of classic papers on ocean-floor spreading and plate tectonics, many data have been accumulated by detailed mapping of the magnetic patterns of the various oceanic regions. Initially, this work was only sufficiently detailed to determine oceanic crustal growth from the late Cretaceous through the Cenozoic to the present day. This early work led Le Pichon (1968), one of the founders of the 'theory' of plate tectonics, to conclude that no appreciable expansion of the Earth had occurred since the Miocene. This, together with the current concept of rigid oceanic crustal plate formation and the inevitable objections of theoretical geophysicists has led most workers on continental displacement to discount the global expansion hypothesis altogether. However, if the oceanic anomaly patterns and crustal generation are examined in detail upon a globe, certain major problems of areal fit become apparent as one progressively removes oceanic crust of a given age in a series of reconstructions of past continental positions back towards Pangaea. The uncritical dumping of surplus oceanic crust down marginal subduction zones or spurious stretching of continental margins *back* in time, as used by some workers, unfortunately will not solve the problems.

The Earth is spherical in shape and not a flat map

This is an observation that is obvious, but one that is too often forgotten when reconstructions of continental displacement are made. Although the Earth is not a perfect sphere, it is sufficiently close to one to make it possible to apply spherical geometric formulae and constants to its form. Any sphere of given radius has a finite surface area. All great circles of this sphere (i.e. those, like the equator, whose plane includes the centre of the sphere) have an equal finite length. Lesser circles of this sphere in which the planes do not pass through the centre (e.g. a high latitude say 60° north or south) are of strictly proportional length to a great circle. So far as the Earth is concerned, if we know the area and shape of the continents at any given time during the last 180 million years, we can apply a test of sphericity and global dimensions to the reconstruction of continental displacement by plotting the ocean-floor spreading pattern and area generated up to that selected point in time, including any subduction of ocean-floor that might have occurred subsequently. In practice, these tests are best carried out on suitable globes.

The Earth, like any sphere, does not lend itself to an undistorted representation of its surface area upon a flat map or page of a book. It is necessary to project the image of the surface features from a point of reference either inside the Earth, usually the centre, or from an external point, according to a selected mathematical formula (Steers, 1962). In any projection, if one retains the modern co-ordinate positions of the intersection of circles of latitude and longitude and moves the position of a continent relative to that graticule, it will change its apparent shape on the map. This is because of the mathematical properties inherent in the projection concerned and is well seen in the reconstructions given here using Winkel's 'Tripel' projection. Most detailed maps of ocean-floor spreading employ Mercator's projection. The fit obtained, for example, between South America and Africa at their common continental margins, made merely by rotating them together without correspondingly changing their shape (a feature commonly seen in illustrations of continental displacement e.g. Rabinowitz & La Brecque, 1979), appears to be very close, but this is a geometrical artefact. Thus continental areas are being illegitimately stretched, distorting their true shape upon the globe. These two continents do not fit perfectly together at their common continental margins on a globe representing modern dimensions (Figs. 12.1B; 12.3). Before considering the nature and age of the crust in the narrow triangular gap extending south from between Angola and northern Argentina, it is pertinent to examine the geometry of the fit between Africa and South America as a function of changes in the value of surface curvature.

In Figure 12.1, the fit of South America and Africa together is tested on three surfaces: (A) is on a flat surface, and marked V-shaped gaps are seen to widen westward and southward from the Benue Gulf–Angola margin region; (B) is on a curved surface subtended by the Earth's current radius, and it can be seen that the V-shaped gaps have narrowed; (C) is on a curved surface subtended by an Earth with a radius 80 per cent of its current mean value, and there are no intervening gaps. It can be seen from this diagram that the accuracy of the fit is dependent on the value of surface curvature as Carey has pointed out. But what is the nature and age of the crust represented by the gap between South America and south west Africa shown by the constant dimensions globe reconstruction (B)?

Off the continental margins, oceanic crust of lower Cretaceous age (about 135 million years old) is present, becoming younger away from the margins. Yet the reconstruction requires oceanic crust older than 180–200 million years to be present; where has this crust gone? These two continental margins, like most 'passive' margins, show rift faulting and crustal thinning; structures that indicate initial extension of the margins before they were pulled away from each other. No marginal subduction zones have been detected down which the older oceanic crust geometrically required to be present could have been thrust before the generation of the crust actually present now. Extensional margins such as these are termed 'passive'

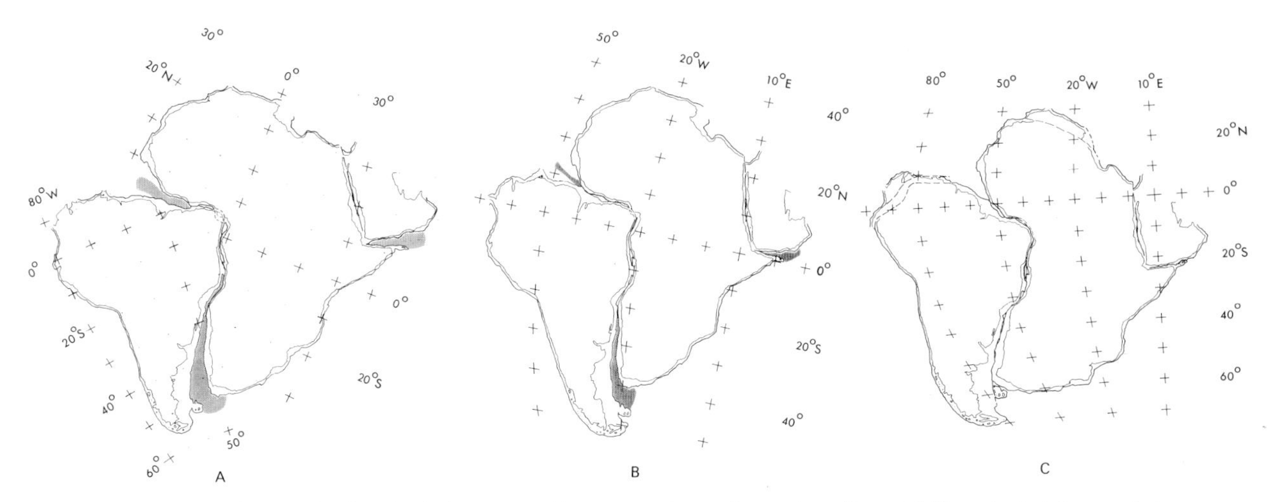

Figure 12.1 A concise exercise in spherical cartographic geometry: (A) body-centred equidistant projections of South America and Africa with modern co-ordinates fitted together as in Pangaea on a flat surface; note the V-shaped gaps, which are shaded, subtended from the common margin of Brazil and Nigeria: (B) using Winkel's projection, the two continents are fitted onto a globe of modern dimensions (co-ordinates from figure 12.3); note that the gap between South America and southern Africa, which is shaded, is of the same order as that of the Hauterivian reconstructions (figures 12.5 & 12.6): (C) also using Winkel's projection, but the two continents are fitted together on a globe 80 per cent of the Earth's modern diameter in a manner which accords with the field data (co-ordinates from figure 12.4).

and provide a relatively complete history of the break-up and separation of continents. Thus if South America and Africa fitted together perfectly without any intervening oceanic crust at the time of Pangaea, which is apparently the case, there has been a decrease in the value of surface curvature in this region since the late Triassic. This decrease is consistent with an increase in the Earth's diameter from 80 per cent of its current mean value to its modern value. If one closed the gap between South America and southern Africa on a modern diameter globe in our Pangaea reconstruction, it would merely transfer the spherical triangular gap to the common Guinea Coast–north Brazilian margins with exactly the same problem of getting rid of old pre-late lower Cretaceous oceanic crust which is not present now.

Reduction in the diameter of the globe solves similar problems in other 'passive' margined areas of poor fit which occur more extensively the further away from the centre of the Pangaea reconstruction one goes. Another example of this is the fit between Greenland and Europe on the one hand and Greenland and the Baffin and Ellesmere Islands continental margins on the other, all of which are passive. To accord with the ocean-floor spreading history of the Norwegian–Greenland Sea, Greenland and Europe have to be fitted together, leaving a gap between Greenland and Baffin and Ellesmere Islands when the reconstruction is made on a modern diameter globe. However, the Labrador Sea, Davis Strait and Baffin Bay spreading region is of late Cretaceous until early Tertiary age without any preceding subduction of pre-Pangaea crust. The Nares Strait separating northern Greenland from Ellesmere Island is a tear fault trace and has apparently never been the site of oceanic crust. Many other examples can be cited in passive-margined areas and are shown in Figure 12.3 as anomalous crust required to be present by the reassembly of the continents on a modern dimensions globe, but which is not present today. Those authorities who consider that these gaps can be explained away by deep-foundered continental crust overlook the problems, both geological and geometrical, implicit in such solutions.

The above situation refers to problems of fit prior to the break-up of Pangaea, which was not a synchronous event, but extended from the early Jurassic into the Tertiary according to a definite pattern. In any test of expansion, it is necessary therefore to take into account also the area and pattern of oceanic crustal generation during this long period of time and indeed up to the present day. The advocates of the constant dimensions Earth hypothesis rely heavily upon subduction zones to dispose of the vast areas of oceanic floor in the Pacific region, which geometrically has to be a feature of their reconstructions of continental displacement. But as Steiner (1977) has indicated, in all oceans the area of oceanic crust generated from the early upper Jurassic onward has increased exponentially, and in the Pacific this is shown as a series of concentric isochrons the western margin of which is truncated at the subduction trenches (Fig. 12.9). It is just as important to the expanding Earth hypothesis that the eastern half of the Pacific ocean-floor spreading pattern has been largely subducted by the relatively westward displacement of the Americas, caused by the development of the Atlantic Ocean, as it is to the constant dimensions hypothesis. However, the expanding Earth model described by the writer (Owen, 1976) takes into account the exponential increase in the generation of oceanic crust from the upper Jurassic and certain anomalies in the distribution of marginal Pacific subduction zones that are ignored by most advocates of a constant dimensions Earth. Although the subduction zones needed to support the constant dimensions argument are present in the western and eastern margins of the Pacific to allow for the subduction of crust in those directions, that on the northern margin is insufficient for the required north–south foreshortening of the Pacific required by their model. The subduction trenches present at the Pacific margins terminate at about 40° south latitude. Further south there is nothing to compensate for the insertion of a lune-shaped area of oceanic crust generated in its most southern region from the late Cretaceous to the present day, extending from the Carnegie ridge triple junction southward to separate Australia from Antarctica and then north-westward across the older pattern in the Indian Ocean to its other apex in the Gulf of Aden. This can best be explained by an expanding Earth model.

A wealth of new geological data from the continents and oceans has been accumulated since the writer presented the ocean-floor spreading evidence for global expansion in 1976 and this is incorporated in the outline maps used here (Figs. 12.3 to 12.9). This new information serves only to confirm that expansion of the Earth has occurred with an increase in its diameter from 80 per cent in the late Triassic-early Jurassic some 180–200 million years ago to its present dimensions, whatever the physical reasons and possible changes in core state that might be implied. There appear to be four options:

1 that the dating, both 'absolute' and relative of the earliest oceanic crust separating any two areas of continental crust is consistently incorrect. This would affect the geometry indicated by the manner of break-up of Pangaea in oceans bounded by passive margins. However, the dating of these initial oceanic crustal areas appears to be correct enough.
2 that there is a mechanism other than subduction for the removal of oceanic crust at *passive* margins which has not yet been detected. Geophysical studies of passive continental margins show them to be extensional in character, with a relatively complete history of their separation, and there is no evidence of re-assimilation of oceanic crust.
3 that we have incorrectly deduced the ages of the identified magnetic anomaly lineations and mis-read the patterns *world-wide*, which does not appear to be the case. If this were so, the whole history of continental

displacement and plate tectonics would be rendered uncertain.

4 that the Earth has increased its crustal surface area substantially with the development of oceanic regions as a result of global expansion. This is confirmed by the ocean-floor spreading patterns produced during the last 150 million years of the Earth's crustal history.

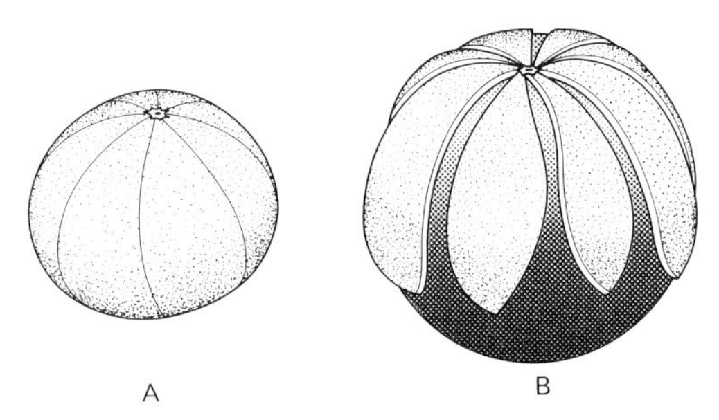

Figure 12.2 The orange skin 'law' which illustrates that the correct fitting together of a spherical shell can only occur on a core of the same diameter as that upon which the skin formed originally.

How does continental displacement work on an expanding Earth?

Continental displacement on an expanding Earth does not follow the pattern shown in Figure 12.2 which illustrates some earlier explanations of the cause of continental displacement by merely pumping up the size of the Earth and, thereby, splitting the continental skin. The double or nearly double generation of oceanic crust each side of a mid-oceanic ridge precludes such a simplistic view. The orange-peel analogue, however, does serve to illustrate a spherical geometric law. If one fits a surface shell (e.g. the peel of the orange) over a spherical core of a size larger than that of the original, it will not fit onto that core without it splitting. Although one perhaps can mould the skin to the different value of surface curvature, it will not fit completely together on the larger core. This principle can be applied to the refit of Pangaea on an Earth of modern dimensions.

It is sometimes assumed that all workers who advocate global expansion discount subduction and do not accept the concept of plate tectonics. The ocean-floor spreading patterns at the Pacific marginal trench systems are clearly truncated, and the seismic evidence strongly supports the down-thrusting of the oceanic crust at these sites. There is nothing wrong with the plate tectonic concept except the use of a constant – the modern radius of the Earth – in the determination of plate growth and motions on the surface of the globe using Euler's theorem. When the Earth's modern radius is used to determine the pole about which a plate has formed by spreading, inconsistencies appear in the form of shifts in the pole of rotation which are usually explained as being due to an interaction with an adjacent spreading region. The changes in the position of the pole of rotation can be explained just as readily, and accord more with the field data, if the radius is increased functionally with time. When this is done the pole apparently migrates, producing transform fault traces that are more strongly curved towards their continental terminations and are more open in the vicinity of the current generating axis. This is particularly evident in the central Atlantic with its long early upper Jurassic to modern ocean-floor spreading history.

The maps given in Figures 12.3 to 12.9, which employ Winkel's 'Tripel' projection, have been produced from models which assume both an Earth of constant dimensions and global expansion. Upon these have been plotted the crustal information from both the continents and oceans up to the time of each reconstruction. The maps themselves are, unfortunately, too small to show the details of the ocean-floor spreading patterns. In the three stages of continental displacement which assume an expanding Earth (Figs. 12.4, 12.6 and 12.8), the continental fits and subsequent oceanic crustal areas accord strictly with the ocean-floor spreading data now available and incorporate revisions to a previous series (Owen, 1976). In the constant dimensions reconstructions (Figs. 12.3, 12.5 and 12.7), the fit of the continents and the configuration of the subsequent oceanic crust is only accurate at certain selected points in order to produce the nearest possible approximation to the actual mapped ocean-floor spreading patterns. If the reader feels that these reconstructions have not been made impartially, he should try to solve the jig-saw puzzle himself; there is sufficient information now available to test them.

The earliest formed oceanic crust so far detected marking the initiation of the break-up of Pangaea (Fig. 12.4) is in the southern region of the North Atlantic. In the common passive marginal areas of the United States east coast embayment and north-west Africa, oceanic crust began to be generated by the Callovian Stage (late middle Jurassic) following upon a long history of rift faulting extending back into the Triassic. During the upper Jurassic, the North Atlantic spreading axis migrated northward towards the region of the Newfoundland–Gibraltar fracture zone as North America and Africa rotated away from each other. During the Kimmeridgian Stage (towards the end of the upper Jurassic) the axis had penetrated sufficiently north to split the Newfoundland continental shelf away from the Iberian Peninsula. The latter commenced to rotate anticlockwise away from the French Biscay margin in response to the relatively eastward movement of North Africa. During the late Jurassic, Antarctica and Australia as a single unit started to rotate clockwise away from India and South East Asia, the hinge of the movement being the common South African–Queen Maud Land margin.

In the early lower Cretaceous, South America began to rotate clockwise away from Africa with the inception of spreading from the initial South Atlantic generating ridge.

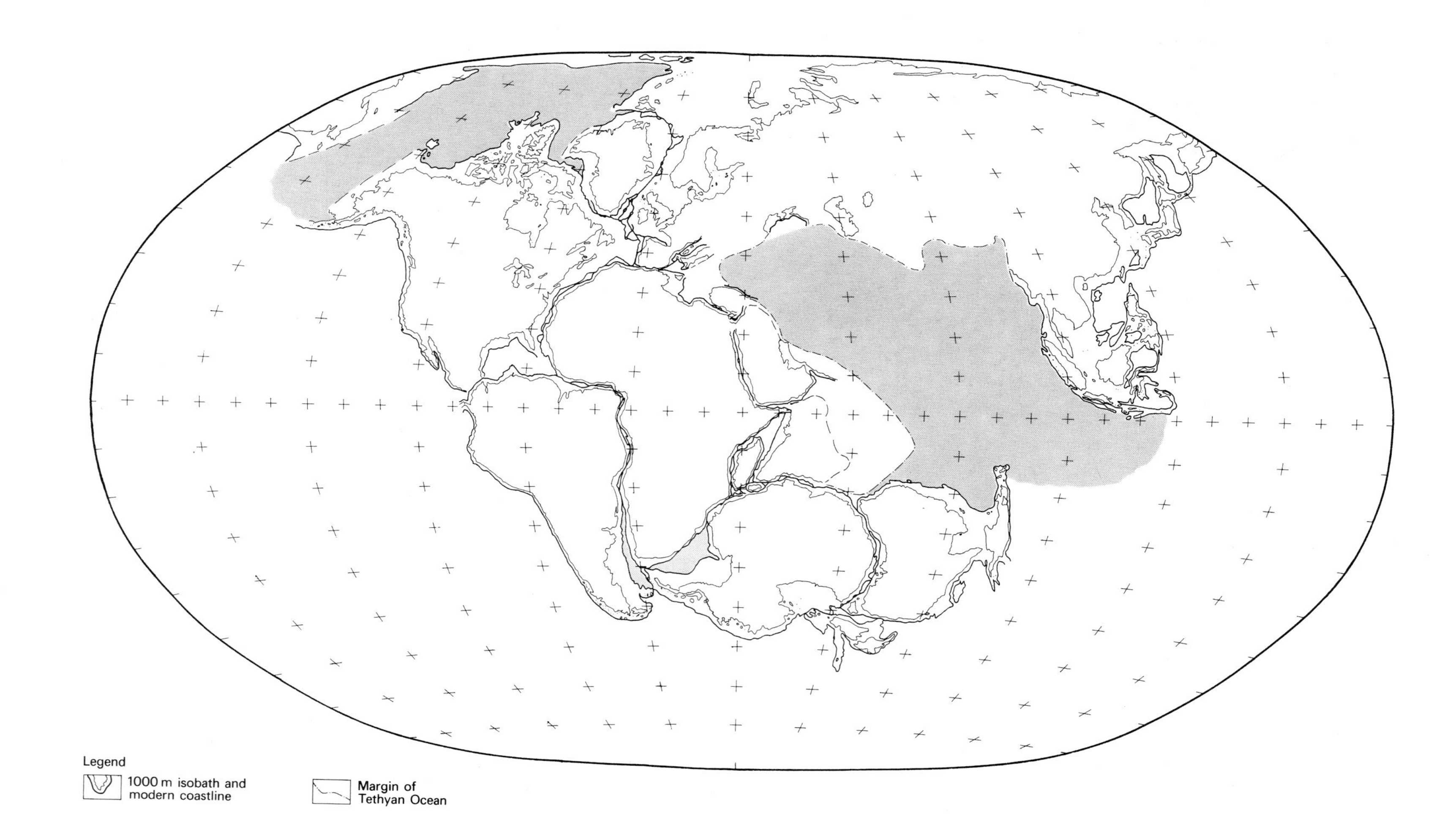

Figure 12.3 Conventional reconstruction of Pangaea (at about 180 million years ago) assuming an Earth of modern dimensions. Winkel 'Tripel' projection with centre meridian 10° E. longitude. Co-ordinates are at 30° intervals of longitude and 10° intervals of latitude. Oceanic crust required to be present in the reconstruction, but of which no evidence exists, is shown shaded except in the Pacific.

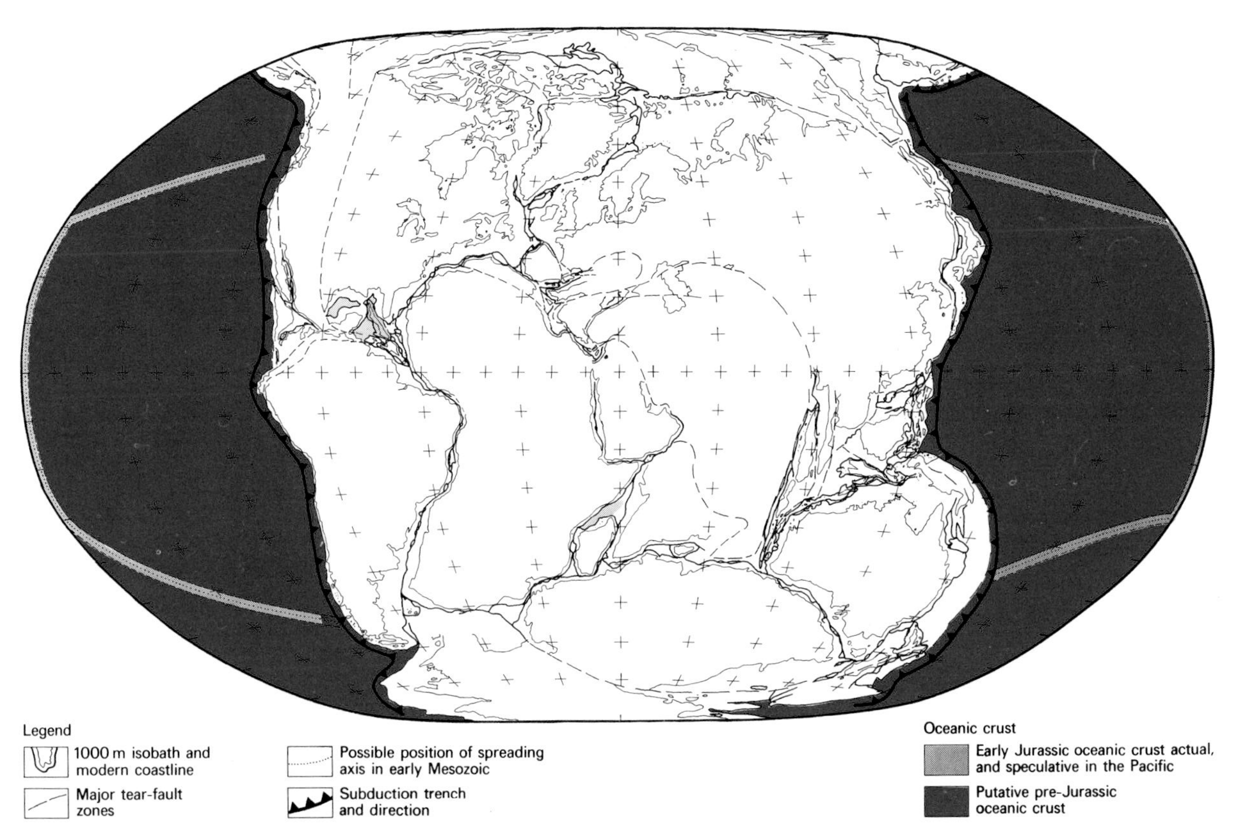

Figure 12.4 Reconstruction of Pangaea (at about 180 million years ago) assuming an Earth with a diameter 80 per cent of modern value. Winkel 'Tripel' projection with centre meridian 10° E. longitude. The fit of the continents together is in accord with the geological data from the continental margins and the subsequent ocean-floor spreading.

Spreading started at the southern end of their common continental margin with rift-faulting extending northward to between Brazil and Angola. The break-up of Pangaea had reached the configuration shown in Figure 12.6 by magnetic anomaly M7 (Hauterivian 120 million years ago). During the lower Cretaceous, the North Atlantic continued to widen, its axis splitting northward towards the British Isles. In a similar manner, the South Atlantic also widened, its axis splitting northwards to reach the Benue Gulf region of Nigeria by the Albian Stage. Slight anticlockwise rotation of Africa caused Madagascar, still joined to India, to appear to move southward with respect to the East African margin. Antarctica and Australia together continued to rotate away from Africa, India and South-east Asia, the hinge area now being at the southern end of South America and the connected West Antarctica peninsula.

During the upper Cretaceous, the North and South Atlantic spreading zones united. The North Atlantic axis penetrated between the British Isles and Rockall Bank, dividing at a triple junction to send a spreading axis to separate Greenland away from Canada by the development of the Labrador Sea. Spreading in the South Atlantic and Indian Ocean was more rapid than in the North Atlantic and the spreading pattern in the Indian Ocean changed to a west–east trending axis offset by the initial Ninety East fracture zone. The separation of Madagascar from India by the relatively northward movement of the latter was probably complete at the end of the Turonian Stage. South America and the West Antarctica peninsula were still connected. The displacement position of the continents by magnetic anomaly 24 (Palaeocene) close to the Cretaceous–Tertiary boundary is shown in Figure 12.8.

The bulk of the area of the Amerasian Basin of the Arctic Ocean was generated during the Cretaceous and early Tertiary; part of it from the spreading axis which had penetrated northward from the Labrador Sea between Canada and Greenland to form the Davis Strait and Baffin Bay. The opening of the Amerasian Basin was due to the rotation of North America against north-east Asia, produced by the developing North Atlantic. Spreading ceased in this basin and in the zone between Canada and Greenland during the Eocene. A few million years earlier (anomaly 24), the main spreading axis in the North Atlantic shifted eastwards to begin splitting Greenland away from Europe as the North Atlantic to the south continued to widen. This new axis penetrated northward to initiate the oceanic crust of the Norwegian–Greenland Sea, reaching into the Arctic to start the generation of the Eurasian Basin from the Nansen Ridge. This spreading

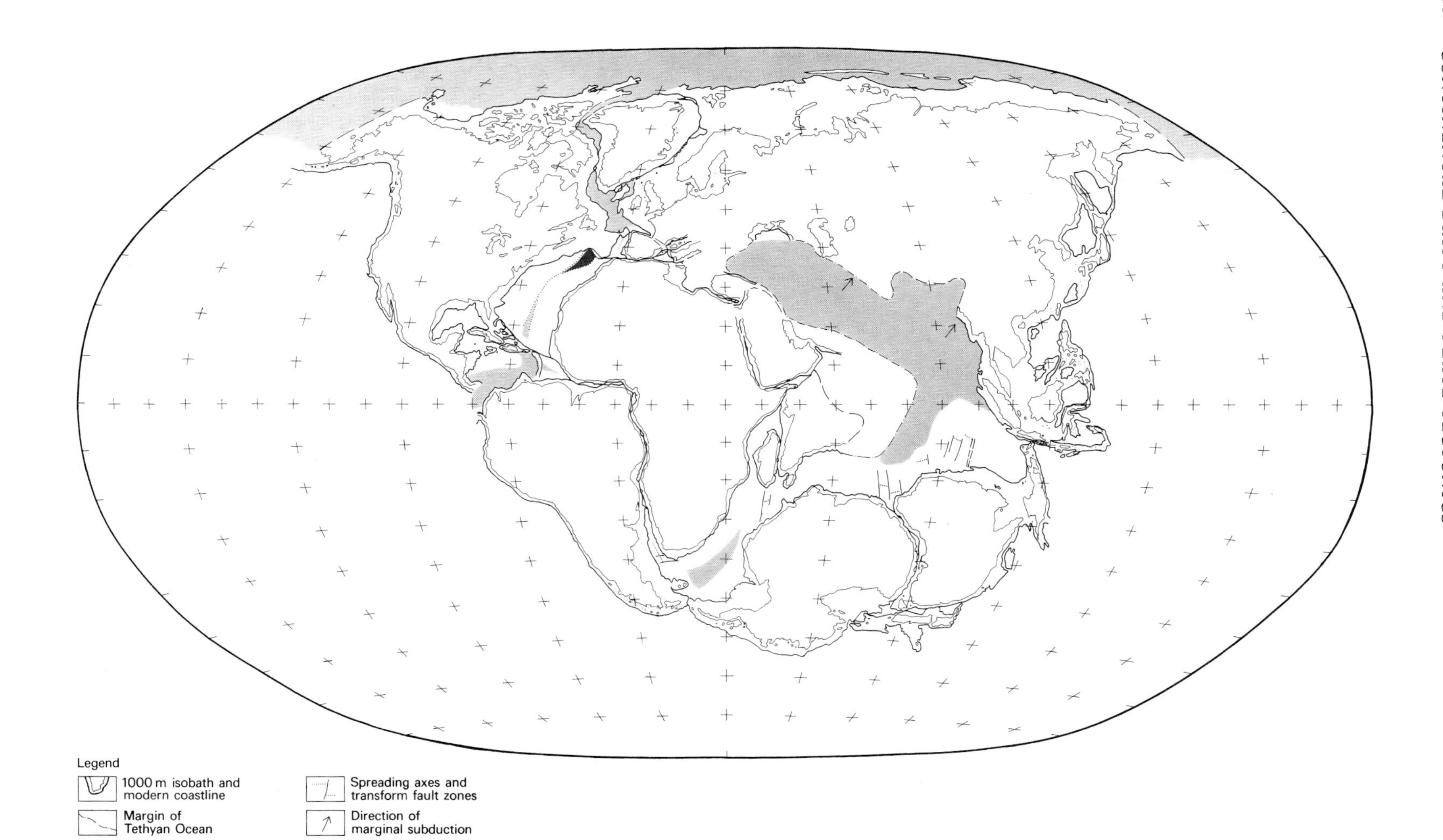

Figure 12.5 Reconstruction of continental displacement at anomaly M7 (Hauterivian at about 120 million years ago) assuming an Earth of modern dimensions. Winkel 'Tripel' projection with centre meridian 10° E. longitude. Anomalous oceanic crust required by the reconstruction, but not present today, is shown shaded except in the Pacific. Area of overlap in the north Atlantic required to provide the nearest fit in the Boreal region is shown in black.

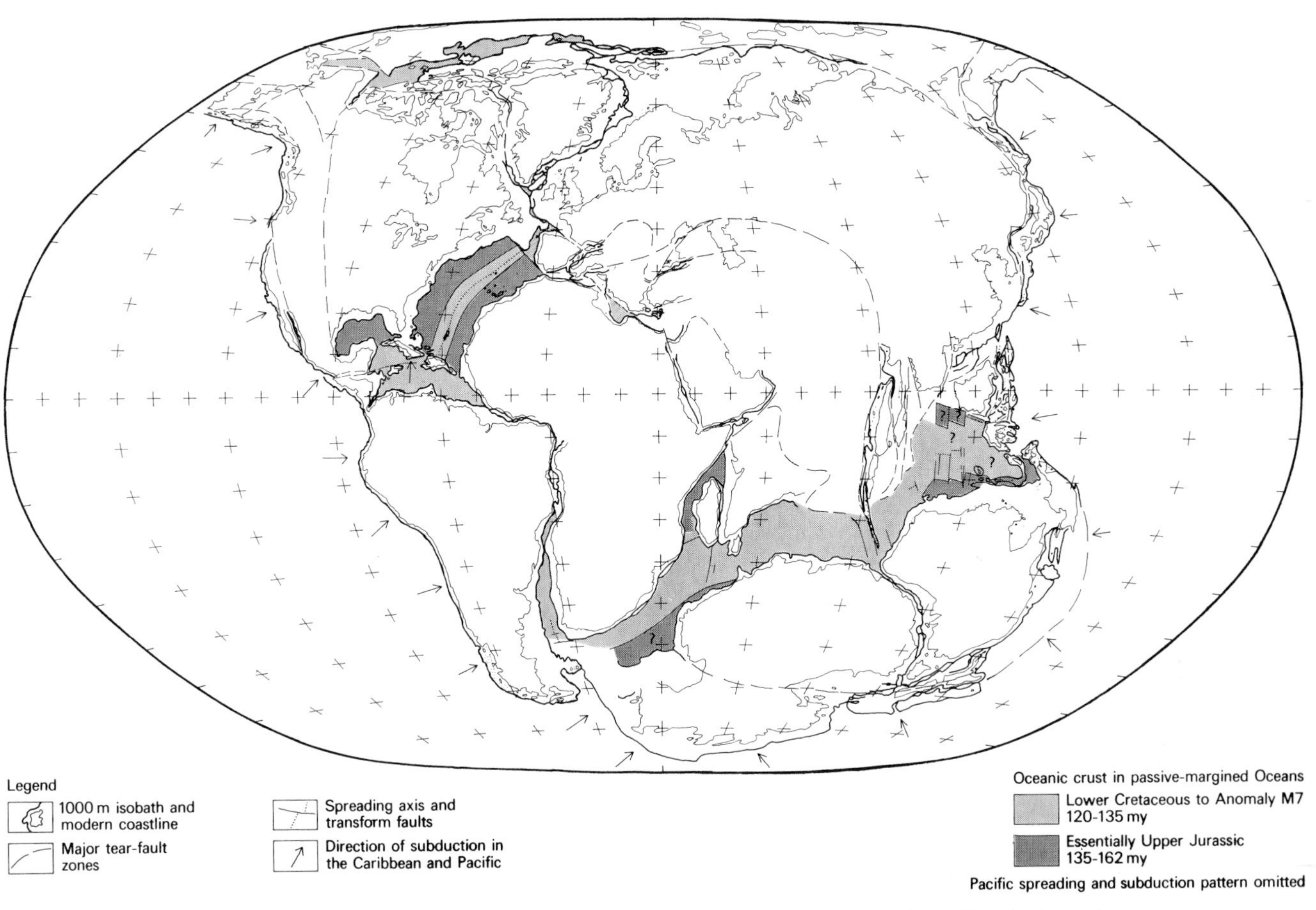

Figure 12.6 Reconstruction of continental displacement at anomaly M7 (Hauterivian at about 120 million years ago) assuming an Earth with a diameter approximately 87 per cent of modern value. Winkel 'Tripel' projection with centre meridian 10° E. longitude. No anomalous oceanic crust is present.

axis, with its 'hot spot' that has produced Iceland, is still producing new crust today. The South Atlantic, like the North Atlantic, has continued to widen, but increased generation of crust in the more southerly regions has caused South America to rotate clockwise relative to North America, producing compressional structures at the southern Caribbean margin, the Puerto Rico Trench and the Lesser Antilles volcanic arc. Such features are required by the expanding Earth model described here, but the major foreshortening of the region including the adjacent Atlantic during the late Cretaceous and Tertiary, required by the constant dimensions model, is not supported by the field evidence.

The principal Cenozoic feature in the Indian Ocean is the penetration westward of a spreading axis that had previously generated oceanic crust in the southern Pacific from within the upper Cretaceous to separate New Zealand both from Antarctica and Australia. This axis started to split Australia away from Antarctica at about the time of anomaly 24, displacing Australasia northward with the associated development of the Mentwai–Java subduction trench, the Banda Arc and the Cenozoic phase of deformation in Indonesia. Antarctica moved relatively southward, but not at the expense of subduction of the south Pacific margin which would be required by the constant dimensions model. The spreading axis split the older Indian Ocean floor across to reach the Gulf of Aden well within the Tertiary. This axis is marked by two triple junctions. One of these has sent off a limb to form the South West Indian Ridge, which links with the South Atlantic ridge and together generate crust which continues the displacement of Antarctica away from Africa and Madagascar. The other has sent off limbs, one of which has produced the Red Sea and the other the East African Rift Valley system. Displacement of India relatively northward into southern Asia had commenced well within the Cretaceous and the expanding Earth model explains the nature of the double thickness of continental crust as a reflection of this long-term continental collision. No Tethys Ocean (as opposed to a Tethyan epicontinental sea) is required in the expanding model, it being merely a geometric artefact required by the spherical geometry of the Pangaea reassembly in the constant dimensions Earth model (compare Figures 12.3 and 12.4). Spreading from

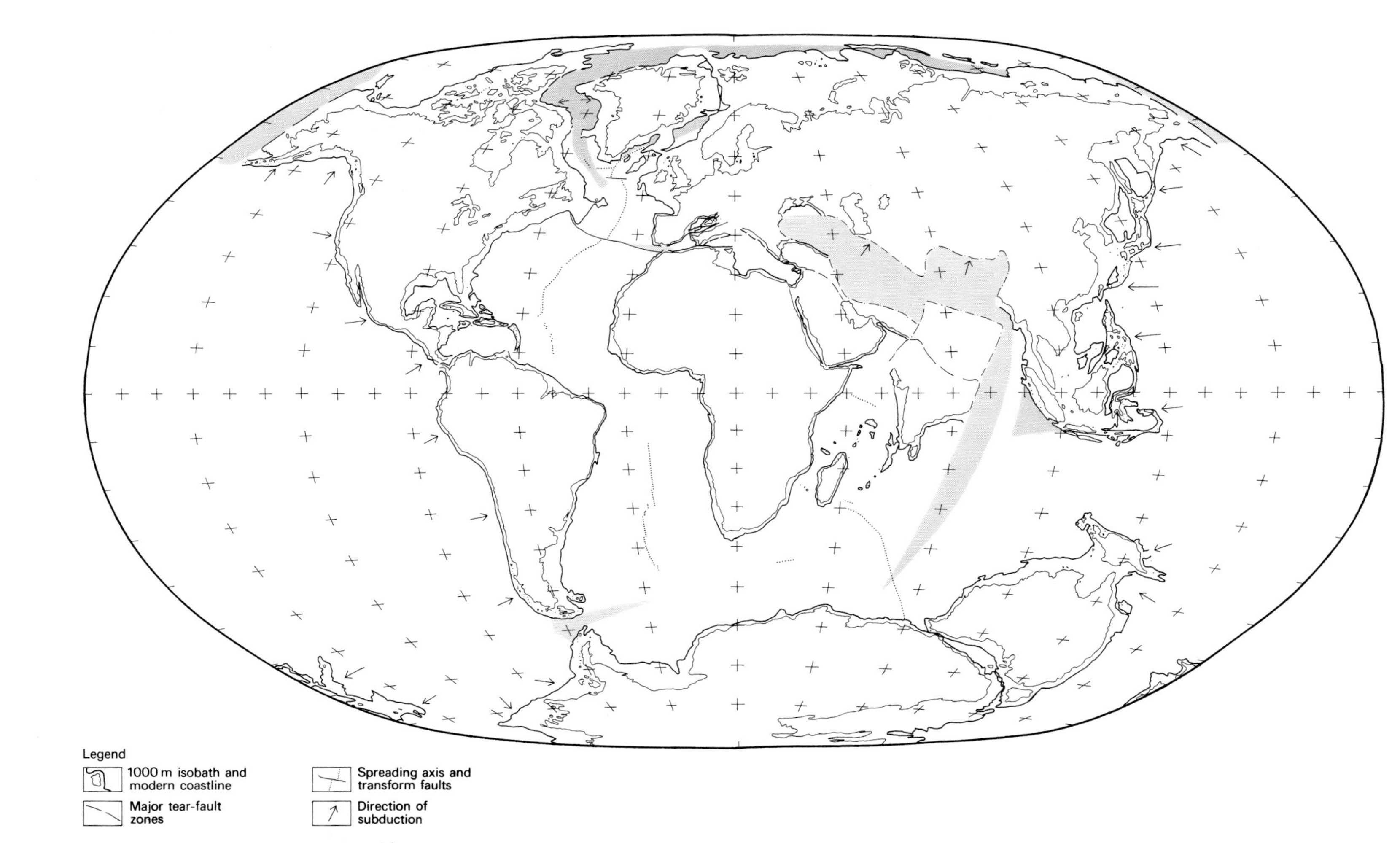

Figure 12.7 Reconstruction of continental displacement at anomaly 24 (Palaeocene at about 56 million years ago) assuming an Earth of modern dimensions. Winkel 'Tripel' projection with centre meridian 10° E. longitude. Anomalous oceanic crust required by the reconstruction, but not present today, is shown shaded except in the Pacific.

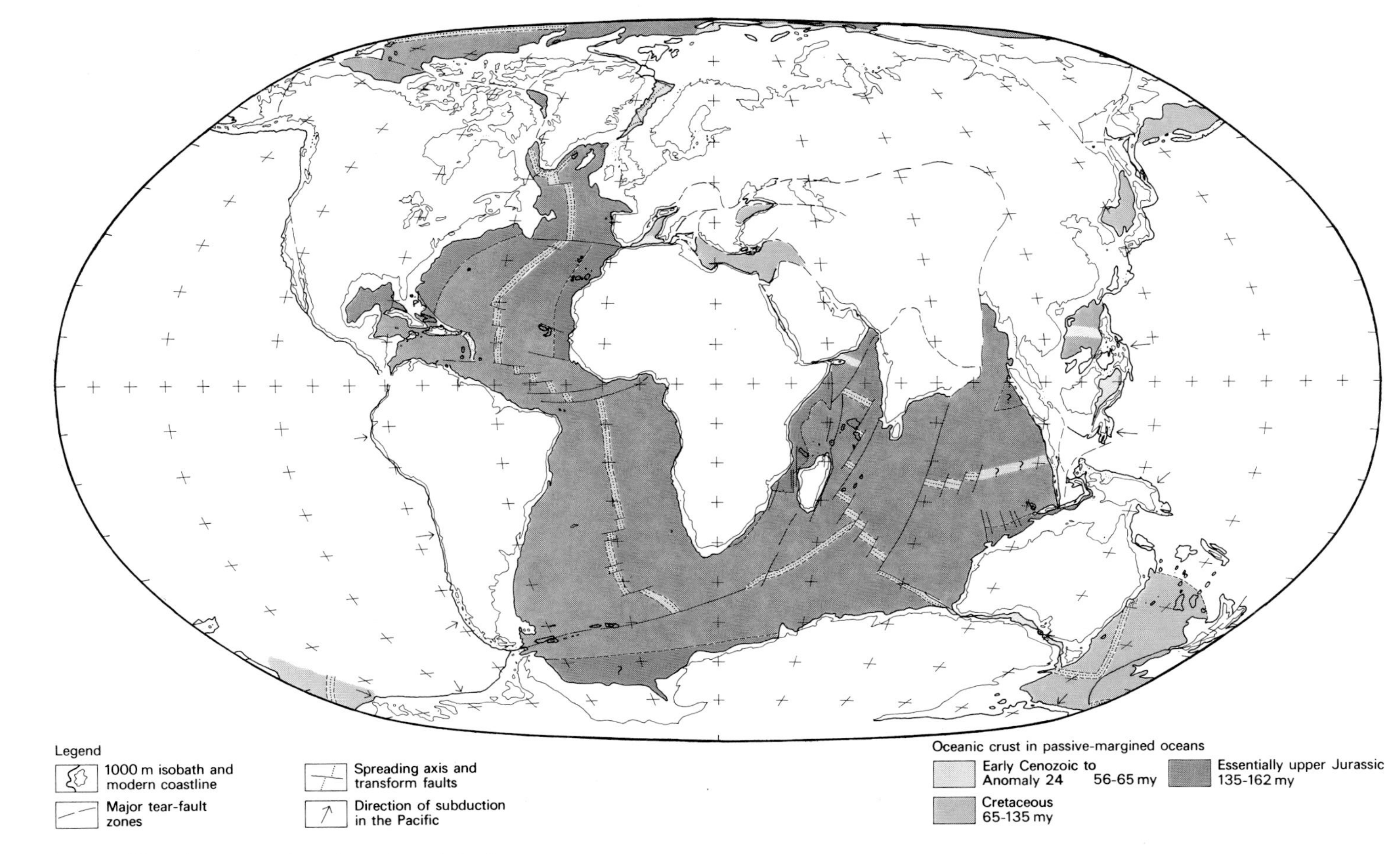

Figure 12.8 Reconstruction of continental displacement at anomaly 24 (Palaeocene at about 56 million years ago) assuming an Earth with a diameter approximately 93 per cent of modern value. Winkel 'Tripel' projection with centre meridian 10° E. longitude. No anomalous oceanic crust is present.

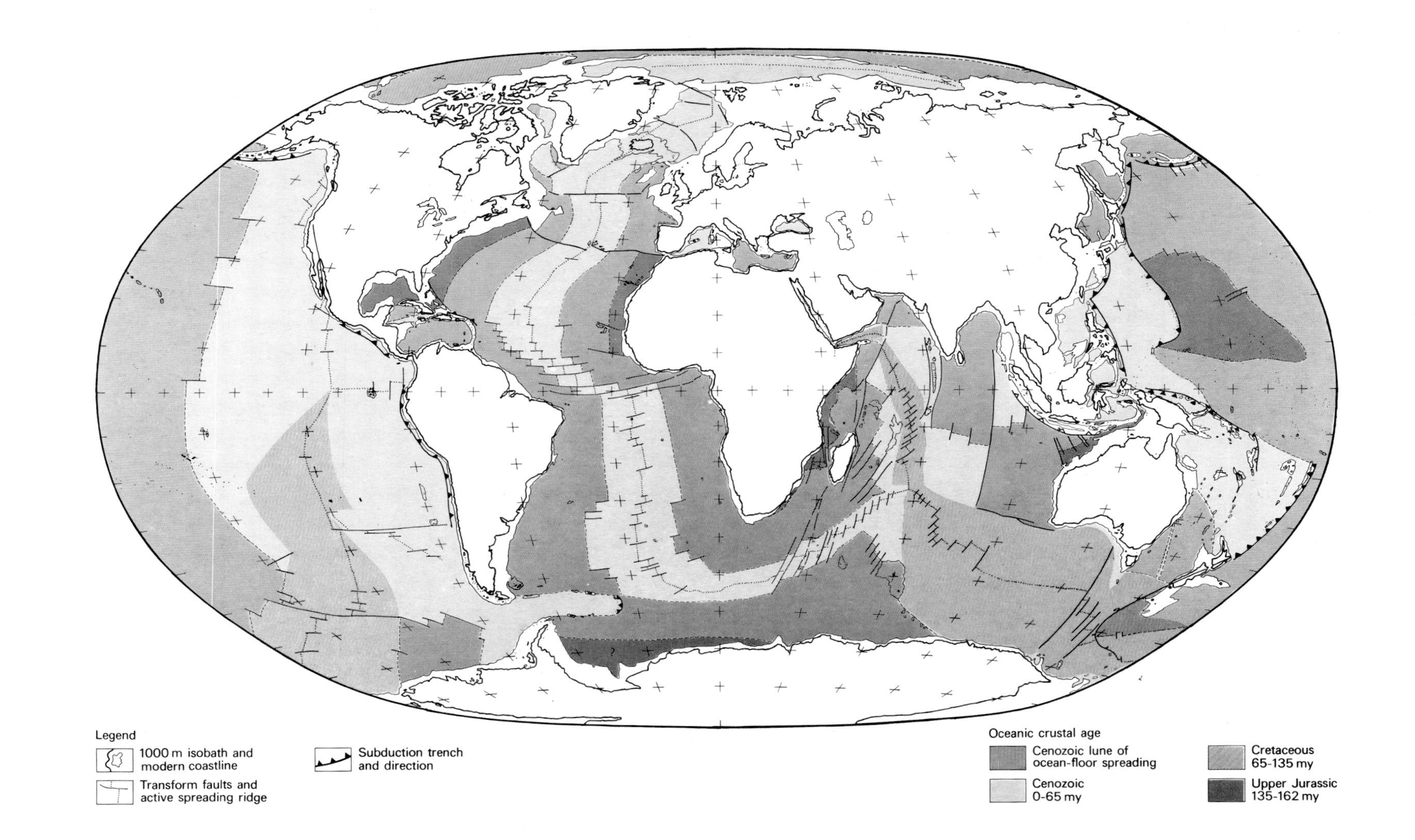

Figure 12.9 The modern Earth with a broad outline of the ocean-floor spreading patterns. Winkel 'Tripel' projection with centre meridian 10° E. longitude. Note that the areas of anomalous oceanic crust decrease in the series of text figures (numbers 12.3, 12.5, & 12.7) constructed on a modern dimensions globe. This is consistent with an expanding Earth.

the new Cenozoic ridge axis in the Indian Ocean has pushed the Indian subcontinent yet further north, rotating it anticlockwise and producing the Himalayan mountain range along the northern margin.

Associated with the north–south extension of the Indian Ocean and displacement of Antarctica, a new spreading region began in the Eocene, to form the Drake Strait, which now separates the West Antarctic Peninsula from South America. Interaction between this spreading region and that of the growing South Atlantic has produced a small subduction zone, the South Sandwich trench.

The development of the passive margined oceans, outlined above, shows the sequence in the break-up of the supercontinent Pangaea and, if the Earth has been a sphere or very nearly so during the last 200 million years, it indicates its spherical geometry during that period. But what of the history of the Pacific Ocean during this interval? Mid-ocean spreading ridges generate a symmetrical or near symmetrical pattern each side of them. It is reasonable to assume that the Pacific Ocean generating ridge behaved in the same way. Today the spreading ridge has reached the north-western margin of North America and is directed well to the eastern side of the Pacific further south. The vast bulk of the post-Pangaea oceanic crust generated to the east of the principal spreading axis in the Pacific has been subducted, together with all the oceanic crust present at the time of Pangaea. The area involved is close to that of the Earth's total crustal area today, a very large but not necessarily over-large area to be subducted at the Pacific margins. The early Mesozoic inception of the Pacific, advocated most recently by Shields (1979), is impossible on the known spherical geometric fit of Pangaea, expanding Earth or not. Superficially, if one closes the Atlantic and Indian Oceans, the Pacific spreading ridge system returns to a mid-ocean position. If the total rate of generation of oceanic crust has remained constant during the last 200 million years, the constant dimensions Earth concept would be supported. However, in the Pacific Ocean, as elsewhere, ocean-floor generation has increased its area exponentially since the mid-Jurassic at least. This, together with the pattern of the spreading region in the southern Pacific, supports an Earth with a diameter 80 per cent of its current value 180–200 million years ago expanding to its present size, and precludes the constant dimensions hypothesis.

The spreading ridge produces twice as much area than is required by a simple expansion model. As the Atlantic Ocean developed its total area *exponentially*, as also did the Pacific Ocean, subduction would have had to occur at one or both margins. It has been recognised for some time now that the older the oceanic floor, and therefore the further away it is from the generating axis and principal crustal heat source of the upwelling convection current, the lower its elevation relative to the continents. The old Pacific oceanic margins would offer less resistance to the displacing continents and so were preferentially over-ridden. Expansion of crustal area, and therefore of the Earth, is clear in the post mid–upper Cretaceous to present-day spreading pattern in the southern half of the Pacific and its Cenozoic continuation in the Indian Ocean. Together, the spreading from this common axis describes a lune, the widest part to the east and south of New Zealand and tapering northward towards the Carnegie Ridge near the Galapagos Islands on the Pacific side and the Red Sea on the Indian Ocean side. It clearly cuts through the older spreading patterns and its insertion is not compensated for in any way by significant subduction. Not only has the Earth to have expanded to accord with the Pacific spreading pattern, but the Earth's core and equator have in effect moved southward relative to the previously formed crust and this is indicated by greater expansion of oceanic crustal area in the southern hemisphere and the apparent migration of the magnetic and climatic equators northward as one goes back in time to the Jurassic.

The foregoing account has dealt with the displacement and development of crustal units. For the palaeobiogeographer this is insufficient to determine the migration and centres of development of plants and animals. Meaningful studies in this field require the superimposition of a palaeogeography upon this crustal framework that takes account of the distribution of epicontinental as well as oceanic seas, currents of hot and cold water, of possible mountain barriers, and indicators of atmospheric conditions and climate.

The real problems

McElhinny *et al.* (1978, p. 317) have dismissed expansion of the Earth on the lack of crustal evidence of expansion on the Moon, Mars and Mercury. The Moon's crust was formed by 3500 million years ago and appears to have been rigid since then. If changes in the value of Universal gravity are invoked as the cause of expansion of the Earth at least during the last 200 million years, why has the Moon not expanded as well during this period of time? There is no evidence for anything but a minute gain in the Earth's mass resulting from the accession of debris from space and the solar wind during the last 200 million years and this cannot, therefore, be the cause of the Earth's expansion. Indeed, we appear to be losing hydrogen ions into space which are liberated by the dissociation of water molecules in the upper atmosphere. The appearance of the surface of Mercury and Mars suggests that their crusts are also largely rigid, as in the case of the Moon, in contrast to the relatively mobile crusts of the Earth and Venus.

Perhaps there is another possible cause of planetary expansion in the terrestrial planets of the Solar System. If one compares the mass of the Earth, Moon, Mercury, Mars and Venus separately against their current dimensions, they fall into two distinct categories as shown in the table below.

The Earth and Venus are planets of high mass in comparison with the others listed, and both possess mobile terrestrial crusts and well-developed atmospheres. Mars,

	Equatorial diameter (km)	*% Mass (Earth 100%)*	*Mean density* ($g\,cm^{-3}$)
Earth	12756	100·0	5·54
Venus	12100	81·5	5·18
Mars	6790	10·7	*c.* 4·00
Mercury	4880	5·5	5·45
Moon	3476	1·2	3·34

Mercury and the Moon, are very light in comparison with the Earth and Venus and probably all three have relatively rigid crustal shells. A thin atmosphere only lingers on Mars. There is no direct evidence of the nature of the Earth's inner and outer cores, but it is assumed from the study of meteorites and the retardation of seismic *P* waves that the core is composed of nickel-iron and/or sulphur, the outer core being molten while the inner one is thought by some to be solid (Jacobs, 1975). But, what if the Earth's inner core is in a plasma state, consisting of matter under such extreme pressure and at a temperature in which the atomic shells are dissociated into densely packed random particles? The behaviour of the *P* waves in the core in such a state would be closely similar to that in a 'solid' iron-sulphur core at high temperature and pressure. The potential for expansion of a plasma core is considerable until such time as the complete core reaches an atomic state, as may now be the case with the molten outer core of the Earth and the inner core of the Moon. This explanation of the development of the Earth's molten outer core provides a more parsimonious reason for the development of the overlying mantle and crustal shells by convection than does the concept of a 'solid' iron-sulphur inner core. The production of new gases and water from deep sources within the Earth is also more readily explainable than by the concept of recycling currently in favour. Moreover, there are no fundamental problems in the generation of the Earth's magnetic field if the inner core is in a plasma state. The original potential for a planetary body to expand would depend not on changes in Universal gravity (*G*) but on its plasma core size.

References

Bucher, W. H. 1933. *The deformation of the Earth's crust: an inductive approach to the problems of diastrophism.* 518 pp. Princeton: Princeton University Press.

Carey, S. W. 1958. The tectonic approach to continental drift, pp. 177–355 *in* Carey, S. Warren (Ed.) *Continental drift – a symposium.* Hobart: University of Tasmania, reprinted 1959.

Carey, S. W. 1975. The expanding Earth – an essay review. *Earth Science Reviews* **11**: 105–143.

Carey, S. W. 1976. The expanding Earth. *Developments in Geotectonics* **10**: 1–488.

Dana, J. D. 1846. On the Volcanoes of the Moon. *American Journal of Science* **2**: 335–355.

Fisher, O. 1889. *Physics of the Earth's crust.* 2nd Ed. xvi+391 pp. London, New York: Macmillan.

Fisher, O. 1891. Appendix to the 2nd edition of *Physics of the Earth's Crust.* 60 pp. London, New York: Macmillan.

Harland, W. B. 1969. The origin of continents and oceans. *Geological Magazine* **106**: 100–104.

Hess, H. H. 1962. History of ocean basins, pp. 599–620 *in* Engel, A. E. J., James, H. L. & Leonard, B. F. (Eds) *Petrologic studies: a volume to honor A. F. Buddington.* New York: Geological Society of America.

Jacobs, J. A. 1975. *The Earth's core.* 253 pp. London, New York, San Francisco: Academic Press.

Le Pichon, X. 1968. Sea-floor spreading and continental drift. *Journal of Geophysical Research* **73**: 3661–3697.

McElhinny, M. W., Taylor, S. R. & Stevenson, D. J. 1978. Limits to the expansion of Earth, Moon, Mars and Mercury and to changes in the gravitational constant. *Nature*, London **271**: 316–321.

Marvin, U. B. 1974. *Continental drift. The evolution of a concept.* 239 pp. (2nd printing) Washington D.C.: Smithsonian Institution Press.

Owen, H. G. 1976. Continental displacement and expansion of the Earth during the Mesozoic and Cenozoic. *Philosophical Transactions of the Royal Society of London* Series A **281**: 223–291.

Rabinowitz, P. D. & Labrecque, J. 1979. The Mesozoic South Atlantic Ocean and evolution of its continental margins. *Journal of Geophysical Research* **84**: 5973–6002.

Shields, O. 1979. Evidence for initial opening of the Pacific Ocean in the Jurassic. *Palaeogeography, Palaeoclimatology, Palaeoecology* **26**: 181–220.

Steers, J. A. 1962. *An introduction to the study of map projections.* 288 pp. (13th edition) University of London Press.

Steiner, J. 1967. The sequence of geological events and the dynamics of the Milky Way Galaxy. *Journal of the Geological Society of Australia* **14**: 99–132.

Steiner, J. 1977. An expanding Earth on the basis of sea-floor spreading and subduction rates. *Geology* **5**: 313–318.

Taylor, F. B. 1910. Bearing of the Tertiary mountain belts on the origin of the Earth's plan. *Bulletin of the Geological Society of America* **21**: 179–226.

Vine, F. J. & Matthews, D. H. 1963. Magnetic anomalies over oceanic ridges. *Nature*, London **199**: 947–949.

Wegener, A. 1912. Die Entstehung der Kontinente. *Geologische Rundschau* **3**: 276–292.

H. G. Owen
Department of Palaeontology
British Museum (Natural History)

PART V

Maps of the past

Introduction

Reconstructing the past geography of the world is the goal of historical geology. To this end, all the lines of evidence derived from any other field of geology may be brought into play. Knowledge of sedimentary environments links up with the petrology of the different kinds of volcanic rocks, palaeontology with plate tectonics. The resulting palaeogeographical maps afford a way of synthesizing all the information in a broad brush view of the past: they bear the same relation to the hard facts of geology as do reconstructions of whole cities to the fragments of pottery and human waste which is the workaday stuff of archaeology.

As geology has changed, so have the reconstructions. It is not so many years ago that the notion of moving continents seemed preposterous to the majority of scientists. At that time, maps of the Mesozoic would have been superimposed on a static geography – to be sure the distribution of seas over the continents would have changed repeatedly, and land bridges or barriers connected or separated faunas according to their similarities – but it was essentially a modern geography taken back in time. One can understand the reluctance of geologists to abandon an aspect of the uniformitarian principle that had placed geology on a scientific footing. To release the continents from their fixity was to abandon the science to a state of flux, and how many other uniformitarian interpretations might themselves be suspect? Now that the movement of continents is accepted, the explanatory power of the resulting different geographies is found to be considerably greater than the static view. Many facts, hitherto unconsidered or inexplicable, seem to slot naturally into the new geographies. Why did we not realise before that the slices of peculiar ophiolitic rocks might represent the site of former subduction, or that the present day juxtaposition of quite different fossil faunas might be due to the disappearance of an ocean that formerly separated them? As so often in science, what was once inconceivable later becomes almost obvious. With extraordinary agility the collective scientific consciousness adopts a new explanation, which

Figure V.1a Locality map showing the position of present-day continents in early Devonian time, about 400 million years ago. Mollweide projection (after Scotese *et al.* 1979).

soon becomes a tenet of the science. From that point on, the way of perceiving known facts operates in the new context. Any other explanation suddenly seems difficult to accept. But in a sense abandoning the fixed geography of the world *was* an irrevocable step. There is now no reason to think that the current view of past geographies should be the last one; other uniformitarian principles *are* coming under examination. Hence the debate over whether the earth has had a constant diameter, which surfaces at several points in this book. When Wegener cut the continents from their moorings, he also severed the past from the present.

Wegener argued essentially for Pangaea, or rather for its previous existence and subsequent break-up. This story is now known in considerable detail, as the chapters by Howarth and Adams show, regardless of caveats about possible earth expansion. The emphasis in this book is on this relatively recent (200 million years or less) segment of past geography, for which the geophysical evidence is now very good. But once the continents are allowed to split and re-aggregate, there is no reason why the process should have begun with Pangaea. By an almost ironic twist of uniformitarianism continental movement can now be conceived as a continuous process, accelerating at some times, quiescent at others. Pangaea itself accreted in the Palaeozoic by the collision of continental fragments, and these fragments may themselves have been conjoined in the Precambrian. The rather neat jigsaw puzzle of the continents today has been cut through in different patterns. Sometimes this occurred along sutures that have been re-awakened in later times, in other cases the continents have been chopped up in different and unexpected ways.

The legacy of past continental collisions is found in the *linear mountain belts*, chains of folded and metamorphosed rocks now often eroded deep into their roots. The Urals chain, for example, snaking across the middle of Asia, is the evidence for a previous ocean that separated the two halves of Asia early in the Palaeozoic. When Pangaea broke up, the two halves of Asia remained fused. Much of the evidence for these early oceans is ambiguous, and, as in so many geological problems, the further one goes back in time, the greater become the areas of uncertainty. In the Precambrian there are numerous folded areas, some of them more or less linear, but it is far from certain how many of these may be attributed to the effects of subduction, and how many to the effects of crustal remobilisation in areas of high heat flow.

The existence of some of the lower Palaeozoic oceans is now quite well established; for example, it is now certain that the Atlantic was open during the Ordovician almost, but not quite, along the same suture in operation today, before closing again to form the late Palaeozoic supercontinent (Wilson, 1966, was the first to suggest this, which shows how recent some of these ideas are). Geophysical evidence is now being brought more to bear on the Palaeozoic, and there have now been several attempts to portray the distribution of all the continental crust for this era (Smith, Briden & Drewry, 1973; Scotese *et al.*, 1979). Many of the results are in dispute; a wide ocean on one map is hardly evident on another. But gradually a consensus of opinion about large pieces of the earth's

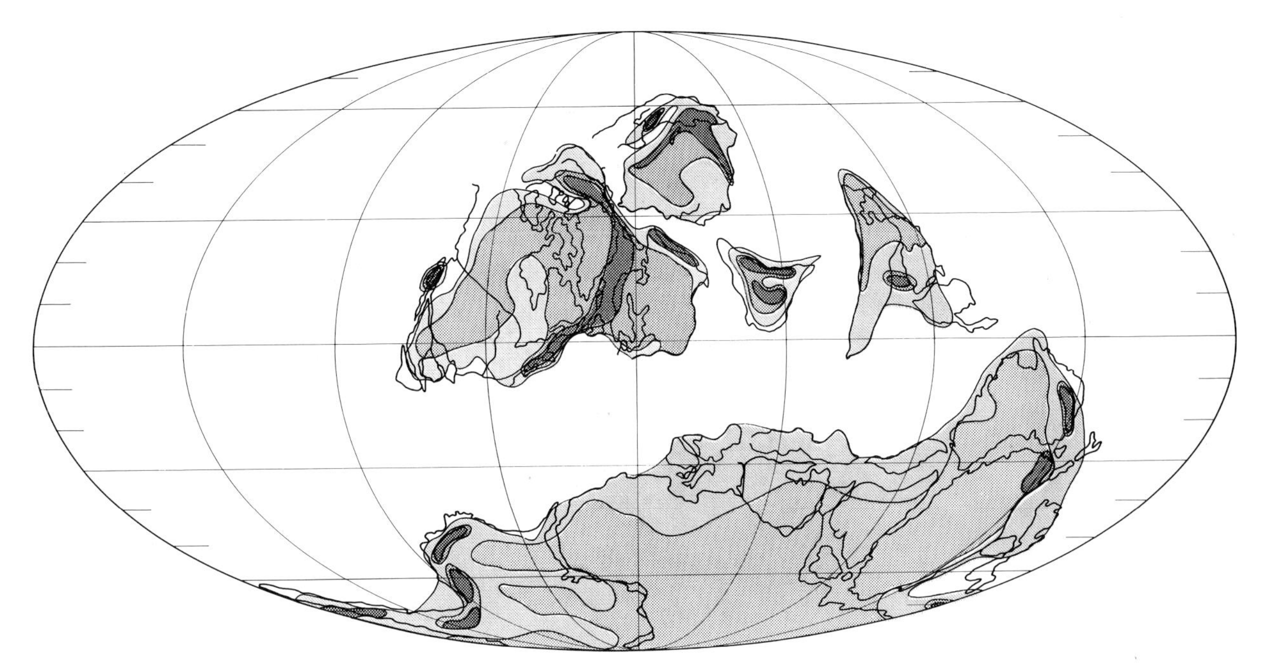

Figure V.1b Palaeogeography of early Devonian time, showing deep oceans, shallow seas, low land, and mountains (after Scotese *et al.* 1979).

crust as far back as the Cambrian is coming into being: for example North America lay more or less on the equator in the Cambrian-Ordovician, and there was an Ordovician pole in North Africa, from which splendid glacial deposits are now known. Figure V.1 shows, as an example, one recent reconstruction for Devonian Time.

Data from fossil animals and plants have always been intimately involved in the construction of past maps. Similarities in terrestrial fossils from India, Africa, South America and Antarctica were used as evidence for their former contiguity from the early days of the theory of Continental Drift. Once fossilised, organic remains are carried as passive passengers through all the subsequent vicissitudes of continental breakup and reconstitution. Some workers have presumed that while plates were separated the organisms on each plate have undergone a separate evolutionary history, the classic example being the Australian marsupials. This is certainly an appropriate working hypothesis for terrestrial life, but with marine organisms the story may be a lot more complicated. An important restriction on the distribution of marine animals is climate, and within the same climatic belt today marine faunas may be relatively uniform *across* oceans. Even so, marine faunal similarities and dissimilarities have been used to reconstruct past continental distributions in periods as far back as the Ordovician (Whittington & Hughes, 1972). No doubt such methods can be used when continental separation is accompanied by latitudinal displacement. The motion of plates, changes in world climate, and changes in position of continents relative to climatic belts will all have influenced the way animals and plants evolved but it is easy to see that the permutations and combinations of these various effects are endlessly complex. There have been bold attempts to write large-scale historical scenarios accounting for the interactions between organic evolution and that of the globe itself (Valentine, 1973). Such attempts are sometimes criticised as "storytelling" – and there is always much contentious material in such panoramic treatments – but the story is, after all, an interesting one, and often stimulating to scientific opinion, even if wrong. Safer ground is to be found in the detailed relations between continental and oceanic distributions based on particular types of fossil organisms, like those portrayed in the following maps. Where the detailed position of plates is known from independent evidence, as in the Mesozoic and Tertiary, then such maps are genuine summaries of what is known.

The Pleistocene is different, for by then the continents have their modern configuration, and the main concern is with climate alone. Furthermore its time scale is in tens of thousands of years, or even less, whereas the rest of the geological column is scaled in a million or more years. Maps of the Pleistocene are concerned primarily with movement: movement of ice, corresponding migration of climatic zones, movement of animals and plants in response to climatic changes. But the maps also serve as summaries of knowledge, in the same fashion as their "older" counterparts. And the Pleistocene has the inestimable advantage that many of the fossils are of species still living, so that cold, temperate and warm floras and faunas can be recognised with greater assurance, the sort of data that have to be inferred indirectly for older fossil faunas. The complexity of the Pleistocene story is enough to strike a chill into the heart of any investigator trying to reconstruct maps for earlier periods. The refinement of dating there is just not fine enough to pick up the kind of changes that have happened over the last 1 000 000 years. There is a tendency to assume that the recent geological past is anomalous, and that events generally happened in a more sedate way in the Jurassic or the Ordovician. However, that impression may partly be because we have not (yet) developed techniques of dating fine enough to see the equivalent fine scale changes in older rocks.

Maps of the past are portraits of the Earth's development. The portrait becomes sketchier the further back in time one goes, and on occasion whole features have to be struck out and rearranged as new facts emerge or our conceptual framework for judging the past changes. There will be no end to the process of reconstruction, but little by little the maps will converge on the historical truth.

References

Scotese, C. R., Bambach, R. K., Barton, C., van der Voo, R. & Ziegler, A. M. 1979. Paleozoic base maps. *Journal of Geology* 87: 217–277.

Smith, A. G., Briden, J. C. & Drewry, G. E. 1973. Phanerozoic world maps. *Special Papers in Palaeontology* **12**: 1–42.

Valentine, J. W. 1973. *The evolutionary paleoecology of the marine biosphere*. New Jersey: Prentice-Hall Inc.

Whittington, H. B. & Hughes, C. P. 1972. Ordovician geography and faunal provinces deduced from trilobite distribution. *Philosophical Transactions of the Royal Society of London*, Series B **263**: 235–278.

Wilson, J. T. 1966. Did the Atlantic close and then re-open? *Nature* **211**: 676–681.

R. A. Fortey
British Museum (Natural History)

CHAPTER 13

Palaeogeography of the Mesozoic

M. K. Howarth

Continental drift has been widely accepted for more than a decade. After many years of debate and controversy, new evidence provided by palaeomagnetism and sea-floor spreading has led to the almost universal acceptance that neither the Atlantic Ocean nor the Indian Ocean existed 200 million years ago. All the present-day continents were united in a single large super-continent known as Pangaea. The northern half of Pangaea was Laurasia, consisting of the present North America, Europe and Asia, while the southern half was Gondwana, a combination of South America, Africa, India, Antarctica and Australasia. A major incursion into the otherwise single land-mass was made by the large wedge-shaped Tethys Ocean coming from the east and separating the eastern half of Laurasia to the north from the eastern half of Gondwana to the south. The whole of the surrounding ocean was the precursor of the modern Pacific.

The best reconstruction of Pangaea that can be made on an Earth of present-day diameter is used as the starting point for the five basic palaeogeographic maps in this chapter. These are for the Trias (240–200 million years ago), the Lower Jurassic Pliensbachian Stage (180 million years ago), the Upper Jurassic Kimmeridgian Stage (145 million years ago), the Lower Cretaceous Hauterivian Stage (125 million years ago), and the Upper Cretaceous Senonian Stage (90–80 million years ago). Maps of continental dispositions and land–sea distributions for geological eras earlier than the Trias are not given, because although a large body of data exists, it is still susceptible to different interpretations. At least the unity of Pangaea from about 240 until 180 million years ago is unlikely to be challenged, and the subsequent break-up of Pangaea can now be traced accurately. The palaeogeographic information has been obtained from numerous sources, of which the most important and most recent key references are listed at the end of this chapter. Stage names are frequently used in the account below, and the stages that are normally used for the Mesozoic are given in Table 1.

Table 1 The Systems and Stages of the Mesozoic

Boundary / System	Stage	Notes
—65 my*		
Upper Cretaceous	Maastrichtian	Senonian
	Campanian	Senonian
	Santonian	Senonian
	Coniacian	Senonian
	Turonian	
	Cenomanian	
—95 my		
Lower Cretaceous	Albian	
	Aptian	
	Barremian	Neocomian
	Hauterivian	Neocomian
	Valanginian	Neocomian
	Berriasian	Neocomian (Ryazanian in Boreal regions)
—135 my		
Upper Jurassic	Tithonian	(Volgian in Boreal regions; Portlandian in Europe)
	Kimmeridgian	
	Oxfordian	
—?155 my		
Middle Jurassic	Callovian	
	Bathonian	
	Bajocian	(Lower Bajocian = Aalenian in some schemes)
—?175 my		
Lower Jurassic	Toarcian	
	Pliensbachian	
	Sinemurian	
	Hettangian	
—200 my		
Upper Trias	Rhaetian	
	Norian	
	Carnian	
Middle Trias	Ladinian	
	Anisian	
Lower Trias	Scythian	
—240 my		

* Absolute ages of some of the boundaries in millions of years (my).

Palaeogeographic maps show the areas occupied by land and sea in former times. They differ fundamentally from maps of continental reconstructions, because seas were present on areas of continental crust, as well as occupying the oceanic basins. In fact the former, epicontinental seas, are geologically more important, for beneath them were laid down the majority of marine sediments studied by geologists. The great thicknesses of sedimentary rocks now forming the major Tertiary mountain ranges – the Alps, Himalayas, Rockies and Andes – were nearly all laid down under epicontinental seas on continental crust. There are similar large thicknesses of relatively undeformed sedimentary rocks on downwarped parts of the continental crust; these are the source, amongst other things, of all the present hydrocarbon deposits. Occasionally large quantities of such sediments spill over the continental edge and fan out to form considerable deposits down on the oceanic crust. However, most sediments deposited on the oceanic crust are very thin in comparison with those on the continental crust, and only rarely do they become incorporated into the continental crust during later Earth movements. So the position of shore lines is the most important feature on the palaeogeographic maps in this chapter. Oceanic seas on oceanic crust are differentiated from epicontinental seas on continental crust, this being the next most important feature on the maps. The edge of the continental shelf, that is the boundary between the continental crust and the oceanic crust, can be seen on the base map (grey) of the reconstructions.

In Chapter 12 H. G. Owen has shown that an even better reconstruction of Pangaea can be made on an Earth 80 per cent of its present diameter. The fitting together of the present continents is better in detail, but the major difference is the great reduction in the width of Tethys (*cf* Figs 13.1 and 13.4). South-east Asia is brought into contact with Australasia, and while the width of Tethys north and east of India is still large, it can now be interpreted as an entirely epicontinental sea on continental crust. The larger Tethys that is necessary in the reconstruction on a present-day-sized Earth has to consist partly of oceanic crust. The possibility that the Earth could have been 80 per cent of the present diameter 200 million years ago is still controversial. Until an acceptable mechanism can be suggested that will explain how the interior of the Earth could double in volume (80 per cent of the present diameter = 51 per cent of the present volume) in the past 200 million years, and yet leave the area and volume of the continental crust unaltered, the author of this chapter remains unconvinced that a substantially smaller Earth in the past is a possibility. It is probably not necessary to postulate a smaller sized Earth in order to eliminate the oceanic crust of the pre-Cretaceous Palae-Arctic Ocean that appears on most conventional reconstructions (e.g. Owen, 1976, p. 225; Smith & Briden, 1977). There is no geological evidence for the presence of oceanic crust in the Arctic before the late Cretaceous, because older oceanic crust is not present in the Arctic today, and there are no Mesozoic subduction zones round the Arctic continental margins down which such older crust could have been consumed. However, the Palae-Arctic Ocean is eliminated by plate tectonics: from about the middle of the Palaeozoic until the Cretaceous the Verkhoyansk-Anadyr Plate (= Kolymski Plate) of eastern Siberia was in contact with the northern continental margin of North America, and the Bering Strait Plate farther east was probably part of the North American continent (Herron, Dewey & Pitman, 1974, p. 379, fig. 4; Christie, 1979, p. 305, fig. 9E, p. 309). These two plates have been placed in this position in all the maps in this chapter (as in Owen, 1976, p. 234, fig. 6*a*), so there is no need to postulate the presence of Mesozoic oceanic crust in the Arctic. The date within the Cretaceous when the plates began to move apart, and started to form the oceanic basins of the modern Arctic Ocean, is still under debate.

Most conventional reconstructions of Pangaea also show an area of oceanic crust off south-east Africa. There is no geological evidence for the presence of pre-Jurassic oceanic crust in that area, and it can also be eliminated by plate tectonics. The West Antarctica Plate was situated farther north-west during the Trias and Jurassic, and was in contact with the continental margin of south-east Africa (Figure 13.1; Owen, 1976, p. 272). Major differences remain in the area around the Tethys Ocean between continental reconstructions on a full-sized or on a reduced-sized globe. In order to illustrate them, a reconstruction of Pangaea on a globe 80 per cent of the present diameter for the Trias is shown in Figures 13.4 and 13.5, and a reconstruction for the Upper Cretaceous on a globe 90 per cent of the present diameter in Figures 13.18 and 13.19. There are considerable differences in the palaeogeography of the Tethys and Arctic areas (compare Figs 13.1 with 13.4, 13.3 with 13.5, 13.15 with 13.18, and 13.17 with 13.19).

The Trias (Figs 13.1 to 13.5)

The most up-to-date account of the relative positions and movements of continental blocks during the Palaeozoic era (Scotese *et al.* 1979), suggests that while Gondwana remained a single unit throughout the Palaeozoic, seven other blocks were initially widely separated from each other. Their slow and irregular approach culminated in the formation of Laurasia and hence Pangaea by the early Trias. The following 60 million years (240–180 million years ago) until the start of the break-up of Pangaea was the only time during the Phanerozoic when all the earth's continental crust was united in a single continent. The whole of the surrounding ocean was 'the Pacific'. Pangaea contained many mountain chains of various ages, and considerable areas of its constituent parts had been flooded by epicontinental seas at various times. In the Trias, however, apart from the major incursion made by the sea called Tethys, epicontinental seas were limited to comparatively narrow areas around the margins.

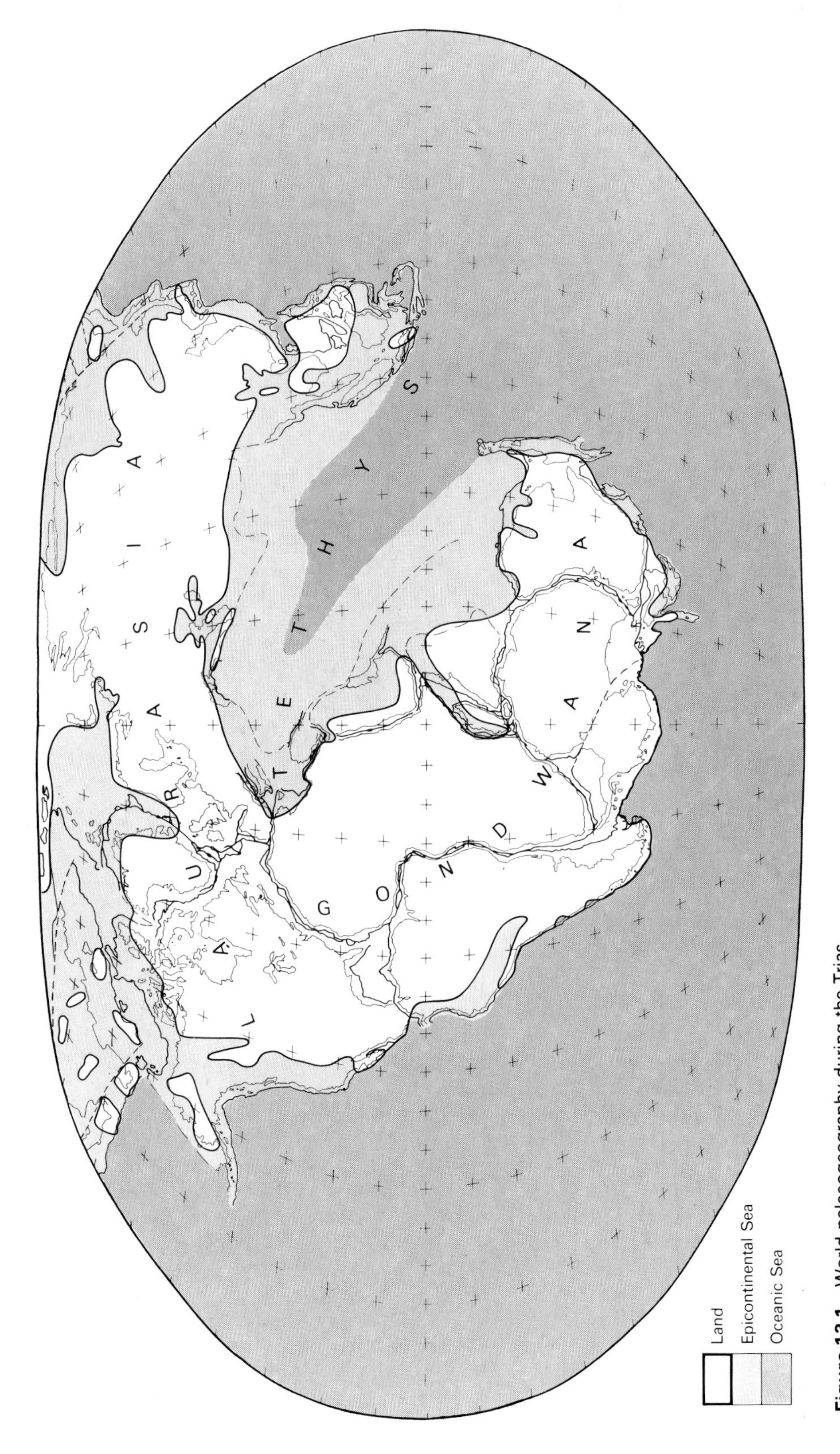

Figure 13.1 World palaeogeography during the Trias (240–200 my) (base map modified from H. G. Owen, new). The single continental land-mass of Pangaea consists of Laurasia to the north and Gondwana to the south, and is divided by the large Tethys Ocean intruding from the south-east. The central part of Tethys is oceanic crust. See the text for more exact dates of the epicontinental seas that are shown at approximately their maximum extent.

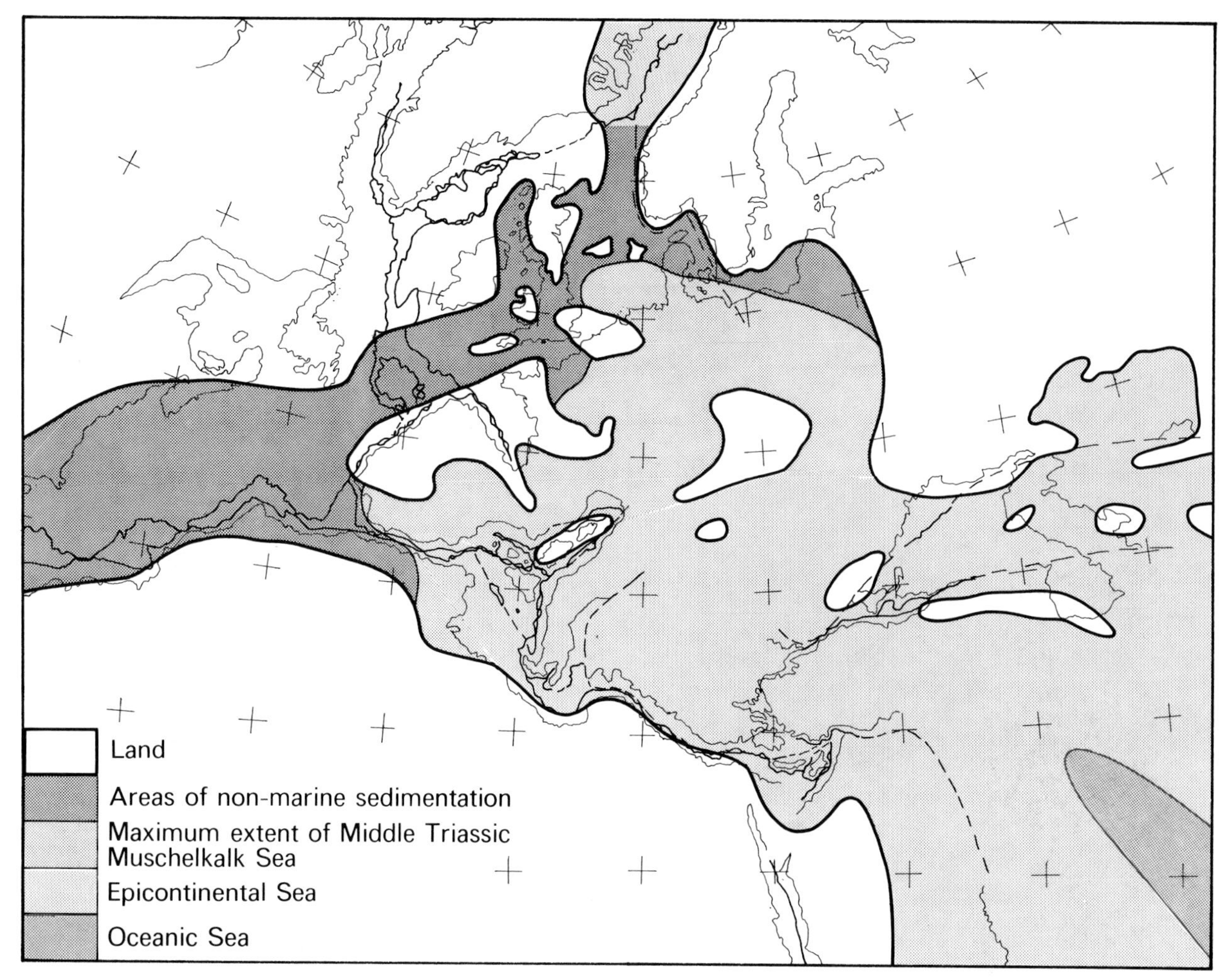

Figure 13.2 Palaeogeography of Europe and adjacent areas during the Trias (240–200 my) (base map from Owen, 1976, fig. 8*a*). The area in which non-marine sedimentation took place for much of the Trias is shown in pale grey. The south-eastern half of that area was transgressed by the brackish Muschelkalk Sea during the Middle Trias.

In order to make the Trias palaeogeographic maps more meaningful, the shorelines are composite and generally show the maximum incursion of the sea onto land during any part of the Trias. So far as Tethys is concerned, much of the northern shore through Europe and central Asia is now incorporated in the Alpine-Himalayan mountain chain, and its exact position at different dates in the Trias is uncertain. The embayments north and east of the Caspian Sea occurred in the Lower Trias, however, and the sea withdrew later in the Trias. The long arm of the sea extending down to cover the western half of Madagascar formed soon after the start of the Trias through rifting and graben structures between Somalia and India. The sea soon withdrew and non-marine rocks were formed later in the Trias. Marine Trias rocks occur in Pakistan and in the Himalayas, but the remainder of India was above sea-level. The marine incursions in the Perth Basin and the Fitzroy Trough of western Australia and the south-east corner of Queensland occurred in the Lower Trias, and the sea withdrew from Australia during the Middle and Upper Trias (Ludbrook, 1978). The shoreline passed through both North and South Islands, New Zealand, and marine rocks are known there for most parts of the Trias except the earliest (Stevens, 1974). No marine Trias is known in Antarctica. In South America no marine Lower or Middle Trias rocks are known, except for a small area on the coast of central Chile which was invaded by the sea in the Middle Trias (Anisian). The shoreline on the map is for the Upper Trias (Harrington, 1962).

The epicontinental seas that were present over large areas of North America during the Carboniferous and Permian had almost completely withdrawn by the start of the Trias, when there was only a short embayment entering through southern California. Spread of the sea from this area during the Middle Trias led to the broad sea on the west of the continent shown on the map at its approximate maximum extent during the Upper Trias (Schuchert, 1955). The sea in east Greenland occurred

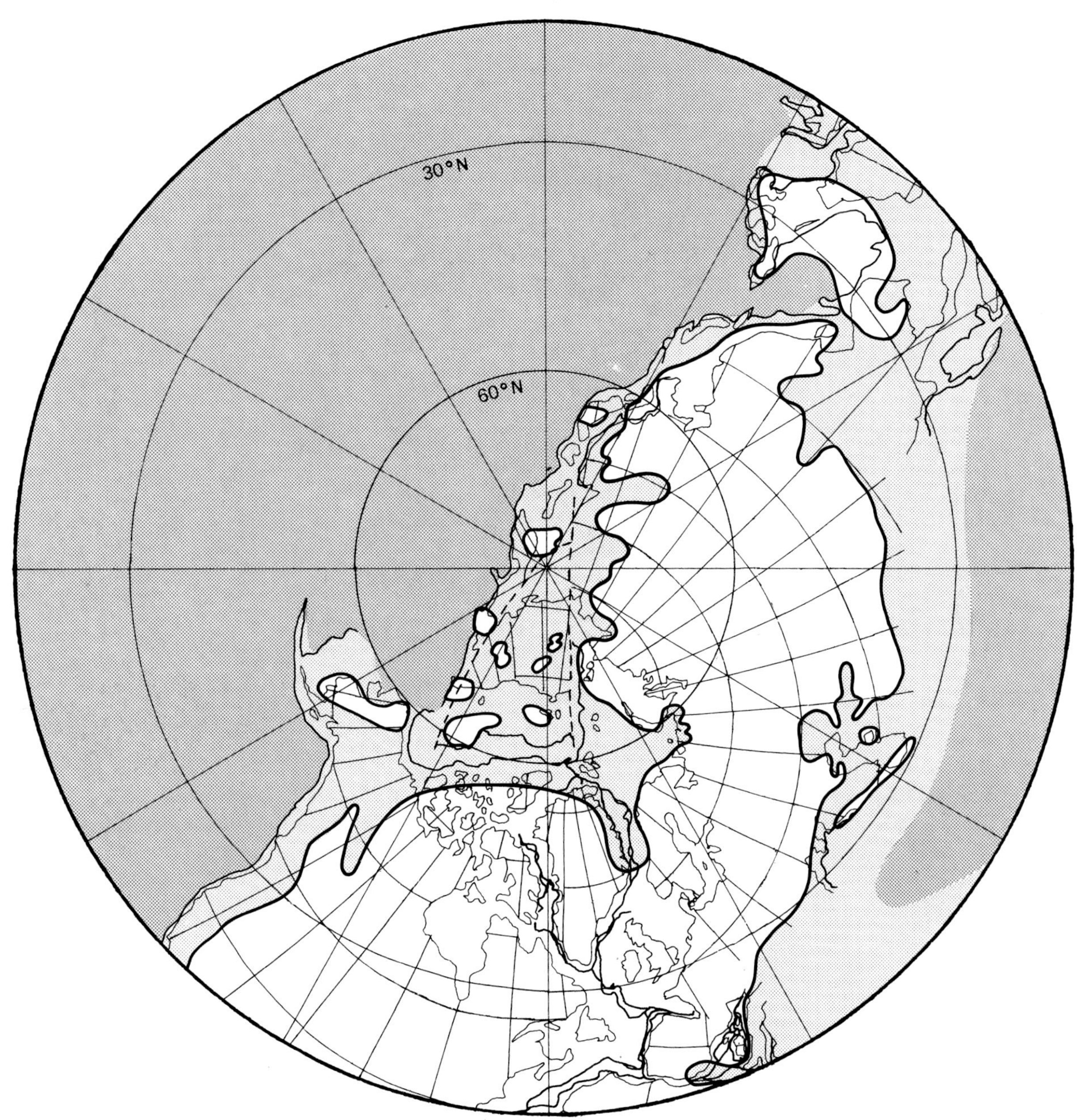

Figure 13.3 Palaeogeography of the north polar regions during the Trias (240–200 my) (base map modified from Smith & Briden, 1977, map 25). The information of Figure 13.1 plotted on a polar projection shows the large epicontinental sea of north-east Siberia.

during the Lower Trias and withdrew later. In eastern Asia the shoreline followed a complicated course through the islands of Japan that varied in detail in different parts of the Trias (Minato, 1965). The arms of the sea in south-west China are shown as in the Middle and Upper Trias, and this is the last epicontinental sea in China during the Mesozoic (Lyu, 1962). In Lower Triassic times the eastern arm extended to the east to find an outlet south of Korea, and cut off a large island to the south. North of Korea the long inlet represented today by the Amur geosyncline reached as far west as north-east Mongolia at its maximum extent in the Lower Trias. North and east Siberia are distorted on Figure 13.1, and the land–sea distribution is shown better on the polar projection of Figure 13.3. A very large area was covered by epicontinental sea and there were at least eight islands. The extent of the sea was similar throughout the Trias (Vinogradov, 1968).

The larger scale map of Europe (Fig. 13.2) shows the

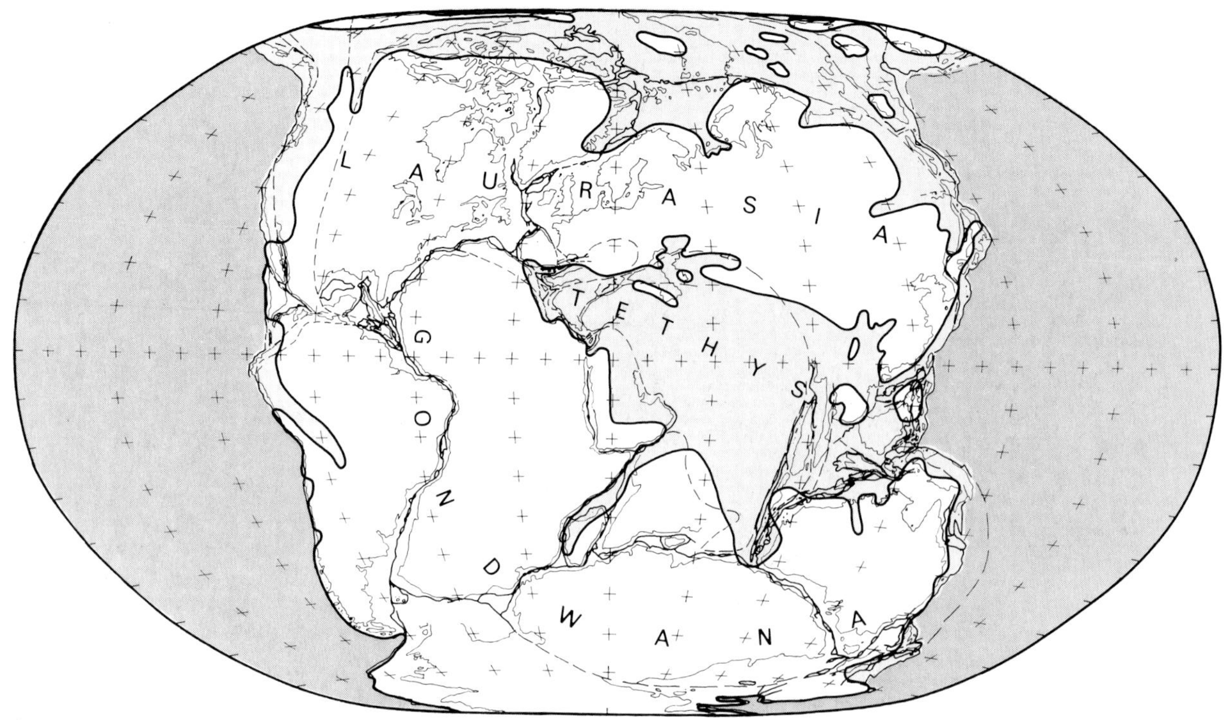

Figure 13.4 World palaeogeography during the Trias (240–200 my) on a base map reconstructed on a globe 80 per cent of the present Earth diameter (from H. G. Owen, new). Compared with Figure 13.1, Tethys is now a wholly epicontinental sea that is nearly closed at its eastern end.

areas of non-marine and evaporite rocks deposited during the Trias because they are such an important feature of the geology of north-west Europe and its continental shelf. The limit of Tethys is shown along the north coast of Africa, then across south-east Spain, between Italy and Sardinia–Corsica and though central Europe to the north part of the Black Sea and the Caspian. The Arctic seas penetrated half way down the east coast of Greenland in the Lower Trias. Between Tethys and east Greenland there was a series of basins in which continental, often 'red bed', sediments were deposited (Audley-Charles, 1970; Warrington, 1970; Wills, 1970). Periodically brackish seas transgressed from the south. The biggest transgression took the Muschelkalk Sea over most of France and Germany during the Middle Trias, and at a late stage, during the deposition of the Waterstones in England, it is possible that a brief marine connection was made with the Arctic sea through a graben in the northern North Sea. This long north–south graben was initiated late in the Trias and separated the two small 'islands' in the middle of the North Sea. Tension faulting and the formation of graben also occurred late in the Trias on the continental shelf off Newfoundland and Nova Scotia, where extensive continental red-bed deposits are found. These Earth movements were accompanied by eruptions of basic volcanic rocks and shallow intrusions that extend westwards along the boundary between Africa and North America (Hallam, 1971*b*). This is an important tectonic event for it is the first sign of rifting and tension that led eventually to the break-up of Pangaea.

The same Trias palaeogeography of Figures 13.1 and 13.3 had been transferred to the different base maps of Figures 13.4 and 13.5 which show a reconstruction of Pangaea on a globe 80 per cent of the modern Earth diameter. Several significant differences can be seen. Tethys is now an entirely epicontinental sea and its eastern outlet is severely restricted, possibly even closed at times. The Laurasian and Gondwana halves of Pangaea are brought into close contact, so that Indonesia touches the north and west sides of Australia. The large Indonesian island of the conventional reconstruction (Fig. 13.1) is drawn as two smaller ones, while the island across eastern Java is joined onto the land-mass of Australia, and the sea to the south is extended further down to cover the Laurasian rocks of Timor which are placed off the south-west corner of Australia in this reconstruction. The polar projection on an 80 per cent diameter Earth (Figure 13.5) has brought Alaska and north-east Siberia into closer contact, so that the Arctic Sea is narrower and there are fewer arctic islands. In particular the islands of Alaska and Siberia immediately west of the Bering Strait merge to form a single larger island.

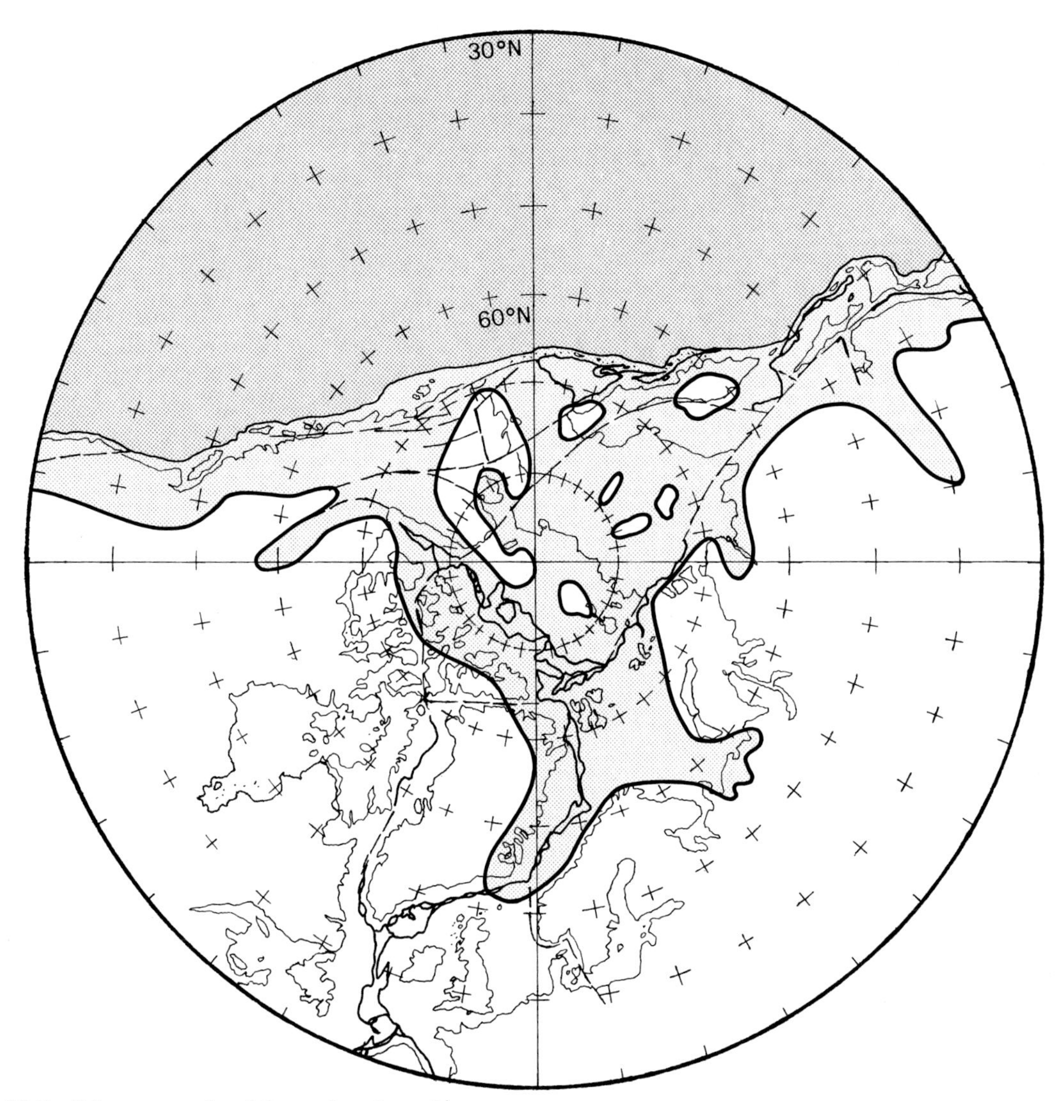

Figure 13.5 Palaeogeography of the north polar regions during the Trias (240–200 my) on a base map reconstructed on a globe 80 per cent of the present Earth diameter (from Owen, 1976, fig. 6*a*). This reconstruction brings north-east Siberia into contact with North America (*cf* Figure 13.3).

The Lower Jurassic, Pliensbachian Stage (Figs 13.6 to 13.8).

The base of the Jurassic was marked by a world-wide transgression of the sea onto the edges and low-lying areas of Pangaea. Thereafter regular cycles of deepening and shallowing of the sea led to transgressions followed by regressions throughout the Jurassic. In the Lower Jurassic the initial transgression was generally maintained through the Sinemurian and into the Lower Pliensbachian. It was followed by a marked regression in the Upper Pliensbachian, then a major new transgression in the Lower Toarcian. The Lower Jurassic transgressions are nowhere better shown than in Europe (Fig. 13.7) where the whole of the area of non-marine deposition in the Trias was flooded by the sea. A complicated pattern of islands and shelf sea was formed, which was constantly changing in detail (Donovan, 1967). Most of Britain was covered by sea for most of the time, except for a large Scottish island and the London–Brabant island which were never covered. The differently shaped islands in the eastern Mediterranean area and the Middle East on Figures 13.6 and 13.7 are due to the same information being plotted on the slightly different reconstructions of the two base maps.

The connection with the Arctic sea past east Greenland was fully marine at least by Lower Pliensbachian times and again in the Toarcian, but it may have been broken at other times, and there is some evidence that the initial spread of the sea was from the south (Hallam & Sellwood, 1975). The possibility of a marine connection between Europe

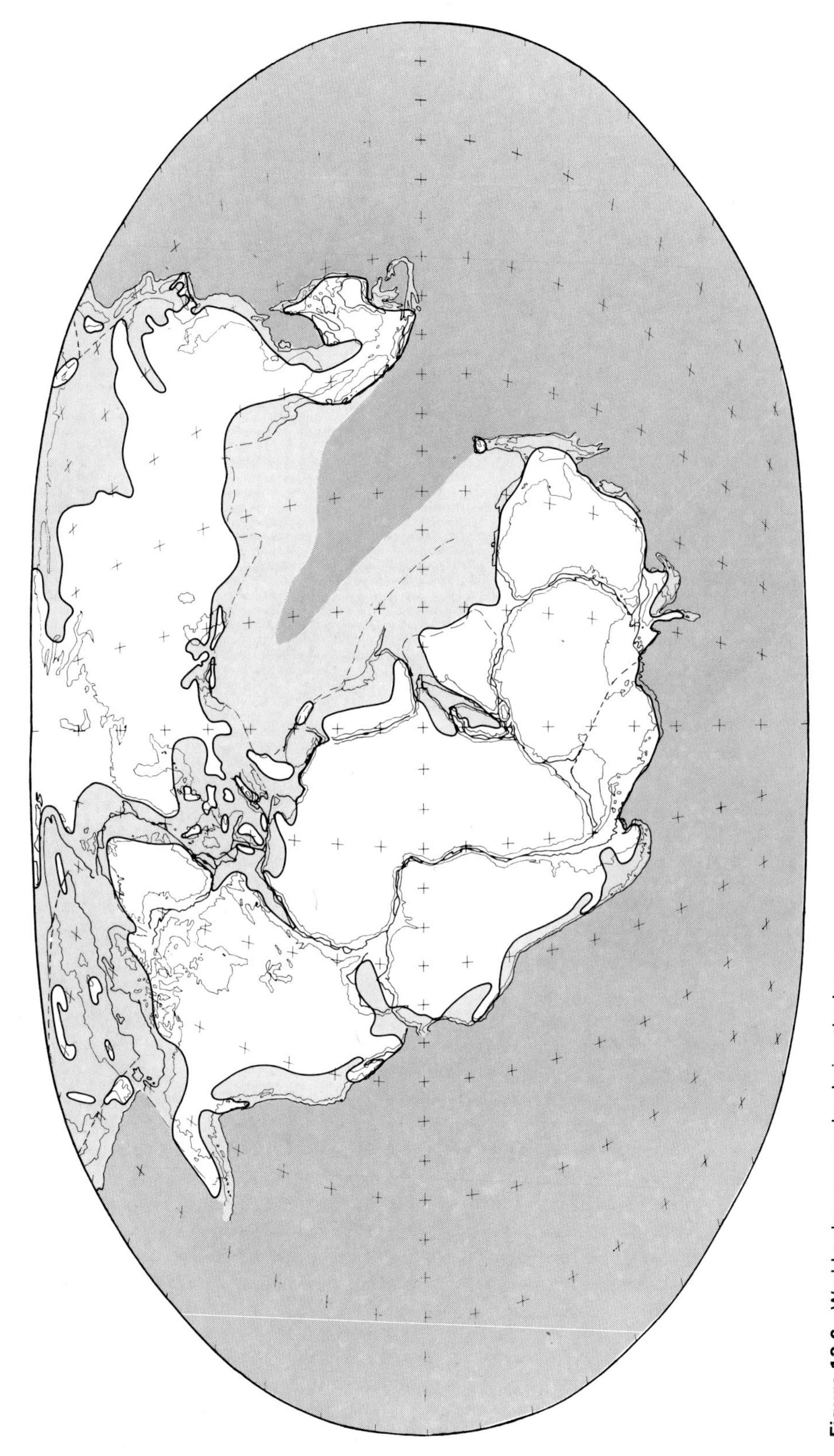

Figure 13.6 World palaeogeography during the Lower Jurassic, Pliensbachian Stage (180 my) (base map modified from H. G. Owen, new). Major differences from the Trias (Figure 13.1) are the epicontinental seas through Europe to the Arctic and through western North America to the Arctic, and the different land-masses in Malaysia and Indonesia.

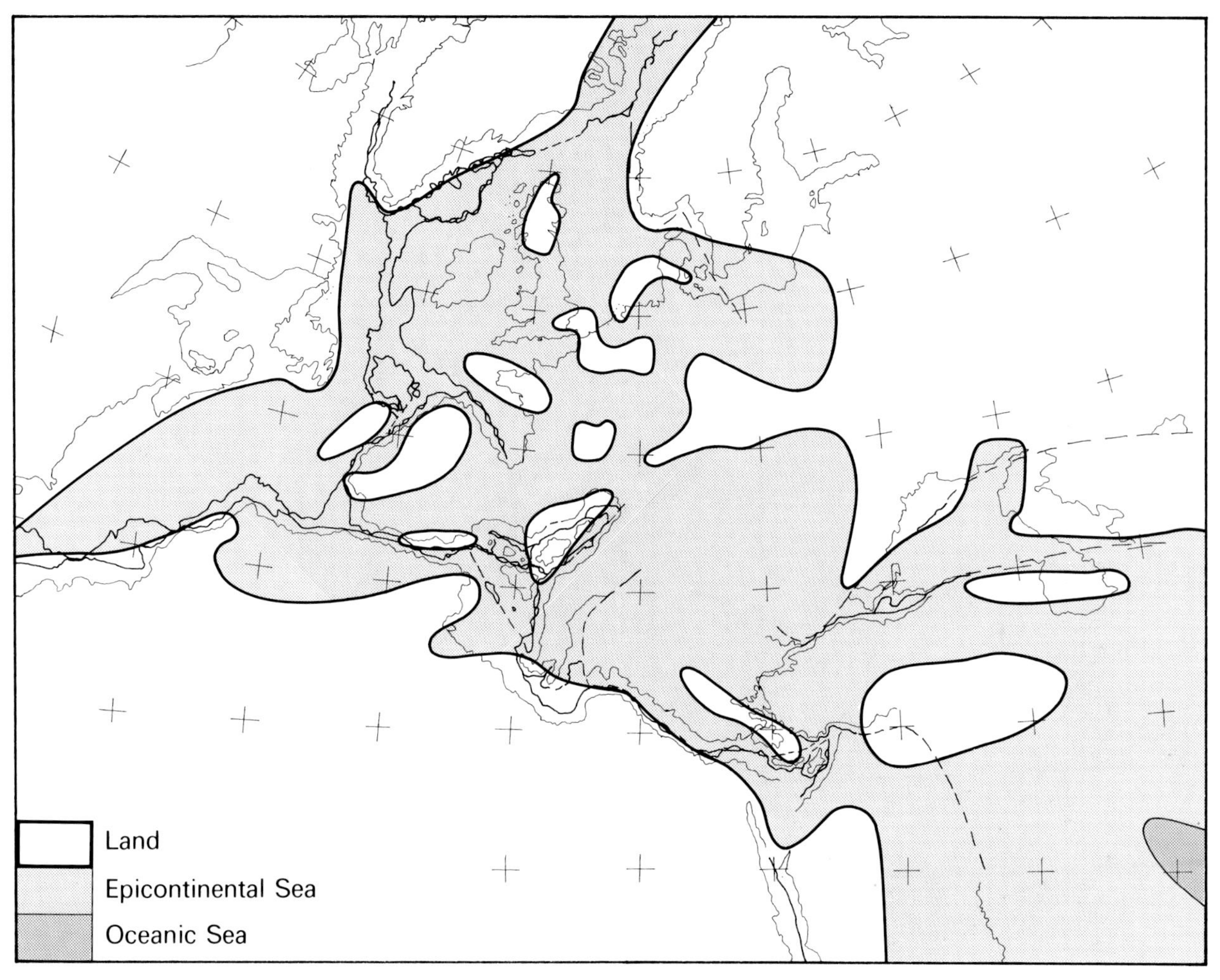

Figure 13.7 Palaeogeography of Europe and adjacent areas during the Lower Jurassic, Pliensbachian Stage (180 my) (base map from Owen, 1976, fig. 8*a*).

and the Gulf of Mexico along the eastern side of North America is more problematical (Hallam, 1977). The first break-up of Pangaea was to occur along this line, and it had already been heralded by tension faulting and volcanicity in the Upper Trias. Temporary epicontinental seaways along the line may have occurred more than once in the Lower Jurassic. The most likely time was during the Toarcian, which would explain the appearance in Chile of the highly distinctive Lower Toarcian ammonite *Bouleiceras*, which otherwise only occurs in central and southern Tethys from Morocco to Madagascar. Other migration routes to Chile are much less likely to have occurred because of the presence of different faunas in the south-west pacific and the Arctic.

After withdrawal of the Triassic sea from Madagascar, the first Jurassic transgression was at the beginning of the Toarcian and was apparently from the east (Blant, 1973). No Lower Jurassic marine sediments are known in Antarctica or in Australia (the first are Middle Jurassic Bajocian rocks in the northern part of the Perth basin). In South America a transgression early in the Lower Jurassic established a sea over much of the northern part of Chile, and there was considerable flooding across part of Patagonia (Harrington, 1962). A large embayment from the Pacific crossed Mexico, where there are several small patches of Jurassic rocks of different ages (Erben, 1957), and it probably extended across the Gulf of Mexico to Cuba. On the west end of Cuba the oldest datable Jurassic rocks are at least Middle and probably Lower Jurassic in age. In western North America the island in western Alaska in the Trias had merged with land further east, and there was probably no direct connection between the seas of north and south Alaska (Imlay & Detterman, 1973). The Toarcian was a time of major transgression in northern Canada and Greenland, and the oldest marine Jurassic of Spitsbergen is of middle Toarcian age.

The Lower Jurassic transgression had little effect on the large area of epicontinental sea over north and north-east

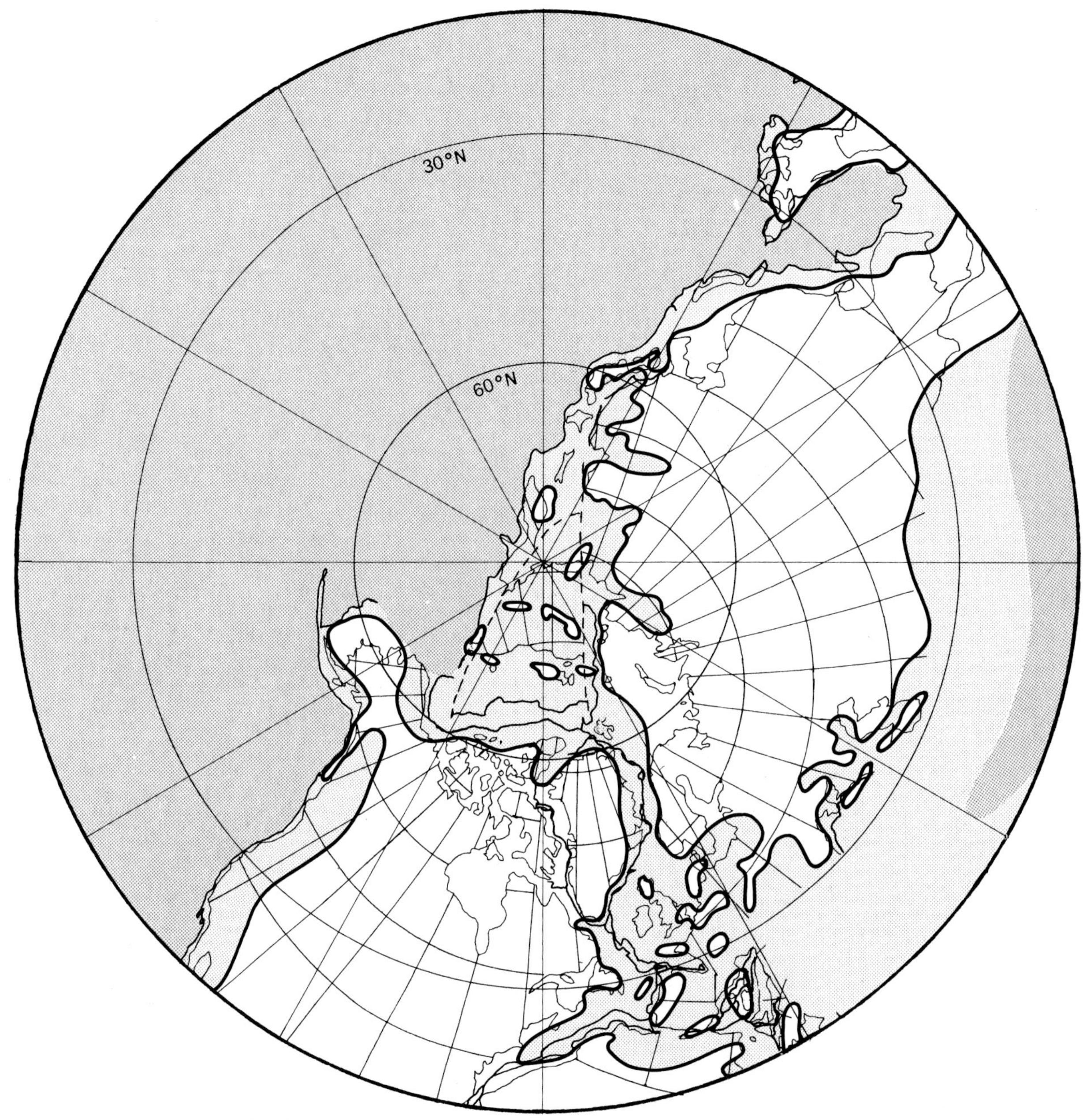

Figure 13.8 Palaeogeography of the north polar regions during the Lower Jurassic, Pliensbachian Stage (180 my) (base map modified from Smith & Briden, 1977, map 24).

Siberia (Figure 13.8) which was at least as large as the European sea (Vinogradov, 1968). The area covered by China, Malaysia and Indonesia is the only other exception to the otherwise world-wide transgression of the Lower Jurassic seas. The sea withdrew from China at the end of the Trias and did not return during the rest of the Mesozoic. A large area of central Malaysia and Indonesia was elevated, leaving only part of Burma, Vietnam, Loas, Hong Kong and the Banda Arc in south-east Indonesia below sea level (Audley-Charles, 1966; Minato, 1965). However, a long arm of the sea reached down to west Borneo, judging by the fossiliferous Lower Jurassic rocks that are known there.

The Upper Jurassic, Kimmeridgian Stage (Figures 13.9 to 13.11)

Although an epicontinental sea through Europe connecting Tethys with the Arctic had divided the land-mass of Pangaea into two large halves from the beginning of the

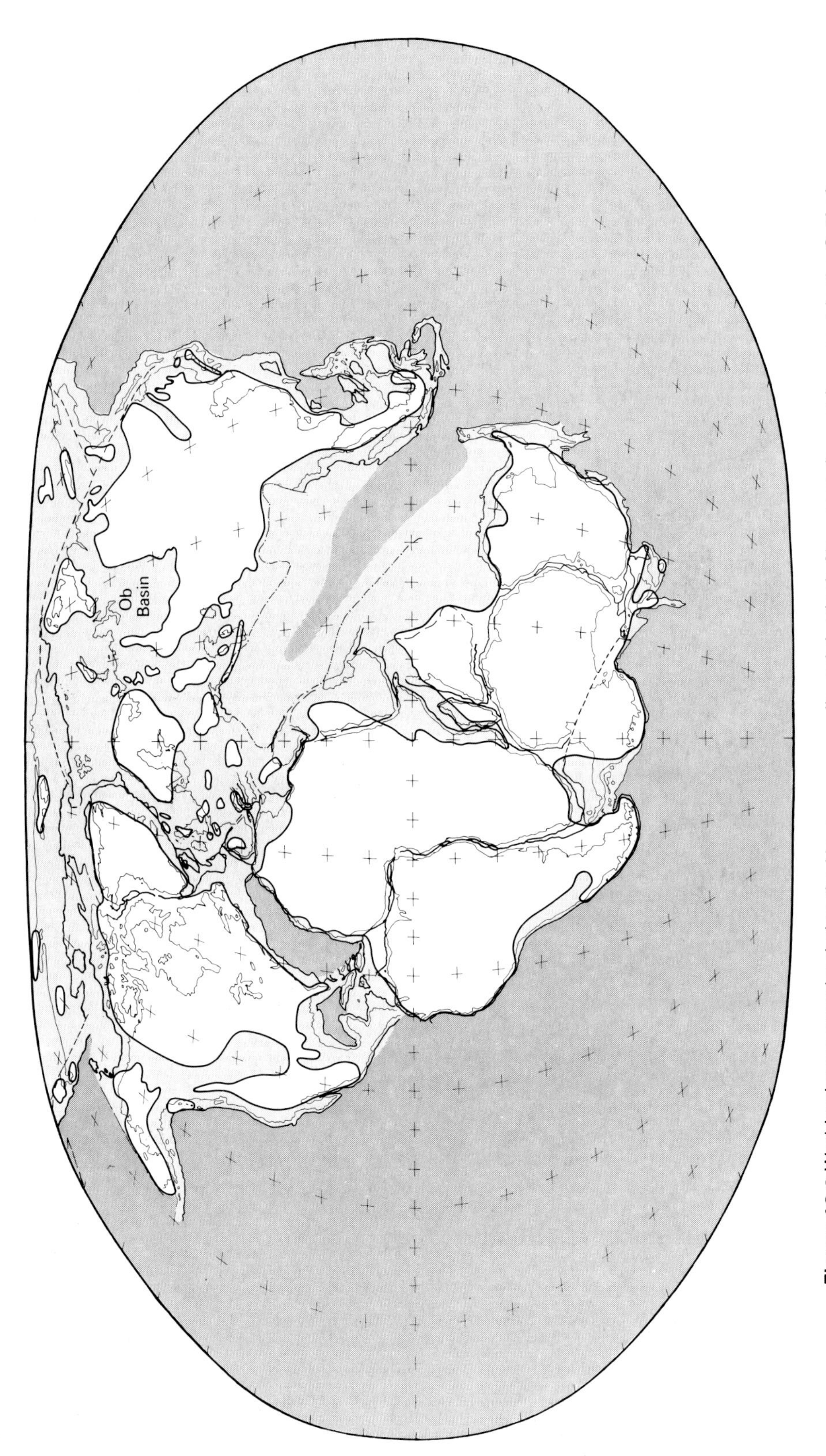

Figure 13.9 World palaeogeography during the Upper Jurassic, Kimmeridgian Stage (145 my) (base map modified from H. G. Owen, new). Compared with the Pliensbachian map (Fig. 13.6), Australia and Indonesia are closer together, and the amount of oceanic crust in Tethys has been reduced (by subduction). New oceanic crust is present in the Gulf of Mexico and the North Atlantic, due to the movement apart of North America and Africa. A large new area of epicontinental sea occurs on the Russian platform west of the Urals and in the Ob basin east of the Urals.

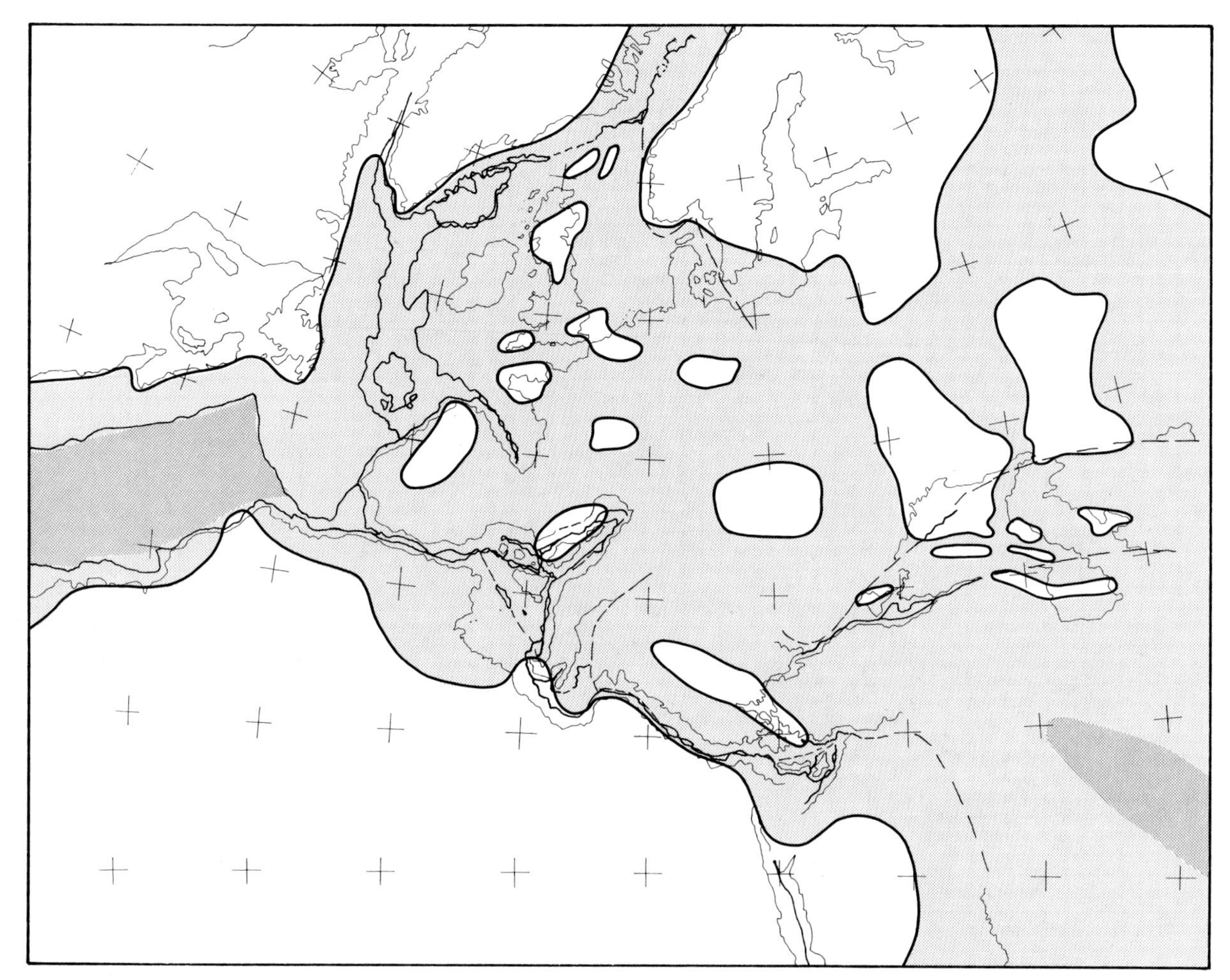

Figure 13.10 Palaeogeography of Europe and adjacent areas during the Upper Jurassic, Kimmeridgian Stage (145 my) (base map from Owen, 1976, fig. 8*a*, modified in the North Atlantic area). New features are the North Atlantic ocean and the large area of sea in the north-eastern part of the map.

Jurassic, the first break in the continental crust of Pangaea came in the Callovian, when North America started to rotate clockwise away from Africa. The northern end remained in contact in the Greenland–Scandinavia region, while sea-floor spreading further south formed new oceanic crust to initiate the North Atlantic Ocean. A permanent seaway now existed between western Europe and the region of the Gulf of Mexico.

The Kimmeridgian was, in general, the time of the greatest transgression of the sea during the Upper Jurassic. Withdrawal in some areas at the end of the Lower Jurassic was followed by renewed transgression during the Bajocian. The Bathonian was, however, a time of marked regression in Britain and north-west Europe, where deltaic, non-marine or brackish-water deposits are extensive. Transgression followed in the Callovian, and much of Europe was covered by an epicontinental sea until the end of the Kimmeridgian (Brookfield, 1973; Hallam, 1975). One of the greatest Jurassic transgressions in the world was across the Russian platform in the Middle and Upper Jurassic. Starting early in the Bajocian, the area of flooding increased steadily until in the Kimmeridgian a vast epicontinental sea existed across most of European Russia, linking Tethys with the Arctic ocean both east and west of the Caspian, and extending round the north end of the Urals into the large Ob basin, then eastwards to link up with the large epicontinental sea of north and east Siberia (Vinogradov, 1968).

North America and Greenland remained a single unit, with little change in palaeogeography except in the west, where the Mesocordilleran Geanticline, a long north–south peninsula of emergent rocks undergoing orogeny at depth, separated the Pacific coast geosyncline to the west from the epicontinental Sundance Sea that occupied a large area of the mid-west United States (Kummel, 1970). The sea transgressed northwards from the Gulf of Mexico into the

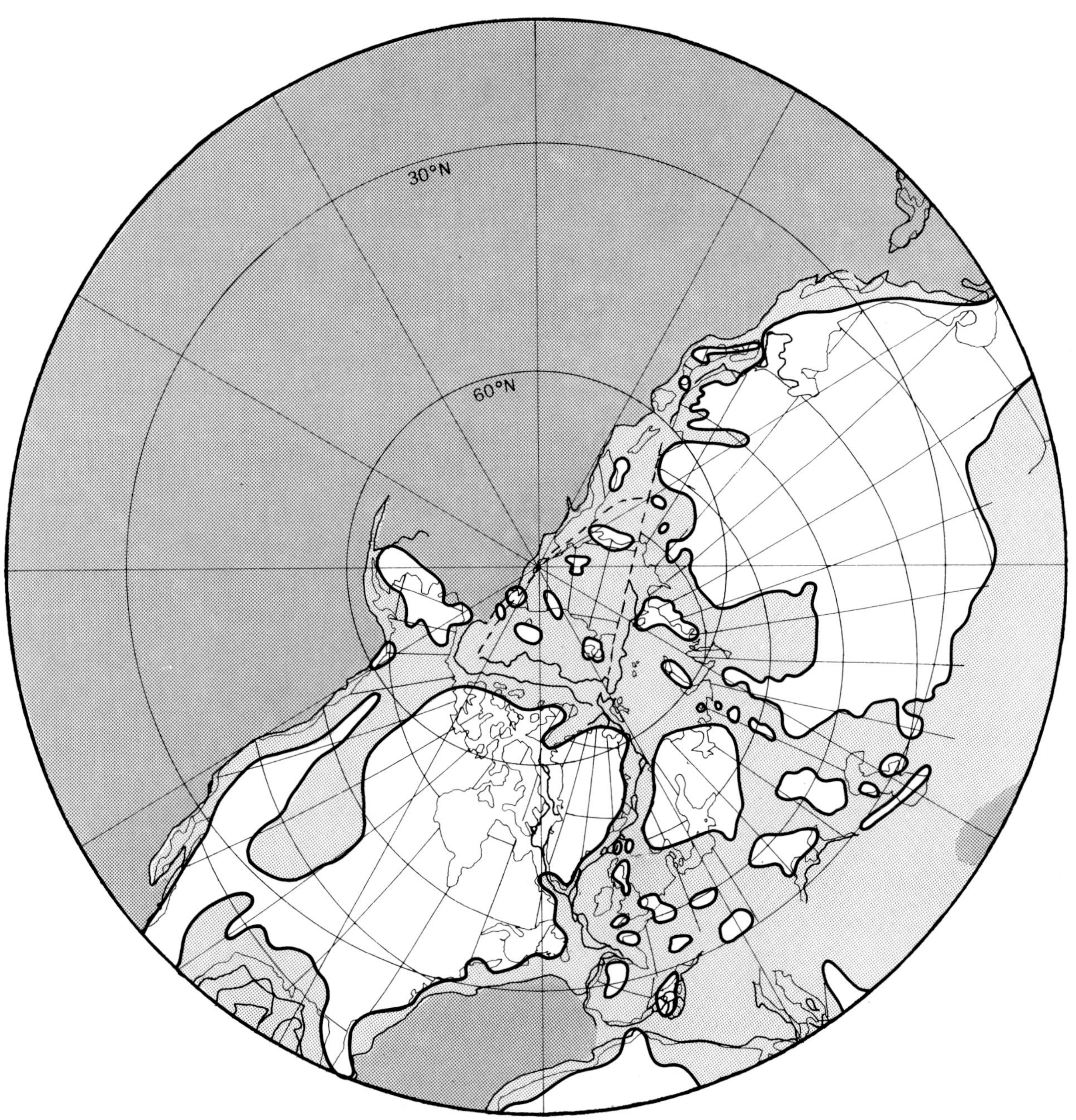

Figure 13.11 Palaeogeography of the north polar regions during the Upper Jurassic, Kimmeridgian Stage (145 my) (base map modified from Smith & Briden, 1977, map 22). The main difference from the Pliensbachian map (Fig. 13.8) is the more extensive epicontinental sea in north-west Siberia.

southern United States where there are extensive concealed deposits of Upper Jurassic rocks (Cook & Bally, 1975). The only area of marine sedimentation in South America in the Kimmeridgian was in the northern half of Chile, and an embayment at the southern end extended into west Argentina (Harrington, 1962).

Middle Jurassic (Middle Bajocian) sediments in the southern part of the Antarctic Peninsula are the oldest Mesozoic rocks found so far in Antarctica (Stump, 1973). Fossiliferous Upper Jurassic rocks are also found in the same area and the sea penetrated right up to the position of the present coast of southern Africa by late Jurassic times (Moullade & Nairn, 1978, p. 428). Rift faulting and graben formation occurred all the way along the line of contact between south-east Africa and Antarctica in the Oxfordian, heralding movement apart that began in the

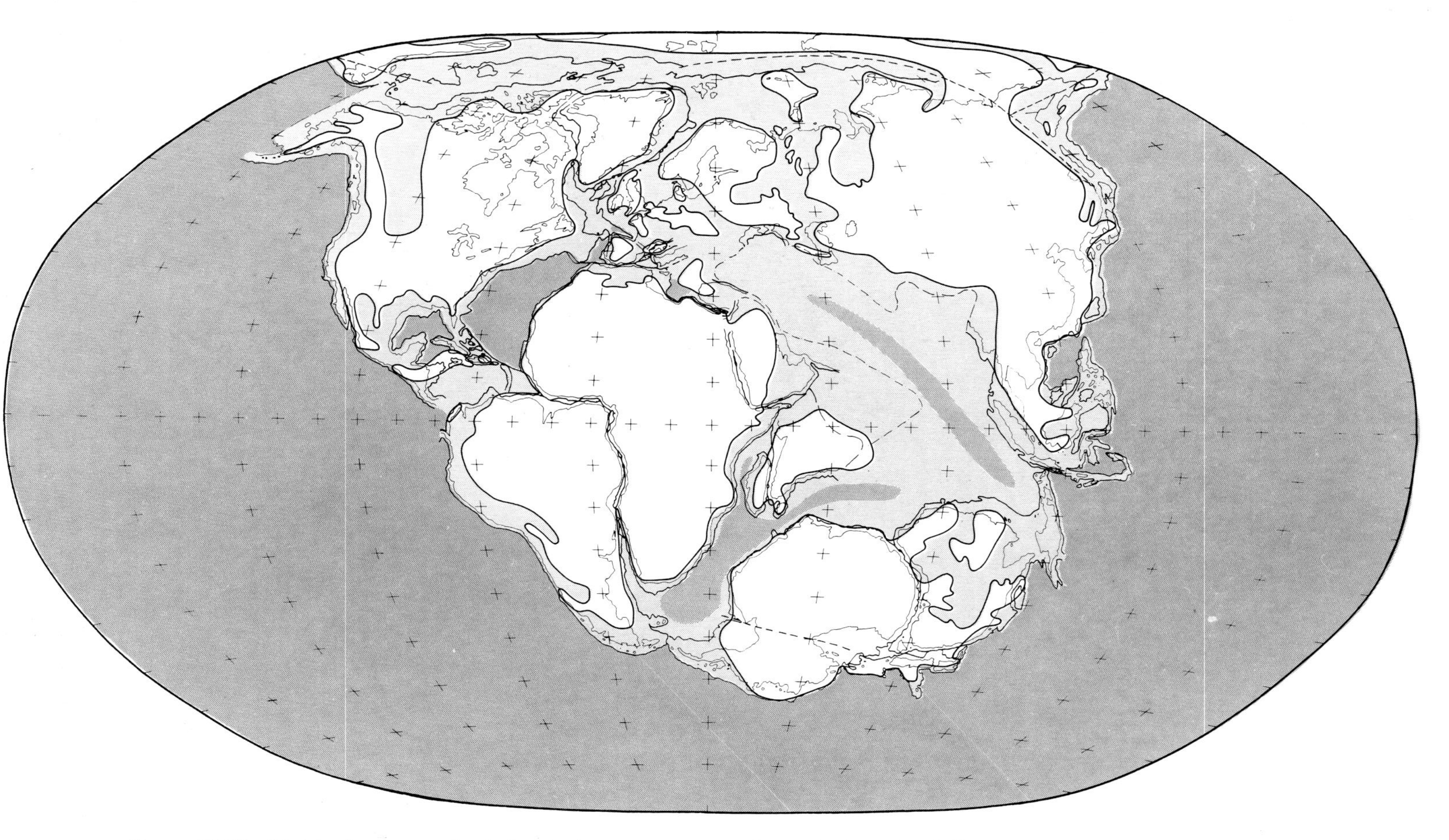

Figure 13.12 World palaeogeography during the Lower Cretaceous, Hauterivian Stage (125 my) (base map modified from H. G. Owen, new). Australia and Indonesia are in contact and the amount of oceanic crust in Tethys has been further reduced (by subduction). The North Atlantic is wider than in the Kimmeridgian (Fig. 13.9), and new oceanic crust has been generated between Africa and Antarctica and both east and west of India. There is an epicontinental sea over much of the interior of Australia.

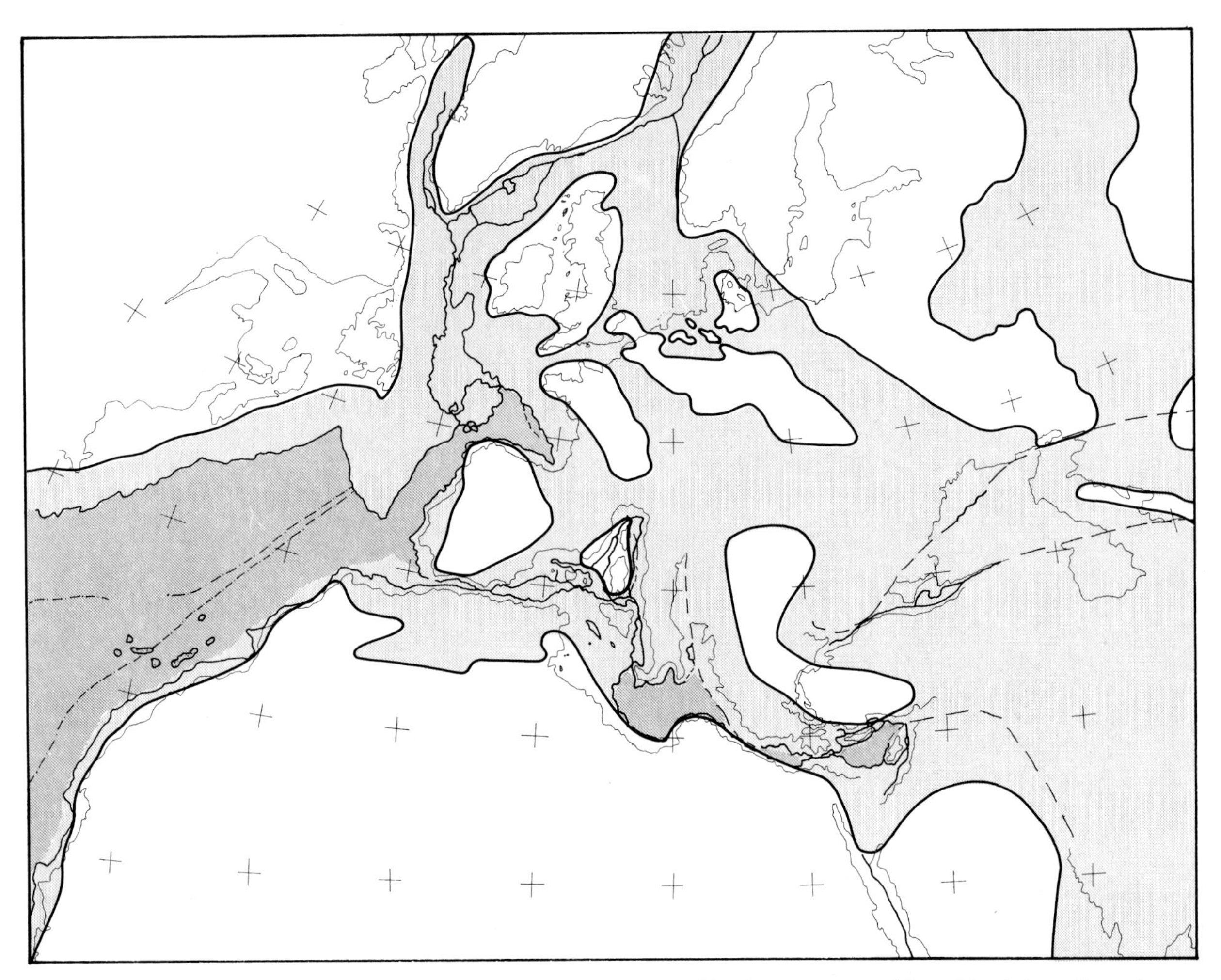

Figure 13.13 Palaeogeography of Europe and adjacent areas during the Lower Cretaceous, Hauterivian Stage (125 my) (base map from Owen, 1976, fig. 8*b*). The extent of the Kimmeridgian epicontinental sea (Fig. 13.10) has been much reduced by the appearance of large islands in north-west Europe. The first oceanic crust of the eastern Mediterranean has been generated, and in the North Atlantic new oceanic crust has appeared north and west of Spain.

Kimmeridgian. Sea probably did not penetrate along the whole line before Kimmeridgian times, but coming from the north it had reached southern Tanzania earlier in the Upper Jurassic. This arm of the sea flooded the west side of Madagascar following tilting of the island to the west in the Middle Jurassic (Blant, 1973). Australia was mainly land during the Upper Jurassic except for small areas of marine deposits in the west and north-west. However, there are extensive non-marine deposits of Upper Jurassic age in Australia, especially in basins in central and eastern areas, where they foreshadow the epicontinental seas that were to appear late in the Lower Cretaceous (Kummel, 1970).

The Middle and Upper Jurassic was a period of increasing differentiation of some marine organisms into Boreal and Tethyan marine faunal realms (Hallam, 1971*a*, 1973). Several groups were affected, and the ammonites have been especially well documented. The Boreal realm was confined to the northern half of the northern hemisphere, and was separated by a varying boundary from the Tethyan realm that occupied the Tethys sea and all regions to the south including much of the Pacific Ocean border, except for the northern part between north of Japan and California. The ammonites of the Boreal realm at first belonged largely to the families Cardioceratidae and Kosmoceratidae and these range from the Bajocian to the Lower Kimmeridgian. Later the dominant group was the Boreal Perisphinctidae, which had moved north from Tethys and evolved independently from the Tethyan branch of the family for the rest of the Jurassic and into the base of the Cretaceous. The Tethyan realm was very rich in Phylloceratidae and Lytoceratidae (which are rare or absent in the Boreal realm) and also large numbers of Perisphinctidae and their derivatives. Reasons for a Boreal–Tethyan differentiation are still hypothetical: climate, salinity and restriction of easy marine connections have all

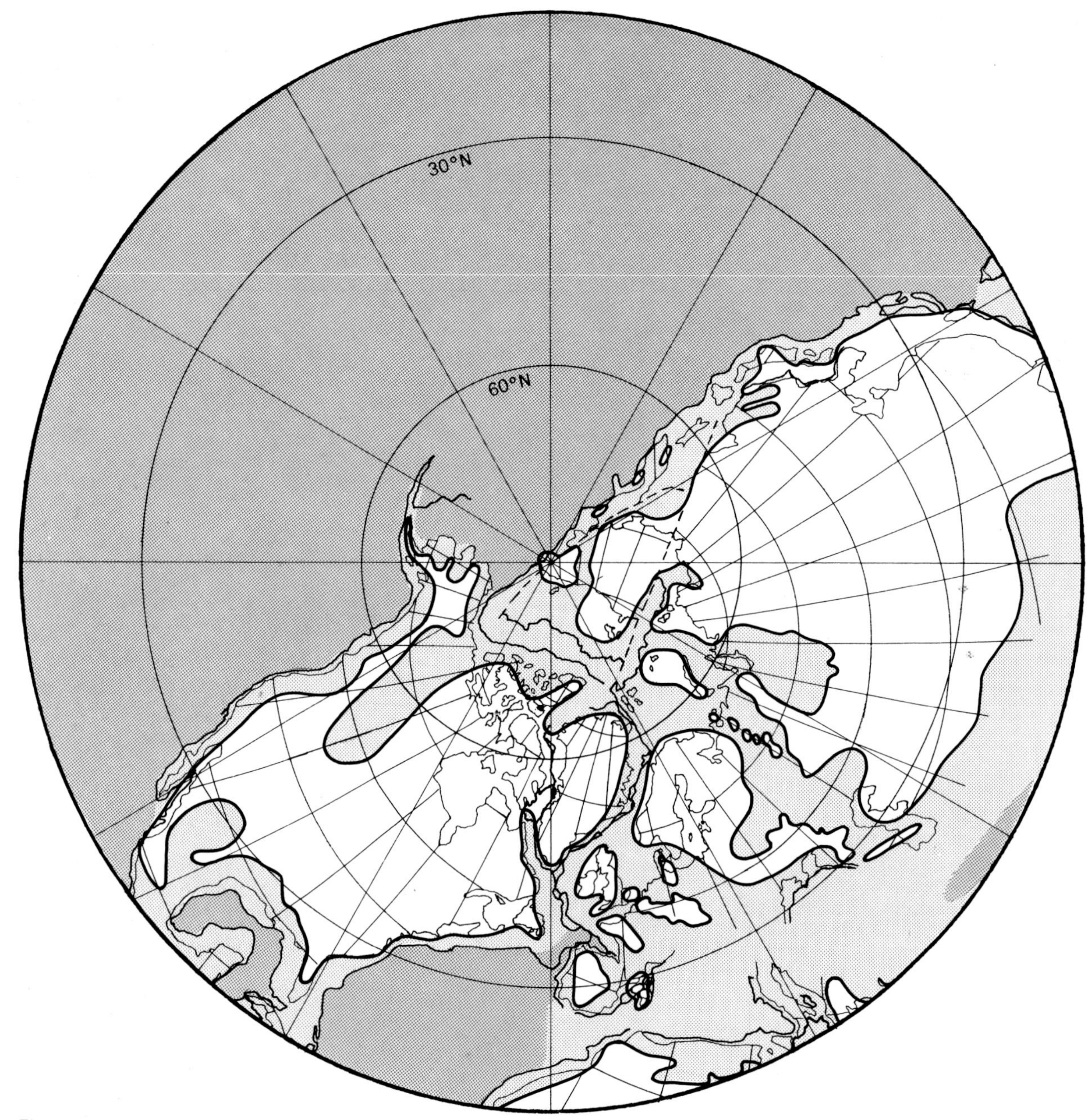

Figure 13.14 Palaeogeography of north polar regions during the Lower Cretaceous, Hauterivian Stage (125 my) (base map modified from Smith & Briden, 1977, map 21). A major withdrawal of the epicontinental sea from north-east Siberia has occurred since the Kimmeridgian (Fig. 13.11).

been put forward, but no explanation provides a wholly convincing answer.

The Lower Cretaceous, Hauterivian Stage (Figs. 13.12 to 13.14)

By the middle of the Lower Cretaceous, new sea-floor spreading between Africa and Antarctica and continuing spreading in the North Atlantic led to the break-up of Pangaea into several areas of continental crust. The largest area to break adrift was the block composed of Antarctica and Australasia which moved away from south-east Africa and India because of new oceanic crust created off the east and south-east coast of Africa and in the channel up the east side of India. Movement had started in the Kimmeridgian, and from that time the sea encroached onto the present seaboard of much of south-east Africa. The only marine rocks of this age known in Antarctica are in the Antarctic Peninsula, where sedimentation took place for most of the period from the late Jurassic Upper Oxfordian to the end of the Aptian.

In Australia, however, the Lower Cretaceous was marked by the greatest incursion of epicontinental seas during its Mesozoic history. Starting in the Barremian, the sea flooded from the north and north-west, and the shoreline shown on the map is for its greatest extent in the Aptian. The Great Artesian Basin of Queensland has very extensive Aptian and Albian marine sediments, and the sea penetrated south of the present south coast of Australia and must have been close to the Wilkes Land coast of Antarctica (Laseron, 1969; Ludbrook, 1978). In New Zealand the area covered by the sea was much less extensive than in the Upper Jurassic, being confined to the eastern part of North Island and the North-east part of South Island (Stevens, 1974).

India and Madagascar together formed the smallest piece of continental crust to be separated from the main mass. It had moved away from east Africa, allowing a wider sea to penetrate between them, but the north-west corner of India was still in contact with the Arabian Peninsula. The sea transgressed slightly onto the east coast of India probably from the Tithonian onwards, though the earliest well-dated marine incursion is Barremian (Robinson, 1967). South America and Africa still formed a single continent. However, the faulted and rifted division between them had started to open from the southern end, because of the rotation of South America away from Africa. The pivot was the north-east corner of Brazil which was in contact with south-east Nigeria. The sea penetrated successively further north up the major rift valley between the two continents and had reached as far north as Angola by the time represented by the Aptian–Albian boundary. Sea had also penetrated from the west as far as Nigeria by the Upper Albian, and probably the whole length of the rifted division between Africa and South America was under marine seas by the end of the Albian (Maack, 1969; Reyment & Taitt, 1972; Reyre, 1966). On the west of South America a major Mesozoic event was the invasion of the sea during the Tithonian, to produce the Venezuelan–Peruvian, Chilean and Patagonian sedimentary basins, which remained as shown on the map until the Cenomanian (Harrington, 1962).

North America–Greenland had moved further away from Africa, and the sea had encroached slightly more extensively onto its eastern and southern coastal areas. In the west the Mesocordilleran Geanticline had probably joined up with the land-mass in Alaska, and the epicontinental sea to the east had withdrawn to the north (Schuchert, 1955; Cook & Bally, 1975).

Much of Europe, the Middle East and the Arabian Peninsula underwent some uplift with regression of the sea at the end of the Jurassic. The main exception was the central Mediterranean area of Tethys which remained marine. This regression is especially well marked in north-west Europe, where much of the present area of the British Isles probably emerged as a large island, and there were other extensive land areas in France and Germany (Fig. 13.13). Continental, deltaic and brackish top Jurassic and Lower Cretaceous sediments occur in southern England and France. Marine sediments of the same age in east and north-east England and north Germany containing a Boreal fauna were formed near the southern end of the seaway that existed between Greenland and Scandinavia and southwards into the present area of the North Sea (Casey, 1973; Kay, 1966). In eastern Europe and western Russia the shelf sea had shrunk somewhat from its maximum extent in the Kimmeridgian, but a north–south connection between Tethys and the arctic and a slightly reduced Ob basin sea remained. Major regression occurred in north and east Siberia from the Valaginian onwards. By the Barremian the whole area was land, except for a small area on the north coast of Siberia and a north–south strait north of the Kamchatka Peninsula, and the sea did not return during the Cretaceous (Vinogradov, 1968). Kamchatka itself and some of the surrounding area was covered by sea, but the Amur geosyncline was greatly reduced, and much of the central part of the Japanese Islands was land (Minato, 1965). The palaeogeography of the Arctic is seen better on the polar projection of Figure 13.14, where the narrowing of the sea and its withdrawal from much of northern Asia can be compared with the wider seas of the Kimmeridgian (Figure 13.11).

The Upper Cretaceous, Senonian Stage
(Figs. 13.15 to 13.19).

By the middle of the Upper Cretaceous sea-floor spreading and continental drift had led to the further break-up of Pangaea. The division between South America and Africa had been rift-faulted since the beginning of the Cretaceous, and opening from the southern end northwards had occurred successively throughout the Lower Cretaceous. The date of final separation in the Benin Basin in the centre of the Gulf of Guinea, and the commencement of movement apart of the two continents, was in the Cenomanian. After that time the sea encroached onto two coastal areas of north-east Brazil, and there is a full history of Upper Cretaceous sedimentation along much of the west coast of Africa (Reyment & Taitt, 1972). India was a separate continental block moving further away from east Africa, though its north-west margin remained in contact with the Arabian Peninsula. Marine sediments were formed along the south-eastern coastal area of India (Robinson, 1967). Malagasy broke away from India in the Coniacian, and there are marine sediments of Maastrichtian age on the east coast of the island (Moullade & Nairn, 1978). Sea gained access to the division between Canada and Greenland by about Albian times, and sea-floor spreading began to move the two apart in the Coniacian. There are marine sediments of Albian and Upper Cretaceous age

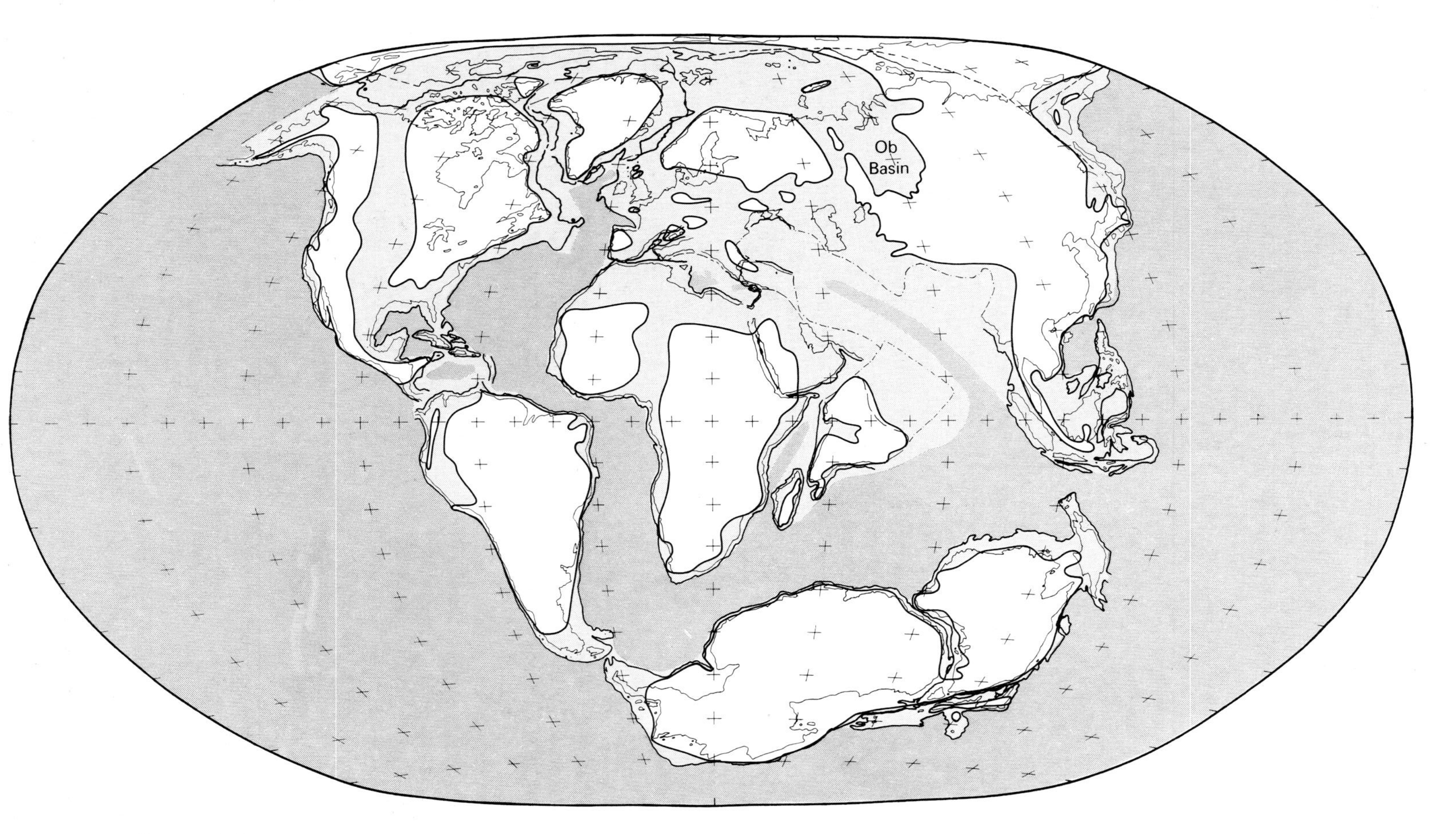

Figure 13.15 World palaeogeography during the Upper Cretaceous, Senonian Stage (90–80 my) (base map modified H. G. Owen, new). The movement of North and South America away from Europe and Africa, and Antarctica/Australia away from Africa and India, and the generation of much new oceanic crust throughout the Atlantic and Indian Oceans are major differences from the Hauterivian map (Fig. 13.12). There is also new oceanic crust in both the north and south Caribbean. The Trans-Saharan seaway and the link between the Gulf of Mexico and the Arctic are new epicontinental seas. The land area is considerably less than in the Lower Cretaceous, though Antarctica probably escaped transgression and the Lower Cretaceous sea over Australia had withdrawn.

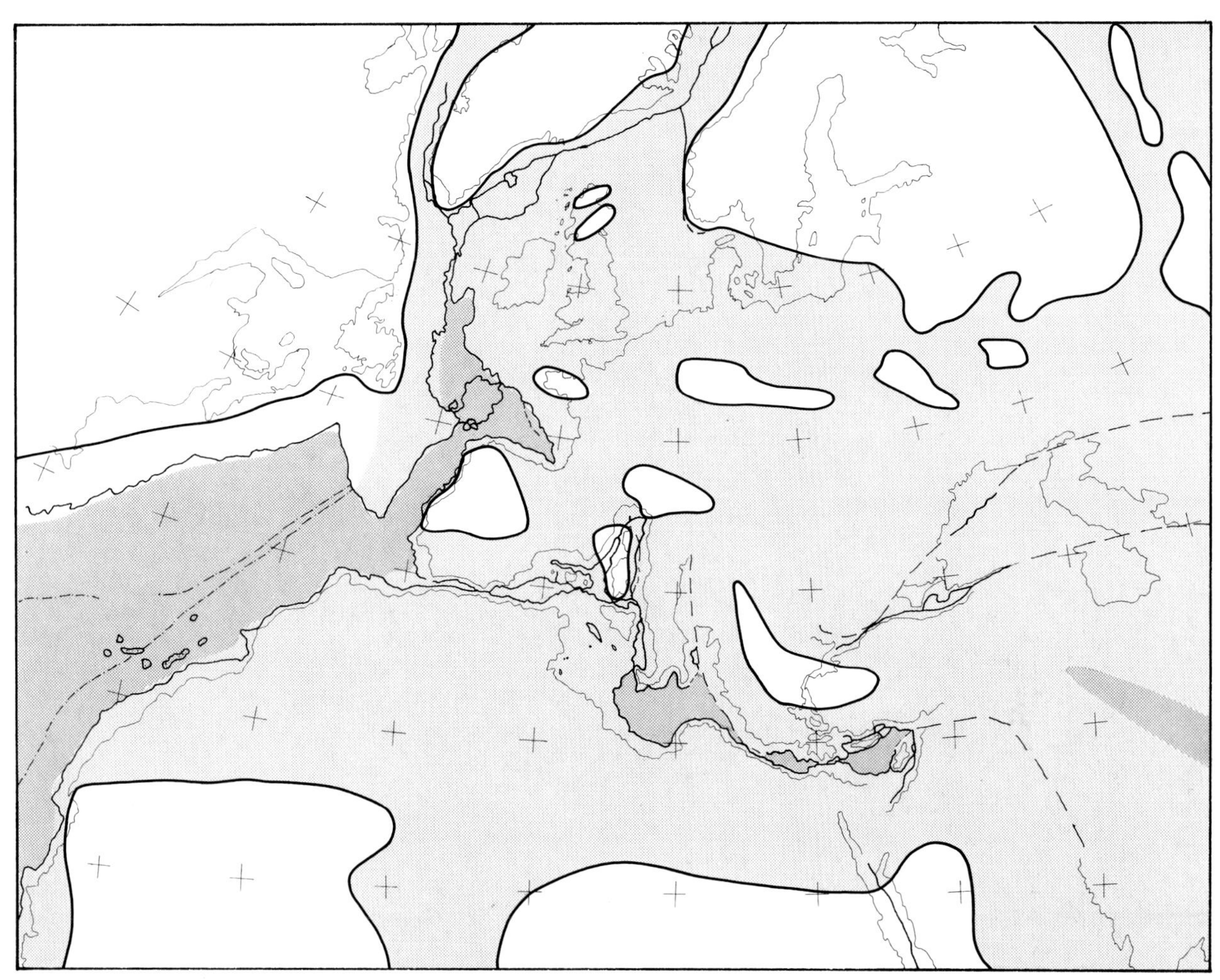

Figure 13.16 Palaeogeography of Europe and adjoining areas during the Upper Cretaceous, Campanian Stage (75–70 my) (base map from Owen, 1976, fig. 8*b*). The epicontinental sea over Europe and north Africa reached its maximum extent high in the Cretaceous at about the Campanian/Maastrichtian boundary, so the date of this map is slightly younger than the others in the Upper Cretaceous series.

in the middle of the west coast of Greenland. By the end of the Cretaceous only two continental blocks had still to commence their separation to complete the break-up of Pangaea. Greenland and Scandinavia were still together, though rifting had been active on the Greenland side at various times in the Mesozoic. Movement apart occurred during the Cenozoic, starting about the middle of the Palaeocene. Antarctica and Australia were still in contact at the end of the Cretaceous, but rifting at what is now their margins had occurred for much of the Cretaceous, and movement apart finally began at the end of the Palaeocene.

The Upper Cretaceous was remarkable for two events, firstly deposition of the unique and remarkably pure limestone, known as the Chalk, over most of Europe, and secondly the rise in sea-level, which reached a maximum height, and consequently the continents underwent their greatest inundation by the sea, at least since the Ordovician and perhaps in the whole of the Phanerozoic. It has been estimated that the Upper Cretaceous sea flooded nearly 40 per cent of the continental area, so that only 18 per cent of the Earth's surface was left as land, compared with 28 per cent today. Estimates of the maximum height reached by the sea during this transgression are 350 m (1150 ft) above present sea-level (Vail *et al.* 1977) and 650 m (2100 ft) above (Hancock & Kauffman, 1979). The lower figure is perhaps a more realistic assessment. (One of the main causes of major sea-level fluctuations, and the transgressions and regressions of the sea that they produce, is generally held to be alteration in the volumes of mid-oceanic ridges, especially during periods of rapid sea-floor spreading. The Mesozoic was largely free from ice ages, the other main cause of alterations in sea-level.)

At the beginning of Albian times sea-level was probably about the same as it is today; detailed analysis reveals a succession of transgressions and regressions in the Albian and the Upper Cretaceous that seem to be sufficiently synchronous in different parts of the world to imply

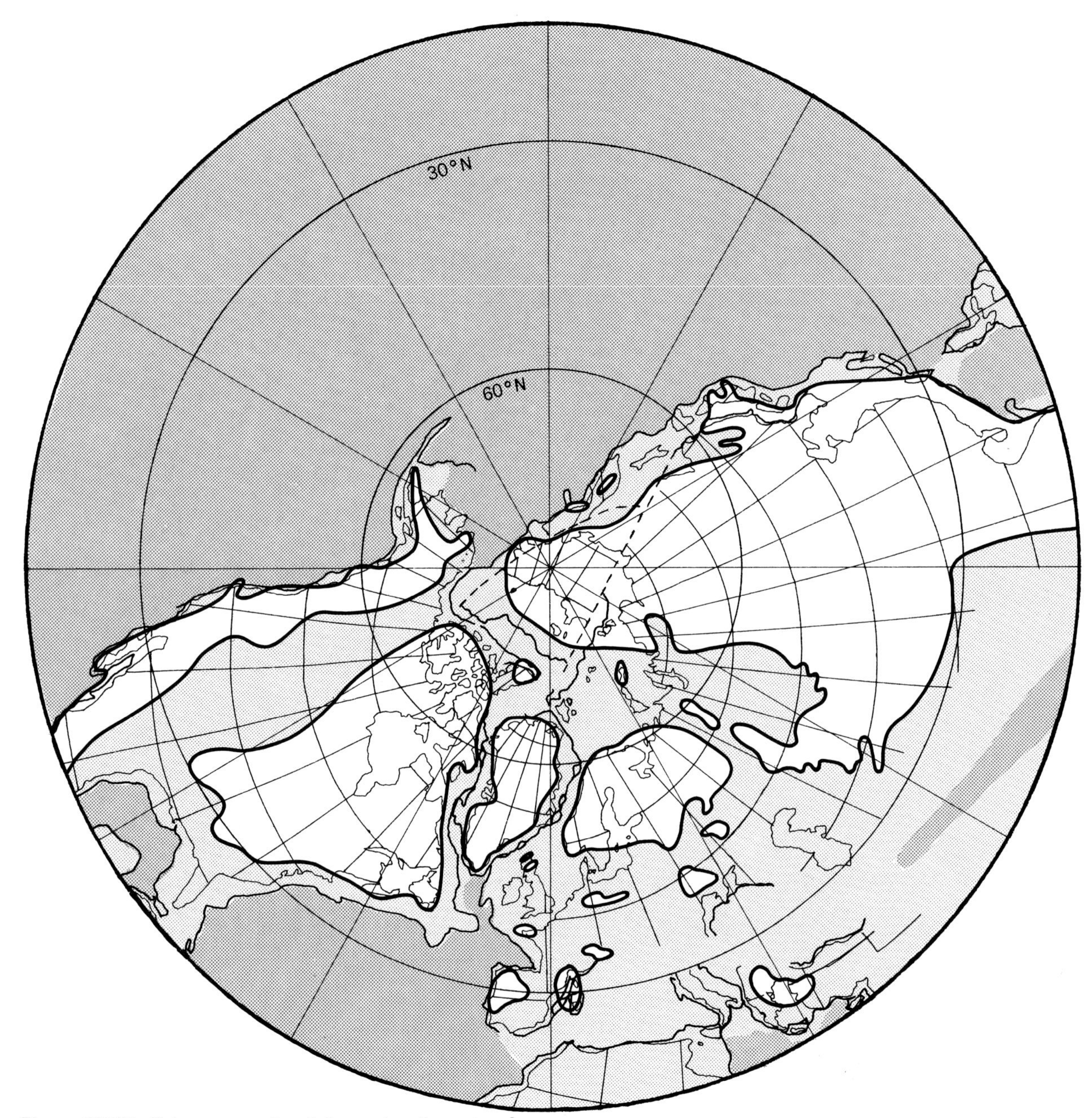

Figure 13.17 Palaeogeography of the north polar regions during the Upper Cretaceous, Senonian Stage (90–80 my) (base map modified from Smith & Briden, 1977, map 19). Since the Hauterivian (Fig. 13.14) the sea had withdrawn from the whole of north and east Siberia.

eustatic (i.e. world-wide) changes in sea-level. The first trangression that took seas onto several stable areas for the first time in the Mesozoic (e.g. west Africa and east India) was in the Upper Albian. Following regression during the late Albian and early Cenomanian, renewed trangression resulted in a major peak in the Lower Turonian. This was the greatest transgression in the mid-west United States, and a wide north–south epicontinental seaway was formed linking the Gulf of Mexico with the arctic (Gill & Cobban, 1973), with sea-level perhaps 300 m above the present level. The Lower Turonian was probably also the time of greatest transgression in western Russia: the sea in the Ob basin was at its largest, and a strait was open down the west side of the Urals, linking it with extensive seas round and to the west of the Caspian area (Vinogradov, 1968). This transgression was also responsible for the most

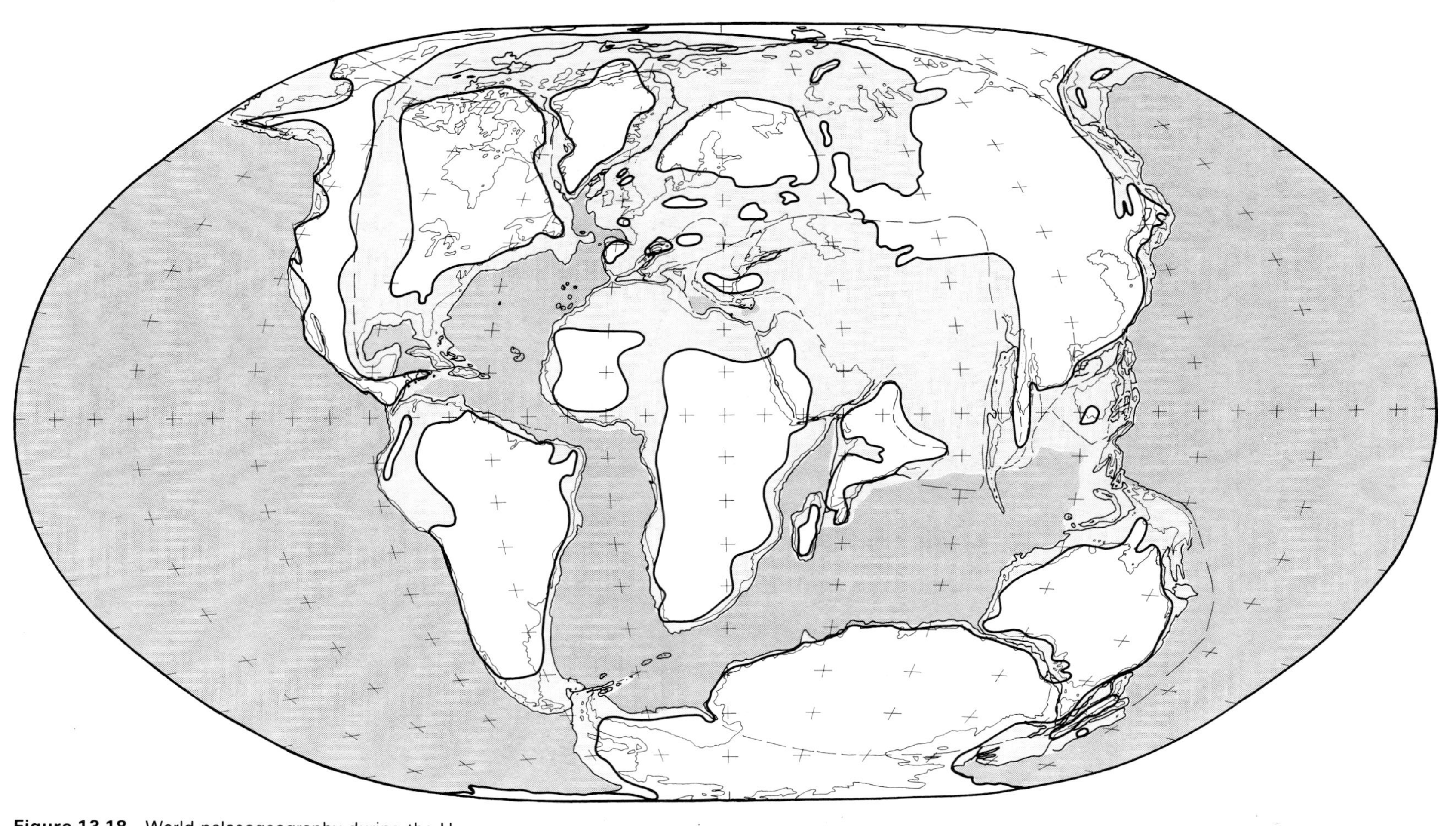

Figure 13.18 World palaeogeography during the Upper Cretaceous, Senonian Stage (90–80 my) on a base map reconstructed on a globe 90 per cent of the present Earth diameter (from H. G. Owen, new). Compared with Figure 13.15, Indonesia and Australasia are nearer together and differently orientated, and there is no oceanic crust in Tethys.

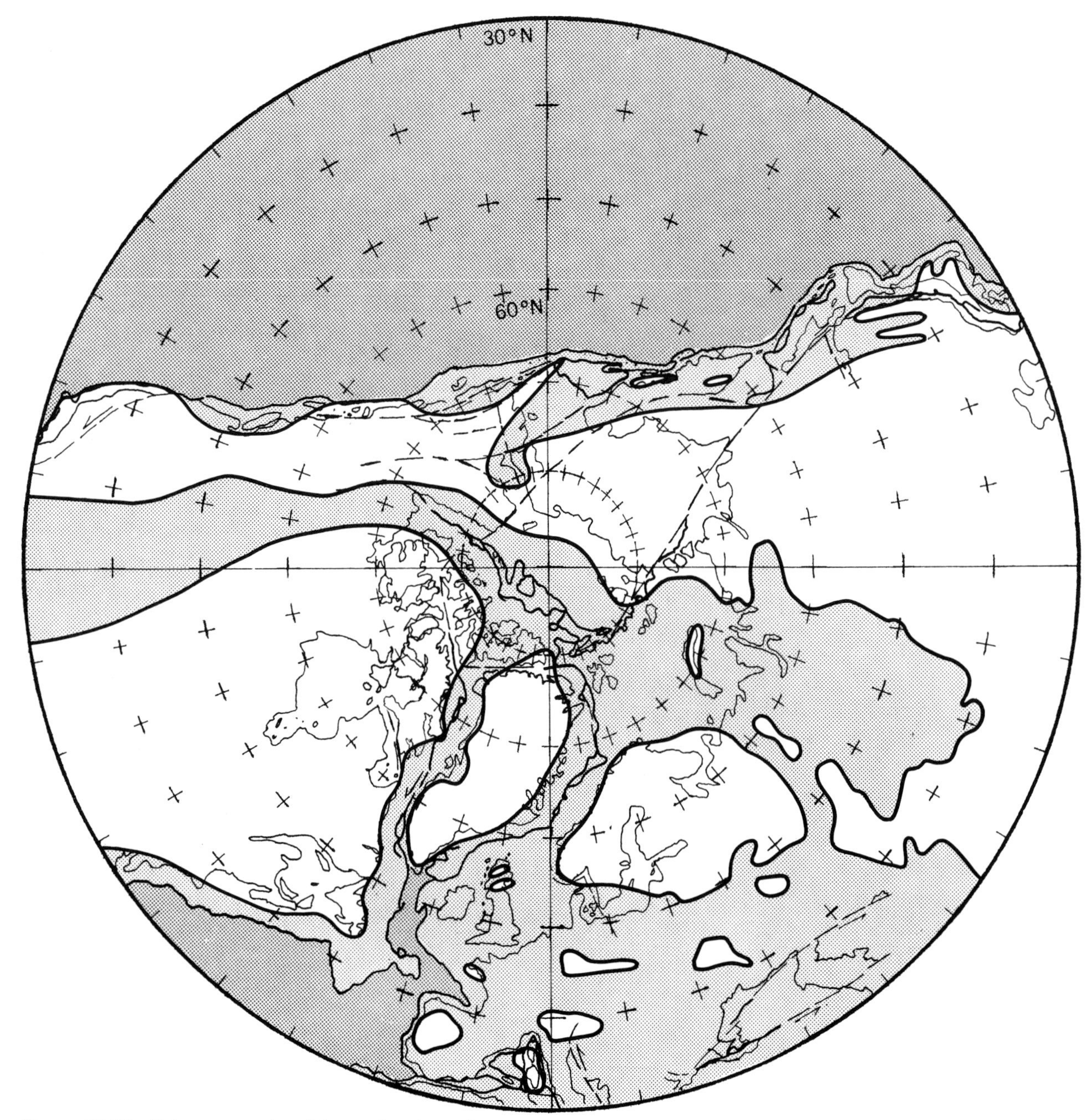

Figure 13.19 Palaeogeography of the north polar regions during the Upper Cretaceous, Senonian Stage (90–80 my) on a base map reconstructed on a globe 90 per cent of the present earth diameter (from Owen, 1976, fig. 6*c*). In this configuration the epicontinental Arctic sea is small, and has no outlet to the Pacific because of the continuous Alaska–Siberia land bridge.

remarkable feature of African Upper Cretaceous palaeogeography – the establishment of the trans-Saharan seaway, which was opened in the early Turonian and led to considerable migration of marine faunas between the Middle East and the Nigeria-Gulf of Guinea area (Reyment & Taitt, 1972). Another result of the early Turonian transgression was the invasion of an epicontinental sea from the west up the Kathiawar–Narmada gulf almost to the centre of India (Robinson, 1967).

After rapid regression in the middle and late Turonian, which broke the trans-Saharan seaway, the Gulf of Mexico–Arctic seaway, and led to withdrawal from central India, renewed transgressions occurred in both Coniacian and middle Santonian times. The latter transgression

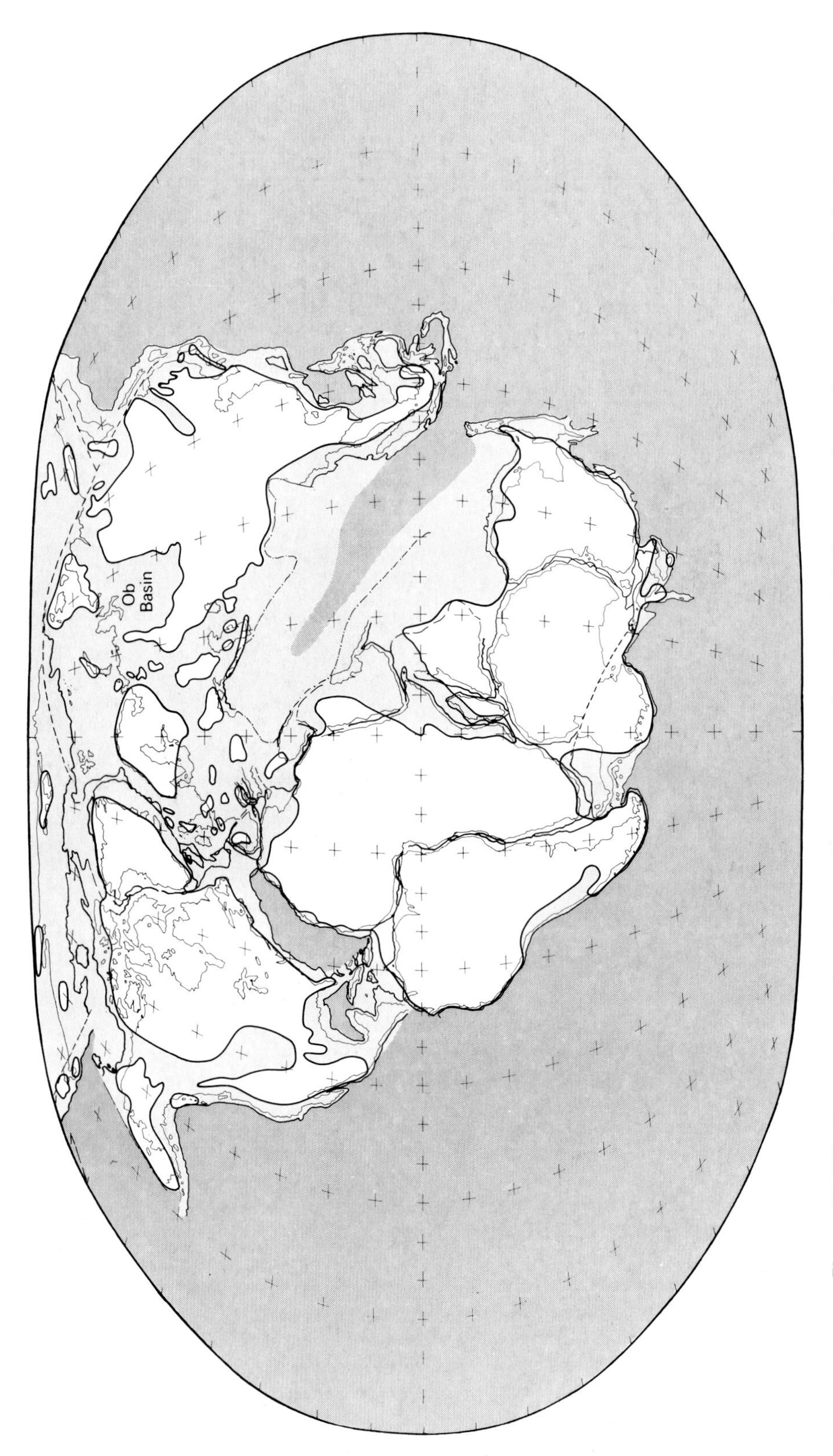

Figure 13.9 World palaeogeography during the Upper Jurassic, Kimmeridgian Stage (145 my) (base map modified from H. G. Owen, new). Compared with the Pliensbachian map (Fig. 13.6), Australia and Indonesia are closer together, and the amount of oceanic crust in Tethys has been reduced (by subduction). New oceanic crust is present in the Gulf of Mexico and the North Atlantic, due to the movement apart of North America and Africa. A large new area of epicontinental sea occurs on the Russian platform west of the Urals and in the Ob basin east of the Urals.

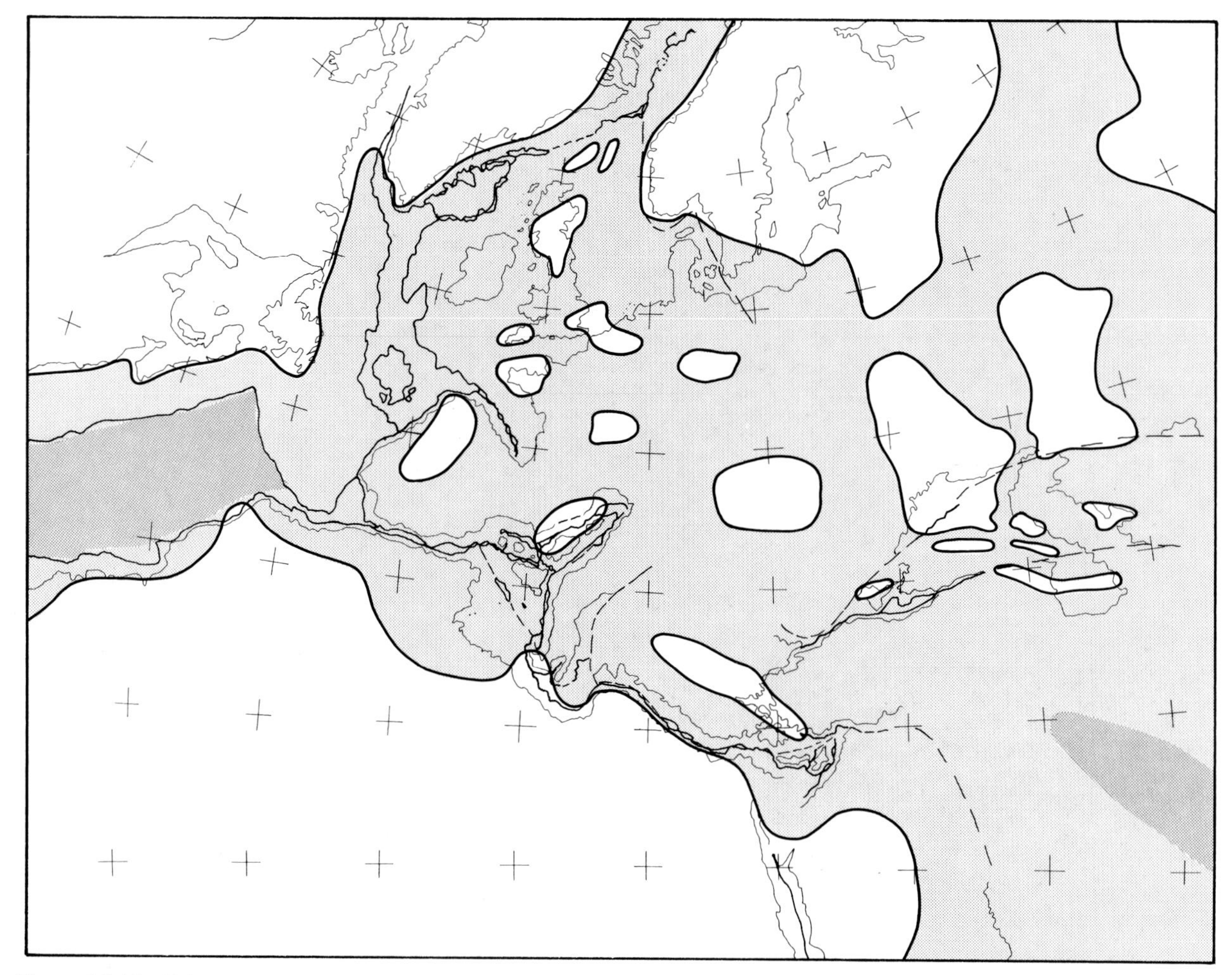

Figure 13.10 Palaeogeography of Europe and adjacent areas during the Upper Jurassic, Kimmeridgian Stage (145 my) (base map from Owen, 1976, fig. 8*a*, modified in the North Atlantic area). New features are the North Atlantic ocean and the large area of sea in the north-eastern part of the map.

Jurassic, the first break in the continental crust of Pangaea came in the Callovian, when North America started to rotate clockwise away from Africa. The northern end remained in contact in the Greenland–Scandinavia region, while sea-floor spreading further south formed new oceanic crust to initiate the North Atlantic Ocean. A permanent seaway now existed between western Europe and the region of the Gulf of Mexico.

The Kimmeridgian was, in general, the time of the greatest transgression of the sea during the Upper Jurassic. Withdrawal in some areas at the end of the Lower Jurassic was followed by renewed transgression during the Bajocian. The Bathonian was, however, a time of marked regression in Britain and north-west Europe, where deltaic, non-marine or brackish-water deposits are extensive. Transgression followed in the Callovian, and much of Europe was covered by an epicontinental sea until the end of the Kimmeridgian (Brookfield, 1973; Hallam, 1975). One of the greatest Jurassic transgressions in the world was across the Russian platform in the Middle and Upper Jurassic. Starting early in the Bajocian, the area of flooding increased steadily until in the Kimmeridgian a vast epicontinental sea existed across most of European Russia, linking Tethys with the Arctic ocean both east and west of the Caspian, and extending round the north end of the Urals into the large Ob basin, then eastwards to link up with the large epicontinental sea of north and east Siberia (Vinogradov, 1968).

North America and Greenland remained a single unit, with little change in palaeogeography except in the west, where the Mesocordilleran Geanticline, a long north–south peninsula of emergent rocks undergoing orogeny at depth, separated the Pacific coast geosyncline to the west from the epicontinental Sundance Sea that occupied a large area of the mid-west United States (Kummel, 1970). The sea transgressed northwards from the Gulf of Mexico into the

caused considerable enlargement of the sea in north-west Canada.

The Cretaceous transgressions had now passed their peak in North America, but in Europe the final transgression in the Upper Campanian and Lower Maastrichtian led to the highest sea-level of all, perhaps 350 m above the present level. Most of Europe (Fig. 13.16) had been covered by sea since the Cenomanian, with only small islands left emergent, and this sea attained its greatest extent late in the Campanian. In fact, with its Chalk limestone facies, much of Europe lacks any good evidence for an Upper Cretaceous shoreline, and the palaeogeography can do little more than depict as land those areas thought to be too high to be inundated (Debelmas, 1974; Matsumoto, 1977). In the Tethys region little land was left, and this final Cretaceous transgression re-established the trans-Saharan seaway down to the Gulf of Guinea.

Areas that do not fit the pattern of Upper Cretaceous transgressions inundating wide areas are mainly those bordering the Pacific Ocean. Much of the Pacific border was tectonically mobile, and the movements, intrusions and volcanism meant that transgressions due to eustatic rise in sea-level did not affect very large areas. In Japan the shoreline was little different from that in Lower Cretaceous times and there was an active orogenic area to the west (Audley-Charles, 1966; Minato, 1965). On the western border of North America a narrow zone of geosynclinal deposition bordered a wide orogenic belt to the east. In South America the Andes were tectonically active. The sea withdrew from the Chilean trough in the Cenomanian and did not return during the Cretaceous. Extensive transgression during the Albian and Cenomanian in Venezuela, Colombia, Ecuador and Peru established a large area of sea by the middle of the Senonian, when volcanic activity and a rising mountain chain appeared in the west (Harrington, 1962). In north-east Siberia the sea had completed the withdrawal that started in the Lower Cretaceous, and the whole area was land except for the Pacific Ocean border.

Australia is particularly anomalous with regard to Upper Cretaceous transgressions, because the large inland sea (see Figure 13.12) withdrew during the period Upper Albian to Middle Cenomanian, and only small coastal areas in the west and north-west have later Cretaceous marine deposits. This appears to imply that during the Upper Cretaceous most of Australia was above the probably 350 m maximum height attained by the sea. So far as is known most of Antarctica also escaped Upper Cretaceous incursions of the sea, for the only sediments of this age known there are clastic rocks of Campanian age in the Antarctic Peninsula (Stump, 1973). The sea gained access to the division between Australia and Antarctica that had been rift-faulted during much of the Cretaceous, and Upper Cretaceous marine sediments were deposited in basins in Victoria and the south of Western Australia (Kummel, 1970; Ludbrook, 1978). It is thought that this sea came from the west. New Zealand was separated from Australia by a continuous channel formed by sea-floor spreading that started in Turonian times. Only small parts of the east side of both islands and the northern tip of New Zealand were covered by the sea. A great regression in the upper half of the Maastrichtian led to withdrawal of the sea from most areas of the world that had been inundated earlier in the Upper Cretaceous.

Finally, the palaeogeography of the Upper Cretaceous has been transferred to the expanding Earth maps of Figure 13.18 and 13.19. The eastern half of Tethys is narrower in this reconstruction and its connection with the Pacific is much more restricted. Tethys is an entirely epicontinental sea, for there is no need to postulate the existence of oceanic crust. Other major differences are in the Arctic, where the sea between Greenland and northern Asia is narrower. This reconstruction brings western Alaska into contact with north-east Asia, and it is possible to merge the land-masses known to occur on each, so that there may have been a continuous strip of land between Alaska and Asia, and no direct connection between the Arctic sea and the Pacific Ocean (Figure 13.19).

References

Audley-Charles, M. G. 1966. Mesozoic palaeogeography of Australasia. *Palaeogeography, Palaeoclimatology and Palaeoecology* **2**: 1–25.

Audley-Charles, M. G. 1970. Triassic palaeogeography of the British Isles. *Quarterly Journal of the Geological Society of London* **126**: 49–89, pls. 7–13.

Blant, G. 1973. Structure et Paléogéographie du littoral méridional et oriental de l'Afrique, pp. 193–233. *in Sedimentary Basins of the African Coasts.* Part 2, South and East Coast. Association of African Geological Surveys, Paris.

Brookfield, M. Palaeogeography of the Upper Oxfordian and Lower Kimmeridgian in Britain. *Palaeogeography, Palaeoclimatology and Palaeoecology* **14**: 137–167.

Casey, R. & Rawson, P. F. 1973. The Boreal Lower Cretaceous. *Special Issue of the Geological Journal* **5**: 1–448.

Christie, R. L. 1979. The Franklinian Geosyncline in the Canadian Arctic and its relationship to Svalbard. *Skrifter Norsk Polarinstitutt*, **167**: 263–314.

Cook, T. D. & Bally, A. W. (Eds). 1975. *Stratigraphic atlas of North and Central America.* 272 pp. New Jersey: Princeton University Press.

Debelmas, J. 1974. *Géologie de la France*, Vol. 1, 296 pp. Paris.

Donovan, D. T. 1967. The geographical distribution of Lower Jurassic ammonites in Europe and adjacent areas. *Systematics Association Publication* **7**: 111–134.

Erben, H. K. 1957. Paleogeographic reconstructions for the Lower and Middle Jurassic and for the Callovian of Mexico. *Twentieth International Geological Congress* Mexico. Section 2: 35–41.

Gill, J. R. & Cobban, W. A. 1973. Stratigraphy and geologic history of the Montana Group and equivalent rocks, Montana, Wyoming, and North and South Dakota. *Professional Papers of the U.S. Geological Survey*, **776**: 1–37.

Hallam, A. 1971*a*. Provinciality in Jurassic faunas in relation to facies and palaeogeography. *Special Issue of the Geological Journal* **4**: 129–152.

Hallam, A. 1971*b*. Mesozoic geology and the opening of the North Atlantic. *Journal of Geology* **79**: 129–157.

Hallam, A. (1973). (Ed.). *Atlas of palaeobiogeography*. 531 pp. Amsterdam: Elsevier.

Hallam, A. 1975. *Jurassic environments*. 269 pp. Cambridge.

Hallam, A. 1977. Biogeographic evidence bearing on the creation of Atlantic seaways in the Jurassic, pp. 23–39 *in* West, R. M. (Ed.) Palaeontology and plate tectonics with special reference to the history of the Atlantic Ocean. *Special Publications in Biology and Geology, Milwaukee Public Museum* **2**.

Hallam, A. & Sellwood, B. W. 1975. Middle Mesozoic sedimentation in relation to tectonics in the British area, pp. 1–32 *in Proceedings of the Jurassic Northern North Sea Symposium*, Stavanger. Part 4.

Hancock, J. M. & Kauffman, E. G. 1979. The great transgressions of the late Cretaceous. *Journal of the Geological Society of London*, **136**: 175–186.

Harrington, H. J. 1962. Paleogeographic development of South America. *Bulletin of the American Association of Petroleum Geologists* **46**: 1773–1814.

Herron, E. M., Dewey, J. F. & Pitman, W. C. III. 1974. Plate tectonics model for the evolution of the Arctic. Geology. *Geological Society of America* **2**: 377–380.

Imlay, R. W. 1967. Twin Creek Limestone (Jurassic) in the Western Interior of the United States. *Professional Papers of the U.S. Geological Survey* **540**: 1–105, pls. 1–16.

Imlay, R. W. & Detterman, R. L. 1973. Jurassic Paleobiogeography of Alaska. *Professional Papers of the U.S. Geological Survey* **801**: 1–34.

Kaye, P. 1966. Lower Cretaceous palaeogeography of north-west Europe. *Geological Magazine* **103**: 257–262.

Khudoley, K. M. & Meyerhoff, A. A. 1971. Palaeogeography and geological history of the Greater Antilles. *Memoirs of the Geological Society of America* **129**: 1–200.

Kristoffersen, Y. 1977. Late Cretaceous sea floor spreading and the early opening of the North Atlantic, pp. 1–25 *in Proceedings of the Mesozoic Northern North Sea Symposium*, Oslo. Part 5.

Kummel, B. 1970. *History of the Earth*. 2nd edn. 707 pp. San Francisco: W. H. Freeman.

Laseron, C. 1969. *Ancient Australia*. 253 pp. Sydney: Angus & Robertson.

Ludbrook, N. H. 1978. Australia, pp. 209–249 *in* Moullade, M. & Nairn, A. E. M. 1978 (Eds) *The Phanerozoic geology of the world*, II. The Mesozoic, A: Amsterdam: Elsevier.

Lyu, K.-Y. 1962. (*Palaeogeographical atlas of China*). 119 pp. Moscow. (In Russian; translation of Chinese original).

Maack, R. 1969. *Kontinentaldrift und Geologie des südatlantischen Ozeans*. 164 pp. Berlin: de Gruyter.

Matsumoto, T. 1977. On the so-called Cretaceous trangressions. *Special Papers of the Palaeontological Society of Japan* **21**: 75–84.

Minato, M. 1965. *The geologic development of the Japanese Islands* 442 pp. 30 pls. Tokyo: Tsukiji Shokan.

Moullade, M. & Nairn, A. E. M. 1978 (Eds). *The Phanerozoic geology of the world II*. The Mesozoic, A. 529 pp. Amsterdam: Elsevier.

Owen, H. G. 1976. Continental displacement and expansion of the Earth during the Mesozoic and Cenozoic. *Philosophical Transactions of the Royal Society of London* Series A **281**: 223–291.

Reyment, R. A. & Taitt, E. A. 1972. Biostratigraphical dating of the early history of the South Atlantic Ocean. *Philosophical Transactions of the Royal Society of London* Series B **264**: 55–95, pls. 3–5.

Reyre, D. 1966. Particularités géologiques des bassins cotiers de l'ouest Africain, pp. 253–304 *in Sedimentary basins of the Africa Coasts, Part 1, Atlantic Coast*. Association of African Geological Surveys, Paris, pp. 253–304.

Robinson, P. L. 1967. The Indian Gondwana Formations – a review, pp. 202–268 *in Reviews for the first symposium on Gondwana stratigraphy*. International Union of Geological Sciences. Mar del Plata, Argentina.

Schuchert, C. 1955. *Atlas of paleogeographic maps of North America*. 177 pp. New York & London: Wiley.

Scotese, C. R., Bambach, R. K., Barton, C., van der Voo, R. & Ziegler, A. M. 1979. Paleozoic base maps. *Journal of Geology* **87**: 217–277.

Smith, A. G. & Briden, J. C. 1977. *Mesozoic and Cenozoic palaeocontinental maps*. 63 pp. Cambridge: Cambridge University Press.

Stevens, G. R. 1974. *Rugged landscape*. 286 pp. Wellington: Reed.

Stump, E. 1973. Earth evolution in the Transantarctic Mountains and West Antarctica, pp. 909–924 *in* Tarling, D. H. & Runcorn, S. K. (Eds) *Implications of continental drift to the earth sciences*, vol. 2. London: Academic Press.

Tanaka, K. & Nozawa, T. 1977. *Geology and mineral resources of Japan*. 3rd edn. Vol. 1, Geology. 430 pp. Japan: Hisamoto, Kawasaki-shi.

Thierry, J. 1976. Paléobiogéographie de quelques Stephanocerataceae (Ammonitina) due Jurassique moyen et supérieur. *Géobios* **9**: 291–331.

Vail, P. R., Mitchum, R. M. Jr. & Thompson, S. III. 1977. Global cycles of relative changes of sea-level. *Memoirs of the American Association of Petroleum Geologists* **26**: 83–97.

Vinogradov, A. P. (Ed.). 1968. *Atlas of the lithological-palaeogeographical maps of the USSR*. Vol. III, Triassic, Jurassic and Cretaceous. 71 maps. Moscow: Ministry of Geology.

Warrington, G. 1970. The stratigraphy and palaeontology of the 'Keuper' Series of the central Midlands of England. *Quarterly Journal of the Geological Society of London* **126**: 183–223, pls. 16–22.

Wills, L. J. 1951. *A palaeogeographical atlas*. 63 pp. London: Blackie.

Wills, L. J. 1970. The Triassic succession in the central Midlands in its regional setting. *Quarterly Journal of the Geological Society of London* **126**: 225–285, pl. 15.

Woodland, A. W. (Ed.). 1975. *Petroleum and the continental shelf of north-west Europe*. Vol. 1, Geology. 501 pp. London: Institute of Petroleum.

Ziegler, A. M., Scotese, C. R., McKerrow, W. S., Johnson, M. E. & Bambach, R. K. 1977. Paleozoic biogeography of continents bordering the Iapetus (Pre-Caledonian) and Rheic (Pre-Hercynian) Oceans, pp. 1–22 *in* West, R. M. (Ed.), Paleontology and plate tectonics with special reference to the history of the Atlantic Ocean. *Special Publications in Biology and Geology, Milwaukee Public Museum* **2**.

M. K. Howarth
Department of Palaeontology
British Museum (Natural History)

CHAPTER 14

An outline of Tertiary palaeogeography

C. G. Adams

The Tertiary sub-era comprises five epochs and represents the greater part of the last 65 million years of Earth history (Table 1). The geographical, oceanographical, climatological, faunal and floral changes which occurred during this time were considerable and closely interrelated. This article briefly describes the principal changes, concentrating on those believed to be of the greatest biogeographical importance. For this reason, geological events in areas such as the Middle East are given rather more attention than similar changes in, for example, Australasia, since the former region was of greater significance than the latter in a global context.

The geographical configuration of the world as it was during Palaeocene times is shown in Figure 14.1. According to the latest geophysical evidence, South America lay to the east of its present position, thus reducing the South Atlantic to 75 per cent of its current width. North America and Greenland were slightly closer to Europe than today, the former being connected to Siberia via the Bering Strait and, perhaps, through Greenland to Europe. Africa lay a few degrees farther south than now, and Arabia was rotated in a clockwise direction so that its present north-west to south-east axis was aligned in a NNW to SSE direction. Arabia was separated from Eurasia by the ancient Tethys Sea which still linked the North Atlantic to the Indian Ocean across the Middle East except, perhaps, during the earliest Palaeocene when the sea was generally regressive. The Indian sub-continent lay across the equator, and Australia formed part of Antarctica.

The arrangement of the continents and oceans in Palaeocene times was thus sufficiently different from today to ensure that, in the southern hemisphere at least, their climates were dissimilar. The absence of polar ice, possibly an indication of greater solar radiation, produced a more equable global climate and ensured that the northern continents in particular enjoyed a warmer régime than they do today. The distribution of land south of the equator created a different pattern of surface currents in the oceans, and modified the local climate accordingly.

In palaeogeographical terms, Cenozoic history charts the movement of the continents towards their present positions and the associated alterations in the ocean basins. It includes the disappearance of the Tethys Sea owing to

Table 1 Divisions of Cenozoic time.
(Compiled from Berggren & Van Couvering, 1974; Haq *et al.* 1977 and Hardenbol & Berggren, 1978).

Era	Sub-era	Period	Epochs	European stages	Time my
Cenozoic	Quaternary		Holocene		0·01
			Pleistocene	Calabrian	1·6
	Tertiary	Neogene	Pliocene	Piacenzian	3·5
				Zanclian	5·0
			Miocene	Messinian	6·6
				Tortonian	10·6
				Seravallian	13·5
				Langhian	15·5
				Burdigalian	19
				Aquitanian	25
		Palaeogene	Oligocene	Chattian	32
				Rupelian	35
				'Lattorfian'	37
			Eocene	Priabonian	40
				Bartonian	44
				Lutetian	49
				Ypresian	53·5
			Palaeocene	Thanetian	60
				Danian	65

Note The Tertiary sub-era comprises the Palaeogene and Neogene periods. Hardenbol & Berggren (1978) recognise only two divisions of the Oligocene, placing the base of the Rupelian at 37 my. The Sparnacian (p. 227) is equivalent to the earliest Ypresian as defined here.

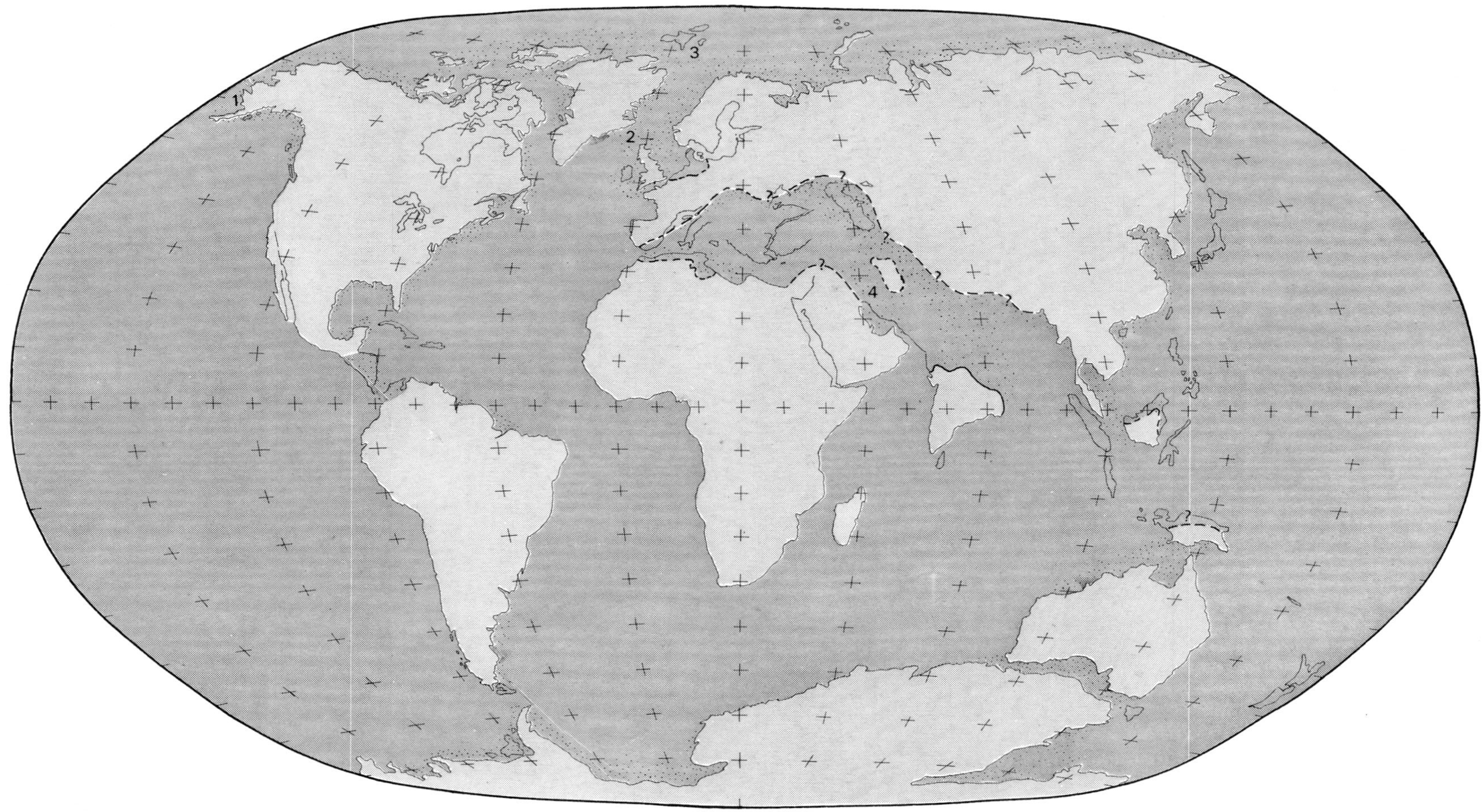

Figure 14.1 Geographical arrangement of the continents in Palaeocene times. Black stipple indicates shelf areas that were probably mainly land. 1 = Bering land bridge. 2 = Greenland-Iceland-Faroes-Britain bridge. 3 = Greenland-Barents Sea-Scandinavia bridge. 4 = postulated Tethys bridge (earliest Danian only). Minor continental margin basins omitted; present Antarctic shoreline shown in absence of other information. Base map by H. G. Owen, Winkel 'Tripel' Projection. Centre Meridian 10° E.

the gradual northward movement of Africa and India, the loss of both eastward and westward land connections between North America and Eurasia, the initiation of the circum-Antarctic Current as Australia broke away from Antarctica, the construction of the Alpine-Himalayan mountain chains, the building of the Andes, the erosion and re-elevation of the Rocky Mountains, and, at a later stage, the separation of the central Atlantic from the Pacific as North and South America became connected by a land bridge.

Cenozoic marine strata can now be dated very accurately using microfossils, and it is theoretically possible to prepare palaeogeographical maps for time intervals as short as 2 to 5 million years. The necessary techniques have been available for only about 25 years, however, and much biostratigraphical work published before the middle of this century needs to be re-evaluated before interregional correlation can be made sufficiently accurate to permit global palaeogeographical reconstructions of that order. It is, of course, possible to study only that part of geological time now represented by rocks exposed at the earth's surface or in boreholes, and the paucity of fossiliferous sediments in some critical areas creates difficulties in the preparation of palaeogeographical maps, as does the uneven quality of the data available from different parts of the world. Information is much more detailed for Europe, the cradle of geological knowledge, than for regions such as Indonesia or South America. On a different scale, data from oil-rich western Iran, a region which has been subjected to intensive geological investigation (see James & Wynd, 1965), are of an entirely different order from those of the geologically no less important, but non-oil producing, eastern part of the same country. Similar contrasts occur in many other parts of the world.

The two principal maps (Figs. 14.2 and 14.5) show the maximum extent of the seas during the Palaeogene and Neogene periods. They are drawn in accordance with the expanding Earth hypothesis but would not look significantly different on a constant dimensions base, since the Earth had attained 95·5 and 97·7 per cent of its final diameter by Middle Eocene and Middle Miocene times respectively. The information on which they are based is taken from modern geological maps and/or palaeogeographical reconstructions (see figure captions for references). The situation in Antarctica is little understood owing to the paucity of information on the Tertiary strata of this continent. It is possible that in some areas, South and West Australia for example, the seas covered slightly more land than is shown here. However, if this was so, the sediments deposited in them have long since been removed by erosion.

The maps are misleading to the extent that at any particular time the total land area would have been larger, and the sea area smaller, than is shown here. Figure 14.2, for example, shows the northern margin of the sea in Europe extending from the Baltic to north of the Caspian. This is correct for the Oligocene but incorrect for the Eocene, when the sea covered only a small part of Europe. In contrast, the shorelines of the Obik Sea (Fig. 14.2) are correct for the Eocene but quite wrong for the Oligocene, when this sea had ceased to exist. The map therefore shows composite shorelines and indicates only that the sea covered an area during part of the period and not that it was there all the time. To overcome this problem without producing epoch and stage maps for the whole Tertiary (the work required would be disproportionate to the value of the results obtained, since changes in areas such as Australia and North America were insignificant in global terms), supplementary maps (Figs 14.3 to 14.7 and 14.6 to 14.8) are included. These show important areas at critical times or provide alternative interpretations of the available information.

Areas of the continental shelf (e.g. the coast of China) that were largely land throughout the Neogene and Palaeogene, or which were only intermittently subjected to indundation are stippled on Figures 14.2 and 14.5. These two maps show the maximum size of the marine barriers to continental migrations and thus the routes available for the dispersal of marine organisms. It has to be remembered that during any one epoch numerous islands would have existed in the shallow shelf seas, particularly in the Tethyan region, the central Americas, and Indonesia. Although some islands were large and existed for millions of years, they are not shown here because they were fundamentally transitory in nature. Some doubtless started their existence as peninsulas and were later transformed into islands.

Repeated connections and disconnections of peninsulas and islands could have permitted some animals and plants to migrate from one continent to another without ever crossing a water barrier (Figs. 14.2 and 14.5 give some impression of the complicated island geography of Europe). The existence of islands between Europe and Africa, or between Arabia and Asia during the Palaeogene does not necessarily mean that they provided migration routes between the continents, but only that they might have done. The line of the Panama Isthmus, although never a land bridge until Pliocene times, was certainly marked by volcanoes throughout much of the Tertiary (Lloyd, 1963). There is, however, no reason to suppose that these ephemeral volcanic islands provided a migration route for terrestrial animals and plants.

Great thicknesses of marine Tertiary clastic sediments were deposited in narrow troughs in front of the rising Alpine chain in southern Europe, but in Britain, and over northern Europe generally, marine sediments accumulated in slowly-subsiding basins where more time may be represented by breaks in the sequences than by the sediments themselves. The successions in these basins are relatively thin, and until fairly recently were difficult to correlate with the enormous thicknesses of sediments which accumulated in the geosynclinal troughs associated with the Alpine–Himalayan chain, the central Americas and the Indonesian Archipelago. In areas where the seas

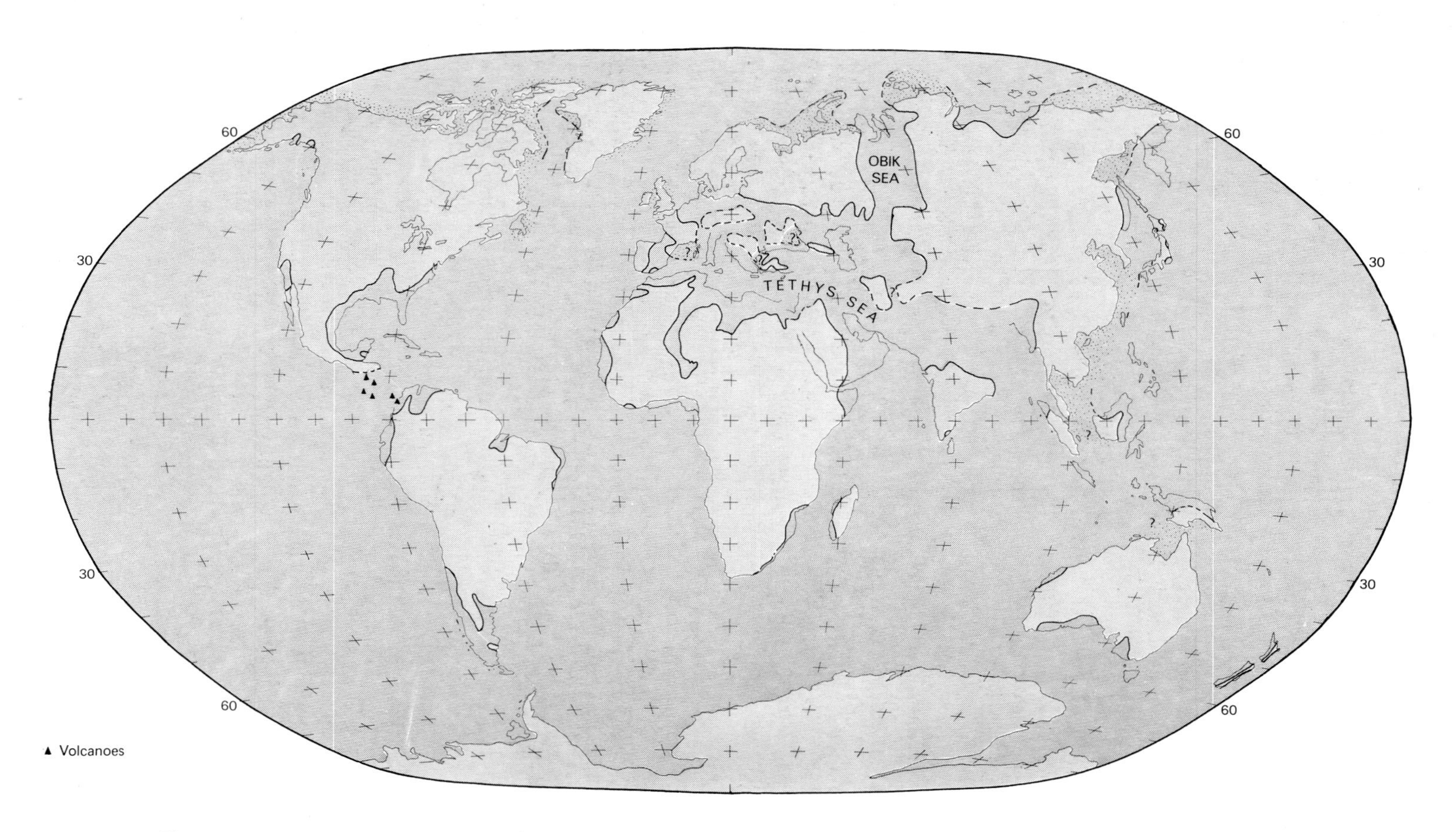

Figure 14.2 Maximum extent of the world oceans during the Palaeogene. A composite map, the shorelines indicating the maximum extent of marine deposition during the Palaeocene, Eocene and Oligocene. Black stipple indicates shelf areas that were probably mainly land. Based on sources too numerous to list, but owing much to Bender (1975), Brinkmann (1960), Cook & Bally (1975), Debelmas (1974), Desio (1971), Dunnington (1958), Gignoux (1955), Harrington (1962), Huber (1978), James & Wynd (1965), Kingma (1974), Lloyd (1963), Lüttig & Steffens (1976), Minato *et al.* (1965), Reyment (1966), Reyre (1966), Said (1962), Vinogradov (1967), Visser & Hermes (1962). Also incorporating information from volumes of the Lexique Stratigraphique International and from numerous geological maps. Base map by H. G. Owen, Winkel 'Tripel' Projection. Centre meridian 10° E. Continental positions shown for the Middle Eocene. Present Antarctic shoreline shown in absence of other information.

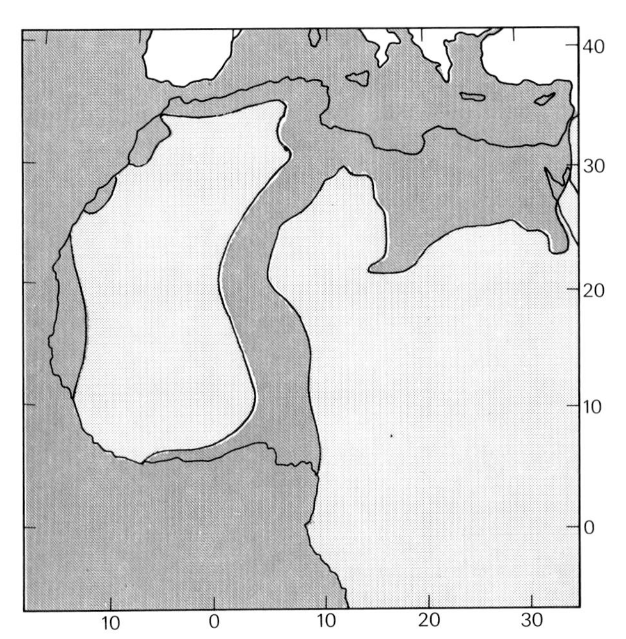

Figure 14.3 The trans-Saharan seaway in the Palaeocene. After Reyment (1966).

were warm, and where terrigenous sediment was in short supply, e.g. over much of North Africa and Pakistan during the Eocene, great spreads of limestone formed from the skeletal remains of marine organisms – mainly foraminifera and calcareous algae.

For some as yet unexplained reason, the Cretaceous Era ended with a fall in the level of the world oceans. The seas, which during Late Cretaceous times had flooded large areas of every continent (see Chapter 13), with the possible exception of Antarctica about which relatively little is known, withdrew some distance down the continental shelves, leaving the last-formed sediments exposed to the destructive effects of erosion. Over the outer parts of the shelves and in the ocean basins, marine sedimentation continued without interruption, but only in a few places on the continents (e.g. north-west Africa), where locally the rate of subsidence outpaced the general fall in sea-level, is there evidence for continuous marine sedimentation across the Cretaceous–Cenozoic boundary. Elsewhere, this is marked by a pronounced erosion surface (disconformity or unconformity). Although the precise cause of the sea-level fall is not known, it can be explained by a change (local deepening) in one of the ocean basins. The only other possible cause would be polar glaciation and there is no evidence for that.

The Palaeogene Period

As already mentioned, the beginning of Palaeocene times saw the continents emergent and the seas standing at an unusually low level. However, this situation did not endure for long. Within a few million years, sea-level again rose and during the Danian some marginal areas of the continents were flooded. Although none of the continents was inundated to the same extent as during the Cretaceous, North Africa was largely submerged and an arm of the sea extended southwards towards an embayment on the north side of the Gulf of Guinea (Fig. 14.2). Some geologists (e.g. Reyment, 1966) believe that a marine connection, the trans-Saharan seaway, existed in Palaeocene times between the western part of the Tethys and the Gulf of Guinea (Fig. 14.3). However, the evidence for this is inconclusive, depending as it does on palaeontological similarities between certain microfaunas in North Africa and Nigeria. There is no continuous band of Palaeocene sediments between the two areas, and the palaeontological evidence is therefore susceptible to more than one interpretation as shown by Petters (1977). Nevertheless, the seas did occupy a greater area of Africa during the Late Palaeocene than at any subsequent time in the Tertiary, although marginal areas, especially in the north, were often submerged.

The Arctic and the Tethys were joined by the Obik Sea which transgressed from the north during Palaeocene times (Fig. 14.4A) and which throughout most of the Eocene extended across the Russian Platform (Figs. 14.4B, C). By Oligocene times (Fig. 14.4D) it had vanished (Vinogradov, 1967). The rest of the huge land-mass of Asia remained above sea-level throughout the Palaeogene, except for a narrow zone in the north east and an embayment in the north (Fig. 14.2). Most of the present-day continental shelf of China was the site of terrestrial and fluviatile deposition throughout most of the Tertiary, and the minor temporary incursions into subsiding basins running inland from the China Sea are too insignificant to be shown on maps of this scale. The palaeogeography of Japan is also difficult to depict on these maps, and for further information the reader is referred to Minato *et al.* (1965).

North and South America stood largely above sea-level during Palaeogene times (Cook & Bally, 1975; Harrington, 1962), but the Gulf Coast region, the southern and north-western parts of South America, and isolated basins on the western seaboard were submerged. South America had a rather more complex Cenozoic history than its northern counterpart owing to the geosynclines that stretched across parts of Venezuela, Colombia, Ecuador and Peru, and to a series of transgressions that affected the southern part of the continent. Surprisingly, perhaps, the Amazon basin was the site of continental deposition throughout the Cenozoic, and was not flooded by the sea even during the Palaeocene (Harrington, 1962).

The Panama Isthmus provides the main focus of palaeogeographical interest in the New World because of its position between the northern and southern continents. Unfortunately, owing to the nature of the present terrain, vegetation, and climate, it is also one of the most difficult areas to decipher. However, it is clear that there cannot have been any continuous land connection between North and South America until mid-Pliocene time, although

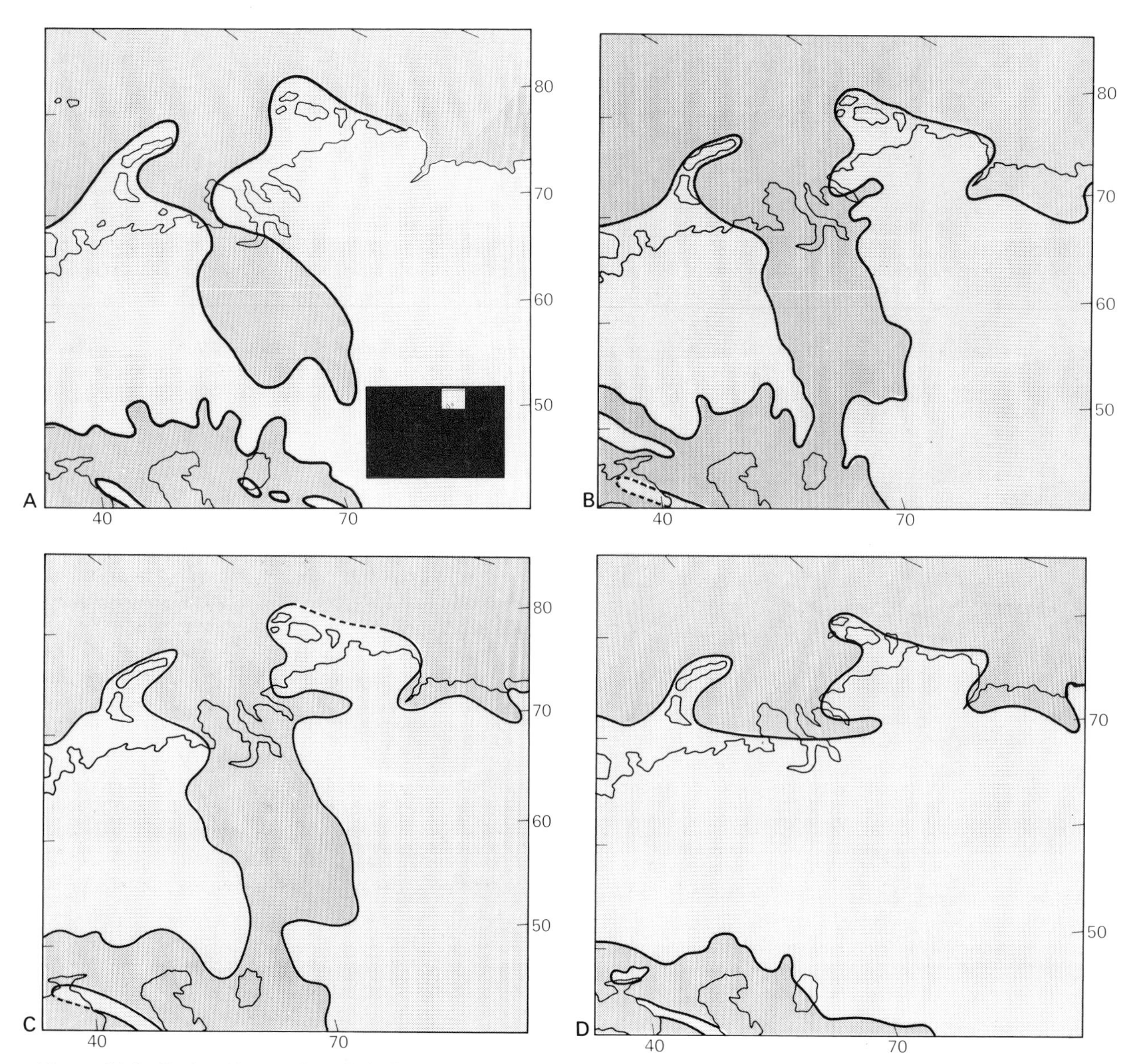

Figure 14.4 Tertiary history of the Obik Sea and Turgai Strait USSR. After Vinogradov (1967): A, Palaeocene; B, Early to Mid-Eocene; C, Late Eocene; D, Oligocene.

chains of volcanic and other ephemeral islands were interposed between the waters of the Atlantic and Pacific. Lloyd (1963) has published a useful modern review of the area.

The Palaeogene history of western and southern Europe is rather complex (see Brinkmann, 1960, Gignoux, 1955) owing to the tectonic movements that produced the Pyrenean and Alpine mountain chains, to areas of subsidence, and to sea-floor spreading which affected the North Atlantic both west and east of Greenland. The east to west (Baltic) and north to south (Rhenish) zones of subsidence which first appeared during the Late Permian were again active during the Palaeogene. In western and central Europe the area of Palaeocene submergence was enlarged during the Eocene and Oligocene, the maximum transgression occurring in the latter epoch when the northern margin of the sea extended from the present-day Baltic coast east-southeastwards towards the Caspian and Aral seas (Fig. 14.2). In southern Europe, the Pyrenean orogeny, which began in the Cretaceous, had a very active phase in the Mid-Late Eocene and there was further folding in the Oligocene. The main effects of the Alpine orogeny also began to be felt in Europe during the Oligocene.

In order to account for the distribution of Palaeogene mammals in North America and Eurasia, three bridges have been postulated between the two land masses. The first, across the Bering Strait, is generally accepted; the second and third are really alternative hypotheses since either or both would account for known distribution patterns. One would have stretched from Greenland, across what are now Iceland and the Faroes into Britain; the other from Greenland across the Barents Sea to Scandinavia. The supposed connections are based upon similarities between the faunas obtained from three or four mammal sites in Europe, a dozen or so in the Rocky Mountains, and one in Asia. Kurtén (1973) lists 17 Palaeocene genera from Europe, 63 from North America, and observes that 10 are common to the two areas. This may seem slender evidence on which to postulate land bridges, especially when allowance is made for our ignorance of Early and Middle Palaeocene faunas and for possible taxonomic errors, but it cannot be ignored. The European faunas are very different from those so far described from Asia which, however, also include taeniolabiid multituberculates, pantodonts, dinocerates and notoungulates, all of which have American affinities (Kurtén, 1966). The Early Eocene (Sparnacian) mammals of Europe also closely resemble those of America, at least at generic level, and McKenna (1975) states that of the 60 described genera from Europe, 34 also occur in North America. Asian and European faunas of this age are dissimilar in that they have few genera in common. From Middle Eocene times onwards, the mammalian faunas of North America and Europe developed differently, and it is possible that sea-floor spreading which occurred in the Early Palaeogene between Greenland and Scandinavia would have broken any land connections existing in Palaeocene and Early Eocene times if they had not been previously disrupted by the rising sea-level.

From the palaeobiogeographer's viewpoint, the Middle East was probably the most important area of Tertiary sedimentation in the world, for although the Tethys Sea was slowly narrowing, it still provided a marine connection between the Atlantic and Indo-Pacific. This link persisted throughout the Palaeogene, although the possibility of a temporary disconnection during the Oligocene cannot be discounted owing to evidence for a regression in Turkey during the later part of this epoch. The south-western shoreline of Asia is difficult to plot accurately (Fig. 14.2) owing to inadequate published information on the Early Tertiary successions of Afghanistan and eastern Iran. Some palaeogeographers (e.g. Termier & Termier, 1960; Minato *et al.* 1965) have postulated a continuous seaway across this region, but the critical strata do not seem to be closely dated or described. Eastern central Iran and the adjacent Lut Block were certainly land (Stöcklin, 1968; Huber, 1978), and the apparent absence of typical Indo-Pacific faunas (especially larger foraminifera) in southern Russia suggests that a land barrier existed between this area and what is now the northern part of Pakistan.

In view of the supposed width of that part of the Tethys Sea which lay between India and Asia during the Early Palaeogene, it is surprising that so little marine sediment has been described from northern India. A few patches of limestone, shale and sandstone, up to 300 m thick in Nepal and Tibet (Sharma, 1977; Ho *et al.* 1976), seem to be all that is known of the Palaeogene sediments that should have been deposited in that part of the Tethys which lay to the north of the Indian subcontinent. In fact, the disposition of Palaeogene sediments around India is consistent with a more northerly position for the subcontinent than geophysicists allow.

Australia was mainly land except for small basins of deposition in the south and west, and for New Guinea in the north. The interior was a flat-lying plain over which volcanoes erupted from Victoria to northern Queensland during the Oligocene and poured out vast quantities of lava (Laseron, 1969). Great thicknesses of lignites accumulated in Victoria at this time. The occurrence of the planktonic foraminifer, *Guembelitria* aff. *stavensis* Bandy, in deep-sea sediments of Oligocene age off South Africa and in the South West Pacific, provides the first palaeontological evidence (rather weak since these occurrences can be explained differently) for the initiation of the circum-Antarctic Current (Jenkins, 1974). However, a shallow sea must have separated the two continents in Eocene and perhaps also Palaeocene times, and an inlet from the west was already present in the Late Cretaceous (Fig. 13.15). New Zealand is of special interest to biogeographers owing to its isolated position. Its importance in terms of global biogeography is, however, minimal for the same reason, and it is sufficient to state here that during the Palaeogene the South and North Islands were occupied by north-east to south-west trending geosynclinal troughs separated by a geanticlinal ridge (Kingma, 1974).

The migration of land animals between the continents of the northern and southern hemispheres (the latter including Africa) was effectively prevented by the presence of marine barriers during the Palaeogene. Migration between Europe and Asia was impeded throughout much of the Eocene by the Obik Sea which stretched north–south across Russia. A similar but shorter-lived barrier may have existed across North Africa during the Palaeocene (Fig. 14.3). It is pertinent to reflect that although references are constantly found in the literature to 'migrations' of animals throughout Europe and from Europe to Asia or America, thus creating the impression that Europe was virtually a single land mass as it is today, this is a misconception. Although a single land area in the early Palaeocene, throughout Palaeogene and Early Neogene times Europe was little more than a changing pattern of islands centred round a few relatively stable blocks. This is important since it affects our view of mammal dispersals. These may have been achieved not so much by steady migrations, or by 'island hopping' involving precarious, and often inexplicable, crossings of open waters, but by opportunistic dispersals as and when temporary land

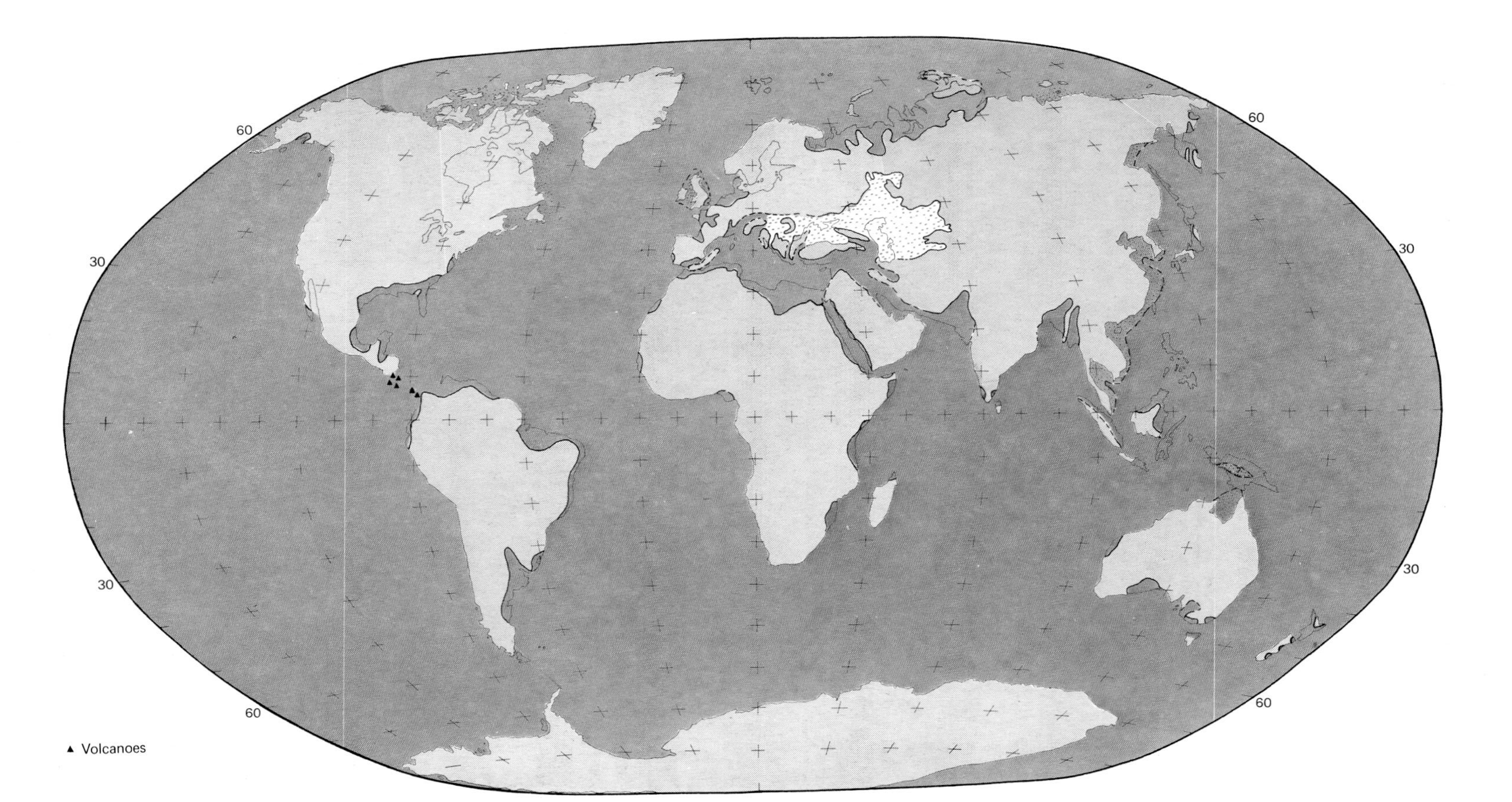

Figure 14.5 Maximum extent of the world ocean during the Neogene. A composite map, the shorelines indicating the maximum extent of marine deposition during the Miocene and Pliocene. Black stipple indicates shelf areas that may have been mainly land. Sources as for Figure 14.2 plus Brown & Jones (1971), Heybroek (1965), Seneš & Marinescu (1974) and Steininger *et al.* (1976). Base map by H. G. Owen, Winkel 'Tripel' Projection. Centre meridian 10° E. Continental positions shown for the Middle Miocene. Present Antarctic shoreline shown in absence of other information.

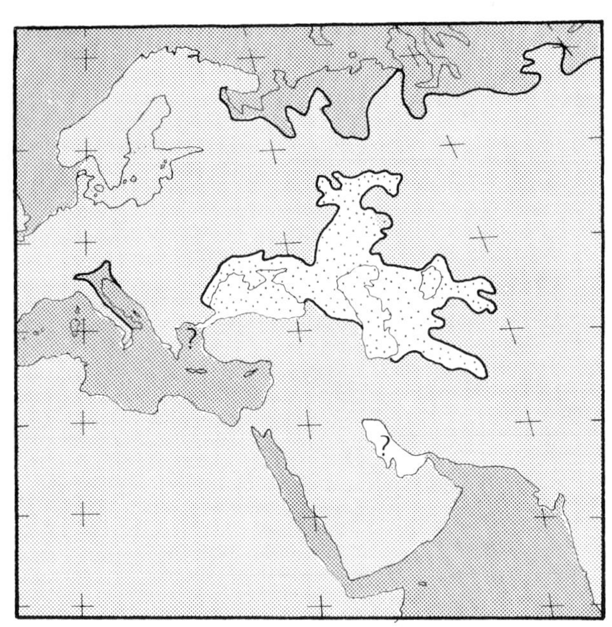

Figure 14.8 Central Europe and the Middle East during the Pliocene, compiled from many sources. Extent of Aralo-Caspian Sea (stippled) in the Late Pliocene from Vinogradov (1967).

During the Late Miocene the eastern Paratethys was a huge area of brackish water which extended north of the Aral Sea during the Late Pliocene (Fig. 14.8). A connection between the Black Sea and the Mediterranean was re-established through the Bosphorus in Quarternary times; this destroyed the Caspian brackish-water fauna and replaced it by that of the Mediterranean in Tyrrhenian times (Smirnov, 1958). The water of the Black Sea today is below normal marine salinity and devoid of oxygen below 525 feet. Lake Balaton, and the Black, Caspian and Aral Seas are all that remain today of the Paratethys.

The final closure of the Tethys in the area now known as the Middle East was preceded by the deposition of limestones and other shallow-water marine sediments. The distribution of these now permits us to plot the course and extent of the seaway which, in Aquitanian times, could have been no more than 200 to 500 km wide in places; the larger foraminifera found in the limestones enable us to date the disconnection of the Mediterranean from the Indian Ocean as occurring during the Burdigalian (14·5 to 19 million years ago). From that time onwards the faunas of the Mediterranean and Indo-Pacific evolved separately. The effect of the disconnection on the Mediterranean faunas, as exemplified by the foraminifera, was profound: their diversity decreased, and the whole character of the fauna changed. Over most of the Middle East, continental deposits were laid down during the Middle Miocene, the only exceptions being in a few coastal areas, for example south-western and southern Iran, where marine sedimentation continued briefly into the Middle Miocene (James & Wynd, 1965). The conjunction of Eurasia and Africa allowed African elephants, bovids and pigs to migrate into Europe.

Recently, Chinese workers (Hsü *et al.* 1978) have suggested that towards the close to the Miocene Epoch, some 5 to 6 million years ago, the Mediterranean Sea temporarily lost its connection with the Atlantic. Geophysical data, coupled with evidence from deep-sea drill holes, indicate that great quantities of evaporites (mainly anhydrite and halite), amounting to one million cubic kilometres, are buried below sediments of Pliocene age on the floor of the present Mediterranean. Since the upper part of the evaporite sequence shows features characteristic of deposition under shallow water conditions, and the Mediterranean is known to have been deep during Tortonian times, it is concluded that the sea must have evaporated almost to dryness during the Late Messinian, leaving only saline lakes in the lower parts of the basins. This increase in salinity prior to the deposition of the evaporites would have killed off most marine organisms, thus accounting for the dramatic faunal changes that have been observed in successions of Late Tortonian–Early Pliocene age in this area. This event is often known as the 'Messinian Salinity Crisis'. There is evidence (Adams *et al.* 1977) that this crisis coincided with a world-wide lowering of sea-level by perhaps 40 to 70 m, and an overall cooling of the climate, associated with polar glaciation. Whether these events were coeval and connected is at present uncertain since our methods of measuring geological time using fossils are rarely accurate to more than one million years, whereas the whole Messinian Event occupied no more than 500000 years. The deposition of evaporites was apparently terminated by the opening of the Straits of Gibraltar in Early Pliocene times, thus allowing the Mediterranean basins to refill from the Atlantic.

The Red Sea rift, although a major physical feature in Middle Miocene times, was not at first connected to the Indian Ocean (Heybroek, 1965; Whiteman, 1968). A land bridge between Arabia and the Horn of Africa made it an arm of the Mediterranean in which large quantities of evaporites (halite and gypsum) accumulated owing to rapid evaporation of the surface waters. In some places these evaporites are several kilometres thick. During Late Miocene times deposition ceased, and the area north of Suez was raised above sea-level. In the Early Pliocene the southern land barrier was degraded, probably by faulting, and the Indian Ocean flowed into the Red Sea; Pliocene sediments found in the Red Sea therefore contain an Indo-Pacific fauna. Although there are patches of unfossiliferous post-Middle Miocene sediments between Suez and the Mediterranean (Said, 1962), there is no direct evidence that the Indian Ocean and the Mediterranean were reconnected until the construction of the Suez Canal. However, the possibility of a brief connection during the Pliocene or during a period of high sea-level in the Pleistocene (Tyrrhenian) cannot be entirely ruled out.

Northern India and the Indonesian Archipelago were

in a state of tectonic turmoil throughout the Neogene. The most intense orogenic phase in the Himalayas is believed to have occurred during the Middle Miocene after the deposition of the Murree Series (Kummel, 1961). On the other hand, Australia (except for New Guinea) was a stable region; only small areas in the west and south being invaded by the sea. Marine deposition virtually ceased in this region with the slow elevation of the south-western part of the continent in early Middle Miocene times. On the eastern side of the continent a gigantic uplift raised the border region to heights of 600 to 2000 m or more above sea-level, thus producing the coastal tablelands from Cape York to Tasmania (Laseron, 1969). It is possible that the sea penetrated into the Ord Basin and into central Australia south of the Gulf of Carpenteria in Middle Miocene times, since sediments in these areas have, according to Lloyd (1968), yielded brackish-water foraminifera. However, no truly marine beds have been found. North Island, New Zealand, was largely the site of marine deposition while South Island was emergent save for isolated coastal basins (Kingma, 1974).

The Neogene was chiefly a relatively warm period. Extensive deposits of shallow water limestone (and coal in terrestrial sequences) of Aquitanian to Burdigalian age testify to the warmth of the climate from Australia to western Europe. The climate remained equable until Tortonian times, when a gradual cooling set in which seems to have reached its climax in the Messinian. The Antarctic ice sheet probably began to form some six million years ago (see Berggren & van Couvering, 1974), and a relatively warm spell in the Early Pliocene was followed by further cooling, with ice beginning to form at the North Pole some 3 million years ago. This produced the Labrador Current which eventually pushed the Gulf Stream southwards to 45° N and altered the climate of northern Europe. This event heralded the Pleistocene glaciation, an event from which the world has not yet recovered.

The migration and distribution of certain Tertiary mammals has already been referred to, but more important than these was the spread of terrestrial grasses during the Neogene. Although these grasses had evolved in the early Tertiary they do not seem to have been widely dispersed until the Oligocene. The importance of this dispersal can scarcely be exaggerated since it was probably a necessary prerequisite for the spread of many herbivores and therefore of the carnivores which preyed upon them.

Faunas and faunal provinces

The Tertiary not only began with the sea standing at an unusually low level, but with a relatively impoverished fauna owing to the number of extinctions which had occurred during the latter part of the Cretaceous Period (for a review see Hancock, 1967). Although the cause of the extinctions is unknown, the effects are obvious; one important consequence was that when the Palaeocene Epoch opened, a considerable number of marine and terrestrial ecological niches were waiting to be filled.

The earliest known Palaeocene deposits are mainly found in deep-sea basins. These sediments contain unspecialised assemblages of planktonic foraminifera, quite unlike those found in Maastrichtian strata, but clearly of Tertiary type. Younger Palaeocene deposits yield progressively more diverse pelagic faunas until, in the Thanetian, they are as varied as during the Late Cretaceous. This is also true of the larger foraminifera which reappeared in the Middle Palaeocene and rapidly became diversified in Late Palaeocene times.

On land the story is the same, although the fossil record is much less complete. Nothing is known about Early and Middle Palaeocene mammal faunas, but from Late Palaeocene times onwards they became steadily more diverse until they were as rich and varied as the terrestrial vertebrate faunas of the Late Cretaceous.

The end of Eocene times saw a decline in foraminiferal faunal diversity, possibly in response to a general deterioration in climate which lasted until well into the Oligocene. In some groups, extinctions during the latest Eocene were hardly less dramatic than in the Cretaceous. However, it is strange that although the larger foraminiferal faunas of the shallow warm waters of the Indo-Pacific were depleted in the same way as in other areas, some species such as *Nummulites fichteli* Michelotti contrived to flourish although many near relatives had disappeared. The Late Oligocene and Early Miocene saw another burst of evolutionary activity (diversification) followed by a gradual decline throughout the rest of the Miocene. Thereafter the faunas stabilised, and, despite the climatic vicissitudes of the Quaternary, the marine biotas changed very little except in regions subjected to serious cooling.

Palaeobiogeographers tend to think in terms of faunal and floral realms and provinces. These deceptively simple concepts were founded on the distribution of Recent mammals. They are based essentially on latitude and continental position, with modifications necessitated by local conditions (mountain ranges, deserts, surface currents etc.).

Terrestrial faunal realms, defined by the distribution of mammals (see Chapter 15), which are themselves controlled by climate (latitude) and other factors already mentioned, are the most easily understood. However, the term 'faunal province' is also applied to marine faunas (an up-to-date review may be found in Davies, 1975), but these provinces are more complex and difficult to define. The problems can easily be illustrated by reference to four different groups of foraminifera – marine protozoans with a worldwide distribution.

The pelagic (often called planktonic) foraminifera are represented today by 37 known species that live mainly in the photic zone of the world ocean. They can travel up and down the water column by altering their buoyancy, but have no other power of movement, and they therefore drift passively in the ocean currents. Plots of the different species show that their geographical distributions are largely related to surface water temperature, and they are therefore found in bands more or less parallel to the equator, for example *Globigerina pachyderma* (Ehrenberg) is a cool water form confined to the polar and sub-polar regions, *Globorotalia hirsuta* (d'Orbigny) is a warm water species, and *Candeina nitida* d'Orbigny a tropical species. Five provinces are currently recognized and have been reviewed by Bé (1977).

A few pelagic species, for example *Globigerinella adamsi* Banner & Blow, are restricted to the Indo-Pacific. In contrast the so-called larger benthic foraminifera are all restricted to shallow warm water in the circumtropical region. They are never found where the minimum summer temperature falls below 18 °C. Although capable of very limited movement by pseudopodia, their dispersal is effected mainly by the drifting of gametes and very young stages, and by occasional rafting on weeds. These methods do not allow the larger foraminifera to match the global distributions achieved by their planktonic cousins, and they therefore show marked longitudinal provincialism as a consequence of the barrier effect of oceans (Adams, 1967).

A third group of foraminifera lives at abyssal depths, where water temperature is uniform from the poles to the tropics. This group shows no provincialism and has a world-wide distribution. Finally, the small group of brackish-water foraminifera, which always occurs in isolated communities, has much the same distribution the world over and is not strongly provincial. (The majority of foraminifera live in shelf seas and are normally able to migrate along the shelves, thus widening their geographical ranges: their distribution is not relevant in the present context.)

Thus, four groups of foraminifera have three different distribution patterns, and it is impossible to think of the entire Order in terms of faunal provinces. We can, however, say that taxa such as *Alveolinella* and *Cycloclypeus* (both larger foraminifera) form part of a generalised Indo-West Pacific faunal province along with molluscs such as *Nautilus*, *Tibia*, and *Tridacna*, and numerous corals including *Fabites* and *Goniopora*. If we consider Late Miocene faunas then we have to recognise an American province, a Mediterranean province and an Indo-West Pacific Province, but none of these is clearly distinguishable on the basis of planktonic foraminifera. Provinces can therefore be distinguished only on the basis of organisms having the same or similar methods of dispersal, and even then there can be difficulties. Benthic foraminifera and echinoderms are largely dispersed by currents, but because the free-drifting stage (if any) of the former is much shorter than the larval stage of the latter, an ocean which constitutes a barrier to one may not be an obstacle to the other.

Summary

The principal palaeogeographical changes in Tertiary times were as follows:

1. The disconnection of North America and Greenland from Europe in the Early Eocene; the gradual widening of the Atlantic Ocean, and its connection with the Arctic Sea.

2. The separation of Australia from Antarctica, a process which began in the Cretaceous, ended in the Palaeogene, and resulted in the establishment of the circum-Antarctic Current.

3. The northward movement of India and Arabia which first narrowed the Tethys Sea and then culminated in its disappearance in Burdigalian times. This brought the Mediterranean Sea into existence as a separate entity from the Indian Ocean.

4. The establishment of the Panama Isthmus during the Pliocene, thus separating the faunas of the Atlantic and Pacific and permitting faunal interchange between North and South America.

5. The building of the Alpine–Himalayan chain, the Andes, and the further uplift of the Rocky Mountains; processes which produced barriers to migration and affected the climates of the continental interiors.

Finally, in palaeobiogeographical terms, the spread of the sea grasses in the Palaeogene and of terrestrial grasses in the Neogene had consequences out of all proportion to their apparent significance, since they provided new ecological niches in which faunal diversification could occur, modified marine sedimentation patterns, and reduced the pace of terrestrial erosion.

Acknowledgements

Thanks are due to my colleagues, P. Kerr, R. L. Hodgkinson and J. E. Whittaker for help in scanning the relevant literature, and to Professor D. Curry for helpful suggestions.

References

Adams, C. G. 1967. Tertiary Foraminifera in the Tethyan, American and Indo-Pacific Provinces, *in* Adams, C. G. & Ager, D. V. (Eds) Aspects of Tethyan biogeography. *Systematics Association Publication* 7: 195–217. London.

Adams, C. G., Benson, R. H., Kidd, R. B., Ryan, W. B. F. & Wright, R. C. 1977. The Messinian salinity crisis and evidence of late Miocene eustatic changes in the world ocean. *Nature*, London **269**: 383–386.

Bé, A. 1977. An ecological, zoogeographic and taxonomic review of recent planktonic Foraminifera, pp. 1–88 *in* Ramsey, A. T. S. (Ed.) *Oceanic micropalaeontology*, Vol. 1. London: Academic Press Inc.

Bemmelen, R. W. van. *The geology of Indonesia*. Vols. 1A, 732 pp. 1B, 60 pp. The Hague: Govt. Printing Office.

Bender, F. 1975. Geology of the Arabian Peninsula Jordan. *Professional Papers of the U.S. Geological Survey*. 560–I; 1–VI+I1–I36.

Berggren, W. A. & Van Couvering, J. A. 1974. The Late Neogene. *Palaeogeography, Palaeoclimatology and Palaeoecology* **16**: 1–216.

Brinkmann, R. 1960. *Geologic evolution of Europe*. Translated by J. E. Sanders. 161 pp. Stuttgart: Ferd. Enke Verlag.

Brinkmann, R. 1976. *Geology of Turkey*. 158 pp. Stuttgart: Elsevier Scientific Publ. Co.

Brown, W. W. & Jones, K. D. 1971. Borate deposits of Turkey, pp. 483–492 *in* Campbell, A. S. (Ed.) *Geological history of Turkey*. Italy: Grafiche Trevisan, Castelfranco Veneto.

Cook, T. D. & Bally, A. W. (Eds) 1975. *Stratigraphic atlas of North and Central America*. 272 pp. N.J.: Princeton University Press.

Davies, A. M. 1975. *Tertiary faunas*, Vol. 2, 2nd ed. revised by Eames, F. E. & Savage, R. J. G. *The sequence of Tertiary faunas*. 447 pp. London: Allen & Unwin.

Debelmas, J. 1974. *Géologie de la France*, vol. 1, 293 pp. Paris.

Desio, A. 1971. Outlines and problems of the geomorphological evolution of Libya from the Tertiary to the present day, pp. 11–36 *in* Gray, C. (Ed.) *Symposium on the Geology of Libya*. Tripoli: University of Libya.

Donovan, D. T. & Jones, E. J. W. 1979. Causes of world-wide changes in sea level. *Journal of the Geological Society of London* **136**: 187–192.

Douglas, R. J. W. (Ed.) 1970. *Geology and economic minerals of Canada*. Vols. 1 & 11; 838 pp.+maps. Canada: Department of Energy Mines and Resources.

Dunnington, H. V. 1958. Generation, migration, accumulation, and dissipation of oil in Northern Iraq, pp. 1194–1251 *in* Weeks, L. G. (Ed.). *Habitat of Oil*. Tulsa: American Association of Petroleum Geologists.

Escher, A. & Watt, W. S. (Eds) 1976. *Geology of Greenland*. Grønlands Geologiske Undersøgelse, 603 pp. Copenhagen.

Gignoux, M. 1955. *Stratigraphic geology*. Translated from 4th ed. 1950 by G. G. Woodford. 682 pp. San Francisco: Freeman & Co.

Hancock, J. M. 1967. Some Cretaceous–Tertiary marine faunal changes, pp. 92–103 *in The fossil record*. Geological Society of London.

Haq, B. U., Berggren, W. A. & van Couvering, J. A. 1977. Corrected age of the Pliocene/Pleistocene boundary. *Nature*, London **269**: 483–488.

Hardenbol, J. & Berggren, W. A. 1978. A new Paleogene numerical time scale, pp. 213–234 *in* Cohee, G. V. *et al.* (Eds) *Contribution to the geologic time scale*. Studies in Geology No. 6. Tulsa: American Association of Petroleum Geologists.

Harrington, H. J. 1962. Paleogeographic development of South America. *Bulletin of the American Association of Petroleum Geologists* **46**(10): 1773–1814.

Hartog, C. den 1970. Origin, evolution and geographical distribution of the sea-grasses. *Verhandelingen der Koninklijke Nederlandsche Akademie Van Wetenschappen, Afdeeling Natur-kunde* **59**: 13–38.

Hecht, A. D. 1976. The oxygen isotopic record of Foraminifera in deep-sea sediment, pp. 1–43 *in* Hedley, R. H. & Adams, C. G. (Eds). *Foraminifera* 2. London: Academic Press.

Heybroek, F. The Red Sea Miocene evaporite basin, pp. 17–40 *in Salt basins around Africa*. London: Institute of Petroleum & Elsevier.

Ho Yen, Zhang Ping-kao, Hu Lan-ying & Sheng Jing-chang **1976**. *A report of scientific expedition in the Mount Jolmo Lungma region (1966–1968)*. (Palaeontology) Fasc. II. Science Press, Peking: 1–76.

Hsü, K. J., *et al.* 1978. History of the Mediterranean Salinity Crisis, pp. 1053–1078 *in* Kidd, R. B. & Worstell, P. J. (Eds) *Initial Reports of the Deep Sea Drilling Project*. Washington, DC: US Government Printing Office.

Huber, H. 1978. Geological map of Iran. Sheets 1–6. Scale 1:1000000. Teheran: National Iranian Oil Company.

Hunting Survey Corporation Ltd, 1961. *Reconnaissance geology of part of West Pakistan*. 550 pp. Toronto: Colombo Plan Co-operative Project.

James, G. A. & Wynd, J. G. 1965. Stratigraphic nomenclature of Iranian Oil Conservation Agreement area. *Bulletin of the American Association of Petroleum Geologists* **49**(12): 2182–2245.

Jenkins, D. G. 1978. *Guembelitria* aff *stavensis* Brady, a palaeo-oceanographic marker of the initiation of the circum-Antarctic current and the opening of the Drake Passage. XL: pp. 687–693 *in* Bolli, H. M. *et al. Initial Reports of the Deep Sea Drilling Project*. Washington: US Govt. Printing Office

Kingma, J. T. 1974. *The geological structure of New Zealand*. 407 pp. London: John Wiley & Sons.

Kummel, B. 1961. *History of the Earth*. 707 pp. San Francisco: W. H. Freeman & Co.

Kurtén, B. 1966. Holarctic land connexions in the early Tertiary. *Commentationes Biological Societas Scientiarum Fennica* **29**, (5): 1–5.

Kurtén, B. 1973. Early Tertiary land mammals, pp. 437–442 *in* Hallam, A. (Ed.). *Atlas of Palaeobiogeography*. Amsterdam: Elsevier.

Laseron, C. 1969. *Ancient Australia* (Revised by R. O. Brunnschweiler). 253 pp. Sydney: Angus & Robertson.

Lloyd, A. R. 1968. Possible Miocene marine transgression in Northern Australia. *Bulletin of the Bureau of Mineral Resources, Geology and Geophysics* **80**: 87–102.

Lloyd, J. J. 1963. Tectonic history of the south central-American Orogen, *in* Childs, O. E. & Beebe, B. W. (Eds) Backbone of the Americas. 320 pp. *Memoirs of the American Association of Petroleum Geologists* **2**, Tulsa.

Lüttig, G. & Steffens, P. 1976. *Palaeogeographic atlas of Turkey from the Oligocene to the Pleistocene*. 64 pp. 7 maps, key & explanatory notes. Hannover: Bundesanstalt für Geowissenschaften und Rohstoffe.

McKenna, M. C. 1975. Fossil mammals and early Eocene North Atlantic land continuity. *Annals of the Missouri Botanical Gardens* **62**: 335–353.

Minato, M., Gorvi, M. & Hunahashi, M. (Eds) 1965. *The geologic development of the Japanese Islands*. Association of Geological Collaborators of Japan. Tokyo: Tsukiji Shotan.

Petters, S. W. 1977. Ancient seaway across the Sahara. *The Nigerian Field* **42**(1): 22–30.

Reyment, R. A. 1966. Sedimentary sequence of the Nigerian Coastal Basin, pp. 115–141 *in* Reyre, D. (Ed.) *Sedimentary Basins of the African Coasts* Pt. I Atlantic Coast. *Association of African Geological Surveys, Paris.*

Reyre, D. 1966. Particularités Géologiques des Bassins Côtiers de l'Ouest africain, pp. 253–273 *in* Reyre, D. (Ed.) *Sedimentary Basins of the African Coasts*. Pt. I, Atlantic Coast. *Association of African Geological Surveys, Paris.*

Said, R. 1962. *The geology of Egypt*. 377 pp. Amsterdam & N.Y.: Elsevier.

Seneš, J. & Marinescu F. 1974. Cartes paléogeographiques du Néogène de la Paratethys centrale. *Mémoires du Bureau de Recherches Géologiques et Minières* **2**(78): 785–786.

Sharma, C. K. 1977. *Geology of Nepal* (2nd edition). 164 pp. Kathmandu: Educational Enterprises.

Smirnov, L. P. 1958. Black Sea Basin, pp. 982–994 *in* Weeks, L. G. (Ed.) *Habitat of oil*. American Association of Petroleum Geologists.

Steininger, F., Rögl, F. & Martini, E. 1976. Current Oligocene/Miocene biostratigraphic concept of the central Paratethys (Middle Europe). *Newsletters in Stratigraphy* **4**: 174–202.

Stöcklin, J. 1968. Structural history and tectonics of Iran: a review. *Bulletin of the American Association of Petroleum Geologists* **52** (7): 1229–1258.

Termier, H. & Termier, G. 1960. *Atlas de paléogéographie*. 99 pp. Paris: Masson & Cie.

Vinogradov, A. P. (Ed.). 1967. *Atlas of lithological-paleogeographical maps of the USSR*. Vol. 4, Paleogene, Neogene and Quaternary, 55 sheets. Moscow: Academy of Sciences.

Visser, W. A. & Hermes, J. J. 1962. Geological results of the exploration for oil in Netherlands New Guinea. 265 pp. *Verhandelingen van het Koninklijk Nederlandsch Geologisch Mijnbouwkundig Genootschap*.

Whiteman, A. J. 1968. Formation of the Red Sea Depression. *Geological Magazine* **105**, (3): 231–246.

C. G. Adams
Department of Palaeontology
British Museum (Natural History)

CHAPTER 15

Pleistocene geography and mammal faunas

A. W. Gentry and A. J. Sutcliffe

We now come to the most recent unit of geological time, the Quaternary Period. All of it is classed as the Pleistocene epoch, except the last 10000 years, which is known as the Holocene, and in which we still are today. The position of the boundary between the Pleistocene and the preceding Pliocene epoch has long been discussed, but since the International Geological Congress of 1948 it has been taken as lying at the base of the marine Calabrian Stage defined in Italy. Haq *et al.* (1977) concluded, from comparison of foraminifera in the stratotype with those in a number of dated deep-sea cores, that this boundary is at about 1600000 years – near the top of the Olduvai event of normal magnetism (Fig. 15.1).

Compared with the earlier periods of geological time, the Quaternary is distinctive. It has a fossil record which is fresher than that of the earlier periods, and sometimes changes of fauna and flora lasting only a few thousand years (in contrast to changes lasting millions or tens of millions of years) can be detected. While the present-day distribution of terrestrial plants and animals still shows the influence of pre-Quaternary continental drift, this process (though still continuing) is not a major factor in Quaternary biotic changes since the period of time involved is so short.

A further character of the Quaternary is that many of its fossil plants and animals are either very closely related to, or (especially in the upper Pleistocene and Holocene) actually conspecific with, living forms, allowing informed ecological comparisons of floras and faunas. In consequence such studies are commonly the province of the botanist and zoologist rather than of the geologist.

Another difference lies in the nature of the field evidence available to the Quaternary palaeontologist. Students of earlier periods often have at their disposal extensive spreads of fossiliferous marine rocks, now elevated above sea-level, providing both long stratigraphical sequences and a means of correlating faunas over long distances. However, most Quaternary marine sediments are still below the sea, where they can be studied only by drilling, and on land the Quaternary palaeontologist must rely mainly on fragmentary deposits – a peat bog in one place, a cave deposit somewhere else and a river terrace at yet another locality. Comparison of such deposits is often difficult since a peat deposit may contain abundant plant remains but no mammalian bones, whereas in a cave bones may be abundant yet plant remains have decayed. With such deposits there is no means of estimating how much time is not represented in the sedimentary record, and it has been suggested that as little as 15 per cent of Quaternary time is recorded on the land.

Greatest, however, of the peculiarities of the Quaternary are its climatic fluctuations. The ice caps and mountain glaciers of the world repeatedly advanced far beyond their present limits and then retreated again as the climate ameliorated. In consequence the Pleistocene is commonly known as the 'Great Ice Age', although it should be remembered that 'interglacial' intervals (some of them warmer than at the present day) alternated with the glacial advances. The first glacial–interglacial fluctuations began to build up during the Pliocene, from about 3200000 years ago, with an increase in the scale of glaciations about 2500000 years ago. Often the Holocene is described as 'postglacial' implying that the 'Ice Age' finished at the end of the Pleistocene, but the Holocene is more correctly regarded as just one more interglacial. A return to colder conditions, even if not to those of a full glaciation, would have serious economic and social consequences. Such a return is likely but not imminent.

The study of the Great Ice Age goes back some time, during which there have been some marked changes of views about its nature and duration. As late as 1840 superficial deposits were still commonly regarded as accumulations of the biblical deluge. Soon afterwards (mainly as the result of the work of Agassiz in the Alps and Archibald Geike in Scotland) it had to be accepted that large areas that are now free of ice had been glaciated in the not-too-distant past, and the idea of a (single) ice age replaced the diluvial theory. Further studies around 1900, mainly in the Alps, led to the idea of four glaciations with three interglacials between them. In the mid-1950s the development of new methods for obtaining cores of submarine sediments at last provided a more precise means of reconstructing Quaternary climatic fluctuations. Since these marine deposits commonly represent long periods of unbroken sedimentation, a much more complete record of Quaternary time is preserved in them than on land. Studies of oxygen isotope ratios in the shells contained in the cores provide an indirect detailed record of past fluctuations of climate back into the Pliocene, and this record is being refined as new cores are examined (see also Chapter 9). It now looks as if, during the last 1700000 years, there have been no less than 17 glacial advances and alternating interglacials, many of which are still unrecognised in the terrestrial record. For the purposes of this chapter it is appropriate to do no more than divide the Pleistocene into lower, middle and upper parts (Fig. 15.1). The Middle Pleistocene begins at the boundary between the Matuyama Reversed and Brunhes Normal Epochs of magnetic polarity at about 700000 years and the Upper Pleistocene at the start of the marine transgression associated with the onset of the Last Interglacial, slightly younger than 130000 years ago (Butzer & Isaac, 1975, pp. 901–903).

Fluctuations in the extent of the polar ice sheets and mountain glaciers were not the only changes during the

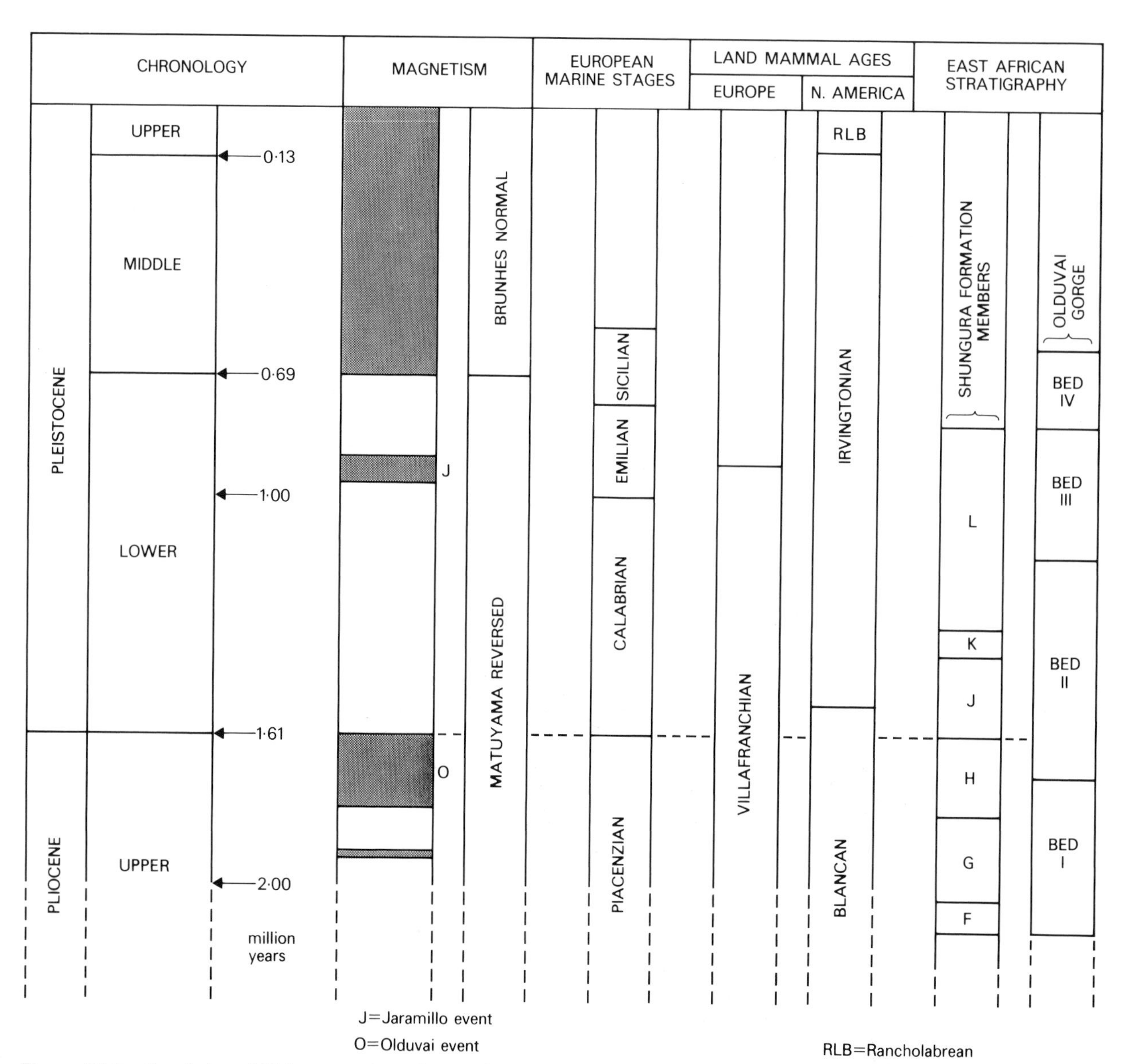

Figure 15.1 Correlation of Pleistocene chronology and magnetism with some of the stratigraphical scales mentioned in this chapter.

Quaternary. Climatic change was worldwide. The periglacial regions migrated to and from lower latitudes and altitudes. In the remaining unglaciated areas, changes of rainfall, temperature, humidity and other factors led to changes in the extent of lakes and desert areas. The availability of absolute dates for many of these events, particularly the more recent ones, now suggests some correspondence between times of glaciation and enlarged deserts. At the time of greatest extent of the last glacial advance, about 18 000 years ago (Figs 15.2, 15.3), the Sahara and many of the other deserts were more extensive than at the present day. During the postglacial optimum, about 9000 to 6000 years ago, the Sahara was less extensive and lakes such as Lake Chad were much larger. There is evidence from rockshelter paintings that such animals as elephant, giraffe and hippopotamus were able to find sustenance in regions that are now almost entirely desert.

A further consequence of these Quaternary fluctuations of climate is that there were changes of sea-level, which greatly affected the extent of the Earth's land surface. During the interglacials the sea stood as high as or higher than at the present day. During the glacial advances, however, much of the Earth's water was locked up as ice, resulting in substantial falls in sea-level, on several occasions by more than 100 metres. For example, at the height of the last glaciation the southern part of the North Sea was dry except for its rivers, as was the Georges Bank off the east coast of America and also a wide expanse of sea off the north coast of Siberia. The Bering Straits disappeared and there was a broad belt of land connecting Siberia to Alaska. In the southern hemisphere New Guinea became joined to Australia.

During the Pleistocene the major marine provincial areas were comparable with those of today; for example, the modern latitudinal sequence of faunas in the west Atlantic was more or less in existence by Pliocene times (Taylor, 1978), but the detailed position of the provincial boundaries fluctuated to the north and south, corresponding to the climatic warming and cooling. Tropical coral reefs were much more extensive during the last interglacial. For example in Kenya and Hawaii, Pleistocene reefs are far more extensive than modern ones; and the giant clam *Tridacna gigas*, now restricted to the western Pacific subprovince, is found in Pleistocene deposits from all over the Indian Ocean.

However, it is the terrestrial faunas which demonstrate in the most striking way the variety of faunal realms in the Pleistocene, and none more so than the mammals; and so the rest of this chapter is devoted to a review of their evolution and distribution during the last few million years. Mammals were more evolutionarily advanced by the end of the Pleistocene than at its start, but the various

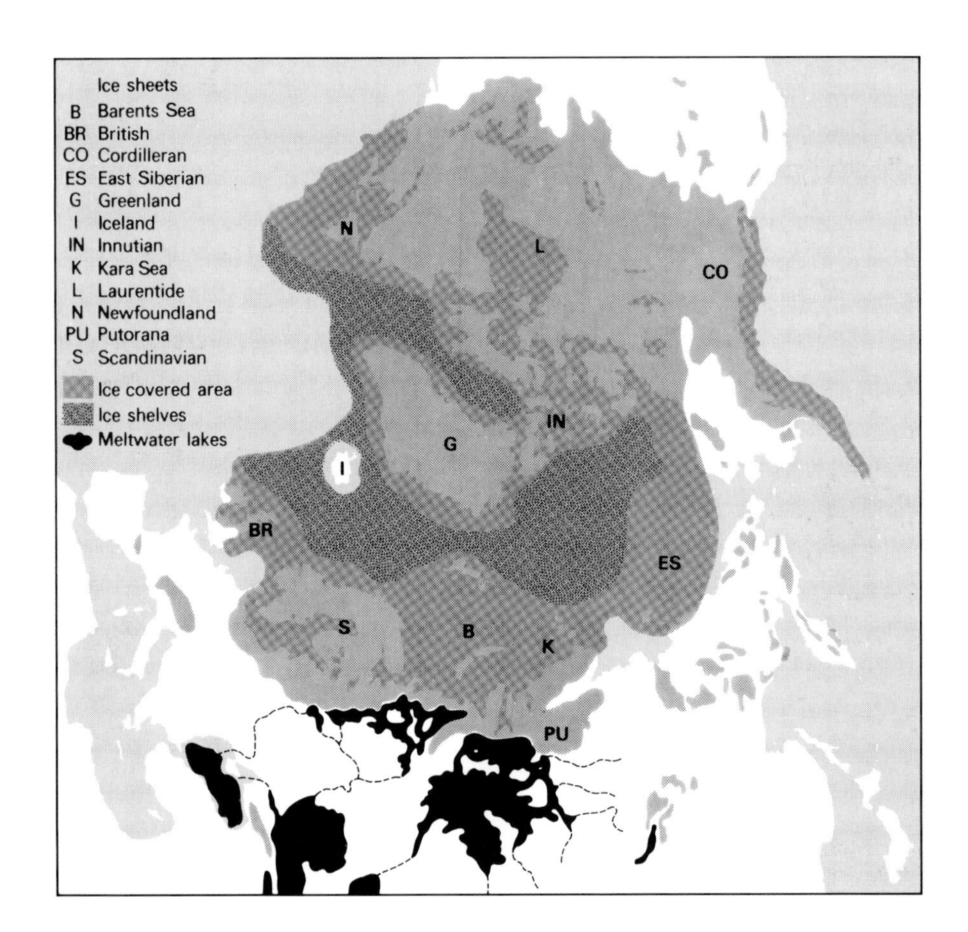

Figure 15.2 The northern hemisphere ice cover during the last glaciation, about 18 000 years ago (after Hughes *et al.* 1977).

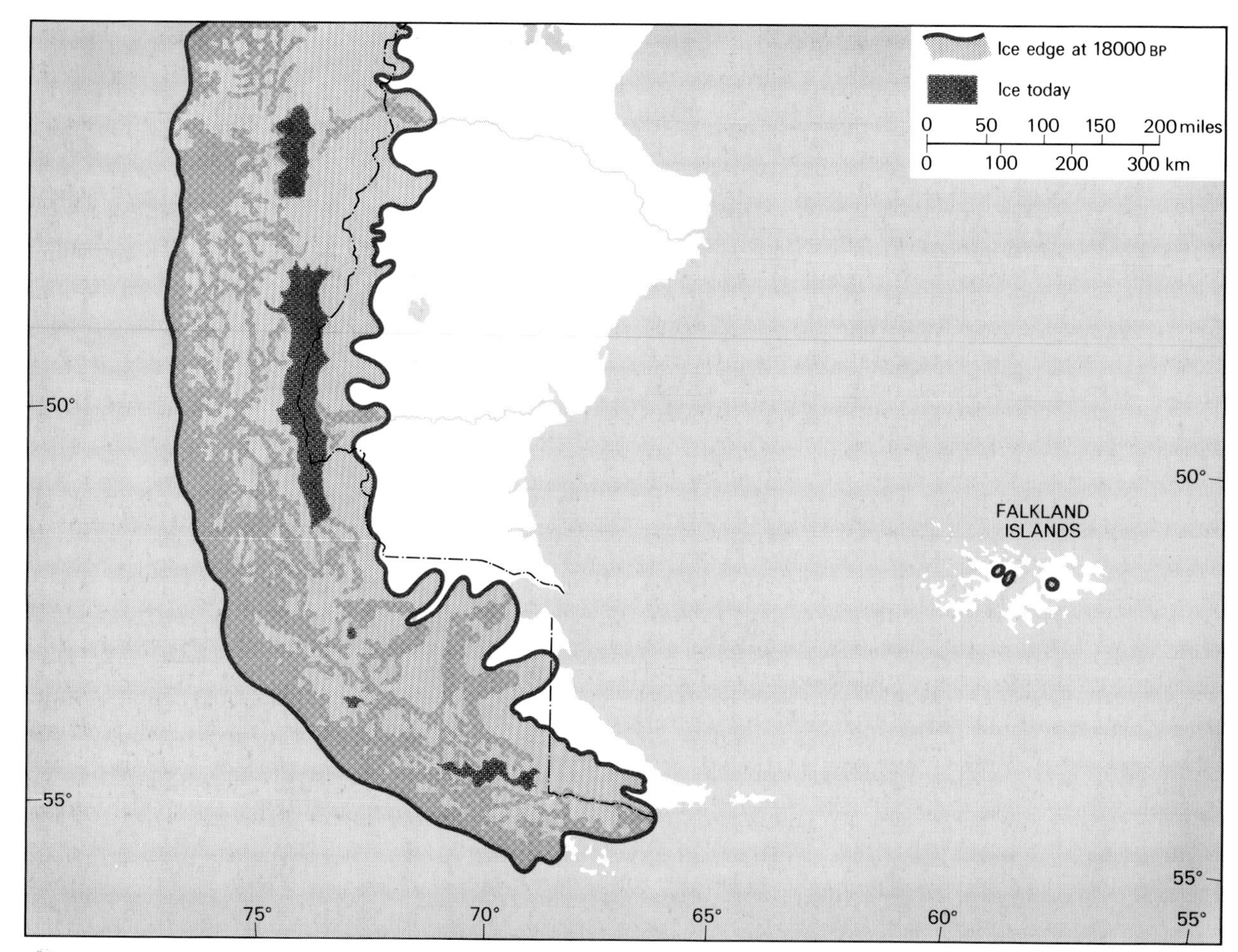

Figure 15.3 Extent of ice cover in the southernmost Andes, South America, today and at the maximum advance of the last glaciation, about 18000 years ago (after CLIMAP).

changes in climate had also brought about dramatic changes in their distribution. Vegetation zones and their dependent faunas moved polewards at times of climatic amelioration, and also toward contracting desert areas. At times of low sea-level during the cold stages such migrations could be extended: easy movement of terrestrial mammals was possible, for example between Alaska and Siberia. However, the development of such land bridges was a barrier to the movement of marine species, and, until the sea rose near its present level, whales and seals could not make their way through the Bering Straits from the Pacific to the Arctic Oceans.

Faunal realms

Alfred Russel Wallace's famous book *The geographical distribution of animals*, published in 1876, documented the generalities that the land animals at present inhabiting the world do not all occur worldwide, nor randomly, nor in ranges that can be wholly explained by reference to climatic or other environmental factors. Species within any particular region are more closely related to one another than to those of other regions, and environmentally similar regions in different parts of the world have different faunas. Edentates (sloths, armadillos, anteaters) occur mainly in South America, marsupials (pouched mammals) mainly in Australia, viverrid carnivores mainly in Africa and southern Asia, and mustelid carnivores mainly in Eurasia and North America. Today six faunal realms can be distinguished (Fig. 15.4):

Palaearctic Europe, North Africa, Arabia, Asia north of the Himalaya, northern China.
Ethiopian Africa south of the Atlas and northern littoral; sometimes taken as Africa south of the tropic of Cancer.
Nearctic North America, Arctic islands, Greenland.
Neotropical Central and South America, West Indies.
Oriental Southern Asia and the islands of Sumatra, Djawa (Java), Borneo and the Philippines.
Australian Australia, New Guinea, Maluku (Moluccas), Sulawesi (Celebes), New Zealand.

These faunal realms exist as a result of the evolutionary history of animals. For example Australia must have its marsupial fauna because it was already isolated by sea barriers before placental mammals arrived there or because early placentals became extinct there. While Wallace's main evidence came from extant faunas he also considered extinct ones. According to Wallace: '...the present condition of the fauna of Europe is wholly new and exceptional. For a long succession of ages, various forms of monkeys, hyaenas, lions, horses, hipparions, tapirs, rhinoceroses, hippopotami, elephants, mastodons, deer and antelopes, together with almost all the forms now living, produced a rich and varied fauna such as we now see only in the open country of tropical Africa'. This is not accurate in detail, since hippopotamuses for example did not live in Europe at the same time as tapirs, but Wallace was certainly right about the contrast between present and past faunas of Europe, and about the necessity for considering past faunas. As a general rule greater numbers of mammal species existed together during the Pleistocene than during historical times. Consequently, many of them must have had more specialised ecological niches than their present-day counterparts, although this was not invariable.

1. The Palaearctic realm

We may start by considering the main features of the Villafranchian fauna of France, which gives us a picture of the mammals living in Europe at the onset of the Pleistocene (Heintz *et al.* 1974 and Ballesio *et al.* 1973). The first appearance of a true elephant (*Archidiskodon* or *Mammuthus*) is in the middle Villafranchian, perhaps 2·5 million years ago, long after the disappearance of the mastodon *Zygolophodon* which had lower crowned teeth with fewer transverse plates. Another mastodon, *Anancus*, which was more primitive in having cusps instead of plates, survived alongside the new elephant for perhaps half a million years. *Hipparion*, the last of the three-toed horses, disappeared after the lower Villafranchian and one or more species of the more advanced one-toed *Equus* appeared in the middle Villafranchian. A new bovine stock, represented by *Leptobos*, appeared near the start of the Villafranchian. This was smaller and more primitive than *Bos*, with females lacking horns; but it did not outlast the Villafranchian.

Other mammals also disappeared during the course of the Villafranchian, for example *Agriotherium* (probably replaced by more advanced and less carnivorous bears of the genus *Ursus*) and the sabre-tooth *Megantereon*, which disappeared at the same level as *Anancus*. Some of the Villafranchian species may be ancestral to those of later periods in Europe, e.g. the bear *Ursus etruscus* to *U. deningeri*, the sabre-tooth *Homotherium crenatidens* to *H. sainzelli*, and the elk *Alces gallicus* to *A. latifrons* and then to extant *A. alces*. In the French Villafranchian assemblage 21 out of 40 species belong to living genera.

Fossil mammals are well known from the Middle

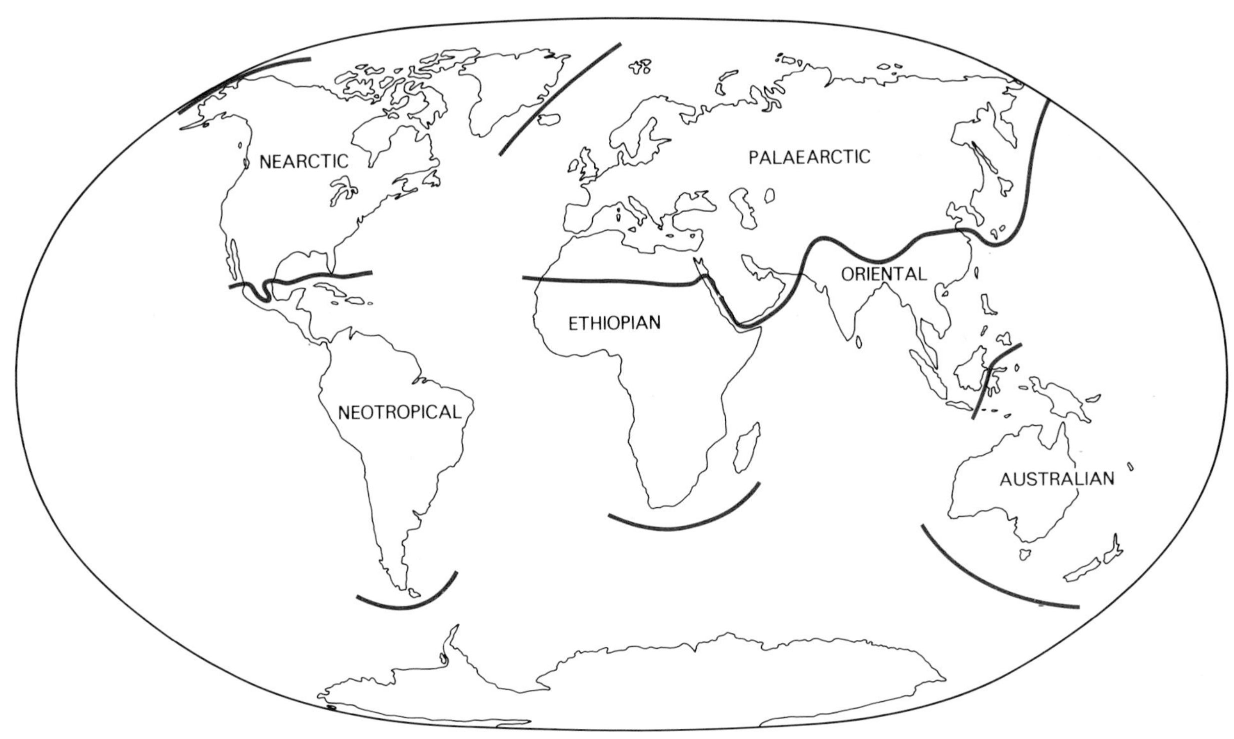

Figure 15.4 Map of present-day faunal realms. The boundary between the Oriental and Australian realms is Wallace's Line. Antarctica does not belong to a faunal realm.

Pleistocene, and Stuart (1974, table 2) provides a list for those from the Cromer Forest-Bed Formation of Britain. We take those above the 'Weybourne Crag' facies as early Middle Pleistocene and certainly later than the Villafranchian mammals just considered. The proportion of species in living genera in Stuart's list (excluding rodents and insectivores) has increased to 37 out of 47, with 38 per cent still living. Similar figures can be derived from Kahlke's (*in* Butzer & Isaac, 1975, pp. 324–347) lists for continental Europe: 83 per cent belong to living genera and 34 per cent to living species. This is a strong change from the Villafranchian faunas, and its most probable explanation is that the newer forms were better adapted to the onset of colder conditions. Some of the European Villafranchian mammals had affinities with southern and south-eastern Asian living forms suggesting that the intensifying ice ages caused a contraction of a formerly more widespread and uniform Villafranchian mammal fauna. European species had included a panda (*Parailurus anglicus*), a tapir (*Tapirus arvernensis*), a serow-like bovid (*Gallogoral meneghini*), a rhinoceros congeneric with the living Sumatran *Dicerorhinus*, a pig (*Sus strozzii*) with resemblances to the bearded pig (*S. verrucosus*) of Borneo, and a species of the raccoon dog *Nyctereutes*. The coexistence of a number of species of deer and bovids also recalls present-day southern Asia.

Among the changes in the Middle Pleistocene are the appearance of the present-day species of wolf and fox, of the extant roe deer (*Capreolus capreolus*), the modern red deer (*Cervus elaphus*), the aurochs (*Bos primigenius*) and two musk oxen – the more primitive *Praeovibos priscus* and the more specialised *Ovibos moschatus*. From its morphology one would expect *P. priscus* to be an older species, and it looks as if this might be so. The Villafranchian elephantid *Mammuthus meridionalis* and the rhinoceros *Dicerorhinus etruscus* survived into this period and were then replaced by *Palaeoloxodon antiquus* the straight-tusked elephant of interglacial periods and *Mammuthus primigenius* the woolly mammoth (Fig. 15.5), and by the rhinoceroses *D. kirchbergensis* and *D. hemitoechus* (and ultimately *Coelodonta antiquitatis*). Similarly the hyaena *Pachycrocuta brevirostris* survives briefly before being replaced by the extant *Crocuta crocuta*.

The Middle Pleistocene is also marked by an evolutionary transition from the early vole *Mimomys*, with rooted cheek teeth, to *Arvicola*, and a gradual increase in numbers of *Microtus* and *Pitymys*. *Lemmus* and *Dicrostonyx* are new appearances among voles with continuously growing teeth. Murid rodents are less common as fossils than the more strictly herbivorous voles. A distinctive large deer *Praemegaceros verticornis* makes a geologically brief appearance in the early part of the Middle Pleistocene record. The southern Asiatic *Bubalus* and the eastern European or Asiatic *Saiga* appear in western Europe from time to time, presumably in response to climatic changes. Man first spread to Europe in the Middle Pleistocene.

From calculations of faunal turnover rates (average of extinction and origination rates of species in successive levels) in relation to the supposed ages of deposits, Kurten

A

B

Figure 15.5 Excavation of two Middle Pleistocene elephants at Aveley, Essex, A, showing the relationship of the two skeletons, mammoth (above) and straight-tusked elephant in the foreground below. There was a short time interval between the burial of the two skeletons, probably on the surface of a marsh. B, view of the mammoth bones from above.

(1968: 262) estimated that mammalian evolution has progressed three or four times faster since the early Middle Pleistocene than in the Villafranchian.

The Late Pleistocene of Europe was a time when faunas consolidated their adaptations to life during the latest ice ages (Fig. 15.6). There is no single important faunal change or break, but some new species do appear in their extant form e.g. the elk (*Alces alces*), the snow vole (*Microtus nivalis*) and the field vole (*M. agrestis*).

Going east across Eurasia the present-day mammal fauna changes. Dormice become restricted east of the Caspian Sea and disappear between 80° and 90° E. Among antelopes *Saiga* recently ranged from 40° to 100° E and *Gazella subgutturosa* from 35° to 110° E, while in eastern Asia from 80° to 120° E there occurred the gazelles *Gazella (Procapra) gutturosa* and *G. picticaudata*. Other forms characteristic of eastern Asia are the giant panda (*Ailuropoda*), tiger (*Panthera tigris*), snow leopard (*P. uncia*), raccoon dog (*Nyctereutes*), and ruminants such as *Hydropotes*, *Elaphurus*, *Pseudovis* and *Ovis canadensis*. The chief differences from the western part of the Eurasian landmass are the more severe extremes of climate in the east and the presence of high mountains and plateaux. Comparable changes may be traced in the Pleistocene. The peculiar deer of the early Middle Pleistocene of Süssenborn (Germany), *Praemegaceros verticornis* and *Dolychodoryceros* are not known from Asia. *Bos primigenius* is found from Europe into Asia but not as far east as China. *Camelus* and the jerboa *Allactaga* appear at least intermittently in eastern Europe, and further eastwards one finds *Gazella*, *Spirocerus* (a larger antelope related to the Indian blackbuck), *Panthera tigris* and the large rhinocerotid *Elasmotherium*. Sometimes animals in China are the direct equivalents of those further west but are sufficiently different to warrant specific or even generic separation e.g. the eastern 'Irish' elk *Sinomegaceros*.

2. *The Ethiopian realm*

The Pliocene fauna from members A to F inclusive of the Shungura Formation, Ethiopia (Fig. 15.1), has been listed by Howell & Coppens (*in* Coppens *et al.* 1976). It dates from between 3 and 2 million years and may be taken as an African temporal equivalent of an early Villafranchian fauna. Fifty-eight per cent of its species belong to living genera and 10 per cent of the species are themselves still living. These are very different figures from those for the

A

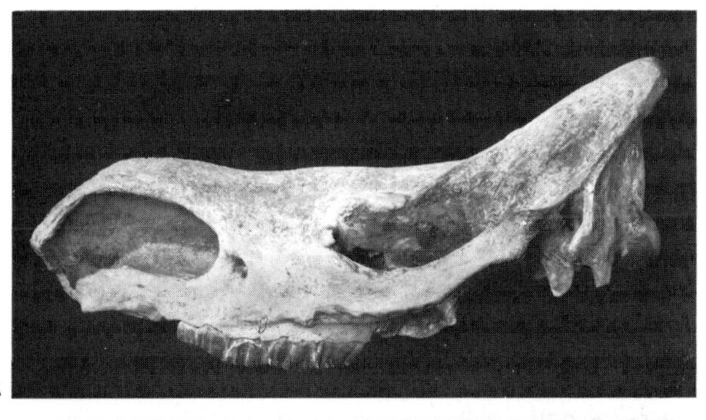

B

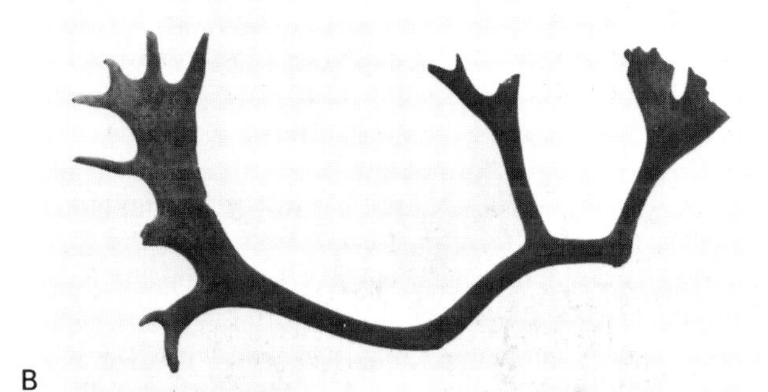

C

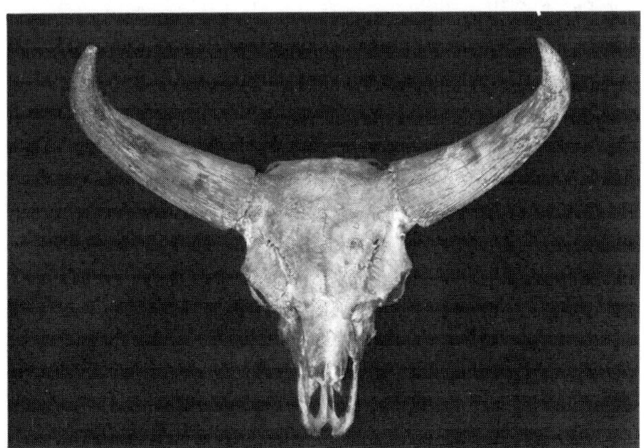

D

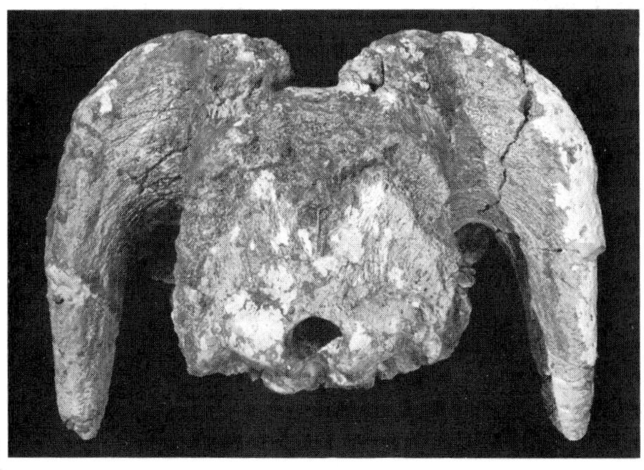

Figure 15.6 Pleistocene mammal remains from the alluvial and other deposits of the London area. A, Skull of woolly rhinoceros, whose distinguishing character is a central bony septum beneath the nasals (on the left). B, Shed antler of a male reindeer from the Floodplain Terrace of the Thames. Today this species is of northerly distribution, and its fossil remains are usually treated as indicators of cold climate. C, Bison skull from the Upper Floodplain Terrace of the Thames. D, Skull of a musk ox (seen from behind) from the Lower Middle Terrace of the Thames and about 150000 years old. The present-day musk ox is the most northerly-penetrating large land mammal, living under severe conditions beyond the tree-line on the Arctic tundra.

Villafranchian of Europe and suggest that older forms have survived longer than in Europe, presumably because they never faced such novel environmental conditions as were produced by the extension of glacial and periglacial areas in higher latitudes. A few species belong to the same genera as in Europe: *Elephas* (although the generic allocation of elephants of this age is not yet stable), *Hipparion*, and some carnivores such as *Homotherium*, *Megantereon* and *Panthera*; but in contrast the diversity of pigs, the large number of antelopes and absence of deer, the survival of *Deinotherium* and the chalicothere *Ancylotherium*, the distinctive rhinoceroses, hyraxes, viverrid carnivores and the hominids are all different from contemporaneous Europe. The pig stocks showed parallel evolution of high-crowned and occlusally-complicated teeth and of some associated skull characters. Among them are *Kolpochoerus* (= *Mesochoerus*) from which the forest hog *Hylochoerus* probably descended, the related *Metridiochoerus* group from among which the wart hog (*Phacochoerus*) appeared, and the large, more distantly related *Notochoerus* which became extinct early in the Pleistocene. An Ethiopian faunal realm must already have been in existence, characterised by indigenous animals, especially herbivores, and also by survivals of animals which had gone extinct earlier in Europe.

Another well-known fauna is that of Middle and Upper Bed II, Olduvai Gorge, Tanzania, later than members A to F of the Shungura Formation but not yet of Middle Pleistocene age. Seventy-three per cent of its species (see Leakey, 1971, pp. 290–294; Gentry & Gentry, 1978, table 11) belong to living genera and 18 per cent of the species are still living. These figures are approached by those for the Middle Pleistocene of Europe, supporting Kurten's observation that evolutionary rates in Europe speeded up after the Villafranchian so that the fauna became more quickly similar to that alive today. *Hipparion* is a notable late survivor in Olduvai Bed IV and another interesting record is a camel in Upper Bed II: the camel had penetrated sub-Saharan Africa as early as member B of the Shungura Formation.

The change from Middle to Late Pleistocene faunas has been little studied in Africa. One difficulty is that although the Pliocene to early-Middle Pleistocene is well-documented from East African sites, the Late Pleistocene is better known in northern and South Africa. A transition which may have occurred around this time is from *Elephas recki* to *E. iolensis*, a non-loxodont elephant which survived into the Late Pleistocene in South Africa. An important extinction which coincides approximately with the end of the Middle Pleistocene is of the large, short faced baboon-like *Theropithecus* over nearly all of its range. The genus survives today only as the relict species *T. gelada* in Ethiopia.

North Africa is properly a southern strip of the Palaearctic faunal realm, only narrowly open to the rest of the Palaearctic, and separated from the Ethiopian realm by a desert barrier. The desert has been intermittently effective in preventing access from the south and some African elements never reached that far north, for example the bovids *Syncerus* and *Megalotragus* and the pig *Notochoerus*. Others have done so but are now totally or locally extinct e.g. the white rhinoceros (*Ceratotherium*), the African elephant *Loxodonta* (used in 218 BC by the Carthaginian Hannibal against Rome), the long-horned buffalo (*Pelorovis*), the reedbuck (*Redunca*) and the wildebeest (*Connochaetes*). Palaearctic elements have been present for varying periods: the brown bear (*Ursus arctos*), the mole-vole (*Ellobius*) of Asia, the gerbil (*Meriones*) and *Bos primigenius* all from the start of the Middle Pleistocene, and *Cervus* possibly from the Middle Pleistocene, *Sus scrofa* and the dormouse (*Eliomys*) from towards the end of the Middle Pleistocene, and *Dicerorhinus kirchbergensis* from the late Pleistocene. The mouse *Apodemus* appears at the end of the Upper Pleistocene. The Barbary sheep (*Capra lervia*) is endemic in North Africa but is a member of a group (Caprinae) with Palaearctic affinities. Some Palaearctic forms have been present from even earlier than the Middle Pleistocene, for example the jerboa (*Jaculus*). The horse *Hipparion* disappeared from North Africa before the Middle Pleistocene.

3. The Nearctic and Neotropical realms

The Quaternary was also a time of rich and diverse faunas and floras in North and South America. Of the mammals some belonged to groups and species peculiar to the New World, whereas others had Asian affinities indicating immigration from that continent. One mammal with a long and well-documented New World history is the horse, which occurred there until the end of the Pleistocene. Its ancestor *Hyracotherium* is known from beds of Lower Eocene age in North America, including probably Ellesmère Island, and also from Europe, showing that at that time there was still a land connection across the region of the present-day North Atlantic (see also Chapter 14). By mid-Eocene times, this land connection had gone as the result of continental drift and the faunas became progressively more different. It was on the North American continent that, during a 50 million year period, *Hyracotherium* evolved into *Equus*.

With Europe separated from North America by the Atlantic the only remaining immigration route to the latter (we will consider the implications of South America shortly) was across the Bering Straits, and there is evidence of successive waves of Asian mammals crossing to the New World until the end of Pleistocene time (Hopkins, 1967). It is remarkable that through the long period that exchanges of fauna have been possible between Asia and North America, many genera have crossed into North America but there has been little equivalent movement in the opposite direction.

The Proboscidea (elephants and mastodons) are an example of an Old World group whose representatives spread to the New World on several occasions. Both bunodont (cusp-toothed) and zygodont (ridge-toothed) mastodons (the latter culminating in *Mammuthus*, which survived into the Holocene in North America) apparently first reached North America in the middle Miocene. Platybelodont ('shovel-tusker') mastodons followed by the same route a little later. True elephants arrived during the Pleistocene but some Old World proboscideans such as *Deinotherium* failed to cross to the New World.

Migrations of the Eurasian fauna into the New World were sporadic with the greatest peaks during the middle Miocene and the Plio-Pleistocene as a whole. The Bering Strait appears to have opened in its present form during the Pliocene, and the land bridge was repeatedly exposed during the glacial advances, facilitating the movement of terrestrial animals, and submerged again during the interglacials.

This land bridge was not the only obstacle to the movement of terrestrial mammals between Asia and North America during the Pleistocene. At times of glacial advance two great ice sheets formed in northern Canada – the Laurentide ice sheet in the Hudson Bay region and the Cordilleran ice sheet to the west. During the glacial maxima they abutted to form a continuous mantle over the whole of Canada and the northern USA. However, at such times central Alaska remained unglaciated and was faunally a continuation of eastern Siberia. Some immigrants from the Old World scarcely spread beyond the North American end of the land bridge, e.g. the early musk ox *Praeovibos*, the yak *Bos mutus*, and the saiga antelope. Immediately before and after a glacial maximum an ice-free corridor through Alberta allowed faunal movements to more southerly areas (Harington, 1978).

As stated earlier, migrations were not so frequent from the New World to the Old World, but *Equus* must have done so if it did indeed evolve in the New World, and the Old World camel is certainly descended from New World ancestors. However, the postulation of migration in specific directions is not always so obvious or implicit in the fossil record as has been claimed, and accepted conclusions are always vulnerable to new finds. At one time the group of cats to which the lynx (*Felis lynx*) belongs was thought to have originated in North America, but an early Old World lynx has been found: *F.* aff. *issiodorensis* from the late Miocene/early Pliocene site of Langebaanweg, South Africa (Hendey, 1974, pp. 162–163).

The North American terrestrial Cenozoic faunas are classified in a succession of biochronological land mammal ages (Fig. 15.1), among which is the Blancan, ending at a date close to or slightly younger than the Plio-Pleistocene boundary. It contains the modern one-toed horse *Equus*, and *Bretzia*, the earliest New World deer. Microtine rodents of the Blancan still have rooted teeth, but late in the period some of them, such as the musk rat (*Ondatra idahoensis*), acquire cement. Near the end of the Blancan the 'northern bog lemming' (*Synaptomys*) and the sabre-tooth (*Smilodon*) appear, and extinctions included some of the microtines, the large dog *Borophagus* (last of a mainly Miocene radiation), the hyaenid *Chasmaporthetes* (also known in Asia and in the European Villafranchian), and the little *Nannippus*, the last three-toed horse of the New World. The bunodont mastodon *Stegomastodon*, however, lasted into part of the succeeding Irvingtonian. Appearances and extinctions of conspicuous mammals in the fossil record are often different from the boundaries of land mammal ages.

The Irvingtonian contains the first large lion (*Panthera atrox*) very like or identical with the cave lion of Europe, and the hare *Lepus*. Another sabre-tooth (*Homotherium*) is first noted in the Irvingtonian but is always a rarer fossil than *Smilodon*. The large llama (*Palaeolama*) appears in Florida. Microtine rodents, for example the *Microtus*-like *Allophaiomys*, acquired teeth which had not only cement but also open roots. A true elephant (*Mammuthus*), the red deer or wapiti (*Cervus elaphus*) and bovids like *Euceratherium* and *Symbos* are all immigrants, or immediate descendants of immigrants, from Eurasia. There are parallels with the European succession, although the length of the Irvingtonian hinders close comparisons. *Allophaiomys* appears in Europe at the beginning of the Middle Pleistocene expansion of *Microtus* and *Pitymys*, but it is difficult to judge whether this is a parallel evolution or whether the change is a single Holarctic event. The genus *Cervus* is known from the Villafranchian or earlier in Europe but the living *Cervus elaphus* only appears in the Middle Pleistocene. The ovibovine *Euceratherium* is no more advanced in its skull characters than *Soergelia* of the European Middle Pleistocene. However, sometimes migrants have better success on new continents, and *Euceratherium* lasted in North America until the end of the Pleistocene.

The Rancholabrean is marked by some new appearances, for example *Bison* and the brown bear. Both appear later than in Europe and are clearly immigrants. There are also transitions in evolving lineages, e.g. from *Felis inexpectata* to the extant puma *F. concolor*, and from the peccary *Platygonus cumberlandensis* to *P. compressus*, which is smaller and has non-expanded zygomatic arches. During the Rancholabrean the long-horned *Bison latifrons* gave way to the shorter-horned *B. antiquus*, probably at the start of the last glaciation.

A carbon date of 27000 years on an artificially worked bone scraper found in the Yukon suggests that man had already arrived there long before the Last Glacial maximum 18000 years ago. But that locality is situated north of the great ice barrier (referred to above), which did not melt until about 12000 years ago. Did man manage to cross what is now Canada and begin to populate the area to the south before the ice barrier formed, or did he have to wait until after it? Alternatively, could he have by-passed the ice by making his way along the coast or by boat? Several carbon dates from both North and South America (not all of them without controversy) suggest that man was present south of the ice before the reopening of the corridor through the ice barrier.

So far we have not considered the fauna of South

America or its influence on that of North America. Today these two continents are joined by the Panamanian land bridge so that (if not prevented by climatic and vegetational factors) an exchange of species is possible; and indeed some species are present on both subcontinents.

However, from late Cretaceous to late Pliocene time North America and South America were separated by a seaway so that faunal exchange was very severely restricted (see Chapter 14). For nearly all of Tertiary time the mammals of South America evolved almost independently and developed into a remarkable indigenous fauna, the detailed ancestry of many components of which is still obscure (Simpson, 1978; Webb, 1978). After a land connection was re-established many North American mammals spread to South America and there was a similar, though less extensive, movement in the other direction – an event known as the 'Great American Interchange'. Its effect on the ecological balance of the mammalian faunas of the two sub-continents was immense and has long been a favourite topic of study among palaeontologists (Patterson & Pascual, 1972; Webb, 1978).

Although the pre-Pleistocene history of the mammalian fauna of South America is not our main concern here, some consideration of the mammals which inhabited that continent before the Great Interchange began is nevertheless fundamental to an understanding of that event. Simpson (1950) divided the pre-Interchange fauna of South America into a number of 'faunal strata' according to the length of time that the various groups had been present. He named the earliest stratum, which includes the marsupials and a variety of unusual placental herbivores, the 'Old-Timers'. It is still uncertain where the marsupials originated and different authorities have favoured North America, South America, Antarctica and Australia, but whatever their origin, they had already evolved in or reached South America by late Cretaceous times. The herbivores, known from the Palaeocene or Eocene onwards were mostly unlike mammals from other continents. The main orders are the extinct condylarths (also in North America and Eurasia but existing as an endemic family in South America); edentates (including sloths and armadillos); notoungulates (also in North America and China and including the rhinoceros-like *Toxodon*); litopterns (including the Miocene *Thoatherium*, which had reduced its side toes even more than modern horses); pyrotheres (which paralleled the elephants); and the astrapotheres whose mode of life has been described by A. S. Romer as 'beyond reasonable conjecture'.

The origin of these herbivores is far from clear. The litopterns, pyrotheres and astrapotheres may have arisen from condylarths, with subsequent differentiation in South America. The early record of the edentates in South America is unfortunately defective and both the phylogenetic and geographical origin of this group are unknown. The earliest known edentates are Palaeocene armadillos. The other groups (glyptodonts, ground sloths, ant eaters and tree sloths) appear subsequently, apparently as the result of evolution within the subcontinent.

The next stratum of mammals, appropriately named by Simpson the 'Island hoppers', includes a small number of families that managed to reach South America while it was still separated from North America. These were the caviomorph rodents and platyrrhine monkeys, first recorded in the early Oligocene, and the procyonid *Cyonasua* in the late Miocene. It is generally believed that these mammals were of North American origin and that they crossed to South America through islands, although the failure of all the other North American species to spread at the same time suggests that crossing was difficult and that no continuous land bridge existed.

Thus until the Middle Pliocene, South America retained its own fauna, supplemented only by the small number of immigrants just described. By this time the condylarths, astrapotheres and pyrotheres had all become extinct, but the other herbivore groups still survived and the native carnivores were almost entirely marsupials.

With the re-establishment of the land bridge across the Panamanian isthmus at the end of Pliocene times there began the Great Interchange. Among the mammals which spread from North to South America in the main Blancan and Irvingtonian interchange were mastodon, horse, tapir, deer, llamoid camelids, peccaries and various placental carnivores including puma, sabre-tooth and bear. Some of these were themselves only recent arrivals in North America from Eurasia. Before the end of the Pleistocene, puma and horse were in Patagonia. Among the South American species which went north were ground sloths, a glyptodont, armadillos, opossum and some rodents. By the end of the Pleistocene the ground sloth *Megalonyx* had reached Alaska. The opossum was a long-delayed replacement for North American opossums which had disappeared at the end of the Oligocene, and it is still spreading north at the present day. No notoungulate, litoptern or marsupial carnivore crossed to North America and they all became extinct in South America during the late Pliocene or Pleistocene. Some mammals survived or differentiated only in their new homes, for example the porcupine *Erethizon* in North America and the spectacled bear *Tremarctos* and horse *Hippidion* in South America, although the last is now extinct. By 13000 years ago man had reached Venezuela. The end of the Pleistocene was a time of extinction for many of the large mammals on both American continents. By the early Holocene the mastodon, ground sloths, glyptodonts, sabre tooths and even the horse were all extinct in both Americas.

4. The Oriental realm

The Oriental realm consists predominantly of tropical forests from India to southern China. These forests also extend through the islands of the East Indies into northern Australia, but by then they have passed into another faunal realm. The Oriental realm is characterised by a variety of forms, for example, bats, insectivores, primates, and deer such as sambar and muntjac.

In the Pleistocene of southern China is the so-called *Stegodon–Ailuropoda* fauna (Kahlke, 1961), which contrasts with northern Chinese fossil faunas by including *Stegodon* (a proboscidean with low-crowned teeth and elephant-like transverse plates), *Ailuropoda* (the giant panda), a large extinct tapir (*Megatapirus augustus*), and some forms still characteristic of the Oriental realm today, such as the badger *Arctonyx*, the porcupine *Hystrix*, and the deer *Muntiacus* and *Elaphodus*. The orang-utan (*Pongo*) and the hippopotamus *Hexaprotodon* are southern forms and do not occur in the *Stegodon–Ailuropoda* localities as far north as Szechuan. They suggest that there was not a sharp boundary across Pleistocene China between the Palaearctic and Oriental realms. Colbert & Hooijer (1953) reasoned that the *Stegodon–Ailuropoda* fauna was probably of Lower or Middle Pleistocene age (as understood here); however, it is probable that fossils have accumulated over a time span, at least at the Yenchingkou fissures, and one would like more definite evidence that an archaic form like the chalicothere *Nestoritherium* was contemporaneous with so many modern species.

The Pleistocene fauna of the Oriental realm extended through the East Indies as far as Wallace's Line (see next section below). Elephant, tiger and tapir extended to Borneo, and the extinct giant pangolin *Manis palaeojavanicus* inhabited the Niah Caves in Sarawak until 40000 years ago (see Hooijer, 1975). Elephants crossed Wallace's Line to Sulawesi, where the pigmy species *Elephas celebensis* evolved, and *Stegodon* reached further east of Wallace's Line to Flores and Timor islands.

5. *The Australian realm*

The Australian realm comprises mainly Australia and New Guinea, which are on the same continental shelf and were periodically united during the low sea-level stages of the Pleistocene. Although separated from the Oriental realm by only a close-knit chain of islands, apparently suitable for island hopping, the Australian mammal fauna is nevertheless the most distinctive in the world. Several boundaries have been proposed by various authors between these two realms (Simpson, 1977), the best known of which is Wallace's Line between Bali and Lombok and then north along the deep Macassar Strait between Kalimantan (Borneo) and Sulawesi (Fig. 15.4).

Whereas all the mammals of the Oriental realm, fossil and living, are placentals, the most characteristic Australian mammals are marsupials and a few monotremes. The Marsupialia have come to be treated as a superorder or even subclass. The Dasyuroidea include the native 'cats' and other insectivores and carnivores. The more numerous herbivorous forms are put in the Order Diprotodonta (wombats, koala, phalangers, kangaroos, possums and others); they have diprotodont incisors (three pairs of unequally sized uppers and one pair of enlarged lowers) and syndactylous hind feet (second and third metatarsals and toes reduced and united in a common ligament). A third group, the Perameloidea or bandicoots, are small invertebrate-eaters or omnivores with syndactylous hind feet but unspecialised incisors. Lastly, we have *Notoryctes*, the marsupial mole, of unknown relationships, and *Thylacinus*, the Tasmanian 'wolf', which used to be placed with the dasyuroids, but is now held to be related to the extinct South American Borhyaenidae (Archer & Bartholomai, 1978, p. 5 and references). The monotremes comprise the platypus and echidna.

Before Europeans began importing domestic animals into Australia, relatively few placental mammals were present there, the most important being bats and some murid rodents. These were apparently late Tertiary immigrants that arrived after the marsupials and monotremes had already occupied the ecological niches filled elsewhere by placentals. In some instances the marsupials have come to resemble their placental equivalents quite closely, for example the mole-like *Notoryctes*, wolf-like *Thylacinus* and ant eater-like *Myrmecobius*.

This Australian fauna has long been a cause of puzzlement among zoogeographers. However, the New World fossil record is important since, though no marsupial remains earlier than Upper Oligocene have been found in Australia, the earliest American finds are Upper Cretaceous. Studies of the North American evidence suggest that the placentals and marsupials separated some time during the Cretaceous. Could the Australian marsupials have originated in America?

At present Asia is the only continent anywhere near Australia, which is otherwise separated from Africa and ice-covered Antarctica by wide expanses of sea. Formerly it was widely believed that the Australian mammalian fauna had somehow reached Australia from North America through Europe and Asia, without the placentals also managing to follow. However, such an explanation presented difficulties. Firstly, although marsupials (presumably of American stock) were present in Europe during Eocene and Miocene times, there is no evidence of marsupials in Africa or Asia. *Deltatheridium*, from the Cretaceous of Mongolia, has some apparently marsupial characters, but these can be interpreted as convergent. Furthermore, although marsupials occur on many of the islands lying between Wallace's Line and Australia, there are fewer species there than on the Australian mainland, suggesting that the marsupial colonisation has been from Australia to the islands, rather than in the opposite direction. Placental mammals, on the other hand, are present only on the most western islands, with little overlap of the marsupial area, and this suggests dispersal from Asia.

A further argument against an Asian source for the Australian marsupials comes from studies of continental drift. In the early discussions of Australian marsupials it was assumed that the continents have always been in the same relative position. As early as 1924 Harrison argued that the marsupials reached Australia from South America by land connections through Antarctica, but this idea did not then receive ready acceptance. With continental drift established, an Asian origin becomes even more unlikely, since in Cretaceous times Australia lay far south of its present position and far from the Asian continent. Island hopping through Indonesia would not then have been possible. As Australia moved northwards, so the distance between the two continents decreased. The occurrence of the first rodents in Australia in deposits of Pliocene age suggests that by that time crossings were occasionally becoming possible by this northern route.

Today most palaeontologists accept an origin (i.e. differentiation from a primitive mammal stock) for the Australian marsupials somewhere within North or South America, Antarctica or Australia itself, although not all are agreed (see essays in Stonehouse & Gilmore, 1977). Some workers favour a North American origin, since the earliest marsupial fossils are known from that continent, and certainly an American origin. Others (Cox, 1973; Tedford, 1974) prefer South America, where late Cretaceous marsupials have been found, whereas Kirsch (1977) has claimed that Australia itself was the place of origin and that either the placentals never arrived there, or died out very early. The earliest marsupial fossils from Australia are late Oligocene diprotodonts (Tedford *et al.* 1975), but no fossil mammals have yet been recorded from Antarctica, where thick ice cover obscures most potential collecting areas. However, there has been discovered in Australia a fossil flea of late Cretaceous age (Riek, 1970) of a type associated elsewhere in the world with furred animals and not birds. The Australian Pleistocene poses special problems; firstly there are few chances to relate terrestrial sites to marine sequences, secondly there are few glacial deposits, making correlation difficult, and thirdly the mammal fauna is unlike that from anywhere else. Australian Pleistocene marsupials show the same trends as Pleistocene placentals in other realms: greater numbers of species apparently living at the same time than today, and a tendency for lineages to progressively increase in body size. Thus in the Darling Downs fauna from alluvial deposits in south-east Queensland there were 23 or more macropodid species, whereas no more than 12 would have been present in the same areas at the time of European colonisation. Part of the excess may arise from fossils of different biomes being gathered into the same accumulating deposits, or from changes in species distributions over a period of time, but these factors are unlikely to be the whole explanation.

The relationship between a Pleistocene and a living form can be intriguing but conjectural, for example as between the common Pleistocene *Macropus titan* and the extant grey kangaroo *M. giganteus*. The teeth of *M. titan* had linear dimensions about 30 per cent larger than *M. giganteus*, and there are other morphological differences between them. It is not known whether *M. giganteus* replaced *M. titan* or evolved from it by post-Pleistocene dwarfing. Possibly *M. titan* survived after the appearance of *M. giganteus* and was sympatric with it. Problems of size and identity may be complicated by sexual size dimorphism within *M. titan*.

The tally of large Pleistocene forms includes *Sthenurus* and *Procoptodon*, both belonging to an extinct macropodid subfamily. *P. goliah* was the largest known macropodid, perhaps one and half times to twice as large in linear dimensions as the living *Macropus giganteus*. They had deep and short faces, thick mandibles, and skulls whose general form is reminiscent of the ground sloths and glyptodonts of South America. They were probably specialised browsers, but not enough of their postcranial structure is known to be sure whether or not they were heavily built, slow-moving animals. At least some had syndactyly of the hind foot so advanced that even the fifth metatarsal was beginning to participate in the reduction, leaving a functionally monodactyl foot.

The large wallaby-like *Protemnodon* had relatively short and massive metatarsals. The diprotodontid *Diprotodon* was the largest known marsupial and was a heavily built browsing mammal with much enlarged incisors. *Palorchestes*, in the same superfamily, probably had a tapir-like proboscis. *Thylacoleo* was the most specialised form. About the size of a hyaena, it had a short face, a pair of enormous blade-like premolars and very reduced molars. It may have been a carnivore, but its affinities are with the otherwise herbivorous Diprotodonta, and members of its family are known back to the Miocene.

Even the monotremes reached a larger size in Pleistocene Australia. The femur of a late Pleistocene echidna (*Zaglossus hacketti*) is more than 1·5 times as long as that of the living *Z. bruijni* and nearly 2·5 times as long as its relative *Tachyglossus aculeatus*.

Extinctions

One of the most spectacular events of the Quaternary is the extinction of so many of the larger mammals, for example the woolly mammoth, mastodon, woolly rhinoceros, giant sloth and sabre-tooth (Martin & Wright, 1967). This occurred about 12000 to 10000 years ago, just as the last ice age was coming to an end, and improved climatic conditions might be expected to have made mammalian survival easier. Many possible explanations have been put forward, those receiving greatest support being climatic change and the coming of man, although catastrophes, biological competition, epidemics, parasites, degeneration and overspecialisation are other suggestions. Sometimes one cause stands out as the likeliest, for example the thylacine probably disappeared from continental Australia following the arrival there of the dingo in Holocene times. Perhaps several factors worked together in other cases. The advent of methods of absolute dating, improved understanding of past climatic change, studies of the ecology of living mammals, and palaeoecological studies (such as of fossil stomach contents) are all bringing the reliable assessment of causes and effects within reach, but many conclusions are still speculative.

Any species has its place in a web of complex ecological chains, which are difficult to decipher, and its survival depends on a balance of circumstances. At times of stress some mammals are more vulnerable to extinction than others. A mammal with specialised food requirements or a long gestation period is more at risk than one with a more catholic diet or rapid reproduction. Catastrophes may cause local extinctions, but these are unlikely to be worldwide. The mixing of once-separated faunas, resulting from the establishment of land bridges or the melting of ice barriers, could lead to epidemics, but some resistant animals are likely to survive and breed to restore previous numbers. The waning of an ice sheet would not necessarily mean improved conditions if, for example, an entire vegetation zone was pinched out. Likewise, in some semi-desert areas of the present day, increased rainfall leads to increased salinity and the destruction of existing trees. At times of such changes an animal has a better chance of survival if it can migrate within a moving biome than if it finds itself stranded on an island, or if the movement of the biome is halted by a mountain range. Examples of changing ranges can be seen by comparing the Pleistocene and modern distributions of the collared lemming (Fig. 15.7) and the hippopotamus (Fig. 15.8). Predation, in so far as it eliminates unfit animals, maintains a vigorous breeding stock and is unlikely to cause extinction, but predators themselves could be exterminated as the result of an extinction of prey species through other causes. In stable communities, interspecific competition is avoided by the exploitation of a variety of ecological niches. A mixing of previously separated communities, following the establishment of a land bridge, may lead to intensified competition and the eventual extinction of some less well adapted species.

Man's influence was of two sorts – direct (principally by hunting) and indirect (for example by destruction of habitat). There can be no doubt that in Europe Palaeolithic man sometimes killed vast numbers of mammals, but he appeared to live in balance with them and did not cause extinction, except locally. By analogy the North American Indians were in no danger of exterminating bison before the arrival of Europeans with their horses and firearms. In an area where man has newly arrived, however, an existing mammalian population might be at a disadvantage if it lacked a defence strategy against the new predator. Man, while spreading across the American continents at the end of the Pleistocene, may have destroyed some of the large mammals, though it is impossible to pin down any specific instances. In Australia the arrival of man (who was already in New South Wales at least 33 000 years ago) seems not to have led to any immediate extinctions. A number of large marsupials flourished for several thousand more years before becoming extinct. In the Russian plains Palaeolithic man's considerable pressure on the mammoth has been strikingly demonstrated on

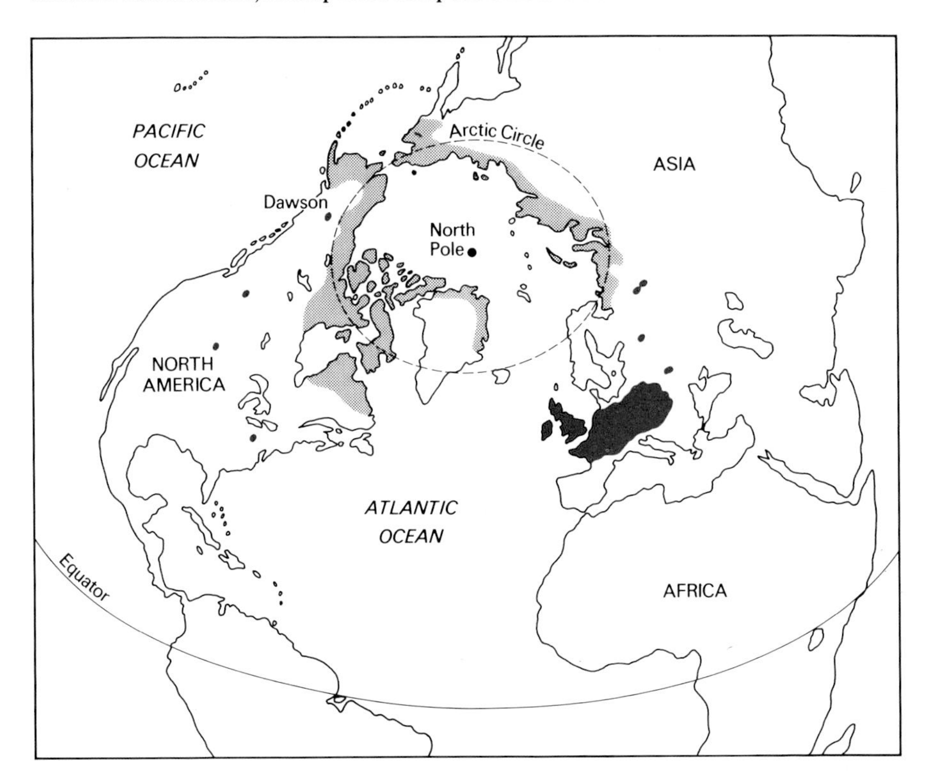

Figure 15.7 An example of biotic displacement: *Dicrostonyx*, the collared lemming, today the most northerly rodent in the world.
Grey areas: present-day circumpolar distribution.
Blue areas: fossil localities dating from the Last Glacial advance, when this lemming was displaced southwards.

Figure 15.8 Distribution of an animal which extended its range during periods of climatic amelioration in the Pleistocene: the hippopotamus, *Hippopotamus amphibius*. Dark blue: distribution in historic and Pleistocene times. Solid blue: dwarf races on various Mediterranean islands. Light blue: additional Pleistocene distribution, not all of which need have been occupied at any one time.
Other species of hippopotamus lived on Madagascar and in the Oriental faunal realm during the Pleistocene.

archaeological grounds, but there is little evidence of his hunting in the north, where the mammoth apparently died out without his intervention. The disappearance from sub-Arctic regions at the end of the Pleistocene of a biome known as Arctic steppe or steppe tundra, which had previously supported mammoths, and its replacement by boggy tundra, is a possible cause of mammoth extinction which is currently receiving wide support among palaeontologists.

The giant deer of Ireland became extinct about 10000 years ago, but not by the hand of man since he only arrived there about 2000 years later. A slight climatic cooling, which led to the formation of a small ice sheet in western Scotland and the formation of corrie glaciers in Ireland, may account for the extinction. However, the species apparently also became extinct in the rest of Europe at about this time and, if it is to be argued that it became extinct in the British Isles for climatic reasons, it is difficult to understand why it was unable to survive under more congenial conditions further south.

In Madagascar there is archaeological evidence that the disappearance of some of the larger lemurs about a thousand years ago was due to man's activities. In Africa larger mammals survived in more variety until the present day, but at least in South Africa late Pleistocene extinctions do not appear to be lacking. A long-horned buffalo and the largest known alcelaphine antelope were present until near the end of the late Pleistocene in the southern Cape Province (Klein, 1974), and other lineages also became extinct or underwent size diminution. A better understanding of Pleistocene mammalian extinctions can only come from more detailed studies. A good start towards this objective has already been made. Initially the extinctions that took place on islands are likely to be the simplest to explain, and it is from islands that the first firm deductions will probably come in the future.

References

Archer, M. & Bartholomai, A. 1978. Tertiary mammals of Australia: a synoptic review. *Alcheringa* Sydney **2**: 1–19.

Ballesio, R., Guerin, C., Meon-Vilain, H., Miguet, R. & Demarcq, G. 1973. Observations et propositions biostratigraphiques sur la limite Pliocene-Quaternaire. *International Colloquium on the Problem 'The boundary between Neogene and Quaternary,'* Moscow **4**: 44–75.

Butzer, K. W. & Isaac, G. Ll. (Eds) 1975. *After the australopithecines: stratigraphy, ecology and culture change in the middle Pleistocene.* 911 pp. The Hague & Paris: Mouton.

Colbert, E. H. & Hooijer, D. A. 1953. Pleistocene mammals from the limestone fissures of Szechwan, China. *Bulletin of the American Museum of Natural History* **102**: 1–134.

Coppens, Y., Howell, F. C., Isaac, G. Ll. & Leakey, R. E. F. (Eds) 1976. *Earliest man and environments in the Lake Rudolf basin.* 615 pp. University of Chicago Press.

Cox, C. B. 1973. Systematics and plate tectonics, pp. 113–119 *in*: Hughes, N. F. (Ed.) Organisms and continents through time. *Special papers in Palaeontology* **12**.

Gentry, A. W. & Gentry, A. 1978. Fossil Bovidae (Mammalia) of Olduyai Gorge, Tanzania. Parts I & II. *Bulletin of the British Museum (Natural History)* (Geology) London **29**: 289–446; **30**: 1–83.

Haq, U. B., Berggren, W. A. & Van Couvering, J. A. 1977. Corrected age of the Pliocene/Pleistocene boundary. *Nature*, London **269**: 483–488.

Harington, C. R. 1978. Quaternary vertebrate faunas of Canada and Alaska and their suggested chronological sequence. *Syllogeus*, Ottawa **15**: 1–105.

Harrison, L. 1924. The migration route of the Australian marsupial fauna. *The Australian Zoologist* **3**: 247–263.

Heintz, E., Guerin, C., Martin, R. & Prat, F. 1974. Principaux gisements villafranchiens de France: listes fauniques et biostratigraphie. *Mémoires du Bureau de Recherches géologiques et minières* **78**, 1: 169–182.

Hendey, Q. B. 1974. The late Cenozoic Carnivora of the south-western Cape Province. *Annals of the South African Museum* **63**: 1–369.

Hooijer, D. A. 1975. Quaternary mammals west and east of Wallace's Line. *Netherlands Journal of Zoology* **25**: 46–56.

Hopkins, D. M. (Ed.) 1967. *The Bering land bridge.* 495 pp. Stanford University Press.

Hughes, T., Denton, G. H. & Grosswald, M. G. 1977. Was there a late-Würm Arctic Ice Sheet? *Nature*, London **266**: 596–602.

Kahlke, H. D. 1961. On the complex of the *Stegodon-Ailuropoda* fauna of southern China and the chronological position of *Gigantopithecus blacki* V. Koenigswald. *Vertebrata Palasiatica* **5**: 104–108.

Kirsch, J. A. W. 1977. The six per cent solution: second thoughts on the adaptedness of the Marsupialia. *The American Scientist* **65**: 276–288.

Klein, R. G. 1974. A provisional statement on terminal Pleistocene mammalian extinctions in the Cape biotic zone (southern Cape Province, South Africa). *Goodwin Series, South African Archaeological Society* **2**: 39–45.

Kurten, B. 1968. *Pleistocene mammals of Europe.* 317 pp. London: Weidenfeld & Nicolson.

Leakey, M. D. 1971. *Olduvai Gorge. 3, Excavations in Beds I and II, 1960–1963.* 306 pp. Cambridge University Press.

Martin, P. S. & Wright, H. E. (Eds) 1967. *Pleistocene extinctions: the search for a cause.* 453 pp. New Haven & London: Yale Univ. Press.

Patterson, B. & Pascual, R. 1972. The fossil mammal fauna of South America, pp. 247–309 *in*: Keast, A., Erk, F. C. & Glass, B. (Eds) *Evolution, mammals and southern continents.* Albany.

Riek, E. F. 1970. Lower Cretaceous fleas. *Nature*, London **227**: 746–747.

Simpson, G. G. 1950. History of the fauna of Latin America. *The American Scientist* **38**: 361–389.

Simpson, G. G. 1977. Too many lines; the limits of the Oriental and Australian zoogeographic regions. *Proceedings of the American Philosophical Society* **121**: 107–120.

Simpson, G. G. 1978. Early mammals in South America: fact, controversy, and mystery. *Proceedings of the American Philosophical Society* **122**: 318–328.

Stonehouse, B. & Gilmore, D. (Eds.). 1977. *The biology of marsupials.* 486 pp. London: Macmillan.

Stuart, A. J. 1974. Pleistocene history of the British vertebrate fauna. *Biological Reviews*, Cambridge **49**: 225–266.

Taylor, J. D. 1978. Cenozoic and present day. pp. 323–364 *in* McKerrow, W. S. (Ed.) *The ecology of fossils.* London: Duckworth.

Tedford, R. H. 1974. Marsupials and the new paleogeography, pp. 109–126 *in*: Ross, C. A. (Ed.), Paleogeographic provinces and provinciality. *Special Publication of the Society of Economic Paleontologists and Mineralogists* **21**.

Tedford, R. H., Banks, M. R., Kemp, N. R., McDougall, I. & Sutherland, F. L. 1975. Recognition of the oldest known fossil marsupials from Australia. *Nature*, London **255**: 141–142.

Webb, S. D. 1978. A history of savanna vertebrates in the New World. Part II: South America and the great interchange. *Annual Review of Ecology and Systematics*, Palo Alto **9**: 393–426.

A. W. Gentry and A. J. Sutcliffe
Department of Palaeontology
British Museum (Natural History)

Index

A

B

C

D

E

F

G

H

I

J

K

L

M

N

O

P

Q

R

S

T